AF806036

Nenad Filipović
Editor

Applied Artificial Intelligence 4: Medicine, Biology, Chemistry, Financial, Games, Engineering

The Third Serbian International Conference on Applied Artificial Intelligence (SICAAI)

Springer

Editor
Nenad Filipović
Department of Applied Mechanics and Automatic Control
University of Kragujevac
Kragujevac, Serbia

ISSN 2367-3370 ISSN 2367-3389 (electronic)
Lecture Notes in Networks and Systems
ISBN 978-3-031-99200-1 ISBN 978-3-031-99201-8 (eBook)
https://doi.org/10.1007/978-3-031-99201-8

© The Editor(s) (if applicable) and The Author(s), under exclusive license to Springer Nature Switzerland AG 2025

This work is subject to copyright. All rights are solely and exclusively licensed by the Publisher, whether the whole or part of the material is concerned, specifically the rights of translation, reprinting, reuse of illustrations, recitation, broadcasting, reproduction on microfilms or in any other physical way, and transmission or information storage and retrieval, electronic adaptation, computer software, or by similar or dissimilar methodology now known or hereafter developed.
The use of general descriptive names, registered names, trademarks, service marks, etc. in this publication does not imply, even in the absence of a specific statement, that such names are exempt from the relevant protective laws and regulations and therefore free for general use.
The publisher, the authors and the editors are safe to assume that the advice and information in this book are believed to be true and accurate at the date of publication. Neither the publisher nor the authors or the editors give a warranty, expressed or implied, with respect to the material contained herein or for any errors or omissions that may have been made. The publisher remains neutral with regard to jurisdictional claims in published maps and institutional affiliations.

This Springer imprint is published by the registered company Springer Nature Switzerland AG
The registered company address is: Gewerbestrasse 11, 6330 Cham, Switzerland

If disposing of this product, please recycle the paper.

Contents

Community Event Discovery Using 'X' Data Stream

O. Adedotun Ogundare[1(✉)], U. Abdullahi Saleh[2], and R. Ayodele Kayode[3]

[1] Department of Computing and Control Engineering, Faculty of Technical Sciences, University of Novi Sad, Dr Zorana Đinđića 1, 21000 Novi Sad, Serbia
adedotun@uns.ac.rs

[2] Department of Computer Engineering, Faculty of Engineering, Istanbul Ticaret University, Örnektepe, İmrahor Cd. No: 88/2, 34445 Beyoğlu/İstanbul, Turkey
abdullahi.saleh@istanbulticaret.edu.tr

[3] Department of Economics, Afe Babalola University, Ado-Ekiti, Nigeria
ayodelekayode100@abuad.edu.ng

Abstract. In recent years, significant research has been carried out to discover current events and the number of events from X's data stream and to reveal the contextual dimension of events. One of the major challenges in this context is that most traditional methods have to estimate the number of events in order to contextually analyze events. Another problem is that some methods tend to detect events that often trigger a significant volume of communication. In this study, we propose a community-based event discovery system that reveals events by applying community detection on the graph representing the co-occurrence relations of n-grams that are frequently observed in X's posts. We evaluate the effectiveness of the proposed system over a dataset used as a comparative dataset. Our system shows the success of discovering as many events as the number of existing events in an unsupervised manner. It also represents each event with words and multiple-word distributions, helping to analyze the subject of events contextually.

Keywords: Community detection · event detection · event discovery · N-gram · X · Twitter

1 Introduction

Social media is a new media system that enables individuals to communicate and share information on the internet. It is a system that has managed to spread rapidly, thanks to its availability at any place and time. In social media, people can share content and easily access the content they are looking for. One of the most popular social media applications is 'X' formally Twitter. 'X' is an online social networking site which is used for connecting people and allowing the people to share their thoughts on different issues using short messages called tweets. Today, 'X' has been a major source of news, which covered both major and minor events. Currently, X has over a billion users worldwide. In a recent survey conducted by Hasan et al. [1], it was presented that people are more

© The Author(s), under exclusive license to Springer Nature Switzerland AG 2025
N. Filipović (Ed.): AAI 2024, LNNS 1446, pp. 1–8, 2025.
https://doi.org/10.1007/978-3-031-99201-8_1

dependent on X as their main source of news compared to Facebook, Google + and other sources of news. Today, politicians, businesspeople, journalists, youths and all categories of people use X for different purposes, such as advertising, politics and weather forecast. For instance, X tends to be a tool used by most journalist to spread news quickly than the traditional media because X does not only serve as a socializing outlet but also an avenue of expressing personal opinions about any real-world event. In recent years, journalists and media outlet use X to spread news to the public. A recent survey conducted by Pew Research Center [2] shows that 59% of Americans get news regularly from X, 31% from Facebook and the other media platform is less than 10%. This makes X the most used social media platform in 2021. After comparing X with other social media platforms, the research shows that X has outpaces all other social media sites when it comes to obtaining news.

Tweet data is very short text due to the 140-character limit. This creates difficulties for many applications such as text classification, sentiment analysis, and event discovery implemented on the X dataset because the data is often sparse and has an irregular format because of a wide variety of tags attached to tweets. Moreover, there exist other challenges for many pre-processing and feature extraction methods due to the volume of data from X, which generates over 340 million tweets per day [3]. Given the volume and diversity of data, extracting useful information from it can be a time-consuming task to easily enable users to discover key events shared on X. Feature extraction and categorization is also another main challenge when it comes to event detection. Extracting the right set of features is important in event discovery because it helps in representing the event and capturing the most valuable information. Revealing the dependency between selected attributes is crucial to represent events with a set of meaningful attributes. It is important to represent events with a set of meaningful and distinctive attributes to reveal the dependency between the selected attributes.

2 Related Works

Many works have been presented for event discovery on social media, most especially on social network like X. Over the years, researchers have developed different methods in discovering both major and minor events in media stream. These approaches are generally classified as term weighting-based methods, Topic modelling-based methods and clustering-based methods.

2.1 Term Weighting Based Methods

Term weighting-based method is a supervised method that assigns appropriate weights to each term observed in all documents in order to enhance the performance of text classification. Term weighting-based event detection methods generally aim to group the high frequency terms which co-occur in large number of tweets. However, describing a complex event using individual words is difficult and insufficient because of the ambiguity it holds due to the lack of context [4]. One of the methods using the term frequency is Mention-Anomaly-based Event Detection (MABED) method which is statistical method which relies on tweets. The method calculates the frequency of dynamic mentions (links) that users use in tweets in order to detect important events and to understand their impact on the crowd [5]. In addition, MABED provides dynamic event discovery and dynamically predicts the time intervals in which events occur. Alvanaki et al. [6], presented En Blogue method for Web 2.0 Streams to detect emergent topics. It focuses on identifying topics that occur in a certain time period on blogs and X social network through the statistical values of tag pairs and correlations between them. Gaglio et al. [7] presented X Live Detection Framework (TLDF) that uses Soft Frequent Pattern Mining (SFPM) algorithm to identify specific topics in a macro event of a X data stream. TLDF uses a dynamic temporal window to detect real-time events based on term associations. TLDF utilizes applicable mechanisms for reducing the number of terms to be considered, such as the term selection method by applying SFPM for generating top-k terms from the set of tweets.

In addition to the TLD, other implementations of term weighting-based methods have been developed, such as the X Monitor [8], Tweet Info [9], Trending Score [10]. All these methods aim to discover discriminative keywords and keyword pairs within a specific time window. The various implementations may differ on the term selection methods that they employ which ultimately affect the variations in the tracking and computation of term correlations.

2.2 Topic Modeling Based Methods

Topic-modelling-based methods focus on extracting latent word distributions of latent events. The general approach is that tweets are represented as a multinomial distribution over the latent events, and each event is characterized by a distribution of the terms contained in those tweets. Tweets consist of different topics and for each topic, they have a probability distribution over the terms in contained [1]. Some well-known topic models such as Latent Dirichlet Allocation (LDA) and Latent Semantic Analysis (LSA) have been used for discovering events on social media. LDA is the most used probabilistic topic model [11]. LDA is an unsupervised approach to finding groups of words called "topics" in a corpus. These topics are made up of words that often occur together and often share a common theme. Therefore, these topics with a predefined set of words can be used as phrases to best describe the entire document. LDA reveals latent word distributions using a Dirichlet process. The other popular model is known as Latent Semantic Analysis (LSA) that extracts word distributions of topics from documents. Differently from LDA, LSA converts the texts into word-topic and document-topic matrices. After then, it applies truncated singular value decomposition. However, capturing good topics from a tweet can be problematic due to the limitation placed on the length of tweets. For event discovery in social media, there exist some methods developed based on these topic models. One of the most popular implementations known as TopicSketch was carried in [12]. The method models the entire tweet stream as a mixture of multiple latent streams and identifies the top k hidden topics whose ratio is greater than a predefined threshold in order to detect bursty events based on an acceleration on the X stream. Latent Semantic Analysis (LSA)-based sound event detection system was presented in [13] by taking the co-occurrence of events which is represented as the degree of their overlapping in a fixed length. The events are clustered using K-means and an observation matrix that is decomposed by using the singular value decomposition.

2.3 Clustering Based Methods

Clustering based methods are unsupervised methods which take a series of objects such as tweets, then organize or group them based on their similarity to find clusters [14]. Clustering-based methods for event discovery can use incremental clustering (partitional clustering algorithm) approaches or community detection algorithms [15] presented ReDites system which discover events related to security domain such as violence, terrorism, and natural disasters.

3 Method and Material

In this chapter, we provide a detail description about our proposed method within the framework of our research. We propose a community-based event discovery system that discover events by applying community detection on the graph representing the co-occurrence relations of n-grams that are frequently observe in X posts. Figure 1 demonstrates the main steps of the proposed method. We firstly created our corpus after having access to X data stream using X API. We downloaded the tweets about different events and pre-processed them. After creating our dataset, we extracted the n-grams, constructed co-occurrence word graph, applied community detection algorithm to detect events and obtained word distributions of the events.

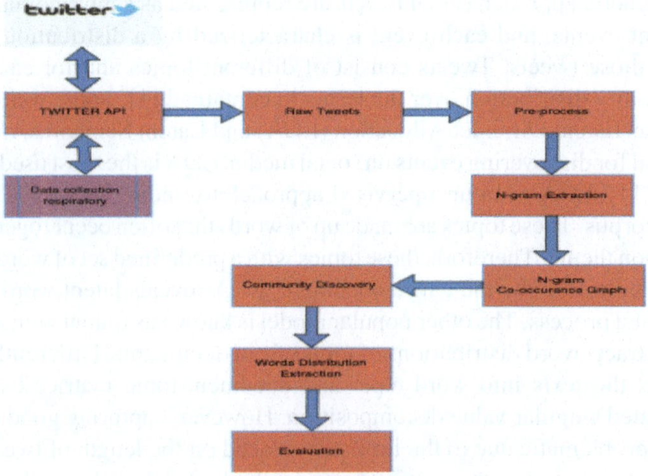

Fig. 1. The illustration of our proposed model

3.1 Creating Corpus

We created our dataset D using tweets ID from the popular X 2012 corpus which contains tweets IDs of different events that were collected between the years 2012 to 2016. This corpus consists of 30 different events which are associated with real world events such as "Ebola" in west Africa, "Brexit" which happened in the UK, "US election". These IDs are only intended for academic purposes and the tweets are downloaded using X API. X API is an application interface that allows access to X in a unique and advance way. Some functions of X API include, sorting, viewing, retrieving and filtering of data on X. For example, a user can create a personal project to observe the sentiment of his own tweets. Build an application which can retrieve user location information of tweets. Users can also filter different events base on their importance.

3.2 Pre-processing Step

In pre-processing step, we cleaned our dataset by removing special characters, numbers, emoji, url link and any other unnecessary character in our dataset. Since our corpus D is obtained from X, there were many unnecessary characters from the tweets. Though some words that were not in the NLTK dictionary (written in English) were deleted while some words that are relevant to our research like "Ebola" were preserved in order to obtain a better performance and to deal with sensitivity.

3.3 Community Detection Process

After constructing our graph structure, we applied Louvain algorithm in order to discover events in our corpus. Louvain algorithm is a greedy algorithm which works in an iterative manner, capable of having a series of community hierarchy. Louvain algorithm shows that every clustering steps of nodes are reduced by a single node which belong to same cluster (Lu et al., 2015). This is an unsupervised algorithm which is divided into two phases namely Modularity Optimization and Community Aggregation. These two steps are repeated until there are no more changes left in the network and the main aim is to achieve a maximum modularity.

Modularity is a measure of the structure of a graph within a community. Modularity optimization is a method used to detect communities in each graph based on their modularity scores. The Maximum modularity quantifies the quality of obtained communities. The formula for calculating modularity optimization is given below:

$$M = \frac{1}{2m}\sum\nolimits_{i,j}(A_{ij} - P_{ij})\delta(c_i, c_j) = \frac{1}{2m}\sum\nolimits_{i,j}(A_{ij} - \frac{k_i k_j}{2m})\delta(c_i, c_j) \qquad (1)$$

where, A_{ij} is the adjacency matrix which represents the weight of the edge connecting the two nodes i and j, k_i represents the degree of the node i, c_i refers to the community of node i. If c_i and c_j are same, $\delta-$ function $\delta(u, v)$ outputs 1 and 0 otherwise. m corresponds to the number of the edges in the graph [16].

Louvain algorithm firstly assigns each node to its own community. Each node is removed from its own community and moved to the community of each neighbor of this node. Then, the modularity is calculated and the new communities are obtained based on the highest modularity increase. This process is repeated for all nodes until an increase in the modularity value is not achieved.

The next step after completing modularity optimization is community aggregation. Here all the nodes which belongs to the same community are merged into a single big node. The links which connect the big nodes are the sum of the ones from the previously connected nodes from the same different communities. This step generates self-loops which are the sum of all the links inside a given community before they are merged into one node (Fig. 2).

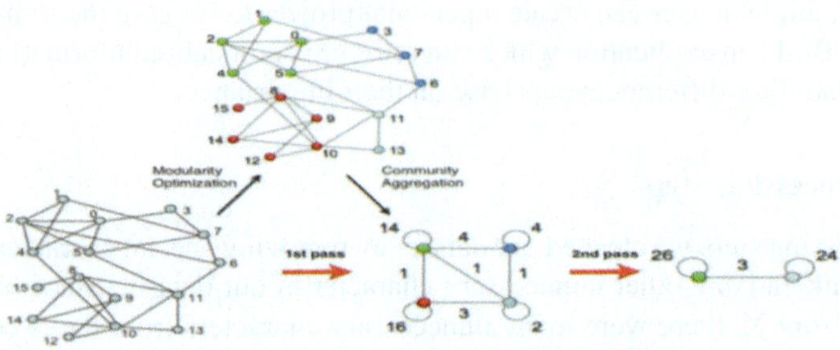

Fig. 2. The illustration of Louvain algorithm (Blondel et al. 2008)

Figure 3 shows the main steps of Louvain algorithm. We apply Louvain algorithm to our n-gram co-occurrence graph to detect communities. Each community refers to one event in our system.

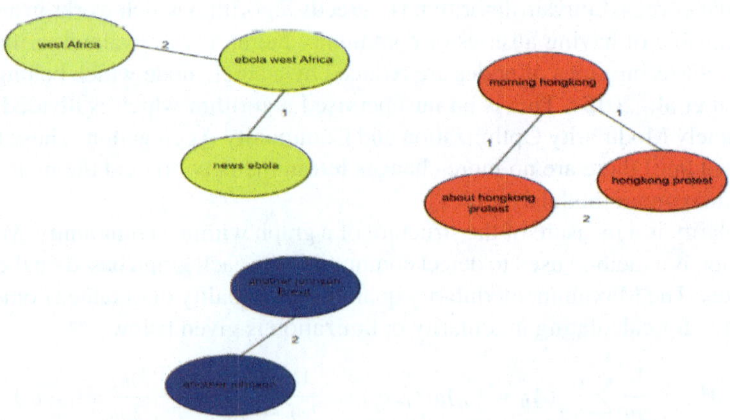

Fig. 3. The obtained communities after applying Louvain algorithm on n-gram co-occurrence graph

The above figure explains how louvain algorithm is used to detect three different events which are represented in a colorful way. These three events are Ebola, Hongkong protest and Brexit.

3.4 Extracting Word Distribution of Events

After community detection process, we extract the word distributions describing contextually these events. Each community obtained by the Louvain algorithm contains different n-grams. Each word that occurs in the n-grams observed in a community is listed. A word can be observed in more than one n-gram. In this community, the frequency of how many n-grams this word appears is calculated. The frequency value calculated for each word is normalized by dividing the total frequency value. This value represents the probability that a word belongs to this community. In other words, we calculate the probability of each word observed in each event.

4 Discussion and Results

4.1 Dataset

This research is carried out using the X Event 2012 dataset. It is a collection of 30 different real-world events that were discussed and collected from X from the years 2012 to 2016 using the X API. This dataset is strictly for non-commercial purposes. Table 1 shows the classes (tags) of 30 different events that occurred between 2012 and 2016 (Fig. 4).

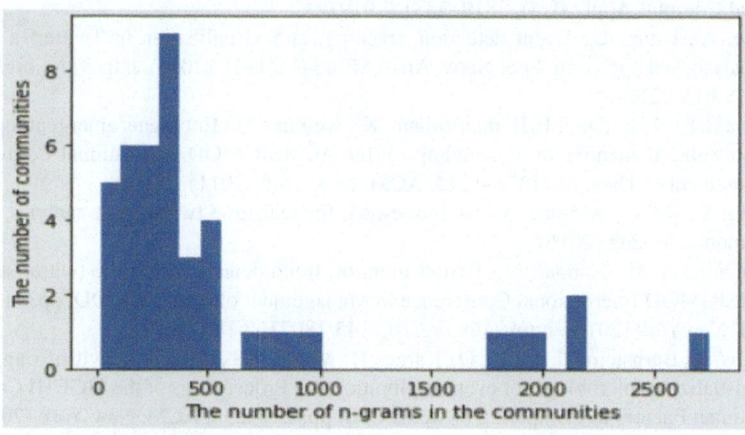

Fig. 4. The distribution of communities produced by community detection algorithm for the dataset consisting of six events.

The distribution of the words included in the six communities is presented in Table 4. In this experiment, the events were "Hongkong Protests", "Ebola", "Hurricane Sandy", "Euro 2012", "Ferguson" and "Obama-Romney". The word distributions in the obtained communities were examined and compared with the selected real event classes. As a result, it is observed that the six selected communities were discovered correctly.

5 Conclusion and Implications

We propose an event discovery system on X using n-gram based graph structure and community detection algorithm. One of the major challenges in event discovery is that most traditional methods have to estimate the number of events in order to be able to contextually analyze events. Another problem is that some methods tend to detect events that often trigger a significant volume of communication. Our proposed method discovers the events in an unsupervised manner. Moreover, it also has the ability to reveal events that do not trigger much interaction compared to other events. Our proposed method shows a very promising performance on unbalanced datasets compared to the previous methods.

References

1. Hasan, M., Orgun, M.A., Schwitter, R.: Real-time event detection from the Twitter data stream using the TwitterNews+ framework. Inf. Process. Manage. **56**(3), 1146–1165 (2019). https://doi.org/10.1016/j.ipm.2018.03.001. (Pew Research Center, 2021)
2. Pew Research Center. News Consumption Across Social Media in 2021, s.l.: Pew Research Center (2021)
3. Al-Dyani, W.Z., Yahya, A. H., Ahmad, F.K.: Challenges of event detection from social media streams. Int. J. Eng. Technol. **7**(2.15), 72–75 (2018)
4. Alsmadi, I., Hoon, G.K.: Term weighting scheme for short-text classification: twitter corpuses. Neural Comput. Appl. **31**(8), 3819–3831 (2019)
5. Guille, A., Favre, C.: Event detection, tracking, and visualization in Twitter: a mention-anomaly-based approach. Soc. Netw. Anal. Min. **5**(1), 1–18 (2015). https://doi.org/10.1007/s13278-015-0258-0
6. Alvanaki, F., Sabastian, M., Ramamritham, K., Weikum, G.: Enblogue: emergent topic detection in web 2.0 streams. In: Proceedings of the ACMSIGMOD International Conference on Management of Data, pp. 1271–1274. ACM, New York (2011)
7. Gaglio, S., Re, G., Morana, M.: A framework for real-time twitter data analysis. Comput. Commun. 236–242 (2016)
8. Mathioudakis, M., Koudas, N.: Twitter monitor: trend detection over the twitter stream. In: ACMSIGMOD International Conference in Management of data SIGMOD, pp. 1155–1158. ACM, NewYork (2010). https://doi.org/10.1145/1807167.1807306
9. Marcus, A., Bernstein, M., Badar, O., Karger, D., Madden, S., Miller, R.: TwitInfo: aggregating and visualizing microblogs for event exploration. In: Proceedings of the SIGCHI Conference on Human Factors in Computing Systems CHI, pp. 227–236. ACM, New York (2011)
10. Benhardus, J., Kalita, J.: Streaming trend detection in Twitter. Int. J. Web Based Commun. 122–139 (2013)
11. Blei, D., Ng, A., Jordan, M.: Latent dirichlet allocation. J. Mach. Learn. Res. 993–1022 (2003)
12. Xie, W., Ren, L., Li, J., Wu, F., Meng, W.: TopicSketch: Real-time bursty topic detection in social media. In: Proceedings of the 2014 ACM SIGMOD International Conference on Management of Data (SIGMOD'14) (2014)
13. Mesaros, A., Heittola, T., Klapuri, A.: Latent semantic analysis in sound event detection. In: 2011 19th European Signal Processing Conference, pp. 1307–1311. IEEE (2011)
14. Karandikar, A.: Clustering short status messages: a topic model based approach. M.S. thesis, Dept. Comput. Sci. and Electr. Eng., Univ. of Maryland, Baltimore County, MD (2010)
15. Osborne, M., Moran, S., McCreadie, R., Von Lunen, A., Sykora, M., Cano, E.: Real time detection, tracking and monitoring of automatically discovered events in social media. In: Proceedings of the 52nd Annual Meeting of the Association for Computational Linguistics (2014)
16. Blondel, V.D., Guillaume, J.L., Lambiotte, R., Lefebvre, E.: Fast unfolding of communities in large networks. J. Stat. Mech: Theory Exp. **2008**(10), P10008 (2008)

Advanced Risk Management Practices - The Application of the Exact Solutions Methods for Resilience Factors Improvement

Aleksandar V. Aleksić, Michael Huber(✉), Danijela P. Tadić, and Snežana B. Nestić

Faculty of Engineering, University of Kragujevac, Sestre Janjic 6, 34000 Kragujevac, Serbia
aaleksic@kg.ac.rs, mail@hubermichael.de

Abstract. In the context of organizational resilience, branch and bound algorithms can be leveraged to determine the most suitable management techniques from a set of options, considering factors such as cost and time constraints. This paper presents a novel approach for assessing the suitability of methods aimed at improving organizational Resilience Factors (RFs). The proposed algorithm integrates fuzzy logic and exact solutions methods. The algorithm comprises several steps, including fuzzy rating of method suitability, determination of aggregate suitability values, and assessment of method application costs. Furthermore, a multidimensional optimization problem is formulated to maximize the overall effectiveness of selected methods while considering budgetary constraints. To efficiently navigate the solution space and identify near-optimal solutions, a Branch and Bound algorithm is employed. The proposed framework offers a comprehensive methodology for decision-makers to make informed choices regarding method selection and resource allocation for business process improvement initiatives. Experimental results demonstrate the efficacy and practical applicability of the proposed approach in real-world scenarios.

Keywords: Resilience Factors · Optimization · Branch and Bound

1 Introduction

In today's rapidly changing and uncertain business landscape, the ability of organizations to adapt and thrive in the face of adversity is paramount. This ability, often referred to as organizational resilience, encompasses the capacity to anticipate, respond to, and recover from disruptive events while maintaining core functions and preserving essential services [1]. As organizations seek to enhance their resilience in the face of complex challenges, the application of innovative methodologies becomes increasingly crucial [2].

This paper explores the potential of branch and bound method application as a novel approach to enhancing organizational RFs. Branch and bound algorithm, as a class of optimization techniques, offers a systematic and efficient way to explore solution spaces and identify optimal or near-optimal solutions [3]. While traditionally applied in fields such as operations research and computer science, their application in the realm of organizational resilience remains relatively unexplored.

© The Author(s), under exclusive license to Springer Nature Switzerland AG 2025
N. Filipović (Ed.): AAI 2024, LNNS 1446, pp. 9–16, 2025.
https://doi.org/10.1007/978-3-031-99201-8_2

The primary objective of this paper is to bridge this gap by investigating how branch and bound algorithms can be leveraged to strengthen organizational resilience. As organizational RFs can be identified and assessed [4], the main objective of this research is to propose a model to improve RFs which are denoted as the weakest link in the system of organizational resilience. By integrating principles of optimization with established theories of resilience, we aim to provide insights into how organizations can effectively navigate uncertainty and build adaptive capacity.

While there is a wealth of literature on organizational resilience and exact optimization methods independently, there is a notable lack of research at the intersection of these two domains. Existing studies have primarily focused on traditional approaches to resilience enhancement, such as risk management strategies and organizational learning processes, overlooking the potential of advanced optimization methods [5]. Furthermore, the few studies that have examined the application of exact optimization methods or metaheuristic methods in organizational resilience have often been theoretical, lacking empirical validation and practical insights [6].

By exploring the intersection of optimization executed by exact optimization methods and organizational resilience, this paper aims to contribute to a deeper understanding of how organizations can effectively navigate uncertainty and build sustainable competitive advantage in today's dynamic business environment. For the optimized enhancement of identified RFs, the branch and bound algorithm is employed.

The paper is structured to include five sections. Section 2 contains the literature review where organizational resilience and exact optimization methods are mainly discussed, while Sect. 3 is used to present the theoretical framework and methodology. Section 4 presents a case study with a discussion of the obtained results, and Sect. 5 provides the conclusion of the research.

2 Literature Review

Organizational resilience has garnered increasing attention in the literature as organizations grapple with the challenges posed by globalization, technological advancements, and environmental uncertainty. Researchers have defined organizational resilience as the ability of an organization to anticipate, respond to, and recover from disruptions while maintaining core functions and adapting to changing circumstances [7]. Various theoretical frameworks have been proposed to conceptualize and measure organizational resilience, including the Dynamic Capabilities Perspective [8], or the Resilience Engineering Framework [9]. The research presented in this paper relies on the RFs fuzzy Delphi assessment [10].

Branch and bound algorithms, a class of optimization algorithms, are particularly well-suited for solving combinatorial optimization problems by systematically exploring solution spaces and pruning suboptimal branches [11]. While branch and bound algorithms have primarily been applied in operations research, computer science, and engineering [12], their potential application in enhancing organizational resilience remains underexplored [13]. It is worth mentioning that the branch and bound algorithms are widely applied in management specifically for the Allocation of Business Model Components while analyzing different business models [14].

Few studies have explored the application of exact optimization methods or heuristic optimization algorithms to enhance organizational resilience [15]. Specifically, the branch and bound algorithms are employed only for resilient path selection in networks [16]. This gap presents a significant opportunity for researchers to leverage the exact solutions methods and heuristic optimization techniques to address the complex and dynamic nature of organizational resilience. The literature review highlights the significance of organizational resilience in today's business environment and the potential of exact solutions methods and heuristic optimization techniques, to enhance RFs. By integrating insights from both domains, this paper seeks to contribute to a deeper understanding of how organizations can effectively navigate uncertainty and build adaptive capacity.

3 Theoretical Framework and Methodology

It might seem obvious that organizations need to develop a resilience capacity to overcome different impacts and to continue successful business operations, in other words, to handle unexpected events, bounce back from Key Performance Indices downfall, and even improve future success [17]. This paper sets the model for optimized enhancement of the RFs considering different constraints.

Organizational resilience is often characterized by a set of interrelated factors that enable organizations to effectively anticipate, respond to, and recover from disruptions. Generally, RFs may be presented through a set of indices $\{1, .., J, .., J\}$. The total number of RFs is denoted as J and $j, j = 1, .., J$ is an index of RF. Drawing on existing literature [10], we identify three key RFs that are particularly critical for organizational success: (1) Redundancy, (2) Self-organization, and (3) Awareness.

To enhance the three key RFs – RF1 - Redundancy, RF2 - Self-organization, and RF3 - Awareness – this study proposes the methodology with the following steps:

Identification of Management Techniques:
The set of methods that could be used for the RFs' enhancement should be identified by drawing on existing literature [18] and expert consultation. For the presented research, the total number of the methods that can be applied for enhancement RF $j, j = 1, .., J$ is denoted as M_j and $m, m = 1, .., M_j$ is an index of methods.

Assessment of the Management Techniques Suitability:
Each treated RF can be enhanced by executing one or several techniques for analysis and improvements [18]. It is assumed that the Operations Manager has chosen the set of methods at the level of each RF that can be applied to enhance the treated RF. In the general case, the methods are presented as a set $\{1, .., m, .., M_j\}$.

The method suitability $m, m = 1, .., M_j$ for the improvement of RF $j, j = 1, .., M$ is stated as a fuzzy group decision-making problem, so the group of three Decision Makers (DMs) is in charge of the assessment. Formally, DMs that participate in the assessment are denoted by a set $\{1, ..., e, .., E\}$. The total number of DMs is denoted as E. The index of DM is noted as $e, e = 1, ., E$. In this manuscript, DMs are: Chief Executive Officer $e = 1$; Operations Manager $e = 2$; Management System Manager $e = 3$. As is well known, it is closer to the human way of thinking that assessments are expressed using linguistic terms than precise numbers of some measurement scales. The fuzzy sets theory [19, 20] presents an appropriate mathematical tool that provides an opportunity to present words of natural language in a quantitative manner.

In this manuscript, the method suitability for the improvement of RFs is described by pre-defined linguistic expressions which are modeled by fuzzy numbers. The main features are membership function, granularity, and the domain.

In the literature, a large number of authors suggest the use of Triangular Fuzzy Numbers (TFNs) for modeling uncertainties and imprecise [18]. They form their opinion on the fact that TFNs do not require complex mathematical calculations and at the same time capture the uncertainties of natural language in a sufficiently good way.

In works that can be found in the literature, the granulation is most often determined based on the subjective assessment of DMs respecting the size and complexity of the problem. The theoretical insight [22] suggests that DMs can use a maximum of seven linguistic expressions, as in [23] and this research. These pre-defined linguistic expressions and their corresponding TFNs are presented:

extremely low applicability -(1,1, 2.5)
very low applicability -(1,2, 3)
low applicability -(2,3.5,5)
medium applicability -(3.5,5, 6.5)
high applicability -(5,6.5,8)
high applicability -(7,8, 9)
extremely high applicability -(7.5,9, 9)

Domains of the defined TFNs are defined on a common measurement scale [24]. A value of 1 indicates that the applicability of the method is almost negligible. On the other hand, a value of seven indicates that the suitability of the method certainly leads to the improvement of RFs.

Improving the resilience of a company can be achieved through simultaneously improving the value of considered RFs. It is assumed that each RF can be improved at least by applying one or more predetermined methods. Within the introduced assumptions, the problem of selection of improvement methods is stated as a multidimensional knapsack problem with mixed max/min constraints.

Cost and Time Evaluation:
Besides the assessment of the management technique's suitability, there is a task to assess the costs and time needed for the execution of the method. The estimates for each management technique are based on empirical data or expert judgment. Of course, this task is intended to be executed by the operations manager.

Formulation of Optimization Problem:
This step implies the definition of the optimization problem as selecting the subset of management techniques that maximize resilience enhancement while minimizing total costs and time required for implementation, subject to budget constraints. The proposed algorithm is further explained.

The Proposed Algorithm:
The proposed Algorithm can be realized through the steps which are further presented.

Step 1. Fuzzy rating of method suitability $m, m = 1, .., M_j$ for the enhancement of RF $j, j = 1, .., J$ at the level of DM $e, e = 1, .., E$ is described by one of the pre-defined linguistic expressions and modeled by TFN $\tilde{x}_{jm}^e$:

where:

$$\tilde{x}_{jm}^e = \left(a_{jm}^e, b_{jm}^e, c_{jm}^e\right) \tag{1}$$

Step 2. Let us determine the aggregate value of the method suitability $m, m = 1, .., M_j$ at the level of each RF $j, j = 1, \ldots, J$ by applying the fuzzy geometric mean operator, $\tilde{x}_{jm}$:

$$\tilde{x}_{jm} = \left(\sqrt[E]{\prod_{e=1,..,E} a_{jm}^e}, \sqrt[E]{\prod_{e=1,..,E} b_{jm}^e}, \sqrt[E]{\prod_{e=1,..,E} c_{jm}^e}\right) = \left(a_{jm}, b_{jm}, c_{jm}\right) \tag{2}$$

Step 3. Representative scalar TFN $\tilde{x}_{jm}, x_{jm}$, is obtained by applying the center gravity method [25], so that:

$$x_{jm} = a_{jm} + \frac{c_{jm} - a_{jm} + b_{jm} - a_{jm}}{3} \tag{3}$$

Step 4. The assessment of cost for each technique application $\gamma_m, m = 1, .., M$ is performed by Operations Manager. This person bases the assessments on experience, evidence data as well as the results of the best practice.

Step 5. A multidimensional KP problem with max/min constraints is presented:

Fitness function:

$$\max_{m=1,..,M} x_{jm} \tag{4}$$

Subject to:

$$\sum_{m=1,..,M} \gamma_m \leq \mathcal{G} \tag{5}$$

where $\mathcal{G}$ the total available budget for the improvement of RFs during one business year.

Application of Branch and Bound Algorithm:
In the end, a branch and bound algorithm is implemented to systematically explore the solution space and to identify the optimal or near-optimal solution.

4 The Case Study

A case study is implemented in a complex international company that produces high-tech equipment for quality management in different industries. The team of DMs has been appointed in compliance with the research guidelines. They had several meetings to provide an assessment of the methods' suitability for the RFs enhancement. The operations manager has provided an assessment in terms of the costs and time needed for the execution of the method.

The proposed Algorithm (Step 1 to Step 3) is illustrated for method $m = 9$ for RF ($j = 1$). The DMs have assessed the suitability of the method $m = 9$ for the enhancement of RF $m = 9$ as follows:

$$L3, L5, L4 \tag{6}$$

The aggregated value is obtained by the application of fuzzy cubic mean:

$$\tilde{x}_{19} = \left(\sqrt[3]{\frac{1}{3} \cdot (2^3 + 5^3 + 3.5^3)}, \sqrt[3]{\frac{1}{3} \cdot \sum_{e=1,\ldots,E} (3.5^3 + 6.5^3 + 5^3)}, \sqrt[3]{\frac{1}{3} \cdot (5^3 + 8^3 + 6.5^3)} \right) = (3.88, 5.28, 6.72) \tag{7}$$

The representative scalar TFN, $\tilde{x}_{19}$ is obtained by the application of the center of gravity defuzzification method:

$$x_{19} = 3.88 + \frac{6.72 - 3.88 + 5.28 - 3.88}{3} = 5.30 \tag{8}$$

Similar calculations will provide the crisp values of the suitability of the method whose execution should enhance identified RFs.

By applying the adequate software solution that integrates the branch and bound algorithm execution [26], we obtain the optimal set of methods that need to be executed for the RFs enhancement in the treated company. The optimal solution is found by the software execution in the 23rd iteration after 0,041872 s.

The methods to be applied to enhance each of the three observed RFs are given in Table 1.

Table 1. The Optimal Set of Methods.

The Treated RFs	RF ($j = 1$)	RF ($j = 2$)	RF ($j = 3$)
The Optimal Set of Methods	$m = 2$; $m = 6$; $m = 20$; $m = 21$; $m = 23$; $m = 24$	$m = 3$; $m = 5$; $m = 6$; $m = 9$; $m = 14$; $m = 16$; $m = 2$; $m = 6$; $m = 20$; $m = 21$; $m = 24$	$m = 3$; $m = 6$; $m = 9$; $m = 20$; $m = 21$; $m = 23$

Taking into account the information provided by Table 1, it may be concluded that methods that should be executed in the first place, are those which are common for every RF. Namely, those are generic flowcharts, brainstorming, and brain writing techniques. After that, the methods that could improve at least two RFs that should be applied until the defined budget is met are balanced scorecards, generic checklists, mind maps, 5W2H, and importance-performance analysis. Finally, relations diagrams could be defined and/or generic tables.

5 Conclusions

In this paper, the theoretical foundation for the study is established, outlining the principles of organizational resilience and branch and bound algorithms, and exploring their potential synergy. The research is described including the implementation of branch and bound algorithms within an organizational context and the steps involved in enhancing resilience factors.

A case study is presented illustrating the application of branch and bound heuristics to enhance organizational RFs, providing real-world insights into the efficacy of this approach. The results indicate the order of management methods that should be executed to enhance organizational RFs most effectively. The chosen tools for the improvement of RFs are mainly generic in nature so their application may have a wide range of benefits. Taking into account the budget for the overall RFs improvement, it is considered that their execution will improve overall business performance and prepare for different disruptions.

The main constraints of the presented research are the following. The structured methodology demands several DMs that are highly skilled, simultaneously holding appropriate business administrative positions to provide the input data. This part of the algorithm execution is the most sensitive one since low-skilled managers may bring wrong input data. The determination of the budget intended for the RFs improvement can also be difficult to perform.

Future research should be scoped to the conducting of case studies in similar organizations so appropriate statistical tests could be conducted followed by appropriate conclusions. Also, heuristic algorithms and meta-heuristic algorithms could be applied to obtain the solution of methods that could be applied for the RFs enhancement. This would leave space for further comparisons and discussions.

References

1. Hillmann, J., Guenther, E.: Organizational resilience: a valuable construct for management research? Int. J. Manag. Rev. **23**(1), 7–44 (2021)
2. Hepfer, M., Lawrence, T.B.: The heterogeneity of organizational resilience: exploring functional, operational and strategic resilience. Organ. Theory **3**(1), 1–29 (2022)
3. Morrison, D.R., Jacobson, S.H., Sauppe, J.J., Sewell, E.C.: Branch-and-bound algorithms: a survey of recent advances in searching, branching, and pruning. Discret. Optim. **19**, 79–102 (2016)
4. Macuzić, I., Tadić, D., Aleksić, A., Stefanović, M.: A two-step fuzzy model for the assessment and ranking of organizational resilience factors in the process industry. J. Loss Prev. Process Ind. **40**, 122–130 (2016)
5. Nozhati, S.: A resilience-based framework for decision making based on simulation-optimization approach. Struct. Saf. **89**, 102032 (2021)
6. Arsovski, S., et al.: Modelling and enhancement of organizational resilience potential in process industry SMEs. Sustainability **7**(12), 16483–16497 (2015)
7. Ponomarov, S.Y., Holcomb, M.C.: Understanding the concept of supply chain resilience. Int. J. Logist. Manage. **20**(1), 124–143 (2009)
8. Teece, D.J., Pisano, G., Shuen, A.: Dynamic capabilities and strategic management. Strateg. Manage. J. **18**(7), 509–533 (1997)

9. Hollnagel, E.: Resilience - The challenge of the unstable. In: Resilience engineering, pp. 9–17, CRC Press (2017)
10. Aleksic, A., Nestic, S., Tadic, D., Komatina, N.: Determination of organizational resilience level within business processes in production companies. In Proceedings of the 6th International Scientific Conference COMET-a. Jahorina, Bosnia and Herzegovina, pp. 750–757, (2022)
11. Chmiela, A., Khalil, E., Gleixner, A., Lodi, A., Pokutta, S.: Learning to schedule heuristics in branch and bound. Adv. Neural. Inf. Process. Syst. **34**, 24235–24246 (2021)
12. Xiao, Y., Zheng, Y., Yu, Y., Zhang, L., Lin, X., Li, B.: A branch and bound algorithm for a parallel machine scheduling problem in green manufacturing industry considering time cost and power consumption. J. Clean. Prod. **320**, 128867 (2021)
13. Coniglio, S., Furini, F., San Segundo, P.: A new combinatorial branch-and-bound algorithm for the knapsack problem with conflicts. Eur. J. Oper. Res. **289**(2), 435–455 (2021)
14. Nestić, S., Aleksić, A., Lafuente, J. G., Cvetić, T., Đurić, G.: The Allocation of business model components under presence of uncertainties by the branch-and-bound method. Math. Prob. Eng. 2958519 (2022)
15. Hassani, L., Ghanbari, R.: The optimization of resilience and sustainability using mathematical programming models and metaheuristic algorithms. J. Clean. Prod. **228**, 1062–1072 (2019)
16. Casazza, M., Ceselli, A.: Optimization algorithms for resilient path selection in networks. Comput. Oper. Res. **128**, 105191 (2021)
17. Duchek, S.: Organizational Resilience: A Capability-Based Conceptualization. Bus. Res. **13**, 215–246 (2020)
18. Tague, N. R.: The quality toolbox. Quality Press (2023)
19. Dubois, D., Prade, H.: Systems of linear fuzzy constraints. Fuzzy Sets Syst. **3**(1), 37–48 (1980)
20. Lehmann, I., Weber, R., Zimmermann, H.-J.: Operations-Research-Spektrum **14**(1), 1–9 (1992). https://doi.org/10.1007/BF01783496
21. Tadic, D., Gumus, A.T., Arsovski, S., Aleksic, A., Stefanovic, M.: An evaluation of quality goals by using fuzzy AHP and fuzzy TOPSIS methodology. Journal of Intelligent & Fuzzy Systems **25**(3), 547–556 (2013)
22. Lootsma, F. A.: Numerical scaling of human judgement in pairwise-comparison methods for fuzzy multi-criteria decision analysis. In Mathematical models for decision support (pp. 57–88). Berlin, Heidelberg: Springer Berlin Heidelberg (1988)
23. Nestic, S., Lampón, J.F., Aleksic, A., Cabanelas, P., Tadic, D.: Ranking manufacturing processes from the quality management perspective in the automotive industry. Expert. Syst. **36**(6), e12451 (2019)
24. Saaty, T.L.: The modern science of multicriteria decision making and its practical applications: The AHP/ANP approach. Oper. Res. **61**(5), 1101–1118 (2013)
25. Wang, W.J., Luoh, L.: Simple computation for the defuzzifications of center of sum and center of gravity. Journal of Intelligent & Fuzzy Systems **9**(1–2), 53–59 (2000)
26. OR-Tools CP-SAT v9.9. Laurent Perron and Frédéric Didier. URL: https://developers.google.com/optimization/cp/cp_solver, last accessed 2024/06/16

Efficient Sequential Detection: Enhancing Cancer Detection in Sequential MRI Sequences

Aleksandar Cvetković(✉) and Veljko Papić

School of Electrical Engineering, University of Belgrade, Bulevar Kralja Aleksandra 73, 11000 Beograd, Serbia
cvetkovicaleksandar1302@gmail.com, papic@etf.bg.ac.rs

Abstract. Conventional object detection models are typically trained to detect objects in images devoid of sequential correlation. However, in the context of MRI imaging, where data is inherently sequential, valuable information lies within the adjacent slices for cancer detection. While 3D Neural Networks present a logical avenue to exploit this sequential nature, their high parameter counts, and complex training dynamics pose significant challenges. Additionally, mutual attention mechanisms, although effective, often require a large volume of training data, which may not always be readily available in medical settings. In response to these challenges, our proposed method efficiently harnesses sequential characteristics of MRI data by integrating information from preceding, current, and succeeding slices. Notably, this integration is strategically incorporated into specific segments of the network architecture to optimize performance. We provide empiric results and give practical advice for using our method with existing architectures. By seamlessly integrating this approach into the YOLOv5 framework, we achieved a notable 2.8% enhancement in the mean Average Precision (mAP).

Keywords: object detection · sequential data · YOLOv5 · brain cancer · MRI

1 Introduction

Brain cancer is among the most devastating and life-threatening diseases, often presenting significant challenges to both patients and healthcare providers. The urgency of early detection cannot be overstated, as timely intervention significantly impacts treatment outcomes and patient survival rates. Magnetic Resonance Imaging (MRI) stands as a cornerstone technique in the diagnosis of brain cancer, offering unparalleled insights into the intricate structures of the brain with high resolution and non-invasiveness. However, despite the utility of MRI in diagnosing brain cancer, its detection remains a complex and nuanced endeavor, and the shortage of experienced radiologists further compounds the issue, as an accurate interpretation of MRI scans demands specialized expertise and time-intensive analysis.

In recent years, the advent of deep learning has revolutionized medical image analysis, offering a promising avenue for automated and expedited brain cancer detection [1]. Deep learning models can efficiently sift through vast amounts of imaging data, swiftly

© The Author(s), under exclusive license to Springer Nature Switzerland AG 2025
N. Filipović (Ed.): AAI 2024, LNNS 1446, pp. 17–24, 2025.
https://doi.org/10.1007/978-3-031-99201-8_3

identifying subtle anomalies indicative of cancerous growths. However, traditional deep learning approaches, particularly general-purpose object detection models, are primarily designed for static image analysis, overlooking the inherent sequential nature of MRI scans. Unlike conventional images, MRI sequences are inherently sequential, with each slice providing unique insights into the three-dimensional structure of the brain. Leveraging this sequential information is crucial for accurate cancer detection, as valuable contextual cues may span across adjacent slices. While 3D Neural Networks [2] present a natural fit for exploiting this sequential nature, they pose significant challenges due to their immense parameter count and complex training dynamics. Moreover, we show that it is suboptimal to apply joint processing of sequential scans too early in networks layers. Models which make use of mutual attention mechanisms such as Non-Local Neural networks [3] have been shown to greatly improve the capturing of long-range dependencies in sequential image data, which is excellent for video analysis but its application in medical image processing remains limited. Furthermore, training these models usually requires extensive training data, making it a luxury not always available in medical settings.

To address these challenges, our proposed method capitalizes on the sequential nature of MRI data by integrating information from preceding, current, and succeeding slices. This integration is strategically incorporated into specific segments of the network architecture, optimizing performance while mitigating computational overhead. By seamlessly integrating this approach into the YOLOv5 [4] framework, we demonstrate notable enhancements in cancer detection accuracy without compromising model efficiency. Our method not only streamlines the detection process but also lays the foundation for scalable and accessible brain cancer screening, facilitating regular check-ups and early interventions to combat this formidable disease.

2 Proposed Method

We first provide the general architecture of YOLOv5s object detection model and set up the numbering of layers which will be examined as points of merging preceding, current and succeeding slice image data. Then we explain how the merge happens as well as the training mechanisms used to provide the best results.

2.1 YOLOv5s Model Architecture

Figure 1 shows YOLOv5s backbone layers consisting of Conv (Convolution), C3 (Triple Convolution) and SPPF (double convolution followed by a MaxPool operation). On the right side of the image, the layer indices are commented and will be used for reference.

Figure 2 shows the head of YOLOv5s model consisting of Conv, C3, Concat (concatenation), and Upsample operations as well as a Detection head. On the right side of the image, the layer indices are commented and will be used for reference.

When we call a layer with index i a merger point, it means that the merger is occurring just before that layer. The merger operation extracts image features from the preceding, current, and succeeding slice and concatenates them across the channel dimension. It also increases the channel dimension of the following layer 3 times to accommodate for

```
# YOLOv5 v6.0 backbone
backbone:
  # [from, number, module, args]
  [
    [-1, 1, Conv, [64, 6, 2, 2]],          # 1
    [-1, 1, Conv, [128, 3, 2]],            # 2
    [-1, 3, C3, [128]],                    # 3
    [-1, 1, Conv, [256, 3, 2]],            # 4
    [-1, 6, C3, [256]],                    # 5
    [-1, 1, Conv, [512, 3, 2]],            # 6
    [-1, 9, C3, [512]],                    # 7
    [-1, 1, Conv, [1024, 3, 2]],           # 8
    [-1, 3, C3, [1024]],                   # 9
    [-1, 1, SPPF, [1024, 5]],              # 10
  ]
```

Fig. 1. Backbone of YOLOv5s detection model.

```
# YOLOv5 v6.0 head
head: [
    [-1, 1, Conv, [512, 1, 1]],                    # 11
    [-1, 1, nn.Upsample, [None, 2, "nearest"]],    # 12
    [[-1, 6], 1, Concat, [1]],                     # 13
    [-1, 3, C3, [512, False]],                     # 14

    [-1, 1, Conv, [256, 1, 1]],                    # 15
    [-1, 1, nn.Upsample, [None, 2, "nearest"]],    # 16
    [[-1, 4], 1, Concat, [1]],                     # 17
    [-1, 3, C3, [256, False]],                     # 18

    [-1, 1, Conv, [256, 3, 2]],                    # 19
    [[-1, 14], 1, Concat, [1]],                    # 20
    [-1, 3, C3, [512, False]],                     # 21

    [-1, 1, Conv, [512, 3, 2]],                    # 22
    [[-1, 10], 1, Concat, [1]],                    # 23
    [-1, 3, C3, [1024, False]],                    # 24

    [[17, 20, 23], 1, Detect, [nc, anchors]],      # 25
]
```

Fig. 2. Head of YOLOv5s detection model.

the fact that we concatenated channels of 3 images together. This is one and only time that information flows between slices and it only increase in the number of parameters.

3 Experiments

In this section, we present the dataset and explain the training regime used to extract the best results. Next, we provide the mAP scores for vanilla version of YOLOv5s model, YOLOv5m model, our methods with merging channels from successive slices (Merge models), and a dummy model when the same slice is repeated three times (to prove that increase in performance is not entirely caused by the increase in the number of parameters – FakeMerge models).

3.1 Dataset

The dataset used in the experiments is the Brain Tumor on MRI dataset [5] originally labeled for segmentation. This dataset was adjusted for object detection by making bounding boxes out of the segmentation masks. As advised in [6], images that did not include any brain tissue after skull stripping were not used in the training process.

3.2 Training Setup

In our experimental setup, while we leverage successive slices for enhanced detection accuracy, we deviate from patient-level batching for two primary reasons. Firstly, the conventional implementation employs batch normalization [7], a technique that normalizes activations based on statistics computed across all images within a batch. However, when all images originate from a single patient, this normalization process becomes skewed due to the inherent variability among patients' MRI scans. Additionally, shuffling, a crucial step for achieving robust results in experiments, becomes challenging to implement consistently when batching is organized by patient, potentially compromising the integrity of the training process.

That is why for each slice that we need to detect tumor on, we also load its preceding and succeeding slice and we store them in the same batch. As depicted in Fig. 3, after the merger happens, the batch dimension is reduced three times, and these images are concatenated in channel dimension (increase from 3 to 9). This effectively makes the batch size three times smaller but gives the best performance.

3.3 Results

Results were obtained by training the network for 100 epochs and storing the best results on the validation dataset. This process is repeated 5 times in a cross-validation setup [8]. Metric chosen for comparison is mAP50 [9]. We also show the number of parameters for each model.

In Table 1, we present the results of experiments with YOLOv5s, YOLOv5m, and models with merge layers. It is apparent that mAP50 metrics of merge layers are worse than the YOLOv5s baseline model if the merging layer is not inserted into the correct place. The best place for insertion of a merge layer based on the results is in layer 11. If we place close attention to Fig. 1 and Fig. 2, we can see that layer 11 is just between the backbone and the head of the network. This conclusion makes sense: the best place

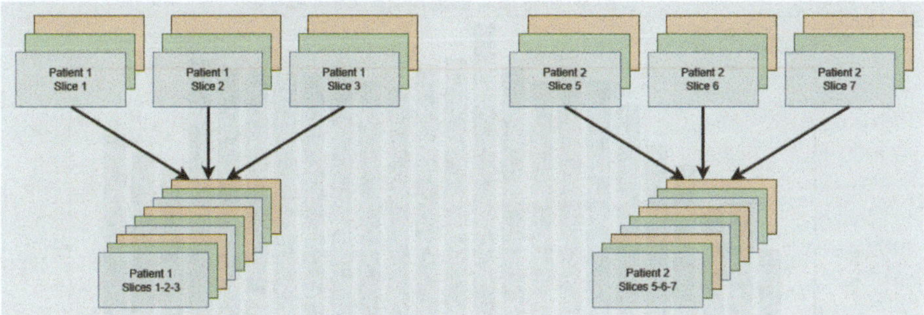

Fig. 3. Merging of slices. Reducing in batch dimension and concatenation along channel dimension.

Table 1. Performance of YOLOv5s, YOLOv5m and models with merge layer.

	YOLOv5s	YOLOv5m	Merge1	Merge2	Merge3	Merge4	Merge5
mAP50	0.770	0.786	0.664	0.691	0.688	0.713	0.725
parameters	7022326	7202294	7202294	7232246	7203574	7342838	7265526
	Merge6	**Merge7**	**Merge8**	**Merge9**	**Merge10**	**Merge11**	**Merge12**
mAP50	0.748	0.762	0.768	0.762	0.770	**0.798**	0.775
parameters	7785206	7457526	9554678	7719670	9293558	**8000246**	7190774
	Merge13	**Merge14**	**Merge15**	**Merge16**	**Merge17**	**Merge18**	**Merge19**
mAP50	0.772	0.758	0.762	0.772	0.767	0.767	0.768
parameters	7190774	7321846	7265526	7062006	7062006	7099382	7317238
	Merge20	**Merge21**	**Merge22**	**Merge23**	**Merge24**		
mAP50	0.770	0.775	0.711	0.721	0.731		
parameters	7092470	7162614	8201974	7293686	7565046		

to reformat the input by inserting the merge layer is between the two distinct parts of the network where a residual connection exists. This way the network can focus on optimizing all the layers in the same way (as opposed to having merged layers at one point of detection head and unmerged at another). This result is also consistent with the experiments we did on image segmentation with UNet where the best place to insert the merger layer is just between the encoder and the decoder modules. These results are graphically shown in Fig. 4 where all models with merging layer are compared to the YOLOv5s baseline. In Fig. 5, we graphically show the number of parameters for each model with merging layer compared to the baseline YOLOv5s model.

Based on the results shown in Table 1, there is a 2.8% increase in mAP50 metric when the merging layer is inserted above the 11th layer of the network. We argue that this increase in performance does not come purely based on the increase in the number of parameters. To prove this, we compare the Merge11 network to the YOLOv5m model

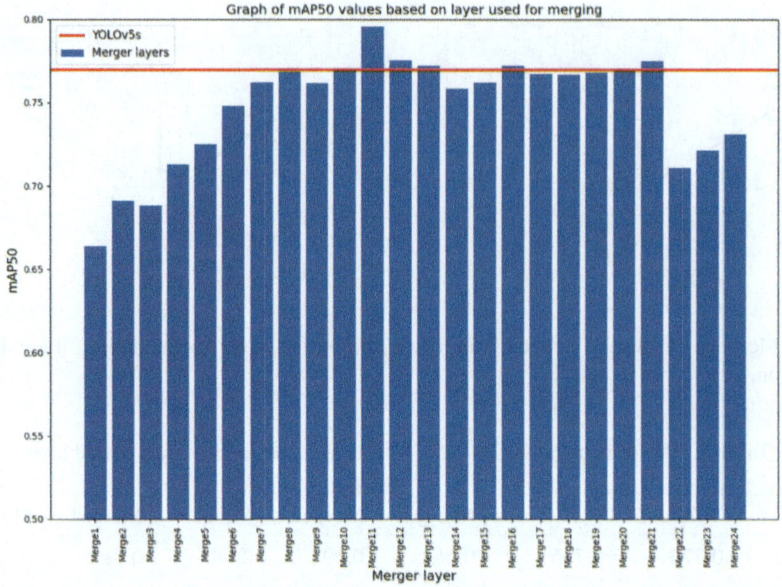

Fig. 4. The mAP50 metric of different merging layers compared to the baseline YOLOv5s model.

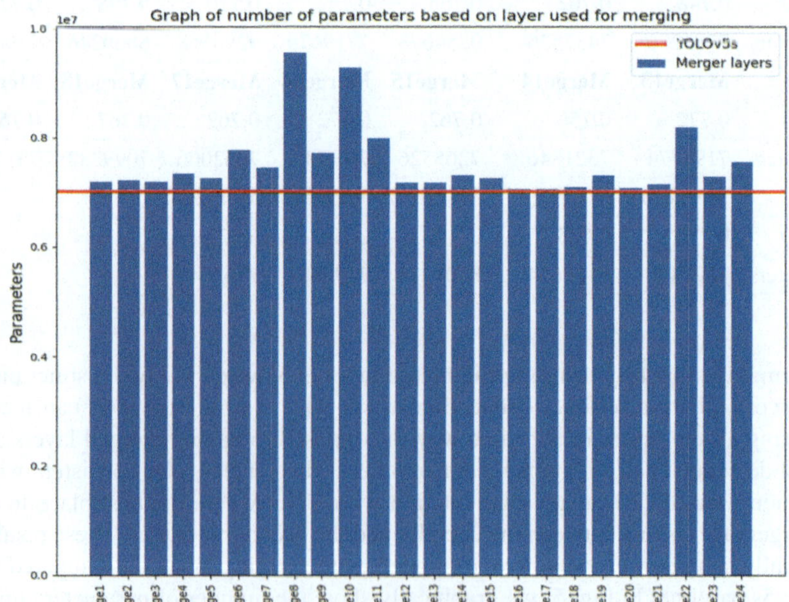

Fig. 5. The number of parameters of different merging layers compared to the baseline YOLOv5s model.

with more than 20 million parameters which dwarfs YOLOv5s (with −7 million parameters) and Merge11 network (with −8 million parameters). Our Merge11 network still outperforms the much larger YOLOv5m model. To further prove this point, we implemented the FakeMerge layer. Instead of merging the preceding, current, and succeeding slice, this network just repeats the current slice 3 times. This means that we do not use the information on preceding and succeeding slices (which we argue is essential for the performance increase) and we only increase the number of parameters. The results of FakeMerge layers are shown in Table 2, where we can see that there is no increase in the mAP50 metric as is the case for models with real merging layers.

Table 2. Performance of YOLOv5s, YOLOv5m and models with fake merge layer.

	FakeMerge1	FakeMerge2	FakeMerge3	FakeMerge4	FakeMerge5
mAP50	0.685	0.704	0.698	0.729	0.730
parameters	7202294	7232246	7203574	7342838	7265526
	FakeMerge6	**FakeMerge7**	**FakeMerge8**	**FakeMerge9**	**FakeMerge10**
mAP50	0. 747	0. 761	0. 764	0. 749	0.756
parameters	7785206	7457526	9554678	7719670	9293558
	FakeMerge11	**FakeMerge12**	**FakeMerge13**	**FakeMerge14**	**FakeMerge15**
mAP50	0. 764	0.763	0.757	0.757	0.760
parameters	8000246	7190774	7190774	7321846	7265526
	FakeMerge16	**FakeMerge17**	**FakeMerge18**	**FakeMerge19**	**FakeMerge20**
mAP50	0.758	0.750	0.748	0.771	0.763
parameters	7062006	7062006	7099382	7317238	7092470
	FakeMerge21	**FakeMerge22**	**FakeMerge23**	**FakeMerge24**	
mAP50	0.768	0.707	0.704	0.721	
parameters	7162614	8201974	7293686	7565046	

4 Conclusions

In conclusion, this study introduces a novel architectural enhancement within the YOLOv5 object detection framework, capitalizing on the sequential nature of data through the integration of merge layers. This architectural modification holds promise for adaptation to various other computer vision architectures, facilitating the exploitation of sequential information inherent in medical datasets. Moreover, our experimentation underscores the pivotal significance of the placement of these merge layers within the network architecture, revealing optimal performance attainment when positioned between the principal modules of the network. Notably, the observed performance improvements are not solely attributed to the augmented model size or parameter count, underscoring the efficacy of our proposed methodology.

Future investigations are warranted to explore the applicability of this approach in processing sequential data streams, particularly in the context of video surveillance systems deployed in self-driving cars, in greater depth. Exploring the integration of merge layers in such scenarios holds the potential to enhance real-time object detection capabilities, thereby advancing the safety and efficacy of autonomous driving technologies.

References

1. Iqbal, S., et al.: Deep learning model integrating features and novel classifiers fusion for brain tumor segmentation. Microsc. Res. Tech. **82**(8), 1302–1315 (2019)
2. Agrawal, P., Katal, N., Hooda, N.: Segmentation and classification of brain tumor using 3D-UNet deep neural networks. Int. J. Cogn. Comput. Eng. **3**, 199–210 (2022)
3. Wang, X., et al.: Non-local neural networks. In: Proceedings of the IEEE Conference on Computer Vision and Pattern Recognition (2018)
4. Ultralytics. YOLOv5 2023. https://github.com/ultralytics/yolov5, unpublished
5. Mateusz, B.: Brain MRI segmentation, Version 1 (2019). https://www.kaggle.com/datasets/mateuszbuda/lgg-mri-segmentation. Accessed 20 Dec 2023
6. Buda, M., Saha, A., Mazurowski, M.A.: Association of genomic subtypes of lower-grade gliomas with shape features automatically extracted by a deep learning algorithm. Comput. Biol. Med. **109**, 218–225 (2019)
7. Bjorck, N., et al.: Understanding batch normalization. Adv. Neural Inf. Process. Syst. **31** (2018)
8. Browne, M.W.: Cross-validation methods. J. Math. Psychol. **44**(1), 108–132 (2000)
9. Padilla, R., et al.: A comparative analysis of object detection metrics with a companion open-source toolkit. Electronics **10**(3), 279 (2021)
10. Ronneberger, O., Fischer, P., Brox, T.: U-Net: convolutional networks for biomedical image segmentation. In: Navab, N., Hornegger, J., Wells, W., Frangi, A. (eds.) Medical Image Computing and Computer-Assisted Intervention – MICCAI 2015. MICCAI 2015. Lecture Notes in Computer Science(), vol. 9351. Springer, Cham (2015). https://doi.org/10.1007/978-3-319-24574-4_28

On Semantic Association Capabilities of GPT LLMs in a Game of Word Associations

Aleksandar Z. Kondić and Vladimir M. Milovanović(✉)

Department of Electrical Engineering and Computer Sciences, Faculty of Engineering, University of Kragujevac, Sestre Janjić 6, 34000 Kragujevac, Serbia
{konda,vlada}@uni.kg.ac.rs

Abstract. Large Language Models (LLMs) excel at language pattern recognition and prediction of words. The current largest models, such as OpenAI's Generative Pretrained Transformer (GPT) models, have attained unprecedented levels of success in their ability to perform a variety of tasks, reaching the level of humans in some areas. This paper focuses on an LLM's core strength—its linguistic predictive capabilities—and how effectively it can be leveraged to find semantic associations between different words. A game of word associations is played with GPT models, in which a set of four words is provided to the player, who has to correctly guess the fifth word that is semantically related to all four words in possibly different contexts. The four provided words are not necessarily semantically related to each other. The testing is done in Serbian, with data obtained from the "associations" game segments of the Serbian national TV quiz "Slagalica" being used to measure the effectiveness of GPT models—notably the currently most powerful GPT-4o model—as players of this game. In a round of the game, ten attempts are given to the model to guess the correct word. It is found that GPT models can discover the correct answer in the majority of the rounds played, with a notable portion of answers being guessed correctly on the first try by the GPT-4 family of models.

Keywords: LLM · word prediction · GPT · semantic associations · game · *TV Slagalica* · quiz

1 Introduction

Artificial intelligence (AI) is a scientific and engineering field providing a rich avenue for research and technical innovation. AI technology has a widespread use across computer systems with a variety of purposes. A commonality shared between them is the need to carry out a task for which an algorithm cannot be easily conceived. In recent years, the sub-field of AI with the most notable advancements has been Natural Language Processing (NLP), whose purpose is the understanding of human language. In particular, the *transformer* architecture [1] introduced significant performance improvements by allowing distant

© The Author(s), under exclusive license to Springer Nature Switzerland AG 2025
N. Filipović (Ed.): AAI 2024, LNNS 1446, pp. 25–33, 2025.
https://doi.org/10.1007/978-3-031-99201-8_4

words in a sentence to be interpreted as being mutually relevant and by enabling parallelism in the processing of an input. Such an architecture lends itself to a higher level of human language comprehension while efficiently using processing resources, resulting in less time needed for training neural networks derived from this architecture. Transformer-based model architectures have been shown to perform significantly better [1] in natural language processing tasks compared to their predecessors, such as Recurrent Neural Networks (RNNs), Gated Recurrent Unit (GRU) and Long Short-Term Memory (LTSM) neural networks.

The performance enhancements have enabled feasible training of ever-larger neural networks. Large Language Models (LLMs) have since emerged as a result of training transformer-based models of unprecedented size, measured in terms of the number of parameters, with vast amounts of data. For instance, OpenAI's GPT-3 is such an LLM whose neural network is composed of 96 layers and 175 billion parameters [2]. Compared to its predecessor—the GPT-2, which consists of 48 layers and 1.5 billion parameters [3]—a result of the GPT-3's development has demonstrated that up-scaling language models can make them effective at various NLP-related tasks without fine-tuning [2], but rather applying bare end-to-end deep learning training approach. As a model's size increases, a general understanding of human language and its various applications seemingly emerges.

Measurement of a language model's effectiveness across a variety of language-related tasks is performed using benchmarking strategies with curated datasets. Some aspects of an LLM's performance tested by such benchmarks are the following: general language understanding [4], question answering [5], common sense reasoning [6,7], and text generation in the forms of summarization [8] and machine translation [9], to name a few. Some of the benchmarks provide a baseline level of human performance which is being increasingly approached with newer generations of LLMs. Humans generally perform well at these benchmarks, with the baseline being approximately 90% or more for most [4,5,7]. GPT-4's benchmark ratings are within 10% of human baselines, and in some tests it even matches human-level performance [10].

1.1 Semantic Word Associations

As LLMs approach humans in performance across language-related tasks, a point of interest arises in how well an LLM could compete in tasks that even humans find challenging. Nontrivial tasks that exercise a person's language-wielding capabilities typically assume the form of quizzes and puzzles. An interesting type of puzzle involves identifying semantic or abstract patterns in a set of words and using them to complete the puzzle. More specifically, given a set of words that are connected by a common semantic or abstract theme, where one word is hidden, the participant is tasked with uncovering it. Without the hidden word, the remaining set of words may not necessarily remain cohesive with respect to a particular semantic or abstract theme.

While providing an adequate level of challenge for humans, such a puzzle is also interesting for testing LLMs as it challenges the core mechanism by which they operate, which is by assigning relevance scores between words [1] and, in the case of GPTs, using them to predict the next word in a sequence. Testing an LLM in this way could provide a window into its semantic and abstract association capabilities. In this paper, the aforementioned puzzle is henceforth referred to as a game of *word associations.*

The specific structure of a word associations game is as follows: four words are provided to the participant, who guesses the fifth hidden word. The words are given in Serbian language. This structure is derived from the segment named "associations" of the popular Serbian national TV quiz "Slagalica" [11], from which the data for testing has been curated [12]. Figure 1 depicts the quiz's "associations" game segment. It is composed of four columns with five fields each. Players can choose to reveal a word hidden behind a numbered field of a column and make a guess for the word hidden behind an unnumbered field, which cannot be revealed unless guessed correctly. Players are awarded points for making the correct guess. They can also make a guess for the central field labeled as "???", which ends the game if guessed correctly. It is the solution to the fifth puzzle consisting of the surrounding unnumbered fields. Thus, this segment of the quiz consists of five word associations games as previously defined, which are depicted in Fig. 1b.

There are previous works using this exact quiz to evaluate neural network performance [13, 14], although they were not LLMs. In this paper, different Generative Pretrained Transformer (GPT) models are tested as players of word associations games in the Serbian language, and their comparative performance is evaluated. The testing is conducted using OpenAI's API [15], and the test dataset is obtained from the actual "associations" game segments of TV quiz "Slagalica."

2 Model Selection and Evaluation Setup

A set of two thousand word associations games is preselected from the test dataset, and the same set is used to test different GPT model generations and versions. Testing is performed using OpenAI's Chat Completions API, through which two messages are sent per instance of the game: a message with the "system" role that describes the game's rules and the response format of the model, and a message with the "user" role containing the four words. System messages are typically used to set the context used by the model in its responses to the user's messages. Here, the models are instructed to return ten guesses for the fifth word as the puzzle's solution and to order them by decreasing probability for each of them being the correct answer. Models then respond with their guesses in a single message that has the "assistant" role.

The following system message, written in Serbian with the Latin script, is sent in every request to the model for playing a word associations game:

> Pričamo na srpskom jeziku. Igramo igru asocijacija, gde ja zadajem četiri reči, a Vi treba da pronđete petu reč koja ih povezuje. Ona može u jezičkom

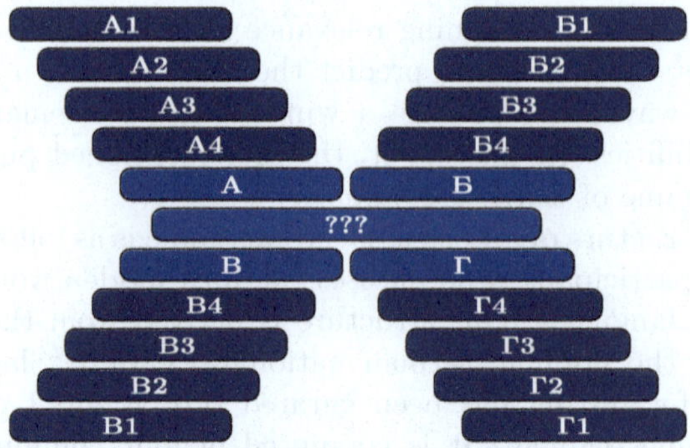

(a) Beginning of the segment where all fields are hidden. Each leaning column corresponds to an instance of the word associations game.

(b) The end of the game segment with all fields revealed. Marked are the individual groups of five words, representing five different games.

Fig. 1. The "associations" game segment of the Serbian national TV quiz "Slagalica".

kontekstu pravilno da se koristi sa sve četiri reči, a da koncepti koje ti izrazi predstavljaju ne budu nužno povezani. Nju ja već znam, a ona nije jedna od navedenih reči. Navedite deset reči koje mogu biti tačan odgovor, u redosledu opadajuće verovatnoće, razdvojene zarezima, bez numerisanja.

In English, this message translates to:

We speak in Serbian language. We are playing an association game where I give four words and You have to find the fifth word that connects them. It can be used correctly in a linguistic context with all four words, without the concepts represented by these expressions being necessarily connected. I already know it, and it is not one of the mentioned words. List ten words that could be the correct answer, in order of decreasing probability, separated by commas, without numbering.

The Chat Completions API offers additional parameters to control model behavior [15]. Default values are used for such parameters. The following models are tested with the prompts: GPT-3.5 Turbo, GPT-4, GPT-4 Turbo, and currently the latest one, GPT-4o.

2.1 Latin and Cyrillic Scripts for Serbian Prompts

Since the Serbian language has both Latin and Cyrillic scripts, the appropriate choice has to be made whenever the language is used for writing. In OpenAI's API and GPT models, textual prompts are partitioned into tokens. The use of API is priced by the number of tokens used in prompts. Through limited testing, it has been determined that using the Latin script in prompts yields fewer tokens than Cyrillic for identical text; an example is shown in Table 1 for the GPT-4 Turbo model. Thus, the Latin script and only two thousand word associations games have been chosen solely for economic reasons to minimize testing costs. Another interesting observation to be made from this example is of the disparity in answers provided for the two scripts, although there is a certain degree of randomness in a GPT model's response to a query when used with the default settings.

3 Testing Results

The criteria used for classifying a correct guess is an exact match with the answer word. Inflected words have the same status as unrelated words, and guesses based on such words do not count as partially correct. The overall performance of GPT models under such criteria is depicted in Fig. 2. Starting from GPT-4, the models can guess the correct answer in more than half of the games when given sufficient number of attempts, i.e., up to ten attempts per each word associations game. A consistent increase in performance with newer model generations and versions is also observable.

Figure 3 depicts the models' accuracies for Top-n guesses, i.e., when the correct guess is within the first n guesses. The models exhibit a significant Top-1 accuracy, with a small and consistent incline when more guessing attempts are allowed. The GPT-4 models perform significantly better than GPT-3.5 Turbo, with slight improvements in performance with each generation and/or version. To reach 50% accuracy, the original GPT-4 needs to be given eight attempts, whereas the GPT-4 Turbo needs four, and the latest GPT-4o model needs only three attempts.

Expectedly, after a certain number of guesses the models' performance begins to saturate. The saturation point can be determined by setting a threshold for the rate of improvement in Top-n accuracy. When an increment in the number of guesses would result in added performance that is below the saturation threshold, the model can be considered to have reached a performance plateau. Table 2 shows the number of guesses for each model at which they reach their performance plateaus for saturation thresholds of 1% and 0.5%. GPT-4o has

Table 1. Example of a disparity in GPT-4 Turbo's responses to Serbian Latin and Cyrillic prompts

	Serbian Latin	Serbian Cyrillic
Prompt	JAGODINA, TELEFON, STRUJA, INSTALACIJA	ЈАГОДИНА, ТЕЛЕФОН, СТРУЈА, ИНСТАЛАЦИЈА
Response	centrala, centralno, centralni, električna, energetska, telekomunikaciona, mreža, sistema, servis, podrška	централа, прикључак, опрема, мрежа, кабл, услуга, пословање, систем, комуникација, информација
Prompt tokens	209	307
Completion tokens	37	59
Total tokens	246	366

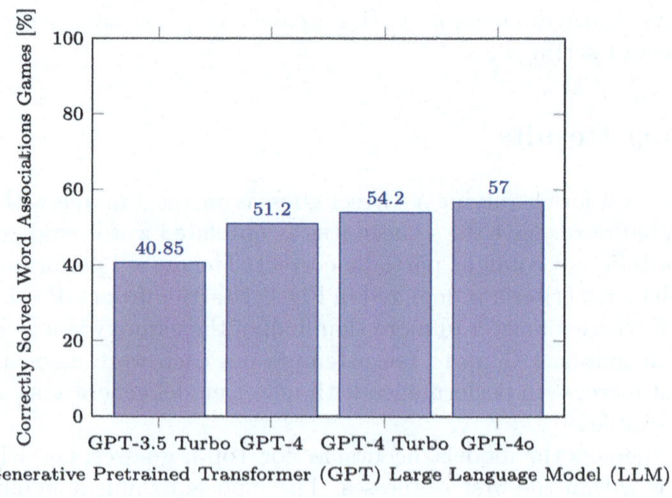

Fig. 2. The exact guess match accuracy for different GPT models when given ten attempts.

a higher saturation point compared to its predecessors in the GPT-4 family of models. With up to ten guesses, the GPT-3.5 Turbo model has not been shown to reach a performance plateau with a saturation threshold of 0.5%.

Another point of interest would be a comparison of model performance when they make correct guesses. In Fig. 4 are shown the mean and median number of guesses taken by each model when the answer was guessed correctly. In most games, the GPT-4 family of models finds the correct answer with the first guess, with mean number of guesses in relative proximity. The GPT-3.5 Turbo on

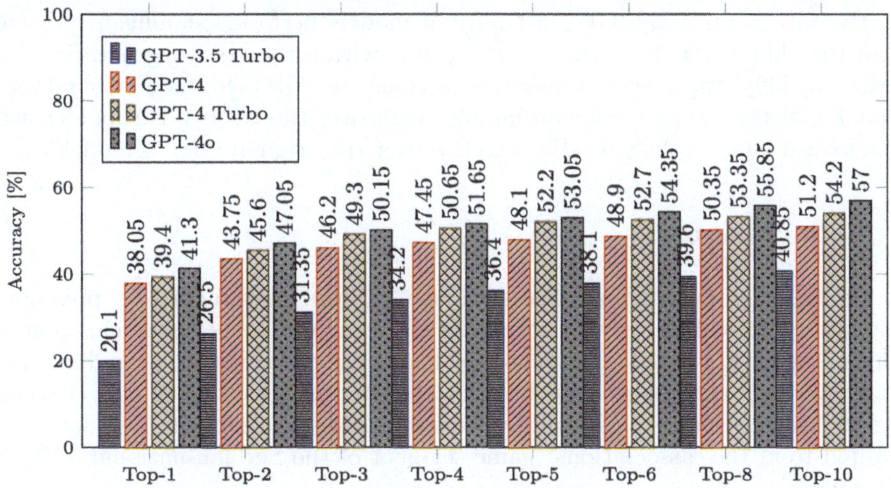

Fig. 3. Top-n accuracies for different GPT model generations and versions.

Table 2. Top-n accuracy plateau points of GPT models for different saturation thresholds.

Threshold	GPT-3.5Turbo	GPT-4	GPT-4Turbo	GPT-4o
1%	6	4	5	6
0.5%	>9	8	6	9

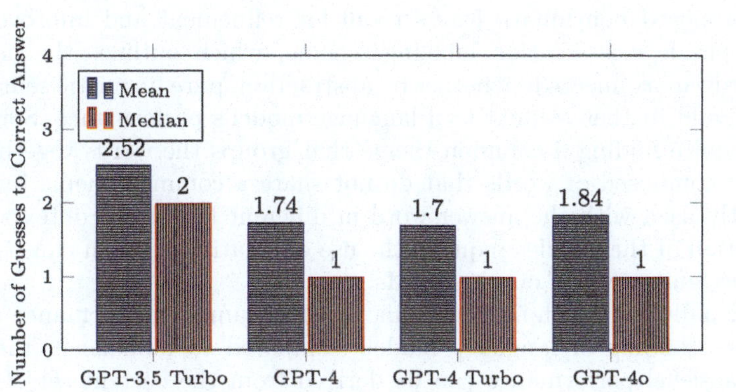

Fig. 4. Mean and median number of guesses for each GPT model within correctly solved games.

average and in most of the cases needs to make an additional guess to find the correct answer.

By all accounts, the GPT-4 family of models performs significantly better than the older GPT-3.5 Turbo. Each version within the GPT-4 generation represents a slight improvement over the previous one. GPT-4o, the newly released model, is demonstrated to be an improvement over the GPT-4 Turbo, although at a rate similar to that of GPT-4 Turbo over the original GPT-4 model.

4 Conclusions

A method of measuring a large language model's performance is presented, potentially providing a way to assess its semantic association capabilities, which is directly related to the core mechanism of the transformer architecture. Unlike already established benchmarks, the proposed method is also challenging for humans, being based on a word associations puzzle. The test dataset is curated from the "associations" game segment of the Serbian national TV quiz "Slagalica."

The tested models have shown to perform well, with the GPT-4 family of models correctly solving the puzzle with an exact guess in majority of the selected test games. When an answer is guessed correctly, it is usually from the first guess by the GPT-4 model generation, and from the second guess by the GPT-3.5 Turbo. Allowing additional guesses to the models slightly, but consistently increases their accuracy, which eventually plateaus as the models' performance begins to saturate. The newly-released GPT-4o model is shown to be the most performant, with the rate of improvement over GPT-4 Turbo in accuracy being similar to that of GPT-4 Turbo over GPT-4.

The proposed benchmark leaves room for refinement and improvement. A sub-type puzzle classification labeling scheme, which outlines the manner in which words are connected—whether by abstraction, pure linguistic semantics, or both—can add further context to a language model's performance. Some of the puzzles require finding a common theme that groups the words together, while others are composed of words that do not share a common theme but can all be correctly used with the answer word in different linguistic contexts. Solving some fraction of the puzzles requires the use of creativity, which can potentially outline the limitations of current LLMs.

Additionally, establishing a baseline level of human performance will help clarify the standing of language models compared to humans in these tests. The human-level performance can be derived from the current set of national TV quiz "Slagalica" recordings which consists of more than three and a half thousand episodes, each having two independent "associations" game segments. If, in the future, large language models start to significantly outperform humans in some areas, this benchmark can potentially help outline that difference since the puzzles are designed to be challenging for humans themselves, unlike current benchmarks.

References

1. Vaswani, A., et al.: Attention is all you need. In: Proceedings of the 31st International Conference on Neural Information Processing Systems, NIPS'17, Red Hook, NY, USA, pp. 6000–6010. Curran Associates Inc. (2017)
2. Brown, T.B., et al.: Language models are few-shot learners. arXiv preprint arXiv:2005.14165 (2020)
3. Radford, A., Wu, J., Child, R., Luan, D., Amodei, D., Sutskever, I.: Language models are unsupervised multitask learners. OpenAI blog **1**(8), 9 (2019)
4. Wang, A., et al.: SuperGLUE: a stickier benchmark for general-purpose language understanding systems. In: Proceedings of the 33rd International Conference on Neural Information Processing Systems, NIPS'19, Red Hook, NY, USA. Curran Associates Inc. (2019)
5. Rajpurkar, P., Jia, R., Liang, P.: Know what you don't know: unanswerable questions for SQuAD. CoRR (2018). http://arxiv.org/abs/1806.03822
6. Levesque, H.J., Davis, E., Morgenstern, L.: The Winograd schema challenge. In: Proceedings of the Thirteenth International Conference on Principles of Knowledge Representation and Reasoning, KR'12, pp. 552–561. AAAI Press (2012)
7. Zellers, R., Holtzman, A., Bisk, Y., Farhadi, A., Choi, Y.: HellaSwag: can a machine really finish your sentence? In: Korhonen, A., Traum, D., Màrquez, L. (eds.) Proceedings of the 57th Annual Meeting of the Association for Computational Linguistics, Florence, Italy, pp. 4791–4800. Association for Computational Linguistics (2019). https://doi.org/10.18653/v1/P19-1472. https://aclanthology.org/P19-1472
8. Lin, C.Y.: ROUGE: a package for automatic evaluation of summaries. In: Text Summarization Branches Out, Barcelona, Spain, pp. 74–81. Association for Computational Linguistics (2004). https://aclanthology.org/W04-1013
9. Papineni, K., Roukos, S., Ward, T., Zhu, W.J.: BLEU: a method for automatic evaluation of machine translation. In: Proceedings of the 40th Annual Meeting on Association for Computational Linguistics, ACL '02, USA, pp. 311–318. Association for Computational Linguistics (2002). https://doi.org/10.3115/1073083.1073135
10. OpenAI, et al.: GPT-4 technical report (2024). https://arxiv.org/abs/2303.08774
11. RTS Slagalica – Zvanični kanal/official youtube channel. https://www.youtube.com/@SlagalicaRTS. Accessed 23 Sept 2024
12. Milovanović, V.M.: Slagalica. https://github.com/milovanovic/Slagalica/tree/master/Asocijacije. Accessed 23 Sept 2024
13. Grujić, N.D., Milovanović, V.M.: Natural language processing for associative word predictions. In: IEEE EUROCON 2019 – 18th International Conference on Smart Technologies (2019). https://doi.org/10.1109/EUROCON.2019.8861547
14. Grujić, N.D., Milovanović, V.M.: Associative word relations in natural language processing. Appl. Artif. Intell. **36**(1), 2034,262 (2022). https://doi.org/10.1080/08839514.2022.2034262
15. API reference. https://platform.openai.com/docs/api-reference/chat/create. Accessed 23 Sept 2024

The Possibility of Bone Fracture Prediction in Osteoporosis Treatment Through Use of AI

Gordana Jovičić[1], Aleksandra Vulović[1,2](✉), Arso Vukićević[1], Miloš Jovičić[3], Nenad Đorđević[4], and Nenad Filipović[1,2]

[1] Faculty of Engineering, University of Kragujevac, Sestre Janjic 6, 34000 Kragujevac, Serbia
gjovicic.kg.ac.rs@gmail.com, {aleksandra.vulovic,fica}@kg.ac.rs, arso_kg@yahoo.com

[2] Bioengineering Research and Development Center (BioIRC), Prvoslava Stojanovica 6, 34000 Kragujevac, Serbia

[3] Institute for Artificial Intelligence of Serbia, Fruškogorska 1, 21000 Novi Sad, Serbia
milos.jovicic@gmail.com

[4] Centre for Assessment of Structures and Materials Under Extreme Conditions, Brunel University London, Uxbridge UB8 3PH, Middlesex, UK
Nenad.Djordjevic@brunel.ac.uk

Abstract. Osteoporosis is a progressive bone disease that is characterized by low bone mass and deterioration of a bone tissue. Patients with osteoporosis have higher risk of fractures, which is a primary complication of this disease. Early identification of individuals that have high risk of fracture, during osteoporosis treatment, is crucial for implementing timely interventions to mitigate fracture risk and prevent debilitating consequences. Artificial intelligence (AI) has emerged as a promising tool in predicting bone fractures in osteoporosis treatment. By employing machine learning algorithms and data analytics techniques, AI models can analyze a number of patient-specific factors, such as bone density measurements and scans, which are commonly used to assess considered problem. The aim of this work is to provide an overview of recent advancements in AI-driven prediction of bone fracture risk in osteoporosis treatment. This work also highlights the challenges and opportunities associated with implementing AI-based fracture prediction models in clinical practice. The integration of AI-based fracture prediction tools into routine clinical care has the potential to improve osteoporosis management by enabling personalized risk assessment. By identifying individuals at heightened risk of fractures early in the treatment process, healthcare providers can tailor therapeutic strategies, optimize treatment efficacy, and improve patient outcomes.

Keywords: artificial intelligence · fracture · osteoporosis · prediction

1 Introduction

Throughout our life, old or damaged bone is being replaced with a new bone in a process known as bone remodeling [1]. During the youth stage of life, more bone is formed than removed, which leads to increase of bone mass. As we age, the speed of bone formation

© The Author(s), under exclusive license to Springer Nature Switzerland AG 2025
N. Filipović (Ed.): AAI 2024, LNNS 1446, pp. 34–41, 2025.
https://doi.org/10.1007/978-3-031-99201-8_5

lowers, while the speed of bone removal increases. This leads to weaker bones and increases the risk of osteoporosis. Osteoporosis is a progressive bone disease that leads to a lower bone density and mass. It occurs when the creation of a new bone cannot keep up with the process of old bone removal, meaning that more bone is removed than produced. As a result of this disease, bones are more fragile and prone to fractures [2]. Since bone loss often occurs gradually and without noticeable symptoms, many patients may not realize they are at risk for osteoporosis until a fracture happens [3].

Fractures are known to cause disability and chronic pain. They can significantly decrease quality of life while increasing mortality rate. The most common fractures experienced by patients with osteoporosis are vertebral, proximal femur, distal forearm and proximal humerus. The distal forearm fractures, known as wrist fractures, are often noticed to be the first fracture that is followed by vertebral or proximal femur fracture. The most common fractures are vertebral, and they lead to the fivefold increase of risk of future fractures [4]. However, the most critical fractures are femoral fractures, as they require hospitalisation and lead to death in about quarter of all cases [5].

Osteoporosis affects more than 200 million people worldwide, with elderly population being at the highest risk [6]. This number is expected to significantly increase over the next few decades due to the number of older population but also related to the changes in the lifestyle of majority of people in the Western world [5].

The aim of this paper is to provide an overview of recent advancements in AI-driven prediction of bone fracture risk in osteoporosis treatment as an initial step for a study that will focus on the use of machine learning algorithms for the estimation of the R-curve slope, toughness threshold, and stress intensity factor based on data collected during clinical examination, such as patients age and crack length.

2 Bone Fracture Prediction

Accurately predicting bone fractures is crucial in the early detection and management of osteoporosis. Current methods rely on both clinical risk factors and bone mineral density (BMD) measurements, which can provide insight into a patient's likelihood of future fractures.

2.1 Risk Factors

Risk factors for osteoporosis can be divided into two groups: fixed and modifiable risk factors. Fixed factors are the ones that cannot be reduced by lifestyle changes, while modifiable ones can be reduced [5]. Fracture risk begins to increase after the age of 50, with the risk doubling with each subsequent decade due to a decrease in bone mineral density and other contributing factors like falls. Women are particularly susceptible to osteoporosis due to a drop in estrogen levels following menopause, while men with primary or secondary hypogonadism, are at higher risks for osteoporosis. It is also known that ethnicity influences osteoporosis, and that Caucasian and Asian populations have higher risks of hip and spine fractures. Family history of osteoporosis as well as previous fractures, significantly elevates the risk of future fractures in both men and women.

Modifiable risk factors can significantly affect bone health by directly reducing bone mineral density and increasing fracture risk. Excessive alcohol consumption raises the risk of fractures due to its harmful effects on microarchitecture and osteoblastic differentiation [7]. Alcohol is linked to poor nutrition, including deficiencies in calcium, protein, and vitamin D, which lead to bone loss and a higher risk of fractures. Smoking also contributes to a higher fracture risk. A low body mass index (BMI), particularly below 20 kg/m^2, doubles the hip fracture risk compared to individuals with a BMI of 25 kg/m^2 [8]. A sedentary lifestyle also raises fracture risk, as physical activity, particularly weight-bearing exercises, stimulates bone remodeling through mechanical load.

2.2 Bone Mineral Density

Osteoporosis is diagnosed primarily through bone mineral density (BMD) measurements and the identification of fragility fractures, especially in the hip or vertebrae. BMD serves as one of the most critical indicators of bone strength, and its decrease correlates strongly with an increased risk of fractures [9]. Low BMD is influenced by factors such as age, family history, and lifestyle, including diet and physical activity [10].

BMD is commonly assessed using dual X-ray absorptiometry (DXA), which provides an estimate of bone strength by measuring the amount of mineral (mainly calcium) in grams per square centimeter of bone (g/cm2). DXA scans are typically performed on the hip and spine since these areas are the most prone to fractures and provide strong predictive value for future fracture risk. According to World Health Organization guidelines, a T-score of –2.5 SD or lower indicates osteoporosis, while a score between –1.0 and –2.5 indicates osteopenia [11]. Every reduction in BMD by 1 SD increases the likelihood of a fracture by roughly twofold, emphasizing the importance of early detection and monitoring [12].

However, BMD does not capture the entire fracture risk profile. Many fractures occur in individuals whose BMD values fall in the range classified as osteopenia (T-scores between –1.0 and –2.5) or even in those with BMD values above -1.0 [13]. This highlights the need for a comprehensive approach to fracture risk assessment that includes other clinical risk factors such as age, previous fracture history, and lifestyle behaviors like smoking and alcohol consumption [11]. As a result, a number of different tools have been developed and applied.

2.3 Tools Used for Bone Fracture Assessment and Prediction

It is common for osteoporosis to be undiagnosed until a fracture occurs. And that delay in early treatment leads to increased morbidity of patients [14] Technology development over the recent decades led to the creation of different tools that are now often used for bone fracture assessment and prediction, which help clinicians identify patients with the highest fracture risk and provide them with appropriate treatment early on.

Fracture risk assessment tools (such as FRAX), utilize clinical variables in order to provide a measure of fracture risk and identify 'at risk' individuals. These tools include a number of different risk factors such as age, BMI, and medical history and they can provide a 10-year probability of major osteoporotic fractures compared to just assessing patients based on BMD measurement [15]. FRAX allows the assessment of individual risk factors even when BMD is unknown [11].

Another commonly used technique is Trabecular Bone Score (TBS). This technique provides insights into the microarchitecture of a trabecular bone, which is useful for the risk fracture assessment for patients who have other conditions. Information about the microarchitecture of a trabecular bone is important as BMD measurements for patients with other conditions can underestimate fracture risk. However, the combination of TBS with FRAX has shown better ability to identify individuals with high-risk of fractures [14].

Although these are the most commonly used tools, they are not the only ones. Tools such as QFracture and the Garvan Fracture Risk Calculator have been used to obtain information about fracture risk.

3 Application of AI for Osteoporosis Management and Treatment

AI has found significant application over the past decades in the field of medicine as a way to help both clinicians and patients. So, it should not be a surprise that AI algorithms have found the application in relation to diagnosis and treatment of osteoporosis. Some of the main applications of AI are related to the following aspects [13]:

- Fracture Risk Assessment
- Treatment
- Treatment Plans
- Fracture Detection

In this paper, we are focusing on the application of AI for Fracture Risk Assessment. One of the most important benefits of predicting fractures is the ability to identify individuals at high risk before a fracture occurs. This empowers healthcare professionals to implement preventative strategies, such as lifestyle modifications, nutritional supplements, exercise programs, and pharmacological interventions, tailored to each patient's specific needs. For instance, strengthening exercises can improve bone density, while medications can help slow bone loss. These proactive measures can significantly reduce the risk of fractures, thereby preventing the associated physical, emotional, and financial burdens.

From a healthcare system perspective, fracture prevention also translates into substantial cost savings. Treating fractures, particularly hip fractures, is expensive and often involves hospitalization, surgery, and rehabilitation. By predicting fractures and intervening early, healthcare providers can reduce hospital admissions, lower the need for invasive treatments, and minimize the long-term costs associated with fracture care. Additionally, reducing the number of fractures lightens the overall strain on healthcare resources, allowing for more efficient allocation of services.

The first research in this field was done in the 90s when artificial neural networks (ANN) were used to identify patients who needed BMD assessment. However, the results were not significantly better compared to conventional statistical methods [16]. Since then, the research and tools in this field have been significantly improved. Over the years ANNs have been used to estimate the occurrence of osteoporosis [17], identify osteoporosis in the population [18], improve the evaluation of X-ray images [19]. Ho-Le et al. [20] have used ANNs to predict hip fractures in postmenopausal women, incorporating BMD, clinical, and lifestyle factors. The model combining BMD and non-BMD factors achieved an accuracy of 87%, outperforming existing statistical models, and demonstrating potential for more precise fracture risk stratification and clinical management. Recently, Ulivieri et al. [21] employed two ANN approaches to create a predictive model for forecasting vertebral fractures at an individual level and a map that provides insights into the complex biological relationships among the densitometric variables analyzed and the two possible outcomes (additional fractures vs. no fractures), with a particular emphasis on the Bone Strain Index.

In 2017, Gourlay et al. [22] assessed the effectiveness of three fracture risk tools (FRAX, Garvan, and QFracture) and two simpler classifiers (one using only the femoral neck BMD T-score and another combining age with femoral neck BMD T-score) for predicting osteoporotic fractures in older men, finding that the age plus femoral neck BMD T-score classifier was as effective as the more complex tools. Ferizi et al. [23] observed that machine learning classifiers, including logistic regression, linear discriminant analysis, and boosted trees, achieve an optimal balance between sensitivity and specificity in predicting osteoporotic fractures. Additionally, MRI and FRAX each contribute unique value to identifying osteoporotic fractures when used with machine learning models. Besides MRI, CT scans have also been used. Muehlematter et al. [24] evaluated bone texture analysis combined with machine learning on standard CT scans to predict vertebral insufficiency fractures. The approach showed high accuracy in identifying at-risk patients, outperforming Hounsfield unit measurements, though predicting fractures in individual vertebrae remains challenging.

Kruse et al. [25] have used unsupervised hierarchical clustering to identify nine patient subgroups with varying fracture risks from data on 10,775 Danish women. Clusters revealed differences in bone mineral density, comorbidities, and treatment compliance, highlighting the potential of clustering to refine fracture risk assessment beyond traditional methods like T-scores, though further validation is needed. A study performed by Engles et al. [26] used the superlearner algorithm to predict hip fractures in older adults using administrative claims data and compared its performance to established methods. While the superlearner showed promising flexibility and similar predictive performance to individual algorithms, its calibration was less effective in the validation set, highlighting its potential for complex datasets but the need for refinement. Genomic data has also been considered as a factor for fracture prediction. Study performed by Wu et al. [27] developed fracture prediction models using machine learning and genomic data from the Osteoporotic Fractures in Men cohort. Their results indicated genetic risk scores as a significant predictor, highlighting the potential of combining genetic profiling with advanced modeling for improving fracture prediction in older men. Study presented by Kong et al. [28] used a fracture prediction model using CatBoost, a machine-learning method, which outperformed the FRAX score in predicting fractures. It identified novel predictors, such as arthralgia scores and homocysteine levels, alongside conventional ones, suggesting a clinically relevant model that needs further validation in larger and diverse cohorts.

The majority of applications of AI for osteoporosis management and treatment are related to the possibility of predicting the risk of osteoporosis. Over the past 2 decades, there have been a lot more research dedicated to fracture risk assessment. However, as noticed from papers covered in this section, there are still a lot of opportunities to improve these results and possibly include additional factors.

4 Conclusion

Osteoporosis is one of the most common diseases, impacting individuals from all racial and ethnic groups. Since it typically presents no symptoms until a fracture occurs, early identification and intervention are critical. Over the past 25 years, significant progress has been made in understanding the various risk factors that contribute to bone mass loss and fracture susceptibility, such as aging and lifestyle choices. The increasing number of people diagnosed with osteoporosis, or expected to be diagnosed in the future, clearly demonstrate the importance of patient identification. Early identification of patients with a higher fracture risk can help mitigate the risk through lifestyle modifications and tailored treatment plans. While BMD remains a critical tool for diagnosing osteoporosis and assessing fracture risk, it should be used in conjunction with other risk factors and technologies to ensure a comprehensive understanding of an individual's fracture risk, support better decision-making and to guide appropriate interventions.

Emerging technologies, particularly AI, are proving to be powerful tools in osteoporosis management and treatment. AI algorithms can analyze patient data, including medical history, and lifestyle factors, to classify individuals based on their fracture risk. This enables healthcare providers to offer more precise and personalized care, focusing on prevention strategies for high-risk patients and adjusting treatments as needed. By using AI for the early detection and management of osteoporosis, healthcare systems can improve the quality of life for affected patients and reduce the long-term financial burden on patients and their families. Preventing fractures and maintaining bone health can significantly reduce hospitalizations, lower medical expenses, and ease the strain on healthcare systems. In conclusion, the integration of AI into osteoporosis management and treatment has the potential to transform both individual patient outcomes and the overall approach to managing this prevalent condition.

Acknowledgements. Authors acknowledge the funding by the Ministry of Science, Technological Development and Innovation of the Republic of Serbia, contract number [451-03-65/2024-03/200107 (Faculty of Engineering, University of Kragujevac)].

References

1. Bolamperti, S., Villa, I., Rubinacci, A.: Bone remodeling: an operational process ensuring survival and bone mechanical competence. Bone Res. **10**(1), 48 (2022)
2. Clynes, M.A., Harvey, N.C., Curtis, E.M., Fuggle, N.R., Dennison, E.M., Cooper, C.: The epidemiology of osteoporosis. Br. Med. Bull. **133**(1), 105–117 (2020)
3. LeBoff, M.S., et al.: The clinician's guide to prevention and treatment of osteoporosis. Osteoporos Int. **33**(10), 2049–2102 (2022)

4. Scheyerer, M.J., Spiegl, U.J., Grueninger, S., Hartmann, F., Katscher, S., Osterhoff, G., et al.: Risk factors for failure in conservatively treated osteoporotic vertebral fractures: a systematic review. Global Spine J. **12**(2), 289–297 (2022)
5. van Oostwaard, M.: Osteoporosis and the nature of fragility fracture: an overview. Fragility fracture nursing: holistic care and management of the orthogeriatric patient, 1–13 (2018)
6. Sözen, T., Özışık, L., Başaran, N.Ç.: An overview and management of osteoporosis. Eur. J. Rheumatol. **4**(1), 46 (2017)
7. Godos, J., Giampieri, F., Chisari, E., Micek, A., Paladino, N., Forbes-Hernández, T.Y., et al.: Alcohol consumption, bone mineral density, and risk of osteoporotic fractures: a dose–response meta-analysis. Int. J. Environ. Res. Public Health **19**(3), 1515 (2022)
8. De Laet, C., Kanis, J.A., Odén, A., Johanson, H., Johnell, O., Delmas, P., et al.: Body mass index as a predictor of fracture risk: a meta-analysis. Osteoporos Int. **16**, 1330–1338 (2005)
9. Jaiswal, R., Zoulakis, M., Axelsson, K.F., Darelid, A., Rudäng, R., Sundh, D., et al.: Increased bone material strength index is positively associated with the risk of incident osteoporotic fractures in older Swedish women. J. Bone Miner. Res. **38**(6), 860–868 (2020)
10. Yang, H., Li, B., Li, H., Zhou, M., Li, B., Guo, J., et al.: The independent and joint association between physical activity, sleep duration and daily sitting time with bone mineral density: a real world study from NHANES 2007–2018. Bone, 117264 (2024)
11. Schini, M., Johansson, H., Harvey, N.C., Lorentzon, M., Kanis, J.A., McCloskey, E.V.: An overview of the use of the fracture risk assessment tool (FRAX) in osteoporosis. J. Endocrinol. Invest. **47**(3), 501–511 (2024)
12. Unnanuntana, A., Gladnick, B.P., Donnelly, E., Lane, J.M.: The assessment of fracture risk. JBJS **92**(3), 743–753 (2010)
13. Dimai, H.P.: New horizons: artificial intelligence tools for managing osteoporosis. J. Clin. Endocrinol. Metab. **108**(4), 775–783 (2023)
14. Baim, S., Blank, R.: Approaches to fracture risk assessment and prevention. Curr. Osteoporos. Rep. **19**, 158–165 (2021)
15. Kanis, J.A., Johnell, O., Odén, A., Johansson, H., McCloskey, E.: FRAX™ and the assessment of fracture probability in men and women from the UK. Osteoporos Int. **19**, 385–397 (2008)
16. Ongphiphadhanakul, B., Rajatanavin, R., Chailurkit, L., Piaseu, N., Teerarungsikul, K., Sirisriro, R., et al.: Prediction of low bone mineral density in postmenopausal women by artificial neural network model compared to logistic regression model. J. Med. Assoc. Thailand= Chotmaihet thangphaet **80**(8), 508–515 (1997)
17. Rizzi, A., Panella, M., Paschero, M., Mascioli, F.F.: Estimation of bone mineral density data using mog neural networks. In: 2004 IEEE International Joint Conference on Neural Networks, vol. 4, pp. 3241–3246 (2004)
18. Chiu, J.S., Li, Y.C., Yu, F.C., Wang, Y.F.: Applying an artificial neural network to predict osteoporosis in the elderly. Stud. Health Technol. Inform. **124**, 609 (2006)
19. De Moura Meneses, A.A., Pinheiro, C.J.G., Schirru, R., Barroso, R.C., Braz, D., Oliveira, L.F.: Artificial neural networks applied to bone recognition in x-ray computer microtomography imaging for histomorphometric analysis. In 2008 IEEE Nuclear Science Symposium Conference Record, pp. 5309–5313 (2008)
20. Ho-Le, T.P., Center, J.R., Eisman, J.A., Nguyen, T.V., Nguyen, H.T.: Prediction of hip fracture in post-menopausal women using artificial neural network approach. In: 2017 39th Annual International Conference of the IEEE Engineering in Medicine and Biology Society (EMBC), pp. 4207–4210 (2017)
21. Ulivieri, F.M., Rinaudo, L., Piodi, L.P., Messina, C., Sconfienza, L.M., Sardanelli, F., et al.: Bone strain index as a predictor of further vertebral fracture in osteoporotic women: an artificial intelligence-based analysis. Plos One **16**(2), e0245967 (2021)

22. Gourlay, M.L., Ritter, V.S., Fine, J.P., Overman, R.A., Schousboe, J.T., Cawthon, P.M., et al.: Comparison of fracture risk assessment tools in older men without prior hip or spine fracture: the MrOS study. Arch. Osteoporos. **12**, 1–11 (2017)
23. Ferizi, U., Besser, H., Hysi, P., Jacobs, J., Rajapakse, C.S., Chen, C., et al.: Artificial intelligence applied to osteoporosis: a performance comparison of machine learning algorithms in predicting fragility fractures from MRI data. J. Magn. Reson. Imaging **49**(4), 1029–1038 (2019)
24. Muehlematter, U.J., Mannil, M., Becker, A.S., Vokinger, K.N., Finkenstaedt, T., Osterhoff, G., et al.: Vertebral body insufficiency fractures: detection of vertebrae at risk on standard CT images using texture analysis and machine learning. Eur. Radiol. **29**, 2207–2217 (2019)
25. Kruse, C., Eiken, P., Vestergaard, P.: Clinical fracture risk evaluated by hierarchical agglomerative clustering. Osteoporos Int. **28**, 819–832 (2017)
26. Engels, A., Reber, K.C., Lindlbauer, I., Rapp, K., Büchele, G., Klenk, J.: Osteoporotic hip fracture prediction from risk factors available in administrative claims data–a machine learning approach. Plos One **15**(5), e0232969 (2020)
27. Wu, Q., Nasoz, F., Jung, J., Bhattarai, B., Han, M.V.: Machine learning approaches for fracture risk assessment: a comparative analysis of genomic and phenotypic data in 5130 older men. Calcif. Tissue Int. **107**, 353–361 (2020)
28. Kong, S.H., Ahn, D., Kim, B., Srinivasan, K., Ram, S., Kim, H., et al.: A novel fracture prediction model using machine learning in a community based cohort. J. Bone Min. Res. Plus **4**(3), e10337 (2020)

Evaluating Bayesian Approaches for Water Quality Classification: A Comparative Study

Ana Dodig[1,2](✉) and Tatjana Lutovac[2]

[1] School of Electrical Engineering, University of Belgrade, Bulevar Kralja Aleksandra 73, 11120 Belgrade, Serbia
ana.dodig@ivi.ac.rs

[2] The Institute for Artificial Intelligence Research and Development of Serbia, Fruskogorska 1, 21000 Novi Sad, Serbia
tlutovac@etf.bg.ac.rs

Abstract. Water quality is crucial for every segment of life and with the increasing importance of reliable and prompt water quality information for environmental monitoring and public health, advanced computational methods are vital. Bayesian approaches, known for their probabilistic foundation and capability to handle uncertainties, provide a promising approach for addressing this need. This study undertakes an in-depth examination of the performance of various Bayesian models in classifying water quality as safe or not safe, using a comprehensive dataset of chemical water parameters. The methodology involves a complete workflow that begins with data preparation through quantile-based discretization and dataset splitting. Next, four distinct Bayesian classifiers—Naive Bayes (NB), Tree Augmented Naive Bayes (TAN), Forest Augmented Naive Bayes (FAN), and a General Bayesian Network (BN)—were implemented and evaluated. Each model's performance was rigorously assessed using a variety of metrics, including Accuracy, Precision, Recall, F1 score and particularly Matthews Correlation Coefficient (MCC) to consider the dataset's imbalance. The results indicate that while NB provides a fundamental approach, its performance, along with TAN and FAN, may be compromised by overfitting. In contrast, the BN model, with its focus on significant features—cadmium and aluminium—exhibited consistent performance across training and testing datasets, indicating effective learning and generalization. Specifically, the BN model achieved an accuracy of 91.731%, a precision of 95.121%, a recall of 95.583%, an F1-score of 95.351%, and an MCC of 0.58. These results underscore the model's balanced and robust performance, especially in the context of an unbalanced dataset. Additionally, the Conditional Probability Distributions (CPDs) extracted from the BN model explained the relationships between specific chemical concentrations and water safety. Taking all factors and results into account this study underscores the integration of Bayesian models into water quality assessment frameworks, particularly advocating for the BN model due to its balanced precision, recall, and reduced bias towards majority class.

Keywords: Bayesian networks · naive Bayes · water quality · classification

© The Author(s), under exclusive license to Springer Nature Switzerland AG 2025
N. Filipović (Ed.): AAI 2024, LNNS 1446, pp. 42–61, 2025.
https://doi.org/10.1007/978-3-031-99201-8_6

1 Introduction

Water quality assessment is crucial for environmental monitoring and maintaining ecological balance, impacting both public health and biodiversity [1]. Communities commonly rely on surface river water as their main source of drinking water. For the same reasons, water professionals are facing growing challenges related to water pollution and the effects of climate change on the amount of available water [2]. With escalating challenges in water pollution and the demand for reliable and prompt information on water quality, advanced computational methods are essential for effective classification. Recently, various machine learning algorithms have been explored for water quality classification. Support Vector Machines (SVM) and Artificial Neural Networks (ANN) have shown effectiveness in handling complex, non-linear relationships in water quality data. In line with these developments, [3] conducted a comprehensive comparison of these models, particularly focusing on their performance in classifying water quality. Their results provide valuable insights into the comparative effectiveness of different machine learning models in this crucial domain. Similarly, methods such as Decision Trees, Random Forests, and XGBoost have gained attention for their robust predictive capabilities in environmental datasets. These techniques, efficient in handling extensive and varied datasets, have been successful in water quality classification. For instance, Decision Trees offer an easily interpretable model [4], while Random Forests provide increased accuracy through an ensemble of trees. In particular, [5] further validates the effectiveness of Random Forest classifiers in predicting water quality. Their research emphasizes the model's proficiency in handling intricate datasets, demonstrating its utility in water quality assessments based on varied parameters. XGBoost stands out for its performance and speed, making it particularly effective in complex data scenarios, as highlighted in the study [6].

Nowadays, studies leveraging Bayesian models in water quality prediction have shown promise. Bayesian approaches, distinct in their probabilistic foundations, are increasingly recognized for their capability to integrate prior knowledge and manage uncertainty, making them particularly suited for the complexities of environmental data. The Naive Bayes classifier exemplifies this, offering a straightforward yet potent mechanism grounded in Bayes' Theorem, known for its practical applicability in various domains including water quality assessment. For instance, the application of Naive Bayes in Serbian watercourses yielded accurate water quality classifications in a significant majority of instances, underscoring its potential as a reliable tool in the transition to digital water management [7]. Building upon the foundation of Naive Bayes classifiers, studies on augmented models such as Tree Augmented Naive Bayes (TAN) and Forest Augmented Naive Bayes (FAN) have shown significant advancements in handling environmental data complexities [8–10]. TAN models introduce a tree-like structure to connect attributes, enhancing the ability to capture dependencies between variables, while FAN models expand on this by incorporating a forest structure for even more complex inter-variable relationships. These augmented models represent an evolving frontier in machine learning for environmental dataset analysis, with research indicating their improved accuracy over traditional models. Similarly, General Bayesian Networks have gained traction in water management, recognized for their adaptability in integrating diverse components and managing data uncertainties, a feature crucial in environmental systems. Concluding the exploration of Bayesian models in water quality assessment, [10] has taken a comprehensive approach by employing and comparing all mentioned Bayesian approaches, for classifying water lead levels in private drinking water systems. This research is particularly notable for its broad application of Bayesian methods in a critical area of public health concern, providing valuable insights into the effectiveness of these models in accurately classifying potential hazards in water supplies. This paper aims to expand on these initial successes, providing a comprehensive analysis of various Bayesian models in water quality classification. It focuses on employing Bayesian models for classifying water quality as safe or not safe, particularly exploring Naive Bayes, Tree Augmented Naive Bayes, Forest Augmented Naive Bayes, and General Bayesian Network structures. By comparing the performance of different Bayesian approaches, the study seeks to identify their unique advantages and limitations for water quality analysis. This comparative analysis is crucial in navigating the challenges posed by rapid environmental changes and escalating water demands, aiming to equip stakeholders with reliable tools for water quality assessment.

2 Methodology

An overview of the methodology is presented in the flow diagram of Fig. 1. It consists of three parts: data set processing, fitting Bayesian networks, and prediction. The first one includes discretization of the data set and dividing into the training and test set. The second one covers fitting Bayesian Networks on the training set. Finally, the third part consists of testing Bayesian Networks on the test set and comparing different Bayesian models.

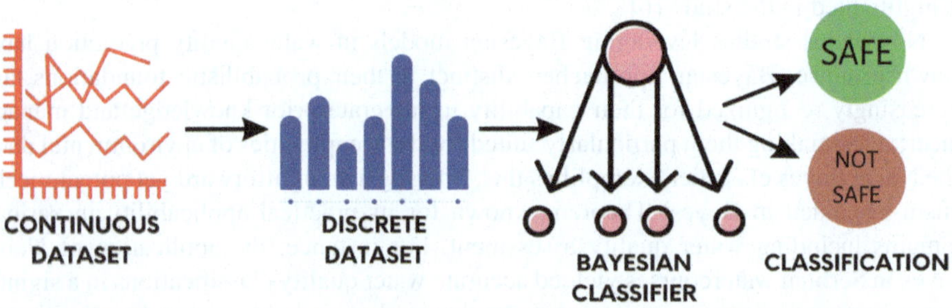

Fig. 1. Flow diagram for water quality classification using Bayesian classifiers.

2.1 A General Approach

This research develops a data discovery framework (Fig. 1) to identify models that classify water samples as safe or not safe. A model is constructed through the processes of discretization, which is applied to continuous variables of dataset and Bayes classification. Each model is evaluated based on a set of metrics that evaluate model performance (Accuracy, Precision, Recall, F1 score, and Matthews Correlation Coefficient). Framework components are described in the following sections.

2.2 Water Quality Dataset

In order to develop good classification models, adequate datasets need to be prepared. For our work, the dataset titled "Water Quality" from Kaggle website [11] is obtained. It comprises various chemical components in water samples and their respective concentrations. Table 1 shows the dataset with all parameters and their values, while the correlation matrix is presented in Fig. 2. Along with these chemical parameters, there is a class variable - is_safe signifying whether the water sample is safe or not. The dataset contains a total of 7,998 water samples. Out of these, 912 samples are marked as safe, and 7,086 are labeled as not safe. As indicated, the dataset exhibits a significant imbalance in its class distribution, with a much larger proportion of samples labeled as 'not safe' compared to those marked as 'safe'. This unbalance can heavily bias the performance of classification models, leading them to favor the majority class and potentially overlook critical insights into the minority class. In this study, a decision was made not to alter the original dataset despite its imbalance. This approach was chosen to observe how Bayesian models handle the challenges posed by an unbalanced dataset in their native state. However, to accurately capture and assess any potential biases in the models' performance due to this imbalance, it is crucial to employ the right evaluation metrics which will be explained in Sect. 2.5.5.

Table 1. Description of the water quality dataset.

Chemical component	Minimum	Maximum	Mean	Standard deviation
Aluminium	0.00	5.05	0.666240	1.265203
Ammonia	0.00	29.84	14.278212	8.878930
Arsenic	0.00	1.05	0.161455	0.252605
Barium	0.00	4.94	1.567849	1.216108
Cadmium	0.00	0.13	0.042811	0.036048
Chloramine	0.00	8.68	2.177094	2.567080
Chromium	0.00	0.90	0.247246	0.270652
Copper	0.00	2.00	0.805950	0.653527
Flouride	0.00	1.50	0.771572	0.435399
Bacteria	0.00	1.00	0.319634	0.329494
Viruses	0.00	1.00	0.328624	0.378102
Lead	0.00	0.20	0.099451	0.058175
Nitrates	0.00	19.83	9.818853	5.541677
Nitrites	0.00	2.93	1.329946	0.573254
Mercury	0.00	0.01	0.005193	0.002967

(continued)

Table 1. (*continued*)

Chemical component	Minimum	Maximum	Mean	Standard deviation
Perchlorate	0.00	60.01	16.461398	17.688307
Radium	0.00	7.99	2.919950	2.322539
Selenium	0.00	0.10	0.049687	0.028771
Silver	0.00	0.50	0.147793	0.143556
Uranium	0.00	0.09	0.044676	0.026904

Since Bayes classifiers are most often used with discrete variables, our dataset needs to be discretized. Discretizing variables that are continuous can improve the performance, accuracy, and interpretation of Bayesian classifiers [12, 13]. Both domain-knowledge and data-driven methods can be applied to transform continuous variables to discrete variables. Due to lack of domain knowledge, we opted for data driven methods. According to previous research, Quantile-Based Discretization (QBD) method is demonstrated as a particularly advantageous and effective choice when dealing with Bayesian classification models [13, 14] and because of this, we also used QBD method as a choice for our work to discretize the water quality dataset. The QBC method bins data into quantiles (Q), ensuring each bin contains roughly an equal number of data points. By setting Q to 200, our dataset is divided into a maximum of 200 quantiles, each representing a similar number of samples. This approach is particularly effective in dealing with skewed data distributions, making it well-suited for Bayesian classifiers that can be sensitive to data distribution, and it helps for achieving a balanced and accurate representation of the dataset.

2.3 Fundamentals

In probability theory, the concept of a 'sample space' S defines all possible outcomes of an uncertain process, like the roll of a die. This space ensures that every potential outcome is unique and mutually exclusive. Subsets of this sample space are termed 'events'. Probabilities assigned to these events adhere to three fundamental axioms:

$\boldsymbol{Axiom}1 : P(S) = 1$, the event S that we will get an outcome in the sample space is certain to occur and is therefore assigned the probability 1.

$\boldsymbol{Axiom}2 : ForallA \subseteq SitholdsthatP(A) \geq 0$, any event A must have a nonnegative probability.

$\boldsymbol{Axiom}3 : IfA \subseteq S, B \subseteq SandA \cap B = \varnothing, thenP(A \cup B) = P(A) + P(B),$, if two events A and B are disjoint, then the probability of the combined event is the sum of the probabilities for the two individual events (see Fig. 3 (a)). On the other hand, if A and B are not disjoint (see Fig. 3 (b)), then it can easily be shown that: $P(A \cup B) = P(A) + P(B) - P(A \cap B)$, where $P(A \cap B)$, is the intersection between A and B and it represents the event that both A and B will occur [15].

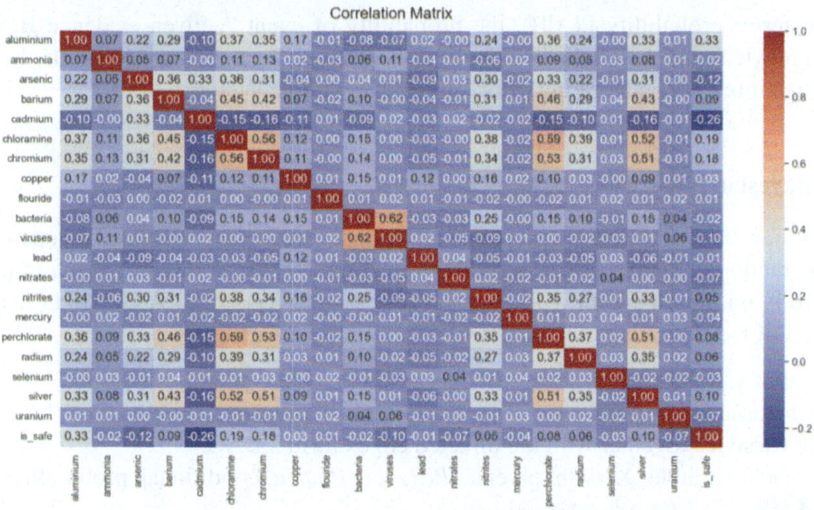

Fig. 2. Correlation matrix for the water quality dataset.

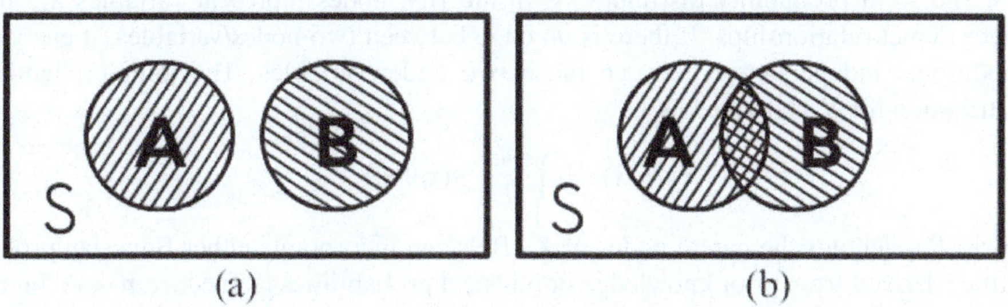

Fig. 3. Graphical presentation of the two disjoint events A and B (a) and the two events A and B that are not disjoint (b).

Next, conditional probabilities and Bayes' Theorem are crucial in probability theory. Conditional probability, denoted as P(A|B), is the probability of event \$A\$ given that event \$B\$ has already occurred. It's fundamental in scenarios where the occurrence of one event influences the likelihood of another. For two events A and B with P(B) > 0, the conditional probability for A given B is:

$$P(A|B) = \frac{P(A \cup B)}{P(B)}, \tag{1}$$

Bayes' Theorem, a natural extension of conditional probabilities, offers a powerful way to update and revise probabilities based on new information. It is mathematically expressed as:

$$P(A|B) = \frac{P(B|A) \times P(A)}{P(B)}, \tag{2}$$

where P(A|B) is the updated probability of A given B, P(B|A) is the likelihood of B given A, P(A) is the prior probability of A, and P(B) is the probability of B. It calculates the posterior probability P(A|B), the probability of event A given evidence B, using prior knowledge of A and the likelihood of B given A. This theorem forms the basis of Bayesian inference, integrating prior knowledge with new data.

2.4 Bayesian Networks

A Bayesian Network (BN) is a graphical model that represents conditional independence among random variables as a directed acyclic graph (DAG). A directed graph is acyclic if there is no directed path $X_1 \rightarrow \cdots \rightarrow X_n$ so that $X_1 = X_n$. A Bayesian network consists of the following:

- A set of variables and a set of directed edges between variables.
- Each variable has a finite set of mutually exclusive states.
- The variables together with the directed edges form a DAG.
- To each variable X with parents $Pa_1, \ldots, Pa_n$, a conditional probability table $P(X\,|Pa_1, \ldots, Pa_n)$ is attached.

A Bayesian Network for a set of variables $X = X_1, \ldots X_n$ includes a network structure and local probability distributions. In the BN, nodes represent variables X_i, and edges depict relationships. If there is no edge between two nodes/variables, it encodes conditional independence between those two nodes/variables. The joint probability distribution for X is derived as:

$$p(x) = \prod\nolimits_{i=1}^{n} p(x_i|\mathrm{Pa}_i), \tag{3}$$

where Pa_i denotes the parent nodes of X_i. BNs can incorporate either Bayesian probabilities derived from prior knowledge or physical probabilities learned from data. In our work, we learn these probabilities from data, which involves using physical probabilities based on empirical evidence to define the conditional probabilities in the network. This approach enhances the realism and applicability of our Bayesian Network in modeling and understanding complex variable interactions.

2.5 Bayes Classifiers

Bayesian classifiers are powerful statistical tools for data classification, built on Bayes' theorem, which enables the computation of predictive probabilities. These classifiers assess relationships and dependencies between variables using probabilistic structures. The strength of the relationship between a pair of variables is quantified by a Conditionally Probability Table (CPT), and each node is assigned a CPT to specify its relationship with a connected node [16]. Bayes' theorem updates probabilities in the CPT as new data are incorporated for the set of variables. In our research, we will explore and compare the performance of four types of Bayesian classifiers: Naive Bayes, Tree Augmented Naive Bayes, Forest Augmented Naive Bayes, and General Bayesian Network, each varying in complexity and structure, to determine their efficacy in classifying data.

2.5.1 Naive Bayes

The Naive Bayes classifier (NBC) belongs to the probabilistic family of classifiers based on Bayes' Theory. The main feature of this classifier is the assumption that all variables are conditionally independent which is the reason for calling it 'Naive'. Because of that, the parameters can be learned separately, more simply and quicker [7]. In general, NBC is structured such that each feature is conditionally independent of every other feature, given the class. That means that each feature variable has the class variable as its only parent. This simplifies the learning process to parameter estimation, usually done by counting methods or employing the Expectation-Maximization (EM) algorithm [15] for incomplete data. To mitigate the issue of rare cases not reflected in the dataset, 'virtual cases' can be introduced, preventing zero probability estimates. Despite the simplicity of its assumption of feature independence given the class, NBCs often perform well in practice, especially in classification tasks where exact probability distributions are less important than identifying the most likely class. In our research, we have utilized a Naive Bayes classifier for our complete dataset without missing values. The Conditional Probability Distributions (CPDs) are calculated using the Maximum Likelihood Estimator (MLE), reflecting the true distribution of our data. The class variable, indicative of water quality being 'safe' or 'not safe', serves as the parent node. All other feature variables are treated as child nodes and are mutually independent, following the foundational assumption of NBC. Figure 4 represents the structure of the NBC for our water quality dataset.

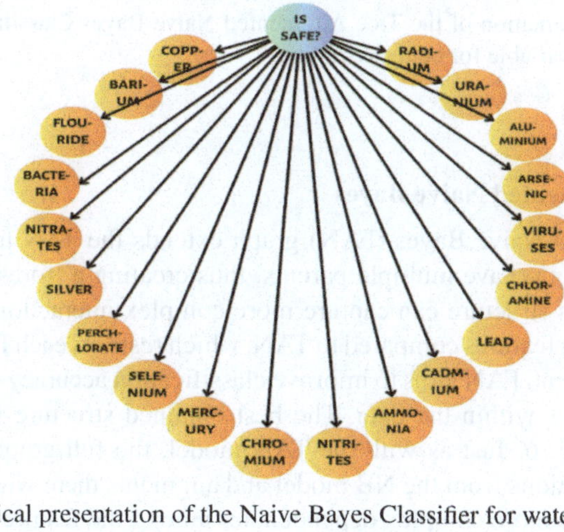

Fig. 4. Graphical presentation of the Naive Bayes Classifier for water quality dataset.

2.5.2 Tree Augmented Naive Bayes

The Tree Augmented Naive Bayes classifier (TAN) expands on the Naive Bayes model by allowing for a more complicated structure where feature variables may have dependencies. Unlike NBCs, TAN does not presume feature independence given the class.

Instead, it permits each feature variable to have another feature as its parent, forming a tree-like structure. However, contrary to NB, the structure is not given here. Structure of the TAN is identified through optimal parameter settings that maximize likelihood, chosen from all possible connections between feature variables. In our analysis, the derived structure for the TAN classifier is illustrated in Fig. 5. For clarity, the structure is presented without the target variable, enhancing the visibility of the tree's connections. The complete graph integrates the Naive Bayes' connections with the additional links identified, which represent the extended dependencies between feature variables.

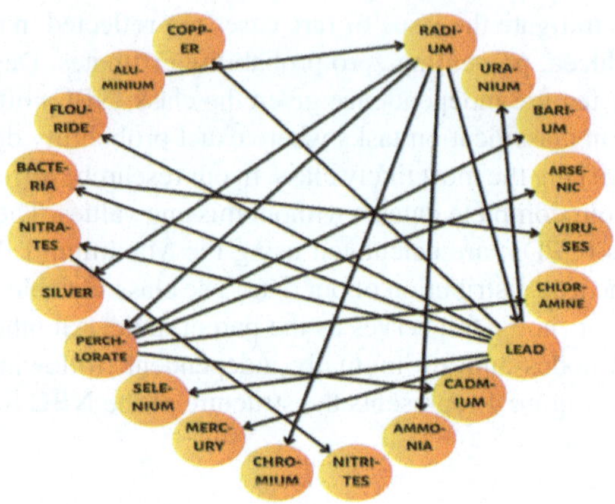

Fig. 5. Graphical presentation of the Tree Augmented Naive Bayes Classifier for water quality dataset without target variable for better visibility.

2.5.3 Forest Augmented Naive Bayes

A Forest Augmented Naive Bayes (FAN) graph extends the TAN model by allowing each feature variable to have multiple parents, thus creating a 'forest' of trees instead of a single tree. This structure can capture more complex interactions and conditional dependencies among features compared to TAN, which restricts each feature to having at most one feature parent. FAN aims to improve classification accuracy by modeling more nuanced relationships within the data. The best obtained structure for FAN classifier is represented in Fig. 6. Just as with the TAN model, the full graph incorporates the foundational connections from the NB model and augments them with newly identified links that capture the more complex dependencies among the feature variables.

2.5.4 General Bayesian Network

General Bayesian Network Classifier (BNC) is Bayesian Network that, unlike NB, TAN and FAN models, does not treat classification feature distinctly from other feature nodes. The classification feature is not assumed to necessarily be the parent of other attributes nodes and may be also a child or parent node in the network. This approach lacks a predefined structure, therefore the discovery of an optimal structure for a given dataset

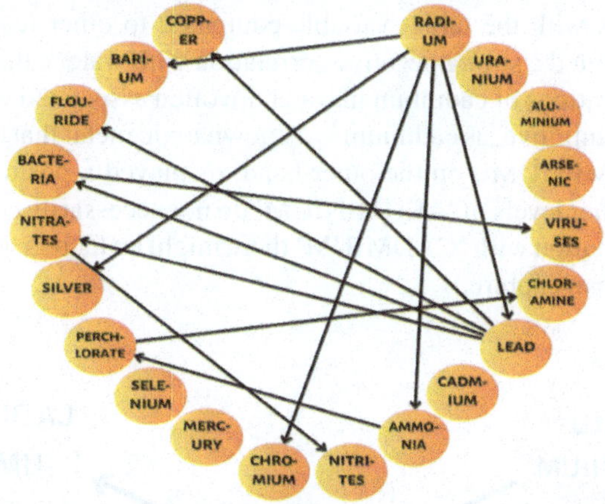

Fig. 6. Graphical presentation of the Forest Augmented Naive Bayes Classifier for water quality dataset without target variable for better visibility.

is necessary. To find this structure, a score-based learning that assigns a number (a score) to each Bayesian network structure is employed. Additionally, to do this, a score function that takes a Bayesian network structure as argument and returns a value needs to be assigned. Based on this, a task of score-based learning can be considered a search problem where we simply look for the model structure with the highest score. A good score function should at least encompass the following properties: it should balance the accuracy of a structure with the complexity of the structure; and it should be computationally tractable to evaluate [15]. An example of a score function satisfying the above two properties is the Bayesian information criterion (BIC), which contains a term measuring how well the data fits the model, as well as a term that accounts for model complexity:

$$\mathrm{BIC}(S|D) = \sum\nolimits_{i=1}^{N} \log_2 P\left(d_i|\widehat{\theta_S}, S\right) - \frac{\mathrm{size}(S)}{2}\log_2(N), \quad (4)$$

where S represents the structure of the Bayesian network, D is the observed data, d_i are the individual data points, $\widehat{\theta_S}$ are the estimated parameters of the structure S, size(S) is the number of parameters to be estimated, and N is the sample size. To find the optimal structure, we used the hill climb search method, a heuristic approach [17]. Other search methods include exhaustive search and greedy algorithms, but hill climbing is often preferred for its balance between efficiency and effectiveness in navigating large search spaces. The optimal structure discovered through this process, as visualized in the accompanying graph (see Fig. 7), reveals a simple yet effective architecture with only two feature variables. The graph illustrates that the variables 'CADMIUM' and 'ALUMINIUM' have direct predictive relationships with the classification node. This minimalist structure implies that, within our model, these two features are sufficient to make a prediction about water safety, thereby enhancing the model's interpretability and computational efficiency. Additionally, this architecture aligns well with insights from the correlation matrix analysis where both 'CADMIUM' and 'ALUMINIUM' showed notable correlations with the target variable compared to other features. Specifically: 'CADMIUM' exhibited a strong negative correlation with water safeness. This suggests that higher concentrations of cadmium in water are often associated with it being unsafe. This relationship is intuitive, as cadmium is a known toxic metal that can pose significant health risks. 'ALUMINIUM', on the other hand, displayed a positive correlation with target variable. Higher levels of 'ALUMINIUM' do not necessarily correlate with unsafe water, but in combination with 'CADMIUM' there might be factors at play that Bayesian Network structure can capture.

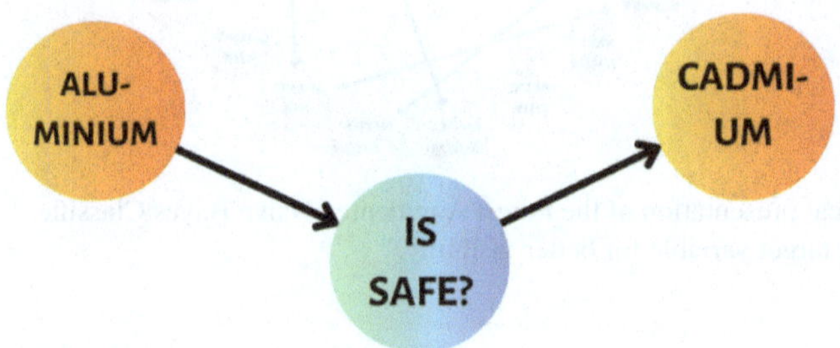

Fig. 7. Graphical presentation of the Bayesian Network Classifier for water quality dataset.

2.6 Performance Metrics

In this research, the performance of classifiers is assessed through a set of widely recognized metrics derived from the confusion matrix, crucial for understanding model predictions in binary classification. As we mentioned before, our dataset is unbalanced and to accurately assess the true performance of the models and detect any biases towards either class, it is imperative to utilize a variety of evaluation metrics. The confusion matrix categorizes predictions into four groups: True Positives (TP), which are correctly identified positive samples, True Negatives (TN), which are correctly identified negative samples, False Positives (FP), which represent negative samples wrongly classified as positive, and False Negatives (FN), which are positive samples incorrectly labeled as negative. In our context, 'positive samples' are those that are not safe (value 0), while 'negative samples' are safe ones (value 1). To evaluate classifier performance, metrics such as Accuracy, Precision, Recall, F-score, and Matthews Correlation Coefficient (MCC) are used. Accuracy represents the ratio of correctly classified samples. While it reflects the overall classifier performance, its effectiveness diminishes in imbalanced datasets, as it can misleadingly suggest high performance due to a majority class bias.

$$\text{Accuracy} = \frac{TP + TN}{TP + TN + FP + FN}, \tag{5}$$

Recall indicates the proportion of actual positive samples correctly identified. It is particularly useful for identifying how well the model detects positive (not safe) samples.

$$\text{Recall} = \frac{TP}{TP + FN}, \tag{6}$$

Precision represents the ratio of true positives to all positively predicted samples. It assesses the accuracy of positive predictions.

$$\text{Precision} = \frac{TP}{TP + FP}, \tag{7}$$

F1-score combines recall and precision through their weighted harmonic mean, which is useful for datasets with imbalanced classes.

$$\text{F1 Score} = 2 \times \frac{\text{Precision} \times \text{Recall}}{\text{Precision} + \text{Recall}}, \tag{8}$$

Additionally, in this study, we introduce the Matthews Correlation Coefficient (MCC) as a robust metric especially effective for unbalanced datasets [18]. Unlike the F1-score, which primarily focuses on positive predictions, MCC considers all four quadrants of the confusion matrix (TP, TN, FP, FN). It delivers a coefficient value between –1 and 1, where 1 indicates perfect prediction, 0 no better than random guessing, and –1 total disagreement between prediction and observation.

$$\text{MCC} = \frac{(TP \times TN) - (FP \times FN)}{\sqrt{(TP + FP)(TP + FN)(TN + FP)(TN + FN)}}, \tag{9}$$

For our purposes, these metrics provide a more comprehensive and nuanced view of the model's performance, highlighting not just its accuracy but also its ability to correctly identify and classify instances from the minority class. By utilizing these metrics, we aim to gain a deeper understanding of the strengths and limitations of Bayesian models when applied to datasets with skewed class distributions.

3 Results

This section presents the results of different Bayesian Classifiers included in the proposed workflow for predicting water quality as safe or not safe. Initially, the data is discretized using Quantile-Based Discretization method, and the discretized data is used for model fitting. Subsequently, the performance of various Bayesian Classifiers is demonstrated. The evaluation of performance is based on five different metrics: Accuracy, Recall, Precision, F1 score and MCC. The results for both training and test sets are provided.

3.1 Naive Bayes

In analyzing the performance of the Naive Bayes classifier on the train set, NBC shows an accuracy of 89.689%, precision of 97.574%, recall of 90.61%, F1 score of 93.963%, and as MCC of 0.609 (see Table 2). On the test set, the following metrics were obtained: accuracy –84.625%, precision –92.554%, recall –89.903%, F1 score –91.209% and MCC 0.303 (see Table 3). Accuracy for our specific problem with an unbalanced dataset may not be always reliable since predicting the majority class for all instances could yield high accuracy. A precision of 92.554% on the test implies that when the model predicts a positive outcome, it is correct about 92.554% of the time. This high precision is particularly valuable in scenarios where false positives are costly. Recall, also known as sensitivity, indicates that the model successfully identifies 89.903% of all actual positive instances. This is crucial in contexts in which missing out on true positives is undesirable. An F1-score of 91.209% suggests a robust model, with a good balance between precision and recall. Finally, MCC is the most informative measure in the context of unbalanced datasets as it considers true and false positives and negatives. An MCC of 0.609 on the train set and 0.303 on the test set, which ranges between –1 and + 1, indicates a moderate positive relationship (especially on the test) between observed and predicted classifications. While not high, it reflects some degree of predictive quality in the model, more so than what the accuracy alone might suggest. The higher metrics on the train set as compared to the test set suggest that the model fits well to the training data, which could be indicative of good learning from the given features. However, the relative difference in performance between the training and test sets, particularly the MCC metric, which is roughly twice as high on the training set, also points to the model being more tuned to the specific distribution and patterns of the training data, also known as overfitting. This difference underscores the importance of using the test set performance as a more reliable indicator of how the model might perform on unseen data. While the model demonstrates high accuracy, precision, recall, and F1-score, the modest MCC value indicates that these metrics might be somewhat inflated due to the unbalanced nature of the dataset. Therefore, it is essential to interpret these results with caution.

Table 2. Performance Metrics of Different Bayesian Classifiers on the train set.

Model	Accuracy	Precision	Recall	F1-Score	MCC
NB	89.689%	97.574%	90.61%	93.963%	0.609
TAN	99.77%	100.0%	99.741%	99.87%	0.989
FAN	95.922%	100.0%	95.395%	97.643%	0.839
BN	91.843%	95.197%	95.614%	95.405%	0.592

Table 3. Performance Metrics of Different Bayesian Classifiers on the test set.

Model	Accuracy	Precision	Recall	F1-Score	MCC
NB	84.625%	92.554%	89.903%	91.209%	0.303
TAN	88.63%	88.74%	99.854%	93.97%	0.018
FAN	85.702%	88.849%	95.922%	92.25%	0.02
BN	91.731%	95.121%	95.583%	95.351%	0.58

3.2 Tree Augmented Naive Bayes

Evaluating the performance of the TAN model on the training set, the TAN model achieved an accuracy of 99.77%, precision of 100.0%, recall of 99.741%, F1 score of

99.87%, and MCC of 0.989 (see Table 2). On the test set, TAN has derived the following results: accuracy –88.63%, precision –88.74%, recall –99.854%, F1 score –93.97%, and MCC 0.018 (see Table 3). As in the case of the NB model, high accuracy can be misleading, and the model might be biased towards predicting the majority class. High values for precision and recall, especially recall, suggests that the TAN model is highly effective at identifying actual positive cases. The F1-score, which balances precision and recall, is quite high at 93.97%. This suggests a good balance between the model's ability to identify positive cases (recall) and its accuracy in these identifications (precision). Despite all of this, the MCC value on the test set in this case is near zero, indicating a negligible correlation between the model's predictions and the actual data. Still, on the train set MCC is notably high (0.989), which reveals a significant performance gap and overfitting. This low MCC on the test set suggests that the model might not be performing as well as other metrics imply, particularly due to the imbalance in the dataset. While the TAN model demonstrates almost perfect classification on the training data, its metrics on the test set are notably lower. It suggests that the TAN model is highly biased towards the class, which is in the majority, in our case positive class and overfitted. Although predicting water as not safe is more important that predicting it as safe, it can be problematic when water is frequently classified as unsafe when it is healthy. Consequently, this water will not be used at all, while resources will be unnecessarily wasted on additional analysis and purification.

3.3 Forest Augmented Naive Bayes

The performance of the FAN classifier shows results similar to those of the TAN model, although with some differences (see Tables 2 and 3). The FAN model demonstrates slightly lower recall (95.922%) and F1-score (92.25%) compared to the TAN model, indicating a minor decrease in its ability to correctly identify true positives and balance precision and recall. However, it shows a marginal improvement in the MCC at 0.02 on the test set, however still very low. This low MCC value, similar to that observed in the TAN model, indicates a potential bias in the FAN model due to the unbalanced nature of the dataset, suggesting that the model might be favoring the majority class.

3.4 General Bayesian Network

The BN model, despite being the simplest form of Bayesian classifier used in this study, has shown great results. On the train set, the BN model displays a robust performance with an accuracy of 91.843%, precision of 95.197\%, recall of 95.614%, and F1 score of 95.405%, with an MCC of 0.592 (see Table 2). The metrics on the train set are very close to those on the test set (see Table 3) indicating a consistent performance and suggesting that the BN model has generalized well without significant overfitting. This demonstrates that the BN model maintains a balance between not missing positive cases and correctly identifying negatives, a significant achievement in the context of an unbalanced dataset. Unlike the other models discussed earlier, the difference between the BN model's performance on the train and test sets is minor, especially in the context of MCC. This slight variance in the BN model's metrics is indicative of good model generalization and robustness to the data's imbalance. It also indicates a good degree of correlation between the model's predictions and actual classifications. This suggests a lower bias compared to other models and a result of the model's strategy to choose the most significant features and incorporate them into the network.

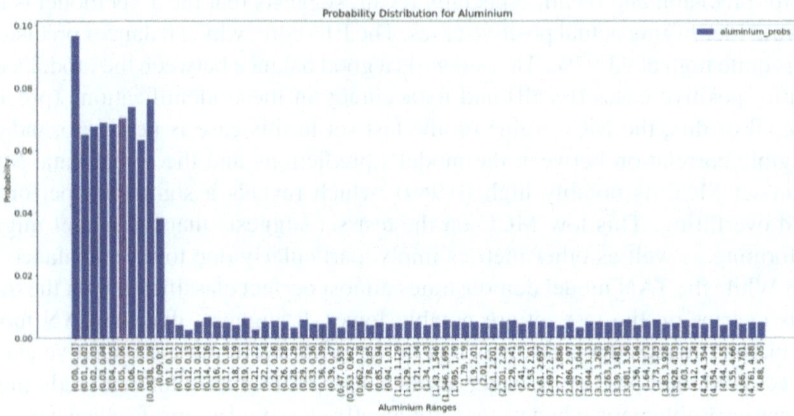

Fig. 8. Probability distribution for Aluminium - BN model.

For a clearer understanding of the practical applications of Bayesian classifiers, we extracted Conditional Probability Distributions (CPDs) from our BN model since it utilized only two features—cadmium and aluminium—considered most significant for predicting water safety.

Figures 9 and 10 illustrate the CPDs for cadmium and aluminium in relation to water safety (target variable), and Fig. 8 shows the overall probability distribution for aluminium. The BN model's CPDs are particularly interesting as they reflect the model's approach to detecting water safety based on specific ranges of cadmium and aluminium concentrations. Figure 9 illustrates the probability of water being safe or not safe given the concentration of aluminium. The blue bars represent the probability of water being unsafe across the various aluminium concentration ranges, while the red indicates the probability of water being safe. Notably, in most cases, the probability of water being classified as unsafe remains consistently high across all aluminium concentrations. This observation correlates with the dataset's imbalance, where a significant majority of the water samples are labeled as unsafe. Figure 10 indicates a more pronounced effect of cadmium concentrations on water safety. It's clear that as the cadmium concentration increases, the probability of water being unsafe tends to increase as well, which is highlighted by the increasing height of the blue bars relative to the red bars. This signals that higher cadmium concentrations are more likely to be associated with unsafe water, a finding that aligns with known health risks posed by cadmium in drinking water. NB, TAN, and FAN classifiers also generate CPDs for all features, but their complexity and the vast number of CPDs produced make it impractical to display them all here. Hence, we chose to focus on the CPDs derived from the BN model for its concise yet effective representation of the data.

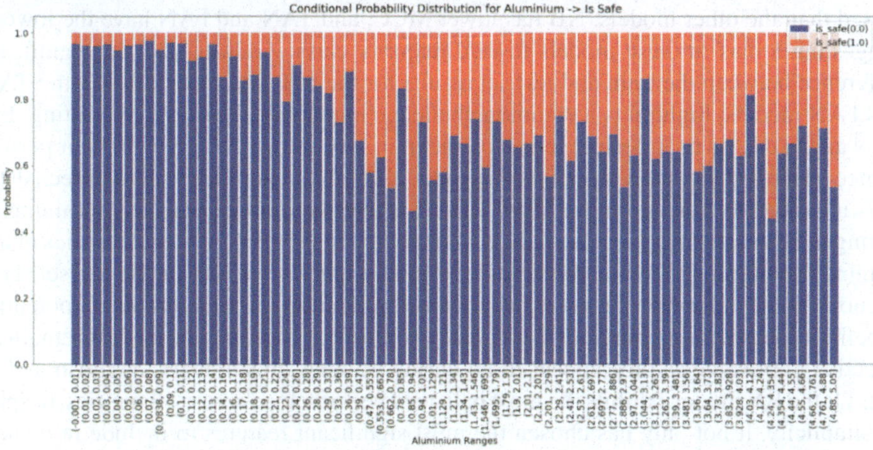

Fig. 9. Conditional Probability Distribution for Aluminium and Target Variable.

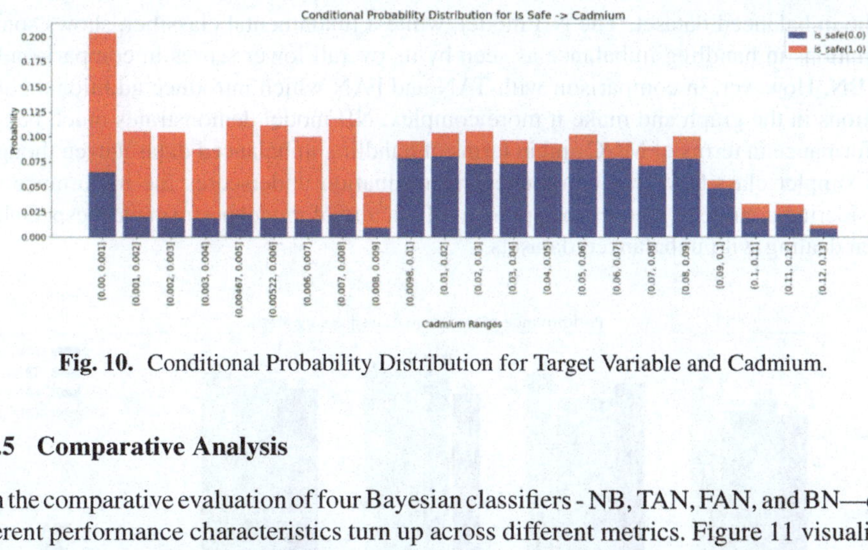

Fig. 10. Conditional Probability Distribution for Target Variable and Cadmium.

3.5 Comparative Analysis

In the comparative evaluation of four Bayesian classifiers - NB, TAN, FAN, and BN—different performance characteristics turn up across different metrics. Figure 11 visualizes the comparison effectively, enabling a clear view of each model's strengths and weaknesses. All models show high accuracy, with BN leading slightly, indicating a strong ability to make correct predictions. Precision is highest also for BN, followed closely by other classifiers. This suggests that all models are particularly reliable in their positive predictions, which is crucial for applications where false positives have severe implications. TAN presents the highest recall. The high recall demonstrates the capability to identify unsafe water samples effectively, which is vital for public health safety. The F1-score is a critical metric which combines precision and recall, and BN exhibits the highest F1-score, suggesting a balanced performance between precision and recall. The TAN model shows a slightly lower F1-score, while NB and FAN have the lowest. MCC is the most telling metric in this analysis, particularly in the context of an unbalanced dataset. BN stands out with a relatively high MCC, indicating that its performance is less biased than the other models. NB has lower MCC, and TAN and FAN have the lowest, suggesting a bias in these models toward majority class. Additionally, the significant difference between the train and test set results for the NB, and especially for the TAN and FAN models, highlights a common challenge in machine learning: overfitting. For TAN and FAN, the near-perfect scores on the train set contrast sharply with their performance on the test set, suggesting that these models do not generalize well to unseen data. This is especially evident in the MCC values, where the dramatic fall from training to testing indicates that the models might be too tailored to the specific characteristics of the training data, failing to maintain the same level of predictive quality on the test set. This phenomenon emphasizes the necessity for strict validation techniques and the potential benefit of incorporating methods to prevent overfitting, ensuring that models remain as effective in practice as they appear to be during training. In summary, as illustrated in Fig. 11, the BN model exhibits the most balanced performance across all metrics, despite its simplicity. It not only has chosen the most significant features to include in its network (graph) but also demonstrates good predictive power and a lower bias towards the majority class, as reflected by its MCC. On the other hand, while TAN and FAN perform well in terms of recall and F1-score, their lower MCC values hint at a greater sensitivity to the imbalanced dataset. The NB model, while a fundamental classifier, shows some limitations in handling imbalance as seen by its overall lower scores in comparison to the BN. However, in comparison with TAN and FAN which introduce additional connections in the graph and make it more complex, NB model demonstrates much better performance in terms of MCC and in terms of handling unbalanced dataset even though it is simpler classifier. This comprehensive evaluation underscores the importance of considering a range of metrics to capture a full picture of model performance, especially when dealing with unbalanced datasets.

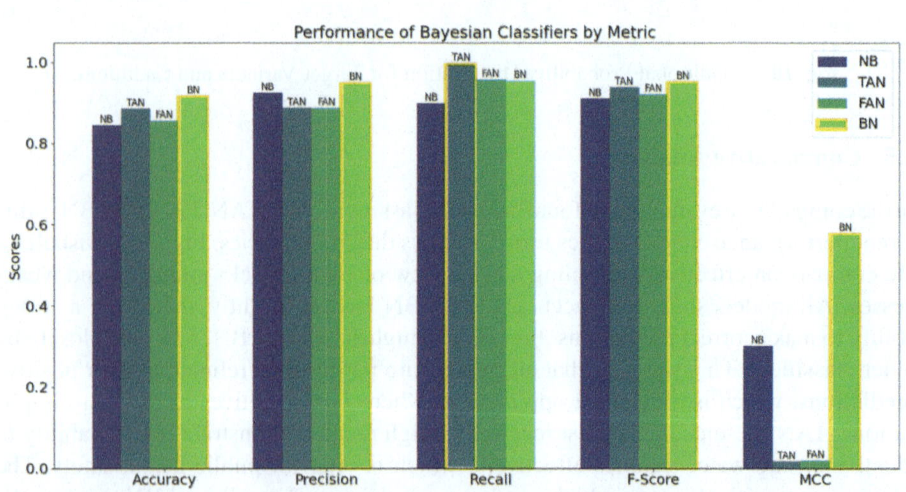

Fig. 11. Performance metrics of different Bayesian Classifiers on the test set. The best classifier for each metric is highlighted.

4 Conclusion

The goal of this paper was to develop a framework that can accurately predict water quality as safe or not safe based on various chemical water parameters (aluminium, cadmium, nitrite, etc.). The objective was to provide a tool that can be used by water resource managers, environmental scientists, and policymakers to monitor and manage river streams and water quality effectively. This paper presented a complete workflow for water quality classification. Firstly, data analysis and preparation steps were performed which represent a very important part because it determines the subsequent success of the prediction method. This step includes quantile-based discretization to discretize the continuous variables, effectively capturing the inherent distributions of the dataset and dataset split into training and testing sets. Next, four different Bayesian Classifiers (NB, TAN, FAN and BN) were implemented to predict water quality. This process consisted of constructing an optimal graph structure for each model, which was then carefully fitted to the training set. Through this fitting process, CPD tables were derived. This methodology was applied on the water quality dataset from Kaggle. The data covers almost 8000 samples and consists of 20 different chemical parameters of the water. The results indicate that while the NB model is foundational, it and other more complex models like TAN and FAN can be susceptible to overfitting, as evidenced by the difference in performance metrics between the training and test sets. On the other hand, the BN model, with its simplified structure focusing on two significant features (aluminium and cadmium), exhibited a robust performance across all metrics and maintained consistency between the training and test sets. This suggests effective learning and generalization capabilities. Additionally, the study underscored the importance of using a range of performance metrics, particularly the Matthews Correlation Coefficient (MCC), to thoroughly evaluate model performance in the context of unbalanced datasets. The MCC provided a more nuanced understanding of the models' predictive quality by considering true and false positives and negatives. The CPDs extracted from the BN model highlighted the relationship between the concentration of aluminium and cadmium and water safety. These CPDs are crucial for understanding the practical applications of Bayesian classifiers, as they summarize the model's approach to predicting water safety based on the specific ranges of these chemical components. In light of the findings, this paper suggests the integration of Bayesian models into water quality assessment frameworks, with a special recommendation for the General Bayesian Network model due to its balance in precision, recall, and lower bias towards majority class representation. The paper also encourages further research into optimizing these models to enhance their predictive accuracy and generalizability. Future work will involve the investigation of some techniques, such as resampling techniques, to provide a more balanced dataset and to reduce any model biases toward the majority class. Furthermore, the comparative aspect of this study will be expanded beyond Bayesian classifiers to include a diverse range of machine learning methods such as Support Vector Machines (SVM), Random Forests, and Neural Networks. Such comparative analysis will provide a broader understanding of the strengths and limitations of each approach in the context of environmental data, which is often noisy, non-linear, and complex Additionally, we plan to apply the insights gained from this study to real-world data from various rivers. Testing the models in a real-world context will validate their practical utility and performance by confronting them with the variances and anomalies of natural water systems. Overall, the comprehensive analysis carried out in this study contributes significantly to the field of environmental data science and offers valuable insights into the application of Bayesian classifiers for water quality classification. The findings hold practical implications for stakeholders in environmental monitoring and public health, aiding in the shift towards data-driven water management and safety assessments.

References

1. House of Commons Environmental Audit Committee et al. Water quality in rivers. Fourth Report of Session 2021, 22 (2022)
2. Jain, N., Yevatikar, R., Raxamwar, T.S.: Comparative study of physico-chemical parameters and water quality index of river. Mater. Today Proc. **60**, 859–867 (2022)
3. Bouamar M, Ladjal M.: Evaluation of the performances of ann and svm techniques used in water quality classification. In: 2007 14th IEEE International Conference on Electronics, Circuits and Systems, pp. 1047–1050. IEEE (2007)
4. Radhakrishnan, N., Pillai, A.S.: Comparison of water quality classification models using machine learning. In: 2020 5th International Conference on Communication and Electronics Systems (ICCES), pp. 1183–1188. IEEE (2020)
5. Alomani, S.M., Alhawiti, N.I., Alhakamy, A.A., et al.: Prediction of quality of water according to a random forest classifier. Int. J. Adv. Comput. Sci. Appl. **13**(6) (2022)
6. Yusri, H.I., Ab Rahim, A.Z., Hassan, S.L., Halim, I.S., Abdullah, N.E., et al.: Water quality classification using svm and xgboost method. In: 2022 IEEE 13th Control and System Graduate Research Colloquium (ICSGRC), pp. 231–236. IEEE (2022)
7. Ilic, M., Srdjevic, Z., Srdjevic, B.: Water quality prediction based on naive bayes algorithm. Water Sci. Technol. **85**(4), 1027–1039 (2022)
8. Friedman, N., Geiger, D., Goldszmidt, M.: Bayesian network classifiers. Mach. Learn. **29**, 131–163 (1997)
9. Jiang, L., Zhang, H., Cai, Z., Su, J.: Learning tree augmented naive bayes for ranking. In: Zhou, L., Ooi, B.C., Meng, X. (eds.) Database Systems for Advanced Applications. DASFAA 2005. Lecture Notes in Computer Science, vol. 3453, pp. 688–698. Springer, Berlin, Heidelberg (2005). https://doi.org/10.1007/11408079_63
10. Fasaee, M.A., Berglund, E., Pieper, K.J., Ling, E., Benham, B., Edwards, M.: Developing a framework for classifying water lead levels at private drinking water systems: a bayesian belief network approach. Water Res. **189**, 116641 (2021)
11. Mssmartypants. Water quality dataset. Kaggle Dataset (2023). https://www.kaggle.com/datasets/mssmartypants/water-quality
12. Yang, Y., Webb, G.I.: Discretization for naive-bayes learning: managing discretization bias and variance. Mach. Learn. **74**, 39–74 (2009)
13. Nojavan, F., Qian, S.S., Stow, C.A.: Comparative analysis of discretization methods in bayesian networks. Environ Model Softw. **87**, 64–71 (2017)
14. Tang, X., Li, J., Liu, M., Liu, W., Hong, H.: Flood susceptibility assessment based on a novel random na¨ıve bayes method: a comparison between different factor discretization methods. CATENA **190**, 104536 (2020)
15. Jensen, F.V., Nielsen, T.D.: Bayesian Networks and Decision Graphs, vol. 2. Springer (2007)
16. Han, J., Kamber, M., Pei, J.: 9 - classification: advanced methods. In: Han, J., Kamber, M., Pei, J., (eds.) Data Mining (Third Edition), The Morgan Kaufmann Series in Data Management Systems, pp. 393– 442. Morgan Kaufmann, Boston, third edition edition (2012). ISBN 978–0–12–381479–1. https://doi.org/10.1016/B978-0-12-381479-1.00009-5. https://www.sciencedirect.com/science/article/pii/B9780123814791000095

17. Tsamardinos, I., Brown, L.E., Aliferis, C.F.: The max-min hill-climbing bayesian network structure learning algorithm. Mach. Learn. **65**, 31–78 (2006)
18. Chicco, D., Jurman, G.: The advantages of the matthews correlation coefficient (mcc) over f1 score and accuracy in binary classification evaluation. BMC Genomics **21**, 1–13 (2020)

AI-Powered Prior Art Search: Towards Enriching Intellectual Property Management?

Ana Hafner[1,4](✉), Dolores Modic[1,2], Nadja Damij[3], Andrej Furlan[4], and Dorian Rampih[5]

[1] Rudolfovo – Science and Technology Centre Novo mesto, Podbreznik 15, 8000 Novomesto, Slovenia
ana.hafner@rudolfovo.eu

[2] Nord University Business School, Nord University, Postboks, 1490, 8049 Bodø, Norway
dolores.modic@nord.no

[3] University of Sunderland, Sunderland Business School, Faculty of Business and Technology, Edinburgh Building, Chester Road, Sunderland, UK
nadja.damij@sunderland.ac.uk

[4] Faculty of Information Studies, Ljubljanska Cesta 31a, 8000 Novo mesto, Slovenia
andrej.furlan@fis.unm.si

[5] University of Nova Gorica, School of Environmental Sciences, Dvorec Lanthieri, Glavni Trg, 5271 Vipava, Slovenia
rampihdarian@gmail.com

Abstract. This paper examines the efficiency of artificial intelligence (AI)-supported patent analytics tools in detecting relevant prior art associated with a specific patent. Our research demonstrates that AI-supported prior art search yields results more rapidly; these results are visually sorted in a transparent manner, presenting titles, abstracts, and images, which facilitates swift decision-making regarding the relevant state of the art. The majority of the results proposed by intellectual property (IP) tools were very broadly connected to the presented invention, but the correlation with patents recognized as relevant by the patent examiner and inventor was weak. Significant differences were observed between tools. Although our initial assumption was that IP tools should provide similar results if data coverage is identical, our research findings do not support it. However, AI-supported IP analytics tools develop progressively and in the next few years, they will certainly successfully complement or even displace traditional prior art searching.

Keywords: artificial intelligence · patents · prior art search · state-of-the-art search · AI-powered patent search · AI-powered IP analytics tools

1 Introduction

With the volume of patent data increasing, filing a new patent application necessitating a thorough state-of-the-art analysis, i.e., a prior art search, is imperative since we must ensure that the proposed invention has not been previously disclosed. However, given the staggering volume of available data, conducting a comprehensive prior art search

© The Author(s), under exclusive license to Springer Nature Switzerland AG 2025
N. Filipović (Ed.): AAI 2024, LNNS 1446, pp. 62–73, 2025.
https://doi.org/10.1007/978-3-031-99201-8_7

becomes a laborious and time-intensive endeavour, which also presents additional complexity for intellectual property rights (IPR) management [1]. This formidable task is typically entrusted to patent attorneys with the expertise to navigate efficiently through patent databases. However, in practice, various patent information users, including inventors, are also compelled to engage in this daunting challenge.

Concurrent with this advent of big data is the understanding that Artificial intelligence (AI) can (and is) transform numerous fields ranging from bioinformatics to image processing [2] and provide new solutions, including those related to patent informatics [3]. In the context of IPR, some researchers, as well as practitioners, point out that AI can penetrate and augment all areas of IP-related work, including the patent drafting, portfolio and management or prosecution [4], analysing concrete patent texts [5], patent classification [6] or patent search supporting prior-art identification [7] [8].

Using an experimental approach using an adapted protocol design as provided by Hafner et al. [7], the purpose of this paper is to examine how efficiently AI-supported patent analytics tools can detect relevant prior art connected to a specific patent. This article thus addresses the question of how semantic search using natural language processing (NLP) can better inform intellectual property (IP) management, i.e. state-of-the-art search, by providing results with greater accuracy than traditional (keyword) search within patent databases. We also not only provide the answer to the question of to what degree the results (prior art patents) using a traditional database search and searches using an AI-powered search match each other but also to how well results within different AI-powered IP tools are consistent across the databases. Regarding the last question, our expectation was that in the case of the same data coverage (i.e., access to a similar set of information), the tools should produce similar results since the algorithms on which the AI is trained are all focused on the optimisation of results. That said, under the same data conditions, different IP tools should perform similarly because they are built with the same intent.

The research contributes to the ongoing discourse on the impact of AI on IPR processes [9], suggesting that while AI technologies show promise, their effectiveness may vary based on algorithmic structure and tool-specific features. We thus address a critical research gap by illuminating whether AI-powered tools produce more accurate and reliable results compared to traditional approaches in patent searches. The analysis builds on existing literature to suggest a nuanced framework for assessing AI's contribution to prior art search, considering factors such as data coverage, algorithmic design, and result optimization. Subsequently, our findings and theoretical framing hold implications for both academic research and practical application in the field of intellectual property management. Empirically, the results offer a benchmark for evaluating AI-driven patent tools, and theoretically, the study enriches the conceptual discourse on AI's transformative potential in managing large-scale patent data.

1.1 Previous Work

AI for Patent Intelligence

Denter (2022) explores the critical role of patent intelligence in a data-driven business environment. It highlights how analysing vast amounts of patent data can provide valuable insights for innovation and strategic decision-making. The study underscores how the analysis of patent information encompassing retrieval, preprocessing and analysis has transitioned from a peripheral activity to a cornerstone of economic success. The paper systematically establishes a foundation for investigation into the application of supervised and unsupervised machine learning (ML) techniques. Supervised ML methods are discussed in the context of predicting outcomes based on the amount of input factors. On the other hand, unsupervised ML is explored for uncovering hidden patterns and trends within patent data without predefined labels [10].

Opportunities and challenges of machine learning approaches are presented in the four key publications, highlighted in his paper. Supervised machine learning is highlighted for its ability to refine decision-making processes by mitigating the noise of human biases in predictive models, minimizing the difference between predicted and true value, thereby enhancing the accuracy and reliability of outcomes derived from data. Conversely, unsupervised machine learning is recognized for its capacity to uncover latent associations and patterns within complex patent datasets, facilitating a deeper understanding of technological trends and innovations. Despite promising advantages, the research identifies significant challenges inherent in deploying ML for patent intelligence. These include the intricate complexity of configuring ML models and their algorithms, which lack transparency and interpretability and are crucial factors for trust and accountability in organizational decision-making. The dissertation emphasized the need to balance high-performance with low explainability, presenting this as a fundamental trade-off for organizations deciding between in house development and external procurement of ML solutions. To achieve effective integration of ML techniques in patent intelligence, strategies that optimize both operational efficiency and inherent interpretability must be advocated for [10]. His work provides a comprehensive analysis of how machine learning can revolutionize patent intelligence, offering practical implications for organizations seeking to harness large-scale data and understanding the balancing of applying advanced ML techniques against the need for transparency and explainability for decision-making.

AI-Powered Prior Art Search in Patent Analysis

Recent studies have highlighted the limitations of traditional prior art search methods within patent databases, such as manual keyword-based search and Boolean logic queries, as well as provided some initial insights into AI-supported patent analytics tools. The integration of AI in patent prior art searches represents a significant advancement in the field of intellectual property (IP) management. AI-powered tools offer the potential to overcome many of the challenges associated with traditional search methods, including the ability to process vast amounts of data rapidly, identify complex patterns, and understand semantic relationships between concepts [11]. Stechi et al. [8] developed an experimental platform for an AI-powered patent prior art search and suggested that the use of AI techniques to retrieve and rank documents could reduce the time and cost of prior art searches. Vaish et al.[12]'s literature review concluded that AI-powered systems can lead to a more streamlined approach to searching that is time-efficient and increases accuracy. Additionally, the integration of AI prior art search extends its capacity to enhance the comprehensiveness of searches, as traditional methods often rely on exact keyword matches, where in contrast, AI systems utilizing natural language processing and machine learning algorithms to understand the context and meaning behind patent descriptions, can potentially bring to the forefront previously overlooked prior art [13].

However, the implementation of AI in prior art search is not without challenges, particularly using it in an ethical manner [14] as well as its impact on the need for human expertise in interpreting search results, assessing the relevance of prior art, and making final determinations on patentability [15]. The optimal approach likely involves a synergistic relationship between AI systems and human experts, leveraging the strengths of both.

Hafner et al. [7] previously conducted an experimental study on a smaller scale, comparing keyword and full-text search results, evaluating their accuracy and time efficiency in prior art searches using a Boolean search within the Espacenet patent database, that served as a benchmark, and AI-supported semantic searches using proprietary IP tools. Their design enabled a robust comparison between traditional search techniques and emerging AI-powered solutions in the context of patent prior art identification. The authors designed the experiment to be replicable, allowing for future studies on the same or different patents, tools, and datasets, which has served as a foundation for our research design. Our study adapts and expands upon theirs to provide a more comprehensive data collection and analysis. Their findings suggested that whilst AI-powered search tools offer significant support to traditional search methods, solely relying on these technologies is not yet recommended for patent prior searches.

2 Research Design

A text of a PCT patent application WO2023211388A1, which was not published yet at the beginning of October 2023 when we made this experiment, was inserted into five different AI-supported patent analytics tools. These tools were proposed in the IP Monitor portal [16] from which we selected 18 relevant tools. However, only five of them expressed their willingness to cooperate in the experiment or were freely available (see Fig. 1).

In these tools, we inserted separately different parts of the patent document: A) abstract, B) claims, C) description, D) complete patent (abstract + description + claims). All five tools calculated similarity scores of patent documents in their databases with the entered text and then sorted the patent documents from the most to the least relevant. We compared these results with:

1) The relevant prior art identified by the inventor in the European Patent Office's (EPO) worldwide database Espacenet before filing a patent application (using keyword searching and Boolean operators) – these patents are mentioned in the patent application as a part of the description (state-of-the-art chapter) part.
2) The relevant prior art identified by EPO's examiner (using EPO's internal tools) during the procedure of PCT application.

We analysed how many matchings appeared regarding 1) and 2) when considering different parts of the patent (A, B, C, D) above. We also compared different IP tools – how well do their results match. The following problem must be mentioned here: patent families can appear under many different names – patent application and publication numbers. We had to be sure that all the different numbers were covered for one patent family. This required some additional work after we obtained the results from the IP tools, as we had to copy the rest of the numbers from the Espacenet database.

After these analyses, we presented our findings to IP tools' providers and asked them to comment on the results.

Therefore, this paper presents a mixed method, combining quantitative and qualitative research approaches.

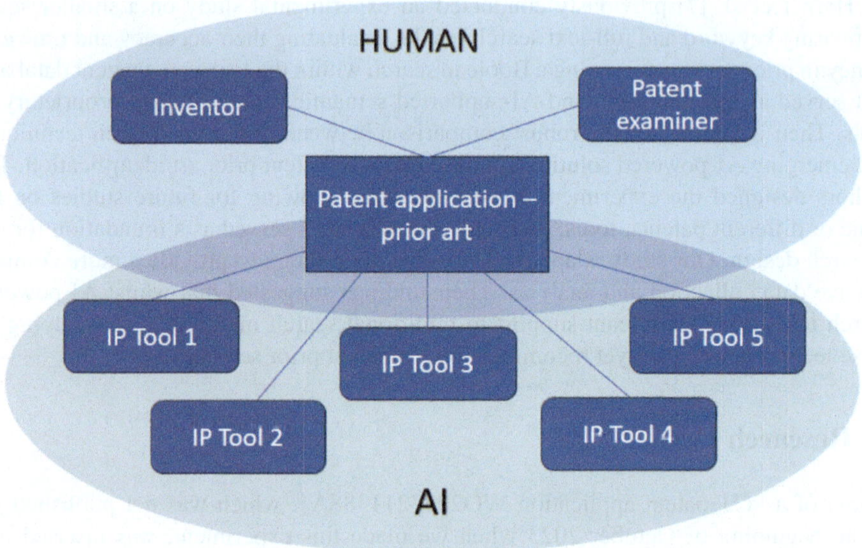

Fig. 1. Research idea.

3 Findings

Firstly, we checked the matches of patents between IP tools and those claimed by the inventor (16 patents) and by patent examiner (7 patents) (Table 1). In both cases, within the first 20 of the most relevant patents recognized by IP tools (see Appendix), these matchings rarely appear (see Table 2). One of the tools (Tool 3) seems to have significantly better performance. For example, it recognized 3 of 7 patents, which were also recognized as prior art by the patent examiner. However, prior art - a patent which compromised the novelty and inventive step of our invention (US10155613) was recognized only by Tool 2.

Secondly, we checked internal consistency: how did recognized relevant prior art in different parts of the patent document (abstract, description, claims, or complete patent) match within one IP tool? Patents marked with red colour in Fig. 2 are those which appear more than once within the same IP tool (uncoloured patents appear only once).

We also checked the matchings of patents within one IP tool between different parts of the patent document (Fig. 3). There were important differences between the

Table 1. Prior art identified by inventor and patent examiner.

Inventor's prior art	Examiner's prior art	Relevance of document identified by examiner
AU201716183	US10155613	X
CN203410834	US2016157647	A
EP3380404	DE6937855	A
WO2010040592	US5848928	A
CN104843329	KR20130039038	A
US2015136642	KR101962680	A
US2010044263	US3035925	A
DE1981252		
USD594323		
CH701734		
EP2132110		
RU2520009		
DE102009014245		
US20160152406		
KR20190134405		
EP2845700		

Note. Relevance of document identified by examiner: Where a document cited in the search report is particularly relevant, it is labelled with the letter "X" or "Y". Where a document cited in the search report represents state of the art that does not prejudice the novelty or inventive step of the claimed invention, it is labelled with the letter "A" [17]

tools, but generally, there was high consistency in identifying the relevant prior art in C (description) and D (complete document), and low consistency between A (abstract) and B (claims). As we can see, Tool 2 and Tool 3 did not recognize that abstract and claims belong to the same invention since there are no equal patents. We can conclude that in terms of internal consistency (Figs. 2 and 3), Tool 1 achieved the highest performance, i.e., it is most capable of determining that different parts of a patent belong to the same invention.

Thirdly, we checked external consistency: how much did the recognized relevant prior art within one IP tool match with other IP tools (Table 3). Overall, there are not many matchings, but there is some matching within some IP tools. Tool 1 and Tool 2 have two patent families in common, while Tool 1 and Tool 3 have seven. Tool 4 and Tool 5 do not have any matchings with any other of the tools.

Table 2. Equal patents.

	Matchings with inventor	Matchings with patent examiner
Tool1	1	1
Tool2	0	1
Tool3	2	3
Tool4	0	0
Tool5	0	0

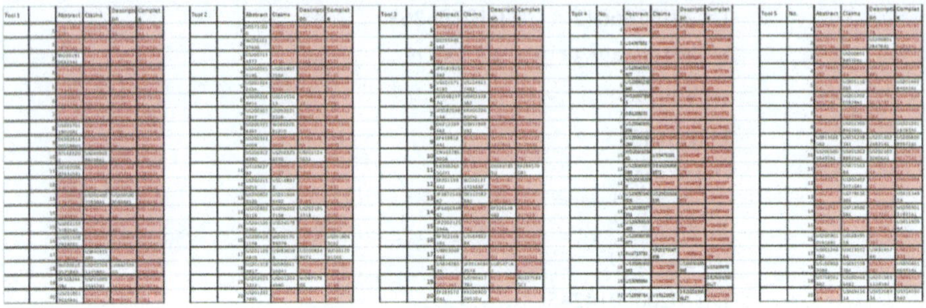

Fig. 2. Internal consistency.

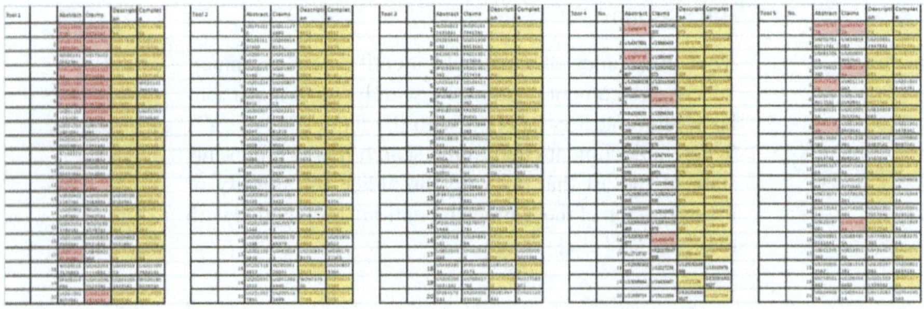

Fig. 3. Internal consistency: abstract and claims/description and complete document comparisons.

Therefore, the highest matching is between Tool 1 and Tool 3 which were the tools that also have the highest matchings with inventor's and patent attorney's selection of relevant patents.

Fourthly, when we analysed the content of the patents proposed by IP tools, we saw that there are important differences: some have considered the field of the invention (chocolate, food, eating, hygiene), while others have a technical implementation (solutions for packaging, grasping, lifting, or picking things up). In both cases, most of the proposed patents were at least partially or broadly connected with the proposed

Table 3. Number of equal patent families.

		Matchings (no. of equal patent families)
Tool1	**Tool2**	2
Tool1	**Tool3**	7
Tool1	**Tool4**	0
Tool1	**Tool5**	0
Tool2	**Tool3**	0
Tool2	**Tool4**	0
Tool2	**Tool5**	0
Tool3	**Tool4**	0
Tool3	**Tool5**	0
Tool4	**Tool5**	0

invention described in a patent application WO2023211388A1. In comparison to our previous research from 2022 [7], we can say that the relevance of the results is higher, so there is a notable progression.

Fifthly, in line with our previous research from 2022 [7], we can again confirm that AI-supported prior art search yields results rapidly, and these results are visually sorted in a transparent manner, presenting titles, abstracts, and images, which facilitates swift decision-making regarding the relevant state of the art.

Lastly, a very interesting fact was that most IP tools – except one – recognized US patents as the most relevant prior art. If the providers of these tools were all US-based companies, we could probably speak about "AI bias": these tools prefer US patents because they are trained that these patents are more important. It is known that AI systems can generate prejudiced outcomes that mirror and sustain the societal biases of those who created the systems. However, only two companies come from the US, while one is EU-based, and the fourth is from Japan. On the other side, of these four companies which prefer US patents, there is only one company – based in the EU – and this provider gave patents which came from all around the world (and strangely, included even more Japanese patents than the provider based in Japan). We need to mention that this is Tool 3, which also provided the largest number of matchings with the patent examiner's search and inventor's proposal.

4 Discussion

We have addressed within this paper several questions. Firstly, can semantic search using NLP better inform intellectual property (IP) management by providing results with greater accuracy than traditional search within patent databases? We can confirm this assumption. Although most of the results were only broadly connected with the described invention, we can report progress in comparison with our research from 2022 where most of the results were irrelevant [7]. We can say that this is significant progress in such a short time.

Secondly, we investigated how well the results (prior art patents) generated by traditional database search and AI-powered search match each other. Matching with the inventor's search, who used the Espacenet database and keyword searching with Boolean operators, was very rare. Only two tools recognized maximally two patents. Furthermore, we also compared the results we got from IP tools with a patent examiner's results. Here, we know only that the examiner used the "EPO internal" database, and we do not know how much this database is supported by AI. In this case, matchings were better but still rare.

Next, we focused on how well results within the same AI-powered IP tools match each other. With this question, we wanted to check to what degree the IP tools recognized the same relevant prior art in different parts of the patent document (abstract, description, claims, or complete patent). In other words, how much are IP tools capable of knowing from different parts of a patent application that we are dealing with the same invention. There were important differences between the tools, but generally, there was high consistency in identifying the relevant prior art in the description and complete document. This is understandable since the description is usually the largest part of a patent document. Abstract and claims did not have many matchings, in our case of two tools, even zero. However, this is also an important feature of AI-based solutions: how quickly can they understand from different parts of patent text that the same invention is described.

Lastly, we investigate how well results within different AI-powered IP tools match each other. This is the most important discovery of our research since we expected that – if the data coverage of IP tools is the same – their results should be very similar. For all tested patent search tools, we could conclude that the search results obtained were very different, with a small number of common patent families (Table 3) and mostly US-registered patents (see Appendix). The exception was one tool (Tool 3), which gave patents from several different jurisdictions. It is not easy to answer why US patents prevail. A representative of a US-based company which manages one of the IP tools responded that because a lot of the inventiveness is coming from the US and the US is such a large innovator, this is why a lot of the results return US patents. However, current data from the Patenscope database [18] shows that in the number of patent applications, the US is in the third position worldwide, behind China and Japan. The quality of patent applications is, of course, another question.

All tested tools come under the domain of proprietary software, with the general functioning principles being covered by the respective companies' confidentiality policies. This makes it impossible to establish the objective reasons for the differences in search results for tested tools. However, we can apply the basic principles of functioning of ML models and methods, and information provided by the providers of tested tools. By these means, it is possible to draw plausible conclusions about the present state of the AI-powered prior art search tools.

From the interview with one of the providers, we know that the ML models supporting the search tool are not trained on the patent data submitted for search by users. For model training, published data in research journals are used, together with the data for the company's internal patent databases. The exact nature of the latter was confidential.

It is reasonable to suppose that each tested tool is being developed independently by the respective company. Due to the nature of the ML models, it is hardly to be expected that two independently developed AI-assisted search tools should give identical results. The major influencing factors for search results are the construction of ML models in both supervised and unsupervised domains [6] [8] [10], data used for training of the supervised domains, and the patent databases searched by those tools. Moreover, the development of AI tools in patent search is very dynamic, being measured on the scale of months rather than years. That has also been confirmed to us by one of the providers we interviewed and by comparing the search results from the previous study [7].

5 Conclusion

AI-supported prior art search can provide results faster, they are visually transparently sorted with titles, abstracts, and images, which enables quick decisions regarding the relevant state of the art. In our case, most of the results proposed by IP tools were only broadly connected with the presented invention, however, the invention is also very specific. In comparison with our research from 2022 [1] we observed significant improvement. There were significant differences between the tools' results. Although our assumption was that IP tools should provide similar results if data coverage is the same, our research cannot confirm this assumption. However, AI-supported IP analytics tools develop progressively and in the next few years, they will certainly successfully complement and in most cases even completely displace traditional (keywords with Boolean operators) prior art searching.

Our research is not without limitations. Firstly, there is only one patent application with which we tested the performance of IP tools. However, even if we had more patents, there is a strong assumption we would come to the same conclusion: that tools gave very different results. Although most of them were to some degree connected with the invention described in a patent application, only one IP tool recognized as relevant patent the only one patent (US10155613) which compromised the novelty and inventive step of our invention. This means that one who is searching for the prior art, should at present concurrently use more AI-supported patent search tools instead of only one to ensure that no relevant prior art is missing. Furthermore, our selection of tools which were taken from IP Monitor [17] could suffer from selection bias - however, to diminish this, our starting point has been a pre-study, during which a review of the tools has been made to ensure their quality. All tested tools are what one could moniker as frontrunners. Hence, we believe the selection bias has not played a major role in our results.

Implications: We believe that utilizing generative AI presents a transformative approach to augmenting prior art searches. By harnessing the power of NLP and machine learning algorithms, generative AI will soon become even more precise in the comprehensive analysis of patent texts, research papers, and other documents, thereby enhancing the efficiency of identifying relevant prior art, which will transform the workflow of inventors, IP attorneys, patent examiners, IP management professionals and other users.

Acknowledgement. This work has been co-funded by the Developing Skills in Intellectual Property Rights Open Data for Sustainability and Circularity (IPR4SC) project funded by the Erasmus + Alliances for Innovation programme, grant number 101056129-IPR4SC.

References

1. Damij, N., Hafner, A., Modic, D.: Activity-to-skills framework in the intellectual property big data era. IEEE Trans. Eng. Manage. **71**, 13251–13265 (2024). https://doi.org/10.1109/TEM.2022.3175738
2. Shuja, J., Alanazi, E., Alasmary, W., Alashaikh, A.: COVID-19 open source data sets: a comprehensive survey. Appl. Intell. **51**, 1296–1325 (2021). https://doi.org/10.1007/s10489-020-01862-6
3. Alderucci, D., Sicker, D.: Applying artificial intelligence to the patent system. Technol. Innov. **20**, 415–425 (2019). https://doi.org/10.21300/20.4.2019.415
4. Adams, S.: Artificial Intelligence–can it deliver improved patent search? 19_2_05.pdf, https://japio.or.jp/00yearbook/files/2019book/19_2_05.pdf
5. Son, J., Moon, H., Lee, J., Lee, S., Park, C., Jung, W., Lim, H.: AI for Patents: a novel yet effective and efficient framework for patent analysis. IEEE J. Mag.| IEEE Xplore, https://ieeexplore.ieee.org/abstract/document/9779775. Accessed 30 Sept 2024
6. Kamateri, E., Salampasis, M., Perez-Molina, E.: Will AI solve the patent classification problem? World Pat. Inf. **78**, 102294 (2024). https://doi.org/10.1016/j.wpi.2024.102294
7. Hafner, A., Damij, N., Modic, D.: Augmented intelligence for state-of-the-art patent search: 2022 IEEE Technology and Engineering Management Conference (TEMSCON EUROPE). 2022 IEEE Technol. Eng. Manag. Soc. Conf. - Eur. TEMSCON-EUR, pp. 61–66 (2022). https://doi.org/10.1109/temsconeurope54743.2022.9801959
8. Setchi, R., Spasić, I., Morgan, J., Harrison, C., Corken, R.: Artificial intelligence for patent prior art searching. World Pat. Inf. **64**, 102021 (2021). https://doi.org/10.1016/j.wpi.2021.102021
9. Modic, D., Hafner, A., Damij, N., Cehovin Zajc, L.: Innovations in intellectual property rights management. Eur. J. Manag. Bus. Econ. **28**, 189–203 (2019). https://doi.org/10.1108/EJMBE-12-2018-0139
10. Denter, N.: Machine learning for patent intelligence: opportunities and challenges. Maschinelles Lernen für die Patent Intelligence: Möglichkeiten und Herausforderungen. (2022). https://doi.org/10.26092/elib/1940
11. The Transformative Impact of AI on Patent Prior Art Searches | Insights | Ropes & Gray LLP. https://www.ropesgray.com/en/insights/alerts/2024/08/the-transformative-impact-of-ai-on-patent-prior-art-searches. Accessed 30 Sept 2024
12. Vaish, K., Rawat, P., Kathuria, S., Singh, R., Joshi, K., Verma, A.: Artificial intelligence reducing the intricacies of patent prior art search. In: 2023 International Conference on Computational Intelligence and Sustainable Engineering Solutions (CISES), pp. 978–982 (2023). https://doi.org/10.1109/CISES58720.2023.10183481
13. GreyB, T.: Top 13 AI-based Patent Search Databases (2024). https://www.greyb.com/blog/ai-based-patent-databases/. Accessed 30 Sept 2024
14. Rai, A.K.: Machine learning at the patent office: lessons for patents and administrative law. Iowa Law Rev. **104**, 2617 (2018)
15. Aristodemou, L., Tietze, F.: The state-of-the-art on intellectual property analytics (IPA): a literature review on artificial intelligence, machine learning and deep learning methods for analysing intellectual property (IP) data. World Pat. Inf. **55**, 37–51 (2018). https://doi.org/10.1016/j.wpi.2018.07.002

16. IPR4SC - IPR made simple with the IP Monitor. https://ip-monitor.eu/. Accessed 25 Sept 2024
17. EPO - 9.2 Categories of documents (X, Y, P, A, D, etc.). https://www.epo.org/en/legal/guidelines-epc/2024/b_x_9_2.html. Accessed 30 Sept 2024
18. WIPO - Search International and National Patent Collections. https://patentscope.wipo.int/search/en/search.jsf. Accessed 30 Sept 2024

Overview of the Integration of Genetic Algorithms and Reinforcement Learning Techniques

Bogdan Milićević[1,2](✉), Vladimir Milovanović[1], Miljan Milošević[2,3,4], and Nenad Filipović[1,2]

[1] Faculty of Engineering, University of Kragujevac, Sestre Janjić 6, 34000 Kragujevac, Serbia
bogdan.milicevic@uni.kg.ac.rs, {vlada,fica}@kg.ac.rs
[2] Bioengineering Research and Development Center (BioIRC), Prvoslava Stojanovića 6, 34000 Kragujevac, Serbia
miljan.m@kg.ac.rs
[3] Institute for Information Technologies, University of Kragujevac, Jovana Cvijića Bb, 34000 Kragujevac, Serbia
[4] Belgrade Metropolitan University, Tadeuša Košćuška 63, 11000 Belgrade, Serbia

Abstract. Genetic algorithms, inspired by natural selection and genetics, employ evolutionary principles such as selection, crossover, and mutation to evolve solutions over generations. Reinforcement learning, on the other hand, focuses on learning optimal decision-making policies through interaction with an environment, guided by reward signals. This paper provides an overview of the integration of genetic algorithms and reinforcement learning, highlighting the benefits associated with their combination. Lately, researchers have combined reinforcement learning and genetic algorithms to improve the optimization process. Reinforcement learning agents can be used to choose genetic algorithm operators or tune hyperparameters, enhancing the performance of large-scale problems. A genetic algorithm can also be used to optimize the set of agents for reinforcement learning tasks. The integration of genetic algorithms and reinforcement learning represents a synergy of two powerful paradigms, each offering distinct strengths in problem-solving methodologies. This hybrid approach has the potential to address complex optimization problems and enhance learning systems.

Keywords: genetic algorithms · reinforcement learning · optimization algorithms

1 Introduction

Natural selection and genetics serve as the inspiration for genetic algorithms, which employ evolutionary concepts such as crossover, mutation, and selection to develop solutions over many generations [1]. Conversely, reinforcement learning is concerned with teaching agents how to make the best decisions by interacting with an environment that is controlled by reward signals [2]. Recently, to enhance the optimization process,

© The Author(s), under exclusive license to Springer Nature Switzerland AG 2025
N. Filipović (Ed.): AAI 2024, LNNS 1446, pp. 74–80, 2025.
https://doi.org/10.1007/978-3-031-99201-8_8

researchers have coupled genetic algorithms and reinforcement learning. By selecting genetic algorithm operators or fine-tuning hyperparameters, reinforcement learning agents can improve the performance of large-scale optimization problems. An additional application of genetic algorithms is the optimization of the agent set for reinforcement learning problems.

The integration of genetic algorithms with reinforcement learning in a variety of research articles [3–7] is overviewed in this paper. In paper [3], the genetic algorithm is guided to determine many parameters simultaneously by use of a controller mechanism based on reinforcement learning. Through activities that alter the probabilities of mutation and crossover, the reinforcement learning agent engages with the genetic algorithm environment [4]. The combination of population diversity and the objective function value is used as the reward function [4]. Another article combines reinforcement learning and genetic algorithm with Q-learning to direct the evolutionary algorithm search [5]. The Q-learning approach selects the crossover and mutation operators used for optimization, and the Q value determines the selection of the operation used for individual evolution. In paper [6], a novel method known as Genetic Algorithm based on Q-learning (QGA) is introduced, utilizing RL to improve genetic algorithm for multi-agent job scheduling [6]. The QGA algorithm optimizes work assignment in production systems by fusing the benefits of both reinforcement learning and genetic algorithm. The flexible job-shop scheduling problem is addressed by the proposed self-learning genetic algorithm (SLGA), in which reinforcement learning agents interact with the genetic algorithm environment to optimize the scheduling process [7].

In our article, we will present genetic algorithms and reinforcement learning and provide an overview of the papers that integrated these two techniques.

2 Methods

In this section, we will briefly describe the two integrated methodologies, namely genetic algorithms and reinforcement learning.

2.1 Genetic Algorithms

Genetic Algorithms (GAs) represent a family of optimization algorithms grounded in the principles of natural selection and evolution [1]. Mathematically, GAs operate on a population of potential solutions, where each solution, denoted as an individual or chromosome, is represented as a vector x in the search space [1]. The optimization problem can be formalized as finding the vector x that minimizes or maximizes a given objective function $f(x)$. The solution space is discretized and encoded into a set of binary strings, real-valued vectors, permutations, or other suitable representations [1]. The total process consists of initialization, fitness evaluation, selection, reproduction, offspring generation and termination.

- Initialization. A population of N individuals denoted as $P_0 = \{x_1, x_2, \ldots, x_N\}$ is randomly generated to form the initial population. Each individual x_i is initialized within the defined solution space S.

- Fitness evaluation. The fitness of each individual x_i in the population is evaluated using a fitness function $F(x_i)$ that quantifies the quality of the solution. The fitness function maps each individual x_i to a real number representing its fitness value $F(x_i) = f(x_i)$.
- Selection. Parents are selected from the current population to undergo genetic operations based on their fitness values. Common selection methods include roulette wheel selection, where individuals are selected with probabilities proportional to their fitness, and tournament selection, where individuals are chosen through tournaments among randomly selected subsets of the population.
- Reproduction. Selected parents undergo genetic operators, including crossover and mutation, to produce offspring. Crossover combines genetic material from the parents to generate one or more offspring, while mutation introduces random changes to individual chromosomes to maintain diversity.
- Offspring generation. The offspring produced by genetic operators replace individuals in the current population, forming the next generation. The offspring population replaces the previous generation to form the next population. Various replacement strategies, such as generational replacement or steady-state replacement, are employed to maintain population size and diversity.
- Termination. The evolutionary process continues for a predetermined number of generations or until a termination criterion is met. Common termination criteria include reaching a maximum number of generations, achieving a target fitness threshold, or observing stagnation in fitness improvement.

Parameters such as population size N, crossover rate p_c, mutation rate p_m, and selection mechanisms are tuned through experimentation to optimize algorithm performance for specific problem domains.

2.2 Reinforcement Learning

Reinforcement Learning (RL) is a computational approach to learning whereby an agent interacts with an environment, aiming to learn a policy that maximizes cumulative rewards over time [2]. Formally, RL can be described using the framework of Markov Decision Processes (MDPs), which is defined by the tuple (S, A, P, R, γ) where S denotes state space, A represents the action space, P is the state transition probability function, defining the probability of transitioning from one state to another given an action, R is the reward function, and γ is the discount factor, which balances immediate and future rewards [2]. A policy π is a mapping from states to actions, determining the agent's behavior in the environment. Formally, a policy can be represented as $\pi : S \rightarrow A$. Value functions are used to evaluate the goodness of states or state-action pairs under a given policy. A basic reinforcement learning agent interacts with its environment in discrete time steps. At each time t, the agent receives the current state s_t and reward r_t. It then chooses an action a_t from the set of available actions A, which is subsequently sent to the environment. The environment moves to a new state s_{t+1} and the reward r_{t+1} associated with the transition (S_t, A_t, S_{t+1}) is determined. The general schematic of reinforcement learning is presented in Fig. 1.

Exploration and exploitation are fundamental concepts in RL that refer to the agent's dilemma of deciding whether to explore new actions or exploit known actions to maximize rewards. Exploration is the process of trying out new actions or exploring unknown

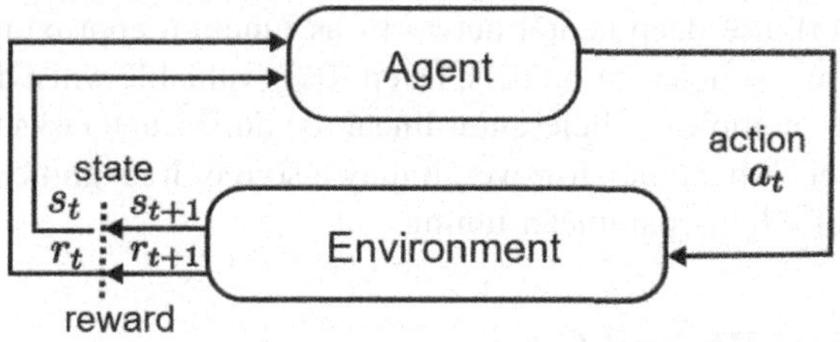

Fig. 1. General reinforcement learning schematic

regions of the state-action space. Exploration is essential for discovering potentially better policies that lead to higher rewards. However, excessive exploration may lead to suboptimal performance in the short term. Exploitation is the process of selecting actions that are known to yield high rewards based on the agent's current knowledge. Exploitation is crucial for maximizing immediate rewards and exploiting known good policies. However, relying solely on exploitation may lead to premature convergence to suboptimal solutions. The environment may be non-stationary, meaning that the optimal policy or reward distribution may change over time. Exploration strategies need to adapt to changes in the environment to ensure continued learning and adaptation. In complex environments with multiple local optima, exploration strategies must allow the agent to escape from suboptimal solutions and explore other regions of the state-action space to discover globally optimal policies. Techniques to balance exploration, trying new actions to discover potentially better policies, and exploitation, taking actions known to yield high rewards, include ε-greedy, softmax exploration, and Upper Confidence Bound (UCB).

Common RL techniques include Q-Learning, Policy gradient methods, Monte Carlo methods, Temporal difference, etc. Q-Learning is a model-free RL algorithm that learns action-value functions directly from experience. It is particularly effective for discrete action spaces and environments with tabular representations. Policy gradient methods directly parameterize the policy and update it in the direction of higher expected returns. Examples include REINFORCE, Actor-Critic methods, and Proximal Policy Optimization (PPO). Actor-Critic methods combine value-based and policy-based approaches by learning both a value function (the critic) and a policy (the actor). Examples include Advantage Actor-Critic (A2C) and Deep Deterministic Policy Gradient (DDPG). Monte Carlo methods estimate value functions or policies by sampling returns from actual trajectories. They are model-free and can handle stochastic environments but may require a large number of samples for accurate estimation. Temporal Difference (TD) learning methods update value estimates based on the difference between successive estimates. They combine the advantages of Monte Carlo methods and dynamic programming and are often more sample-efficient.

Deep RL extends RL by leveraging deep neural networks to learn representations directly from high-dimensional sensory inputs. Deep RL algorithms, such as Deep Q-Networks (DQN), Deep Deterministic Policy Gradient (DDPG), and Proximal Policy

Optimization (PPO), use deep neural networks as function approximators to approximate value functions, policies, or models. Deep RL is valuable when dealing with high-dimensional inputs, complex policies, non-linear dynamics, and tasks that require transfer learning and generalization. However, it may also require significant computational resources and careful hyperparameter tuning.

3 Integration of RL and GA

In this section, we will provide brief summaries of several selected research articles that combined RL and GA. There are multiple ways to combine RL and GA. An RL agent can be used to choose the operators of the GA or to tune the hyperparameters of GA. Both of these approaches show promise in terms of improving the optimization process. Additionally, a GA can be used to optimize the set of agents that can be used for a reinforcement learning task.

In paper [3], a reinforcement learning-based controller mechanism is used to guide the GA to determine multiple parameters simultaneously. Specifically, the RL agent interacts with the GA environment by taking actions that affect mutation and crossover probabilities. The state space is defined as a tuple of two elements, current objective function value and diversity of the population pool. The RL agent learns to improve the GA's performance towards solving large-scale and practical problems by iteratively improving the off-policy method using the Q-learning model [3]. The reward function is defined as the combination of objective function value and population diversity.

In [4], RL and GA are combined by using RL agents to learn a behavior policy that dynamically tunes the operators of Genetic Algorithms at runtime. The RL agents adjust the crossover and mutation probabilities based on the state of the optimization process.

RL and GA are combined by using Q-learning to guide the search for an evolutionary algorithm in paper [5]. The crossover and mutation operators used for optimization are chosen by the Q-learning method and the operation used for individual evolution is selected according to the Q value [5]. The RL agent's states in the RL-GA algorithm indicate whether the fitness function value has improved through the search and are divided into two states, higher and lower. The RL agent's actions are the evolutionary operations selected according to the Q value, which can be one or two evolution operations selected from a total of 15 actions in the Q-value table [5]. The RL-GA uses real numbers to encode individuals, with the decoding process using the task time window selection algorithm, which determines whether to execute each task and the specific execution time of each task.

In the research article by Zhipeng Li et al., RL is used to enhance GA for multiagent task scheduling by introducing a novel approach called Genetic Algorithm based on Q-learning (QGA) [6]. Specifically, the QGA algorithm combines the strengths of both RL and GA to optimize task assignment in manufacturing systems. In QGA, the gene space of the genetic algorithm problem is considered as the action strategy space of the Q-learning algorithm [6]. The fitness function of performing an action within the gene space is treated as the reward obtained from executing the action. Furthermore, the Q-learning algorithm is responsible for selecting good actions, while the genetic algorithm is responsible for finding good structures. By combining these two algorithms, QGA aims to reach the optimal value in a shorter time, resulting in a smaller makespan for task scheduling [6]. Overall, the use of reinforcement learning in conjunction with genetic algorithms in the QGA approach provides a more effective and efficient method for improving task scheduling in multiagent systems.

In the proposed self-learning genetic algorithm (SLGA) for solving the flexible job-shop scheduling problem, RL and GA are combined in a synergistic manner to enhance the optimization process [7]. In the context of the self-learning genetic algorithm (SLGA) for the flexible job-shop scheduling problem, the RL agent interacts with the genetic algorithm (GA) environment by considering states and taking actions to optimize the scheduling process [7]. The state of the GA at a given time step, represents the current configuration or status of the genetic algorithm during the iterative process. It includes information such as the current population of solutions, the fitness values of individuals, and the progress of the optimization algorithm [7]. The actions taken by the RL agent involve adjusting and updating specific parameters of the genetic algorithm, such as the crossover rate and mutation rate. By dynamically adjusting these parameters, the agent aims to guide the genetic algorithm towards more effective exploration of the solution space and improved convergence to high-quality solutions.

4 Conclusions

Optimization is a fundamental concept across various domains, encompassing fields as diverse as engineering, economics, machine learning, and beyond. At its core, optimization aims to identify the best possible solution from a set of feasible alternatives, considering specified objectives, constraints, and criteria.

The integration of genetic algorithms and reinforcement learning represents a promising approach to enhance the optimization process. Drawing inspiration from evolutionary concepts and guided by reinforcement signals, this hybrid methodology combines the strengths of both techniques to tackle complex problems more effectively. Through an overview of several research articles, we have highlighted the diverse applications of this integration. From simultaneous parameter determination to large-scale issue optimization and multi-agent job scheduling, the synergy between genetic algorithms and reinforcement learning has demonstrated significant potential in improving solution quality and convergence speed. The outlined papers showcase innovative approaches, such as utilizing reinforcement learning agents to guide genetic algorithm operations or introducing novel algorithms such as QGA for multi-agent job scheduling. These methodologies leverage the complementary strengths of genetic algorithms and reinforcement learning to address challenging optimization problems more efficiently. Overall, the integration of genetic algorithms and reinforcement learning opens new avenues for advancing optimization techniques and solving real-world problems. In addition to the diverse applications highlighted across various domains, the integration of genetic algorithms and reinforcement learning holds promise for improving the optimization problems in biomedical engineering. In future, by leveraging this hybrid approach, researchers can address complex challenges inherent in biomedical research, such as parameter estimation, system modeling, and drug design optimization. Through the synergy of genetic algorithms and reinforcement learning, enhanced solutions can be developed to improve medical diagnosis, treatment planning, and healthcare management, ultimately leading to better patient outcomes and advancements in the field.

Acknowledgements. The research was supported by the STRATIFYHF project that has received funding from the European Union's Horizon Europe research and innovation programme under Grant Agreement No 101080905. Views and opinions expressed are however those of the authors only and do not necessarily reflect those of the European Union or granting authority - European Health and Digital Executive Agency (HADEA). Neither the European Union nor the granting authority can be held responsible for them. This research was also supported by the Ministry of Education, Science and Technological Development of the Republic of Serbia, contract number [451-03-65/2024-03/200107 (Faculty of Engineering, University of Kragujevac) and 451-03-66/2024-03/200378 (Institute of Information Technologies, University of Kragujevac)]. The authors acknowledge support from the City of Kragujevac, Serbia.

References

1. Mitchell, M.: An introduction to genetic algorithms (1996)
2. Kaelbling, L.P., Littman, M.L., Moore, A.W.: Reinforcement Learning: a survey. J. Artif. Intell. Res. **4**. AI Access Foundation, 237–285 (1996). https://doi.org/10.1613/jair.301
3. Quevedo, J., Abdelatti, M., Imani, F., Sodhi, M.: Using reinforcement learning for tuning genetic algorithms. In: Proceedings of the Genetic and Evolutionary Computation Conference Companion. ACM (2021). https://doi.org/10.1145/3449726.3463203
4. de Miguel Gomez, A., Toosi, F.: Continuous parameter control in genetic algorithms using policy gradient reinforcement learning. In: Proceedings of the 13th International Joint Conference on Computational Intelligence. SCITEPRESS - Science and Technology Publications (2021). https://doi.org/10.5220/0010643500003063
5. Song, Y., Wei, L., Yang, Q., Wu, J., Xing, L., Chen, Y.: RL-GA: A reinforcement learning-based genetic algorithm for electromagnetic detection satellite scheduling problem. Swarm Evol. Comput. **77**. 101236 (2023). https://doi.org/10.1016/j.swevo.2023.101236
6. Li, Z., Wei, X., Jiang, X., Pang, Y.: A kind of reinforcement learning to improve genetic algorithm for multiagent task scheduling. Math. Prob. Eng. Hindawi Limited, 1–12 (2021). https://doi.org/10.1155/2021/1796296
7. Chen, R., Yang, B., Li, S., Wang, S.: A self-learning genetic algorithm based on reinforcement learning for flexible job-shop scheduling problem. Comput. Ind. Eng. **149**, 106778 (2020). https://doi.org/10.1016/j.cie.2020.106778

Utilization of Augmented Reality for Improvement of Balance Disorders

Đorđe Ilić[1,2](✉), Aleksandra Vulović[1,2], and Nenad Filipović[1,2]

[1] Bioengineering Research and Development Center (BioIRC), 6 Prvoslava Stojanovica Street, 34000 Kragujevac, Serbia
djordje.ilic@kg.ac.rs

[2] Faculty of Engineering, University of Kragujevac, 6 Sestre Janjica Street, 34000 Kragujevac, Serbia

Abstract. Falls of the elderly, as well as younger people and children, can cause injuries. In the elderly, falls are one of the leading causes of injury and death, and the second leading cause of accidental and unintentional deaths. Balance and problems related to balance are one of the most common health problems in the elderly and, in most cases, are the primary cause of falls. A measure to improve functionality in maintaining balance and preventing falls is balance rehabilitation. The platform of the TeleRehaB DSS project is developed to actively engage patients physically and help them in the rehabilitation process. It contains Virtual Coach (VC) support that is adapted in such way so that it can effectively enable the user to achieve various goals. The TeleRehab DSS system for patients with balance problems delivers a training program at home with the help of VC support. The active participation of the patient in these games results in significant improvements in the rehabilitation process. "Catching Game" is one of the games within the Cognitive Games section, in which the patient has the task of collecting apples that fall from the tree at a certain speed. Regulating the speed of falling apples as well as their quantity can be achieved by introducing the field of artificial intelligence (AI), which will have the ability to recognize the patient's needs and personalize their experiences, thereby raising the functionality of VC support to a higher level. In the future, the goal is to work on the improvement of these types of systems based on augmented reality with the help of AI that contributes to the rehabilitation of patients by increasing the percentage of their efficiency.

Keywords: Telerehabilitation · Balance · Augmented reality · Virtual coaching

1 Introduction

Traditional therapies typically require physical presence of a therapist. However, advancements in technology have introduced new treatment possibilities, allowing patients to manage their own care under the remote supervision of professionals through digital devices. This type of treatment belongs to the field of telerehabilitation, which involves engaging the patients and monitoring them through digitally scheduled sessions. Telerehabilitation implies a system of telecommunication technologies for controlling or monitoring remote rehabilitation.

© The Author(s), under exclusive license to Springer Nature Switzerland AG 2025
N. Filipović (Ed.): AAI 2024, LNNS 1446, pp. 81–89, 2025.
https://doi.org/10.1007/978-3-031-99201-8_9

Virtual Reality (VR) and Augmented Reality (AR) allow access to simulated areas that would be impossible to reach in person. In addition to psychological conditions, VR treatment options include conditions requiring stroke rehabilitation as well as conditions for physical rehabilitation and balance rehabilitation [1, 2]. For the aforementioned areas of rehabilitation, VR has proven to be an effective treatment tool that has been used for a long time, while AR, on the other hand, is increasingly used in all of these.

Through programs such as motion-enabled games, AR can reproduce precise physical movements and provide additional motivation for exercise. Using these types of applications or platforms that create a virtual environment, patients practice using simulations that are set up to create imbalances, allowing them to adapt and improve their balance [3]. With the help of various motion sensors located on people using AR platforms, they provide experts with various information, such as movement accuracy or execution speed, etc. Based on this information, doctors can monitor the current clinical picture and adjust future sessions. In addition, virtual reality helps patients overcome anxiety by providing a controlled and personalized virtual environment, thereby increasing patients' self-confidence and freeing them from the fear of falling or losing balance [4]. It is important to emphasize that virtual reality does not replace traditional treatments for balance disorders, but complements conventional rehabilitation sessions.

In modern VR and AR applications, artificial intelligence plays a key role in creating realistic and dynamic environments. Artificial intelligence algorithms have taken the development of these applications to a higher level by improving the realism of virtual worlds [5]. Using artificial intelligence, environments with dynamic elements can be introduced that adapt to the behavior of the application user, creating a more attractive experience. From object recognition to content generation, AI can contribute to the authenticity of virtual landscapes, which can significantly improve the realism of applications. Also, the AI area in VR/AR applications realizes many possibilities for additional improvement of realism within them.

The AR platform in this study contains Virtual Coach (VC) platform, which is a part of the TeleRehaB DSS platform. In addition, we explore several approaches that can be used in order to improve VC support for more efficient and easier use. The goal of this paper is to analyze and present ways to improve the VC System of the TeleRehaB DSS project platform through the utilization of artificial intelligence algorithms.

2 Technologies Used for Application Development

The AR platform of the TeleRehaB DSS project is designed and built in Unity, a cross-platform game and application engine based on C++. Running code in Unity automatically uses Mono version 3.5, which has API compatibility roughly equal to that of the.NET Framework 3.5/CLR 2.0. The Virtual Coach Platform is supported on two platforms, PC and Android, and was made in Unity, which is composed of so-called scenes that contain different types of components such as: buttons, input fields, fields for printing text, plain text or images, etc. The application consists of three scenes: StartScene, CognitiveGames and VirtualCoach. StartScenes (Fig. 1) represent the initial screen at the beginning, which contains the basic menu for selecting the language and choice between Virtual Coach segment and Exergames and Cognitive games segment.

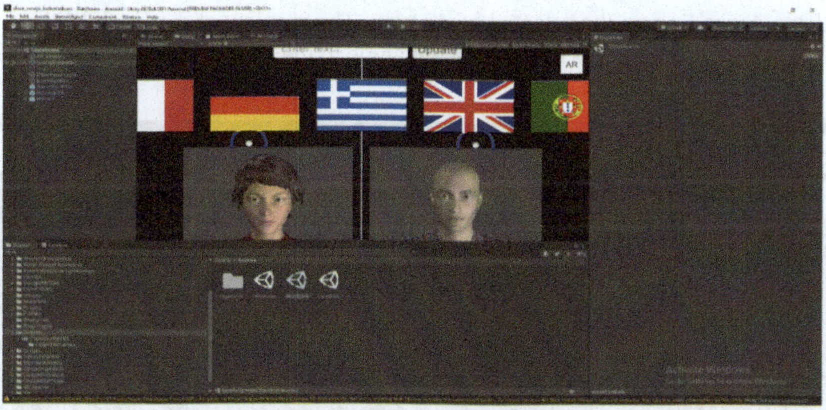

Fig. 1. Unity development environment

The scenes in the project are regulated by C# scripts in the background. In addition to the C# programming language, functionality can be controlled by JavaScirpt (UnityScript) or Boo programming languages. The code written in the programming languages that do not represent Unity engine code runs on Mono or the Microsoft.NET Framework, which is compiled Just-in-Time (JIT). The version of the Unity editor in which the project was developed was 2019.4.18f1. When exporting an application for the Android platform, the editor copies the recommended versions of Android development tools (JDK, SDK, NDK, GRADLE) that are in its range.

3 Application Overview

As previously mentioned, the TeleRehaB DSS platform contains two types of interactive exercises/games for patient rehabilitation: Cognitive and Exergames. Cognitive games represent active games that include only the patient's head movements, while Exergames also include the patient's head movements but in different body positions. The Cognitive games section contains a total of 6 games, including: Memory game (stacking cards in certain positions), Catch food game (collecting apples that fall from a tree), Remember previous game (guessing the type of animal that comes out of the sensor), Animal feeding game (feeding rabbits), Preparing animal food (feeding rabbits by choosing the correct product), Marble game (managing the ball towards the goal). Each game in the Cognitive Games section has 5 different levels, each of which carries a certain weight. The idea is that while the user is playing these games, he interacts and performs the necessary actions by moving his head. The Exergames section contains three different types of games: Sitting (the game is performed while the patient is sitting), Standing (the game is performed while the patient is standing) and Walking (the game is performed while the patient is walking). If a patient chooses the Sitting game type in the Exergames section, he can choose between the two following options: Look left and right or Look up and down. Both types are performed from a sitting position and the goal is to follow the target in the center of the screen that moves by moving the head left and right or up and down depending on the selected game type. The preview of the Look left and right game, where the patient tries to follow a rectangle to the left or right, is shown in Fig. 2.

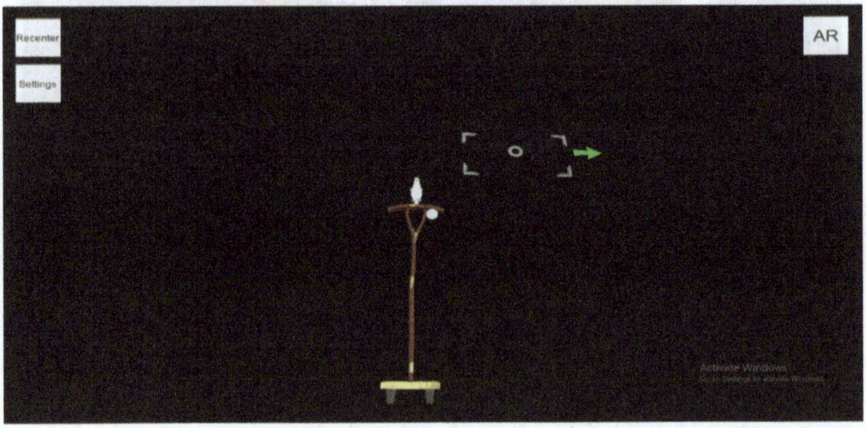

Fig. 2. Sitting games (Look left and right)

The second game of the Exergames section, Standing, has three types: Standing seal, Bending down and Reaching up. When playing Standing Seal, the patient must keep the body straight with hands by the hips in a standing position with eyes closed for one minute, maintaining the balance that way (Fig. 3). After the end of one minute, the main goal is to help the seal control the ball on his head by standing still and maintaining his balance.

Fig. 3. Standing game (Standing seal)

In the Bending down game, the patient is required to bend the body at an angle of 90 degrees from the hips upwards, including the entire upper part of the body. The task is that, by bending and straightening the body at an angle of 90 degrees, the patient commands the dog that he/she sees in front of him/her to stand up and lie down.

If the patient's body is in the lower position, the dog is lying down at that moment, and if the patient's body is in the upper position, the dog is standing at that moment. The third type, Reaching up, has the task of involving the whole body of the patient by moving his/her arms above the head upright. With this movement of the body, the patient commands the dog in front of him/her to jump, when the hand is raised above the head of the dog or to be on the ground when the hand is not raised.

The third game of the Exergames section, Walking, consists of four types: Walking straight, Walk and look left and right, Walk and look up and down, Walk and look diagonally. In the first type, Walking straight, the patient sees an empty space in front of him/her that is surrounded by grass and flowers in which he/she is required to move straight, following the bird that goes in front of him/her to a certain point (flower), trying not to look left or right or at the floor. When the patient reaches the flower, the patient returns to the starting position, concentrating on looking straight ahead following the bird. Figure 4 shows the appearance of the game from the patient's profile, while Fig. 5 shows the appearance of the application from the first person.

Fig. 4. Walking game (Walking straight, side 1)

In the next type, Walk and look left and right, the patient interacts in the same way as in the previous game with a difference of following the bird by moving the head left and right. Also, the types Walk and look up and down and Walk and look diagonally contain the same process of interaction. The difference is that in these games the user or patient follows the bird by moving the head up and down or diagonally. With all the previously mentioned types of games within the framework of the Walking section, the patient is required to keep the body straight while walking, synchronizing the movements of legs and arms.

Fig. 5. Walking game (Walking straight, side 2)

4 Application of AI with Augmented Reality

AR works by combining the real world with elements of the virtual world. Images or sounds are created by programmers where the goal is to include them in the real world. On the other hand, VR creates its own reality that is computer generated and guided. Both areas are similar in the way they provide the user with an enhanced experience. AR and VR have great potential in advancing the medical field by making things like remote surgery a real possibility. Our application is related to the AR field because it uses the external world in combination with virtual elements, but it turned out that there are small differences between AR and VR, and for this reason we can take into consideration the literature related to VR. The way AI contributes to the improvement of an AR application is almost the same for VR. Analyzing and understanding the role of artificial intelligence in combination with VR in medicine was carried out by Wu et al. [6] in their paper categorizing the taxonomy of the application into three categories: *Visualization Enhancement*, *VR-related Medical Data Processing* and *VR-assisted Intervention.* Based on this classification, a comprehensive review of the current state of the art of AI in VR applications adapted for medical needs was performed, where the contribution of each category to health care was defined. The *Visualization Enhancement* category is focused on increasing the user's visual and sensory perception in the virtual space, while *VR-related Medical Data Processing* focuses on the advanced capabilities of VR Systems for analyzing and interpreting complex medical data. The last one, *VR-assisted Intervention*, shows the application of VR combined with AI in real-time and interactive medical scenarios. By categorizing applications based on their functions in the spectrum of patient care, the aim of the paper is to provide a detailed analysis that will facilitate the understanding and identification of current shortcomings in the combination of these two areas, as well as their improvement. The decision-making process in medicine requires or involves the health care provider and the patient, where the clinician offers options and describes their risks and benefits, and the patient expresses his values. In the paper by Antel et al. [7] the way to provide safe and successful doctor-patient communication using AI methodology by incorporating VR technology has been analyzed. Successful communication represents an accurate understanding of the relevant factors surrounding clinical choices about the division of responsibility for surgical decisions among all participants. The inclusion of the mentioned technologies can lead to more personalized communication adapted to the needs, characteristics, and values of the patient, thereby improving communication. The use of AI algorithms can enable doctors to generate customized decision-aid documents for individual patients based on demographic characteristics or individualized pathologies.

Lampropoulos et al. [8] performed a detailed analysis by reviewing the applications of deep learning, knowledge graphs and the semantic web to influence the improvement of augmented reality (AR). They determined three areas - object detection, image processing and computer vision with which deep learning can improve the user experience in AR services. In this paper, AI is expressed as a technology to improve the detection of inputs such as behavior or speech in AR. Ribeiro de Oliveira et al. [9] reviewed papers focusing on how VR combined with AI is applied to specific problems in industry, commerce, logistics, services, processes or systems. The authors point out that AI methods contribute mainly to higher accuracy and higher efficiency of problems in VR (for example, in surgery). Lack of training data or high computational costs are the main drawbacks in applying AI to VR problems. One of the more recent articles in this area is the one by Reiners et al. [10] where they address the combination of AR and AI research and, according to their review, the main applications of AR and AI are training (medical or military), robots, games, autonomous cars and advanced visualization. These existing reviews co-center on the areas where AR and AI are applied. Generally speaking, most research is done in the field of medicine and training. Also, the papers point out that computational costs and limited training data are the two main issues that limit existing methods. AR/VR are relatively new technologies that are increasingly used both in medicine and in other segments. Accordingly, consideration of the application of AI in combination with these technologies has begun. Based on the information from the available literature, we believe that AI can be used with our application in the ways of personalization content and communication. In particular, in the Walk and look left and right game from the Walking section, things can be made more difficult for the user during interaction in order to encourage the user to progress in the rehabilitation process. This method involves increasing the distance between the starting point and the end point, which the patient should cover by following the bird by moving the head left and right. In order to achieve the best possible response in terms of balance, it is important to increase the number of movements of the bird to the left and right, so that the patient moves the head to the right and left several times. Each patient would have their own progress, which would be tracked by AI algorithms, on the basis of which each subsequent attempt would be adjusted to the difficulty level based on the previous attempt. Given that our application communicates with the Edge server and sends it certain information from the application, it can be introduced to send accurate information about the patient's movements that define the condition, based on these parameters AI can automatically determine the condition and proposal for the next therapy to a certain medical worker. In the ways of personalized content and communication that can be improved by AI, other games in the application can be improved as well.

5 Conclusion

VR/AR have the potential to significantly influence the transformation of health care, enabling new types of treatments and diagnostics and the possibility of changing the way and place of providing care. The ability to deliver standard and very new types of content in expressive and realistic ways at a distance that are adapted to different clinical contexts is their main potential in diagnosis and treatment. VR/AR are not only contexts related to the field in medicine, but represent valid solutions in education, venous or surgical visualization, patient relaxation, PTSD treatment, speeding up recovery in physical therapy - or even supporting medical presentations. The development of these applications is a difficult and expensive process. However, the development of these applications can contribute to avoiding doctor's mistakes, saving lives, and also in relaxing the patient, which contributes to a more pleasant experience.

This paper presents an application that is used for medical purposes for the rehabilitation of patients who have problems with balance. Three types of games are presented within the application, which include movements of the patient's head and other parts of the body. We have shown that patients can have an interesting experience using the mentioned application as well as benefits in the process of their balance rehabilitation. The entire methodology proved to be successful, but with the combination of artificial intelligence, it can improve the application by creating a more personalized and adapted environment for the patient, which can lead to more efficient functionality of the entire project.

Acknowledgements. This research is funded by the European Union, under the Horizon Europe programme, GA number 101057747 (TeleRehaB project). Views and opinions expressed are however those of the author(s) only and do not necessarily reflect those of the European Union or the granting authority. European Health and Digital Executive Agency (HADEA). Neither the European Union nor the granting authority can be held responsible for them. Authors also acknowledge the funding by the Ministry of Science, Technological Development and Innovation of the Republic of Serbia, contact number [451-03-65/2024-03/200107 (Faculty of Engineering, University of Kragujevac)].

References

1. Pérez, I.C., Nieto-Escamez, F.A., Obrero-Gaitán, E.: Immersive virtual reality in stroke patients as a new approach for reducing postural disabilities and falls risk: a case series. Brain Sci. (15 May 2020)
2. Sana, V., Ghous, M., Kashif, M., Albalwi, A., Muneer, R., Zia, M.: Effects of vestibular rehabilitation therapy versus virtual reality on balance, dizziness, and gait in patients with subacute stroke: a randomized controlled trial. Randomized Controlled Trial, (16 Jun 2023)
3. Hemme, C.L., Carley, R., Norton, A., Ghumman, M., Nguyen, H., Ivone, R., Menon, J.U., Shen, J., Bertin, M., King, R., Leibovitz, E., Bergstrom, R., Cho, B.: Developing virtual and augmented reality applications for science, technology, engineering and math education. Biotechniques **75**, 11–20 (2023)

4. Lucas, L., Parker, J.: Physiotherapists perceived role in managing anxiety in patients with relapsing-remitting multiple sclerosis: a mixed-methods study (2022)
5. de Oliveira, R., Rodrigues, B., da Silva, B., da Silva, M., Spinass, M., Spinass, A.N., Ludke, R., Ludke, G., Gaudio, G., Gaudio, R.S., Gomes, M., Gomes, I.R., Cotini, G., Cotini, G., Vargens, L., Vargens, Sd., Schimidt, D., Schimidt, Q.: Virtual reality solutions employing artificial intelligence methods: a systematic literature review. ACM Comput. Surv. **55**, 29 (2023)
6. Yixuan, W., Kaiyuan, H., Danny, C.Z., Jian, W.: AI-enhanced virtual reality in medicine: a comprehensive survey (2024)
7. Ryan, A., Samira, R.A., Guadagno, E., Jason, H.M., Dan, P.: The use of artificial intelligence and virtual reality in doctor-patient risk communication: a scoping review. Patient Educ. Couns. **105**(0738–3991), 3038–3050 (2022)
8. Georgios, L., Euclid, K., Konstantinos, D.: Enhancing the functionality of augmented reality using deep learning, semantic web and knowledge graphs: a review. Vis. Inform. **4**, 32–42 (2020)
9. Oliveira, R.d., Silva, T.d., Silva, M.d., Spinasse, M., Spinasse, N., Ludke, G., Antonio, R., Gaudio, G., Gaudio, R. S., Gomes, M., Gomes, I. R., Cotini, G., Cotini, G., SCHIMIDT, L., SCHIMIDT, V., SCHIMIDT, D., SCHIMIDT, Q., Andreao, M., Andreao, V., Mario, R., Mario, M.: Systematic review of virtual reality solutions employing artificial intelligence methods}. In: Proceedings of the 23rd Symposium on Virtual and Augmented Reality, New York, pp. 42–55 (2022)
10. Davahli, R., Davahli, D., Reza, M., Reza, K., Neira, W., Neira, C.: The combination of artificial intelligence and extended reality: a systematic review. Front. Virtual Reality (2021)

Artificial Intelligence as an Instrument of Self-determination: Current Regulatory Frameworks

Dragan Dakić(✉)

Faculty of Law, University of Kragujevac, Jovana Cvijica 1, Kragujevac, Serbia
ddakic@jura.kg.ac.rs

Abstract. This study examines the complex interplay between artificial intelligence (AI) and self-determination against the backdrop of existing regulatory structures. It investigates how advancements in AI technology can transform traditional human rights norms, even as these innovations are shaped by human rights laws. This reciprocal relationship is particularly noticeable in fields like automation and the privatization of military functions, where mechanisms for accountability are facing increasing scrutiny. To address these issues, we utilize a qualitative approach that involves an extensive literature review, case analyses, and evaluation of legislative frameworks, with particular emphasis on the European Union's recent risk-based regulatory initiatives for AI. Additionally, we assess the effects of AI on legislative reforms in Serbia, investigating how AI-driven tools can streamline the legislative process by analysing legal texts, identifying discrepancies, and proposing changes to ensure compliance with EU benchmarks.

Our research indicates that AI has the capacity to markedly enhance judicial systems by reducing case delays, minimizing human mistakes, and boosting public trust in local courts. By assessing judicial decisions, AI can help guarantee adherence to international human rights standards, especially those outlined in the European Convention on Human Rights. Moreover, we contend that AI can promote inclusive governance in ethnically diverse countries, such as Bosnia and Herzegovina, by forecasting election results and amplifying the perspectives of underrepresented populations. The study concludes that the incorporation of AI into legal systems not only advances Serbia's goals for EU integration but also bolsters the tenets of self-determination and human rights. Through the application of AI technologies, legal frameworks can better adhere to international obligations, safeguarding individual rights while fostering societal unity. Ultimately, this paper highlights the critical importance of implementing strong accountability frameworks to ensure that AI enhances human dignity, equality, and self-determination rather than jeopardizes them.

Keywords: self-determination · artificial intelligence · regulatory framework

1 Introduction

An examination of normative approaches concerning the regulation of Artificial Intelligence (AI) reveals a complex interplay between technological advancement and human rights law that is reciprocal rather than unilateral. The advent of new technologies can

© The Author(s), under exclusive license to Springer Nature Switzerland AG 2025
N. Filipović (Ed.): AAI 2024, LNNS 1446, pp. 90–106, 2025.
https://doi.org/10.1007/978-3-031-99201-8_10

reshape established human rights standards, while simultaneously, human rights legislation can steer the path of technological innovation. This intricate relationship is especially pronounced in the realms of automation and the privatization of military operations, as highlighted by Laura Dickinson. She contends that these developments erode mechanisms of accountability within both domestic and international legal systems [1]. In her chapter titled "Safeguarding Human Rights from Problematic Technologies," Lea Shaver underscores the rising recognition of a right to science, which places an obligation on states to ensure that new technologies do not violate human rights. Shaver illustrates this point through the example of legal actions in South Africa concerning prohibitive water meters, demonstrating that human rights advocates must take into account the technological aspects of rights infringements. She calls for a framework that includes the protection of human subjects to ensure accountability from those conducting harmful technological experiments [2]. Dalindyebo Shabalala, in his piece "Climate Change, Human Rights, and Technology Transfer," advocates for a human rights-centered perspective on technology transfer, especially pertinent to climate change issues. He argues that human rights frameworks offer a stronger basis for claims related to technology transfer compared to conventional development models. This viewpoint prioritizes the needs of at-risk populations and emphasizes the importance of focusing on technologies that specifically address their unique needs [3]. Thérèse Murphy's chapter, "Judging Bioethics and Human Rights," delves into the convergence of bioethics and human rights law, particularly with respect to medical and biological technologies. She investigates significant cases from the European Court of Human Rights that pertain to the governance of reproductive technologies. Murphy advocates for enhanced dialogue between bioethics and human rights to ensure that ethical considerations are effectively woven into the legal frameworks that regulate new medical technologies [4].

In contemplating the future of AI regulation, one might wonder whether we require an alternative normative approach. The current landscape is dominated by a risk-based framework [5], particularly evident in the European Union's recent legislative efforts, including the approval of the Artificial Intelligence Act [6]. This approach categorizes AI systems based on their potential dangers [7], aiming to mitigate risks that could threaten individual rights and safety [8]. However, as we delve deeper into the implications of AI on society, it becomes increasingly clear that a more nuanced perspective may be necessary. The concept of new rights, which emerge from the extensive interpretation of existing human rights instruments, offers a compelling example of how we might construct a human right to AI. These new rights do not need to be explicitly stated in human rights texts; rather, they can be recognized through the work of human rights interpreting bodies. The formation of such rights relies on the principles of implied rights and derivative rights, suggesting that if society deems a right necessary, it will find a way to establish it. This notion raises the question: if we recognize the need for a right to AI, how might we go about creating it?

To address this question at some later point, we must first consider whether AI can play a role in safeguarding the most essential human right—the right to self-determination. This right encompasses the ability of individuals to make choices about their own lives, free from external coercion. As AI technologies continue to evolve, they hold the potential to enhance personal autonomy and empower individuals in their decision-making processes. However, this potential also comes with significant risks, as AI systems can be designed to manipulate or influence behavior, thereby undermining the very autonomy they are meant to support. Thus, the exploration of AI as a human right must be grounded in a careful examination of its implications for self-determination. If we can establish that AI can indeed serve as a tool for enhancing individual autonomy, we may find a compelling argument for recognizing a human right to AI. This inquiry not only aligns with the broader principles of fundamental human rights, such as dignity and privacy, but also emphasizes the importance of ensuring that AI development and implementation respect and uphold these rights.

Therefore, as we navigate the complexities of AI regulation, it is essential to remain open to the possibility of alternative normative approaches. The recognition of a human right to AI could emerge from our collective understanding of the need for such a right, driven by the desire to protect and promote self-determination in an increasingly automated world. Ultimately, if we determine that the right to AI is necessary, we will find a way to create it, ensuring that the advancement of technology aligns with the fundamental rights that underpin our society. In that end, we will explore how artificial intelligence (AI) can profoundly impact future self-determination. It will begin by outlining the historical development of the self-determination principle, tracing its path from early acknowledgment in the League of Nations to its formal inclusion in the United Nations Charter. This principle will be emphasized as a vital right that enables peoples to self-govern and encourages harmonious international relations.

In the section dedicated to the influence of AI on legislative reform, we will analyze how AI-powered tools can enhance legislative processes, particularly in Serbia's efforts to comply with European Union (EU) standards. These tools will demonstrate their ability to evaluate legal documents, detect inconsistencies, and suggest necessary amendments, thereby strengthening Serbia's dedication to EU membership. The discussion will then transition to the judiciary, where we will investigate AI's potential to improve adherence to the European Convention on Human Rights (ECHR). By evaluating judicial rulings, AI can help alleviate case backlogs, reduce human errors, and enhance public confidence in local courts, thus ensuring compliance with international human rights norms. In a different section, the paper will explore the intricate interplay between collective self-determination and individual rights, especially in ethnically varied nations such as Bosnia and Herzegovina. We will argue that AI can aid in fostering inclusive governance by predicting electoral results and encouraging public consultations that amplify the voices of marginalized groups, thus harmonizing the demands for collective identity with the principles of universal equality.

2 Self-determination in International Law

One of the most significant yet intricate and contentious issues in international law is self-determination. It is rooted in formal sources of international law at both universal and regional levels. Its historical roots can be traced back to the Covenant of the League of Nations introduced in 1919. While the Covenant of the League of Nations did not explicitly declare self-determination as a right, it is widely interpreted that Article 22 of the Covenant laid the foundation for this principle. This article advocated for the creation of mandates for peoples who were unable to "stand by themselves under the strenuous conditions of the modern world," reflecting an early acknowledgment of the right to self-governance for those peoples. The mandate system was primarily applied to former colonies and territories governed by defeated nations, indicating a recognition by the international community of the self-determination principle. The notion of self-determination was articulated more explicitly following World War II with the founding of the United Nations. Within the UN Charter, self-determination is recognized as a fundamental principle in Articles 1(2) and 55. Article 1(2) outlines that one of the purposes of the UN is to foster "friendly relations among nations based on respect for the principle of equal rights and self-determination of peoples," with the goal of ensuring peace and security. Similarly, Article 55 emphasizes the UN's dedication to "promoting conditions of stability and well-being" and advocating for the respect of "equal rights and self-determination of peoples." These provisions have reinforced self-determination as a key objective within the framework of international law.

While the Universal Declaration of Human Rights (UDHR) does not directly reference self-determination within its text, it implicitly endorses the idea through its focus on individual rights and freedoms that support self-governance. For instance, Article 21(3) asserts that "the will of the people shall be the basis of the authority of government," thereby connecting self-determination to democratic governance. The UDHR's dedication to human rights has also established a global context in which individuals and groups possess the right to select their own political and social systems. Additionally, both the International Covenant on Civil and Political Rights (ICCPR) and the International Covenant on Economic, Social, and Cultural Rights (ICESCR)—which, along with the UDHR, constitute the International Bill of Human Rights—commence with the same declaration regarding self-determination in Article 1(1): "All peoples have the right of self-determination. By virtue of that right, they freely determine their political status and freely pursue their economic, social, and cultural development." This clear acknowledgment was a significant development, establishing self-determination as a legally recognized right and creating a universal principle for all peoples.

The principle of self-determination was later enhanced by two important resolutions from the United Nations General Assembly. The first, Resolution 1514 (The Declaration on the Granting of Independence to Colonial Countries and Peoples 1960), serves as a pivotal landmark in the development of the self-determination principle, particularly in colonial scenarios. This Declaration claimed that "the subjection of peoples to alien subjugation, domination and exploitation constitutes a denial of fundamental human rights" and affirmed that all peoples possess the right to self-determination. It highlighted that colonial rule contradicts peace, security, and international cooperation, thereby sparking a wave of decolonization throughout the 1960s and 1970s. The second resolution, UN General Assembly Resolution 2625 (the Declaration on Principles of International Law concerning Friendly Relations and Co-operation among States in accordance with the Charter of the United Nations 1970), goes beyond decolonization. This Declaration strengthened self-determination as a fundamental aspect of international law, offering interpretations of the UN Charter in relevant sections. In one of it's the most frequently but incompletely quoted paragraphs it clarified that all peoples have the right to freely determine their political status without external interference, while adding that respect for states' territorial integrity is conditioned by states 'conducting in compliance with the principle of equal rights and self-determination of peoples' as well as by the 'government representing the whole people belonging to the territory.' Additionally, it states: 'By virtue of the principle of equal rights and self-determination of peoples enshrined in the Charter of the United Nations, all peoples have the right freely to determine, without external interference, their political status and to pursue their economic, social and cultural development.'

The European Convention on Human Rights of 1950 (ECHR) stands out as the most crucial instrument within the Council of Europe. Although self-determination is not directly mentioned, the ECHR lays the groundwork for numerous rights that are fundamental to this concept. By establishing essential freedoms, the Council of Europe promotes various forms of political autonomy through the ECHR. The convention indirectly supports self-determination by emphasizing political involvement and the individual rights necessary for the democratic manifestation of collective identity. Articles within the ECHR, like Article 3 of Protocol 1—which ensures the right to free elections—and Article 8, which safeguards the respect for private and family life, are interpreted as vital components of democratic self-determination, thereby advocating for individual autonomy and the liberty to preserve personal, familial, and cultural connections. In the case of Sejdić and Finci v. Bosnia and Herzegovina (Application nos. 27996/06 and 34836/06), the European Court of Human Rights (ECtHR) tackled issues surrounding political participation for minority groups, acknowledging the democratic rights that are essential for self-determination. Before this, in the case of Sidiropoulos and Others v. Greece (Application No. 26695/95), the ECtHR affirmed the rights of a minority to establish associations, underscoring the significance of group identity and autonomy. Additionally, the ECtHR has contributed valuable insights into self-determination in relation to individual autonomy through its gender-related rulings, such as B. v. France (Application No. 13343/87), Christine Goodwin v. The United Kingdom (Application No. 28957/95), and A.P., Garçon and Nicot v. France (Applications nos. 79885/12, 52471/13, and 52596/13).

Another significant regional source is the Helsinki Final Act (1975), which was approved by the Conference on Security and Cooperation in Europe (CSCE), currently known as the Organization for Security and Co-operation in Europe (OSCE). This landmark document plays a crucial role in European international law by harmonizing the principles of sovereignty with human rights considerations. It upholds the concept of self-determination indirectly through its principles of non-intervention and sovereignty. The Act affirms that "all peoples have the right to self-determination, allowing them to freely decide their political status and pursue their economic, social, and cultural development without external interference." Regarding cultural self-determination, Article 7 of the European Charter for Regional or Minority Languages emphasizes the necessity to "acknowledge the significance of regional or minority languages in relation to group identity and to encourage their usage in public life." This Convention, adopted by the Council of Europe in 1992, aims to protect the rights of groups to maintain their cultural identities. Furthermore, in 1995, the Council of Europe established the Framework Convention for the Protection of National Minorities, which fosters self-determination by providing protections for national minorities and ensuring their rights to preserve and enhance their distinct identities and to engage in governance.

The Treaty of Amsterdam (1997) plays a significant role within the European Union by incorporating fundamental rights into EU law, thereby promoting democracy and strengthening principles related to self-determination, including the freedoms of assembly and association. Self-determination is reflected in Article 2, which states that the Union is based on the values of liberty, democracy, and human rights, and Article 6, which emphasizes that the EU acknowledges and respects the national identities of its Member States, maintaining local and regional diversity as a core aspect of the Union's foundation. Additionally, Article 11 of the Treaty of Lisbon (2007) upholds freedom of expression, facilitating the unimpeded pursuit of political objectives by various groups. Article 12 further affirms the significance of freedom of assembly and association as vital elements for the actualization of self-determination. In 2017, the Union collaborated with the United Nations to issue a joint declaration stating, "all peoples have the right to self-determination, and this right must be exercised in a manner that respects the territorial integrity and sovereignty of States." Likewise, the European Court of Justice has addressed matters of individual self-determination in gender-related cases, similar to the European Court of Human Rights, with notable rulings including P. v S. and Cornwall County Council (1996), MB v Secretary of State for Work and Pensions (2018), and K.B. v National Health Service Pensions Agency and Secretary of State for Health (2004). Notable case of the European Court of Justice in the context of collective self-determination is the case of Council of the European Union v. Front populaire pour la libération de la saguia-el-hamra et du rio de oro (Front Polisario), recognized self-determination as one of the essential principles of international law. The Court interpreted a treaty in a manner that respected this principle, indicating its foundational role in international law.

Indisputably, self-determination is recognized in both universal and regional legal sources as a complex construct that encompasses both collective and individual dimensions. This concept includes both internal and external aspects: at the collective level, the internal aspect pertains to participatory rights enshrined in state legislation, judiciary, and administration, while the external aspect relates to integration and, in certain instances, separatism. At the individual level, the internal dimension focuses on the protection of rights such as freedom of thought and belief, which predominantly reside in the personal realm and become apparent only when expressed externally through rights like freedom of speech, conscientious objection, or public religious practice. Given that both collective and individual dimensions may involve controversial issues, including separatism or gender transition, we will explore the impact of AI on addressing self-determination challenges. Initially, we will employ AI to address previously resolved issues to evaluate its adherence to established reasoning and identify any discrepancies. This analysis will enhance our understanding of the potential of AI in future self-determination matters. Additionally, we will investigate the innovative capabilities of AI. To conduct this examination, we will define both levels of self-determination using recognized interpretative instruments.

2.1 Artificial Intelligence and Collective Self-determination

In 2019, the Draft Conclusions on jus cogens [9] were endorsed by the International Law Commission, which included a "non-exhaustive list" of potential jus cogens norms: (a) the prohibition of aggression; (b) the prohibition of genocide; (c) the prohibition of crimes against humanity; (d) fundamental principles of international humanitarian law; (e) the prohibition of racial discrimination and apartheid; (f) the prohibition of slavery; (g) the prohibition of torture; and (h) the right to self-determination [9].

The inclusion of the right to self-determination among the peremptory norms of international law carries significant implications, both theoretically and practically. As previously discussed [10], peremptory norms are characterized as essential principles that serve as the foundation of international law. These norms reflect the shared moral values of the international community, promoting core principles such as justice, human rights, and peace. Practically, one potential impact of peremptory norms arises from the 1969 Vienna Convention on the Law of Treaties (VCLT), specifically Article 64, which addresses the notion of jus cogens superveniens. This idea refers to the introduction of new peremptory norms that can render prior treaties invalid if they are inconsistent with these norms. The evolving nature of jus cogens, as emphasized in other discussions [10], demonstrates how international law can respond to shifting moral and ethical standards, thereby enhancing the integrity and coherence of the legal framework. In essence, jus cogens norms are fluid; they transform in accordance with the collective will and moral values of the international community. This adaptability is vital for ensuring that international law remains pertinent and effective in tackling contemporary challenges and upholding accountability for breaches of fundamental human rights [10].

The significant number of opposing comments regarding the International Law Commission's 2022 Draft Conclusions on the Identification and Legal Consequences of Peremptory Norms of General International Law [11] may have been influenced by various factors. For example, Argentina was a strong proponent of including self-determination, asserting that this right is crucial for the legitimacy of statehood and the conduct of international relations. The country stressed that self-determination is vital for safeguarding human rights and fostering peace. Argentina pointed to historical examples where the assertion of self-determination has resulted in the freedom of oppressed groups, bolstering its argument that classifying this right as a peremptory norm would reinforce international law. Similarly, South Africa called for the inclusion of self-determination, referencing its historical fight against apartheid to illustrate the importance of this principle. The nation contended that self-determination is essential for correcting injustices and ensuring that all peoples can choose their political status. In its position, South Africa underlined that acknowledging self-determination as a peremptory norm would demonstrate the international community's dedication to human rights and dignity. Additionally, India voiced its support for the recognition of self-determination, portraying it as a foundational element of international law that encapsulates the hopes of peoples across the globe. The country argued that self-determination is key to advancing democracy and stability, enabling communities to articulate their identity and governance preferences. India maintained that recognizing this right as a peremptory norm would strengthen the legal frameworks surrounding self-determination and encourage compliance among states.

China presented a contrasting perspective, expressing worries over the concept of self-determination, suggesting that it might clash with the principles of state sovereignty and territorial integrity. The Chinese government stressed the importance of self-determination, but insisted that it should not be exploited to support secessionist agendas or disrupt the stability of established nations. They advocated for an approach that honors both the right to self-determination and the necessity of national unity. In a similar vein, Russia conveyed its concerns about labeling self-determination as a peremptory norm, cautioning that such a designation could exacerbate tensions and disputes over territories. They argued that self-determination needs to be framed within the context of international law to prevent possible abuse. Furthermore, Russia highlighted the significance of upholding state sovereignty and warned against any interpretations that might fuel separatist movements [12]. The United States also raised alarms about the repercussions of accepting self-determination as a peremptory norm, especially with respect to its foreign policy and global relations. The U.S. asserted that, although self-determination is an essential principle, its implementation must consider the rights of sovereign states to maintain their territorial integrity. Moreover, the U.S. stressed the necessity for a practical approach that addresses the intricacies of international law and the risks of conflict stemming from unilateral claims of self-determination.

Generally speaking, former colonies advocate for self-determination, whereas former colonial powers and current AI powers resist its recognition as a peremptory norm. The primary argument against acknowledging self-determination as such revolves around its supposed contradiction to the principles of territorial integrity and state sovereignty. These principles, which were excluded from the jus cogens list, depend on states adhering to the principle of equal rights and the self-determination of peoples, as well as having a government that represents the entire population of the territory. In this context, they are effectively subordinate to self-determination. However, this debate concerns external self-determination through secession, which is not the focus of this discussion.

The case of Serbia illustrates a particular facet of collective self-determination that includes both an internal dimension—regarding the alignment of its legal framework with EU acquis—and an external dimension, focusing on its European Union (EU) membership pursuit. The Stabilization and Association Agreement (SAA), serves as a fundamental element in Serbia's efforts towards EU accession, establishing a comprehensive framework for the reciprocal rights and responsibilities shared between Serbia and the European Union. Through a monistic approach, the SAA has been integrated into Serbia's legal system, granting it a unique legal authority. By integrating international treaties into its domestic legislation, Serbia guarantees that the SAA takes precedence over any conflicting national laws, thus reinforcing its dedication to aligning with EU standards. This integration is especially significant for Serbia's internal self-determination concerning legislative and judicial matters, where the implementation of artificial intelligence (AI) has demonstrated significant transformative potential [13].

AI has the capability to play considerable role in Serbia's legislative reform. By employing AI-driven tools to assess legal documents and detect inconsistencies, the Serbian administration could improve the process of aligning domestic laws with EU standards. These technologies might propose specific amendments and track adherence to EU obligations, effectively addressing any deficiencies. In addition, the judicial system could benefit from AI by enhancing its ability to ensure that court rulings are in line with the European Convention on Human Rights (ECHR). AI solutions that analyse judicial decisions for compliance with the ECHR would not only help alleviate the case backlog at the European Court of Human Rights (ECtHR) but also minimize human errors, thereby enhancing public trust in local courts [13]. Such developments not only boost the efficiency of institutions but also signal Serbia's commitment to fulfilling EU membership requirements. Despite over a decade since Serbia achieved candidate status, challenges persist in meeting the Copenhagen criteria, particularly regarding judiciary reform and the protection of fundamental rights [14]. AI offers a strategic means to address these challenges. For instance, AI systems could analyse judicial data to identify patterns of ECHR non-compliance and recommend adjustments [13]. This would not only help Serbia meet accession requirements but also strengthen the rule of law. Additionally, AI could support administrative reform by automating bureaucratic processes, enhancing efficiency, and reducing corruption risks.

On an external level, Serbia's self-determination in the EU accession process is defined by its ability to fulfil obligations set forth in the SAA and accession negotiations. These commitments, especially in key areas like reforming public administration, ensuring judicial independence, and upholding fundamental rights, necessitate tangible results [13]. AI presents an efficient means to achieve these results. By automating data and conducting real-time assessments, AI can simplify Serbia's reporting mechanisms, enabling clear and precise evaluations of compliance with EU standards. Furthermore, such systems could aid EU oversight bodies in assessing Serbia's advancements, bolstering its image as a competent and responsive candidate for EU membership.

As it was previously discussed [13], the technical features of AI systems tailored to Serbia's EU accession should be both innovative and pragmatic. Key functionalities must include normative translation modules, enabling the conversion of legal norms into machine-readable formats. This would facilitate precise applications in legislative and judicial contexts. Furthermore, these AI tools should be dynamic, incorporating machine learning capabilities to adapt to legal and procedural developments. Transparency and public accessibility are equally critical; AI systems must engage not only institutions but also citizens, building trust in the broader accession process.

While the explicit obligation to use AI is absent from the SAA and related agreements, the principle of good faith under international law underscores Serbia's need to adopt effective measures to meet its obligations. The principle of good faith, as affirmed in international jurisprudence such as the Gabčíkovo-Nagymaros Project case [15], emphasizes the importance of implementing treaty obligations in both letter and spirit. AI's potential to accelerate compliance with EU standards aligns directly with this principle, positioning it as a necessary tool rather than a discretionary innovation. Obstacles like the absence of advanced AI infrastructure do not constitute valid impediments under international law. The Vienna Convention on the Law of Treaties (1969) [16] makes clear that internal limitations cannot excuse non-performance of international obligations [17]. Serbia's willingness to develop judiciary-focused AI solutions, such as a proposed system for aligning judicial decisions with ECHR standards, further demonstrates its capability to integrate such technologies. Given the estimated cost of these systems—well within reasonable limits—the doctrine of impossibility of performance cannot apply. AI thus emerges as a crucial component in Serbia's EU accession strategy. By leveraging AI for both internal reforms and external negotiations, Serbia not only fulfils its obligations but also exemplifies a forward-looking approach to international integration. This dual focus on technological innovation and legal alignment ensures that Serbia's path to EU membership remains credible and progressive.

A specific scenario illustrating the dynamics between individual and collective internal self-determination is exemplified by Bosnia and Herzegovina. The ECtHR ruling in the case of Sejdić and Finci v. Bosnia and Herzegovina (Application nos. 27996/06 and 34836/06) reflects the intricate balance between the collective self-determination of the constituent nations and the fundamental right to equality as outlined in the ECHR. The case stemmed from the disqualification of two applicants, a member of the Roma community and a Jew, from running for positions in the House of Peoples and the Presidency of Bosnia and Herzegovina. This exclusion, based on constitutional clauses originating from the Dayton Peace Agreement, underscored the ongoing difficulties of reconciling ethnic representation with universal democratic principles (paragraphs 1–15).

The Constitution of Bosnia and Herzegovina serves as a fundamental element of the Dayton Agreement (Annex IV), aimed at stabilizing the nation in a post-conflict environment by securing the political dominance of the "constituent peoples"—the Bosniaks, Croats, and Serbs (paragraph 6). Nonetheless, it effectively marginalized individuals classified as "Others," which encompasses ethnic minorities such as the Roma and Jews (paragraphs 11–12). Although the ECtHR recognized the necessity of these provisions during the delicate post-war phase, it also noted their increasing misalignment with current human rights standards (paragraphs 13–15). The ECtHR pointed out that while the ethnic-based framework was crucial for maintaining peace, it could no longer excuse the systemic discrimination faced by those not belonging to the constituent groups (paragraphs 26–31). In examining this constitutional setup, the Court assessed its alignment with Articles 14 and Protocol No. 12 of the ECHR, which prohibit discrimination (paragraph 26). The applicants contended, and the Court agreed, that their disqualification from candidacy represented a clear case of direct racial and ethnic discrimination without adequate justification (paragraphs 32–33). Although the responding government attempted to portray these provisions as essential for preserving a fragile peace, the Court stressed the disproportionate effects on individual rights. It urged Bosnia and Herzegovina to progress beyond the strict ethnic quotas previously considered vital for governance in the aftermath of conflict (paragraphs 31–33).

The ruling also considered the wider effects on collective self-determination, highlighting how inflexible frameworks can perpetuate inequality. The Court pointed out the Venice Commission's suggestions, which aimed at refining the concept of vital national interest, simplifying decision-making processes, and ensuring representation is inclusive (paragraphs 22, 37). These recommendations were intended to modernize Bosnia and Herzegovina's governance while also maintaining necessary safeguards for political stability in a diverse society (paragraphs 34–39). Addressing these complexities, future approaches might find value in the incorporation of AI, which can help reconcile collective and individual rights. AI can provide tools for forecasting electoral outcomes based on updated constitutional standards, helping policymakers predict how proposed modifications will affect representation and stability. Additionally, AI-enabled platforms can promote inclusive public consultations, ensuring that a diverse array of viewpoints, particularly those from historically marginalized communities, are taken into account. Predictive analytics could assist governments in evaluating the human rights consequences of legal changes, aiding in the alignment of domestic legislation with international norms. Furthermore, AI-based monitoring systems could improve transparency by identifying and reporting discriminatory practices in real time. The case of Sejdić and Finci v. Bosnia and Herzegovina illustrates the persistent conflict between the need for collective self-determination in ethnically varied societies and the necessity of maintaining universal principles of equality and non-discrimination. The Court's decision acts not only as a critique of the current constitutional system but also as an appeal for innovation and change. By adopting new technologies and encouraging inclusive conversations, societies such as Bosnia and Herzegovina can work towards balancing these conflicting needs, creating governance frameworks that respect their diverse identities while upholding human rights.

2.2 Artificial Intelligence and Individual Self-determination

The exploration of the convergence between AI and reproductive autonomy reveals that the integration of AI systems in decisions related to reproduction can greatly improve individual self-determination [8]. Reproductive autonomy, which refers to an individual's right to make informed choices regarding their reproductive health and family planning, is a crucial element of self-determination. The emergence of AI technologies within this area introduces both possibilities and obstacles, prompting a thorough examination of their effects on personal agency. In a prior paper, we emphasized the capability of fully automated deep learning models to assess pregnancy chances and aid in efficient embryo selection. This functionality not only aims to enhance the safety and effectiveness of reproductive services but also empowers individuals by offering them more precise information for making well-informed decisions about their reproductive choices. By utilizing AI, people can acquire insights into their reproductive health that were once unreachable, thereby improving their capacity to make choices that reflect their personal values and circumstances.

A major advantage of implementing AI in reproductive context is its capacity to analyze vast data on implantation, including genetic details. This capability allows for a deeper appreciation of the health conditions of embryos, which is vital for those confronted with challenging reproductive decisions. As previously mentioned, the legal interpretation of implantation data differs from one jurisdiction to another within the EU, underscoring the need for AI systems to be adaptable and capable of addressing these diverse legal landscapes [8]. By customizing AI applications to accommodate different parameters on a case-by-case basis, individuals can exercise greater control over their reproductive decisions, thereby reinforcing their self-determination. Furthermore, clear insights into the AI's decision-making processes are critical for individuals to anticipate the consequences of their decisions. This kind of transparency not only builds trust in the technology but also enables individuals to participate in shared decision-making alongside healthcare professionals.

The capacity to question AI logic and outcomes empowers individuals, enabling them to vocalize their own preferences and values throughout the reproductive decision-making process [8]. This approach aligns with the principles of informative, interpretative, and deliberative care models that honor the established hierarchy of decision-making in reproductive health and emphasize the importance of individual self-determination. Consequently, it is crucial that AI systems are equipped with efficient mechanisms for human involvement, guaranteeing that individuals maintain final authority over their reproductive choices. In the absence of sufficient human oversight, there is an elevated risk of eroding individual self-determination. The lack of distinct guidelines concerning legal accountability in AI decision-making may create circumstances in which individuals feel powerless or pressured by the technology.

Another dimension of individual self-determination pertains to Serbia's external obligations, which include its commitments to the EU and the Council of Europe. Recently, the Constitutional Court of Serbia put a temporary hold on the enforcement of the national Law on Gender Equality, which will be reviewed later for its constitutional merit. This scenario illustrates how AI might be applied in decision-making processes. The European Court of Human Rights has been instrumental in developing the concept of gender through its jurisprudence. In rulings by the ECHR, gender is closely linked to the right to private life as outlined in Article 8 of the European Convention on Human Rights. The Court has asserted that both gender identity and gender expression are critical components of personal self-determination and dignity. Landmark cases, such as Christine Goodwin v. The United Kingdom (2002), highlight the Court's acknowledgment of gender identity as a protected right. In this ruling, it determined that the UK's failure to legally recognize post-operative gender identity breached Article 8. Additionally, the Court underscored the necessity for member states to adapt to the evolving social understanding of gender to uphold individual self-determination. In a similar vein, A.P., Garçon and Nicot v. France (2017) saw the ECHR reject the mandate for irreversible sterilization as a prerequisite for gender recognition, broadening the Court's interpretation of privacy to encompass the right to gender identity without obligatory physical alterations. The Court differentiated between corporeal rights (physical autonomy) and psychological rights (self-identification), affirming that the right to gender identity exists independently from any physical changes. Together, these rulings delineate that within the ECHR framework, gender rights encompass self-identification, freedom from discrimination, and protection against invasive or coercive practices. The decisions made by the ECHR significantly influence member countries, including Serbia, by necessitating adherence to these human rights standards.

The recognition of gender as a legal construct within the framework of the European Union has been significantly shaped by the ECJ. In the ECJ's legal interpretations, gender is intricately connected to anti-discrimination principles found in directives like Directive 2006/54/EC, which addresses gender equality in employment, and Directive 79/7/EEC, which focuses on equal treatment in social security. A pivotal case in this context is P. v. S. and Cornwall County Council (1996), where it was established that discrimination on the basis of gender reassignment is a form of sex discrimination. The Court's broad interpretation of EU law ensures that transgender individuals are entitled to the same protections against discrimination as individuals whose classification is based on biological sex. In the decision of MB v. Secretary of State for Work and Pensions (2018), the ECJ determined that requiring surgical gender reassignment for the receipt of social security benefits infringed upon EU anti-discrimination laws. This ruling broadened the understanding of gender in EU law, recognizing rights related to gender identity beyond solely physical alterations. The ECJ has continuously emphasized that gender includes aspects of gender identity, expression, and reassignment, asserting that these components are essential to an individual's dignity and equality.

These principles constitute a fundamental part of the EU's legal acquis, which Serbia, as a candidate for EU accession, must incorporate into its national legal system. Additionally, Serbia's monistic approach to international law guarantees that ratified treaties and acknowledged universal principles are integral to its domestic legal order. The ECtHR and its the rulings of the European Court of Human Rights hold direct applicability within Serbia, necessitating that domestic courts, including the Constitutional Court, interpret and apply these principles. It is important to note that while Serbia has not yet joined the EU, its obligations under the SAA require alignment with the EU acquis, including the jurisprudence of the ECJ. Although the rulings of the ECJ do not have immediate applicability, their principles act as important benchmarks for legal reforms in Serbia. The integration of these standards is crucial for Serbia's accession to EU membership.

The capacity of the Constitutional Court to analyze intricate legal matters, especially those surrounding the subtleties of gender identity and its components, can be greatly improved with AI technologies. Utilizing natural language processing (NLP) algorithms, these AI systems can adeptly scrutinize vast amounts of case law from both the ECHR and ECJ, allowing for the identification of pertinent precedents and trends. AI could play a crucial role in assessing how Serbia's domestic laws align with international commitments through comparative studies of both legal frameworks. This functionality would empower the Court to evaluate whether the Law on Gender Equality is consistent with the Constitution as well as Serbia's responsibilities under EU and international law. Furthermore, AI's ability to analyze socio-economic data could forecast the potential repercussions of legal decisions, assisting the Court in formulating rulings that are legally robust while also socially and economically sustainable. Such insights would be vital for making decisions based on merit, ensuring that the Court's judgments protect individual rights and promote societal cohesion.

3 Conclusion

The paper begins by examining the notion of self-determination, tracing its historical development and importance as a fundamental element of global legal systems, highlighting its twofold nature—collective and individual. This vital principle, crucial for fostering autonomy and governance, has seen considerable evolution, originating from its early conceptualization in the League of Nations and being formally enshrined in the UN Charter. The UN Charter articulates self-determination in Articles 1(2) and 55, positioning it as a tool for enhancing global stability and welfare. Subsequent documents, such as the Universal Declaration of Human Rights (UDHR) and the International Covenants on Civil and Political Rights and Economic, Social, and Cultural Rights, affirmed self-determination as a universal entitlement, empowering individuals to make choices about their political status and seek development autonomously. This principle was further strengthened by pivotal UN resolutions, including Resolution 1514, which acted as a catalyst for decolonization, and Resolution 2625, which elucidated the relationship between self-determination and the integrity of territories, contingent upon equal rights and inclusivity being maintained by states. Consequently, former colonial powers adopt a restrictive interpretation of self-determination rights when it comes to their former colonies.

Transitioning from a universal perspective, the focus shifts to regional mechanisms, particularly the European Convention on Human Rights (ECHR) and the corresponding judgments by the European Court of Human Rights (ECtHR). In this context, the relationship between self-determination and individual rights takes center stage. Significant cases, such as Sejdić and Finci v. Bosnia and Herzegovina, underscore the conflict between collective identity and democratic inclusion, illustrating the ongoing difficulty of achieving a balance between ethnic representation and universal equality. Legal frameworks, including the Helsinki Final Act, the Framework Convention for the Protection of National Minorities, and various EU treaties, further weave self-determination into governance structures, underscoring its significance in protecting cultural identity and promoting democratic engagement. These frameworks underscore the need for flexible governance models that harmonize diverse identities with human rights principles. In its conclusion, this section identifies artificial intelligence (AI) as a possible facilitator of self-determination. By tackling existing gaps in legislation and the judiciary, AI has the capacity to promote inclusivity and accountability. It offers resources for predictive governance, public engagement, and the unification of legal standards. As societies confront the intricate issues surrounding multi-dimensional self-determination, AI stands out as a revolutionary tool that integrates established legal principles with modern challenges, thereby strengthening the universality of self-determination within the context of human rights.

Following subsection addressing AI and collective self-determination investigates the potential of AI technologies to enhance collective governance autonomy, particularly in the contexts of legislation and the judiciary. This analysis is contextualized within Serbia's efforts to join the EU and examines how AI can aid in aligning national governance with international human rights standards. The research highlights AI's capacity to effect transformative changes in legislative procedures. With AI tools, countries like Serbia can analyze legal texts, pinpoint discrepancies, and recommend modifications, which aids in synchronizing national laws with EU regulations. This process not only ensures adherence to international commitments but also bolsters the credibility of institutions. Additionally, the application of AI in evaluating judicial rulings can uncover patterns of non-compliance with frameworks such as the European Convention on Human Rights (ECHR), providing suggestions for rectification. By mitigating case delays and minimizing human errors, AI enhances public trust in the judicial system and reinforces Serbia's dedication to its aspirations for EU membership.

On a broader scale, AI's contribution to Serbia's pursuit of self-determination is depicted as a tool for achieving EU accession criteria. Automated systems can produce accurate compliance reports and increase transparency within administrative operations.

Moreover, sophisticated machine learning models designed for Serbia's legal environment can swiftly adjust to changes in legislation and facilitate the incorporation of EU legal standards. Such capabilities ensure that AI-led reforms are consistent with the principles of good faith in international law, reflecting Serbia's proactive involvement in the EU accession journey. Furthermore, the discussion highlights the complex relationship between collective self-determination and individual rights in ethnically diverse nations, referencing Bosnia and Herzegovina as an example. The European Court of Human Rights case of Sejdić and Finci v. Bosnia and Herzegovina illustrates the conflicts between ethnic representation and equality. AI is suggested as a means to bridge these challenges by fostering inclusive public engagement, anticipating the effects of governance alterations, and detecting discriminatory actions in real-time. In summary, this subsection introduced the potential of AI to transform governance by enhancing collective autonomy and ensuring adherence to human rights standards. It suggests that, when incorporated with due diligence, AI can serve as a crucial tool for fostering fair governance, supporting self-determination, and addressing intricate socio-political challenges in accordance with universal and regional standards.

The subsection addressing AI and personal self-determination explores how AI technologies can profoundly influence individual self-determination, specifically concerning delicate matters like reproductive rights and gender identity. The importance of reproductive autonomy is highlighted as a fundamental aspect of self-determination, empowering individuals to make informed decisions about their reproductive health and family planning. AI emerges as an essential resource in this area, capable of processing intricate data sets to deliver accurate and actionable insights. For example, sophisticated AI algorithms can assess the likelihood of pregnancy and aid in embryo selection, allowing individuals to make choices that reflect their values and situations, while also enhancing the safety and effectiveness of reproductive services. One significant benefit of AI is its capacity to process extensive genetic and implantation data, which is vital for those navigating tough reproductive choices. Nevertheless, introducing AI into reproductive health can be challenging, especially in regions with varying legal and cultural contexts, such as within the EU. By customizing AI tools to fit local conditions, individuals can gain more control over their decisions, thereby strengthening their autonomy. It is imperative that AI decision-making processes are transparent; this transparency fosters trust and enables individuals to engage actively in collaborative decision-making with healthcare providers. However, the lack of well-defined accountability mechanisms and legal frameworks governing AI usage may threaten personal autonomy, potentially resulting in situations where individuals feel pressured or powerless against technological systems.

The discussion also encompasses gender identity, pointing out how AI may facilitate legal changes and judicial measures. In Serbia, gender equality has recently garnered significant national attention, particularly after the Constitutional Court temporarily halted the implementation of the national Law on Gender Equality. The court has expressed concerns about the law's compliance with the Serbian Constitution, postponing its final decision for a comprehensive analysis. This postponement highlights the challenges in harmonizing national legal standards with international human rights obligations, especially regarding Serbia's hopes for EU membership.

AI has the potential to significantly aid the Constitutional Court of Serbia in its evaluation of the Law on Gender Equality. Utilizing natural language processing (NLP) technology, AI can scrutinize case law from both the European Court of Human Rights (ECHR) and the European Court of Justice (ECJ), allowing for the identification of pertinent precedents and patterns. Furthermore, AI can effectively analyze domestic laws in relation to international obligations, enabling the court to determine whether the legislation aligns with the Serbian Constitution and its international commitments. Predictive analytics could also provide support in anticipating the socio-economic consequences of possible decisions, thereby ensuring that judicial outcomes are legally sound as well as socially responsible. In summary, AI represents a valuable instrument for promoting individual self-determination in areas of reproductive autonomy and gender rights. Its capability to navigate intricate legal and social frameworks makes it an essential resource for countries like Serbia, which face the challenge of reconciling local constitutional norms with global obligations. Nevertheless, the deployment of AI must occur alongside ethical supervision and strong accountability measures to guarantee that its use fosters, rather than diminishes, the ideals of human dignity, equality, and self-determination.

References

1. Dickinson, L.: Automated Weapons, and Private Military Contractors: Challenges to Domestic and International Legal Regimes Governing Armed Conflict. In: Normative Approaches to Technology and Human Rights, (2023)
2. Shaver, L.: Safeguarding human rights from problematic technologies. In: Normative Approaches to Technology and Human Rights, (2023)
3. Shabalala, D.: Climate change, human rights, and technology transfer: normative challenges and technical opportunities. In: Normative Approaches to Technology and Human Rights (2023)
4. Murphy, T.: Judging bioethics and human rights. In: Normative Approaches to Technology and Human Rights, (2023)
5. Chamberlain, J.: The risk-based approach of the European Union's proposed artificial intelligence regulation: some comments from a tort law perspective. Eur. J. Risk Regul. **14**(1), 1–13 (2023). https://doi.org/10.1017/err.2022.38
6. EPRS_BRI(2021)698792_EN.pdf (europa.eu)
7. Grieman, K., Early, J.: A Risk-based Approach to AI Regulation: System Categorisation and Explainable AI Practices" (2023) 20:1 SCRIPTed 56 https://script-ed.org/?p=4109 https://doi.org/10.2966/scrip.200123.56
8. Dakic, D. Reproductive autonomy conformity assessment of purposed AI System. In: Intelligent Systems for Sustainable Person-Centered Healthcare, vol. 205. Springer, Cham, (2022). https://doi.org/10.1007/978-3-030-79353-1_5
9. The Text of the Draft Conclusions of *Jus Cogens* and the list of possible norms of jus cogens, together with commentaries thereto. UN Doc A/74/10, 2019
10. Dakić, D.: The Resolution on the Srebrenica Genocide and Dayton Peace Agreement: In Which Direction Is It Heading?. Studia Europejskie – Studies in European Affairs 3/2024, 233–247 (2024). https://doi.org/10.33067/SE.3.2024.13
11. United Nations.: Draft Conclusions on Identification and Legal Consequences of Peremptory Norms of General International Law (Jus Cogens). A/77/10. 2022. Accessed [November 13 2024]. [https://legal.un.org/ilc/texts/instruments/english/commentaries/1_14_2022.pdf]

12. Russian legal argument in the case of so-called remedial secession at Pelliconi, Andrea Maria. Self-defence as remedial self-determination: continuity in Russia's legal arguments for the invasion of Ukraine. Netherlands Int. Law Rev. **71**(2), 233–260. https://doi.org/10.1007/s40802-024-00260-w (2024)
13. Dakic, D.: Is there a legal obligation to use artificial intelligence in the European Union accession process?. In: Filipović, N. (Eds.), Applied Artificial Intelligence 2: Medicine, Biology, Chemistry, Financial, Games, Engineering. AAI 2023. Lecture Notes in Networks and Systems, vol 999. Springer, Cham. https://doi.org/10.1007/978-3-031-60840-7_28
14. Serbia Report 2023. European Commission. Available at: https://neighbourhood-enlargement.ec.europa.eu/serbia-report-2023_en
15. Gabčíkovo-Nagymaros Project (Hungary/Slovakia), Judgment, I.C.J. Reports 1997
16. Jovičić, K.: The Relationship between International and Domestic Law According to the Serbian Constitution of 2006. Srpska Politička Misao 29–46 (2017)
17. Vienna Convention on the Law of Treaties 1969. United Nations. Available at: https://legal.un.org/ilc/texts/instruments/english/conventions/1_1_1969.pdf

Applied Artificial Intelligence in Detecting Hate Speech

Jelica Cincović, Uroš Radenković, Marko Mićović, Adrian Milaković, Vladimir Jocović, and Dražen Drašković(✉)

University of Belgrade - School of Electrical Engineering, Belgrade, Serbia
drazen.draskovic@etf.bg.ac.rs

Abstract. The problem this research addresses is detecting hate speech (HS) in texts in Serbian on the Internet. Detecting and reducing HS is crucial for the safety and well-being of individuals, as it can otherwise lead to real-world harm and tragedies. Traditional methods of manually monitoring online content are time-consuming, costly, and ineffective in dealing with the vast amount of user-generated content. Therefore, there is a need for automated tools that can efficiently detect and prevent HS, which constitutes the primary goal of this research. In this paper, we describe the methods of data collection that we intend to implement, present a short analysis of the existing solutions, and provide the conceptual solution of the model.

Keywords: hate speech · serbian language · natural language processing · convolutional neural networks · recurrent neural networks · BERT architecture

1 Introduction

The term "Hate Speech" has been known for a long time. Still, with the development of the Internet and the more significant number of users on the Internet, it has become very popular on a global level. By spreading from traditional media to the Internet, this phenomenon has led to the creation of Cyberbullying (electronic violence), which unfortunately retains all aspects of traditional violence. Many institutions, experts, organizations, and companies have tried to define hate speech with more or less success. The United Nations defines hate speech as any form of communication in speech, writing, or behaviour that attacks or uses pejorative or discriminatory language against a person or group based on their religion, ethnicity, nationality, race, colour, origin, gender, or other characteristics. Hate speech categories are shown in Fig. 1.

Sexism or gender discrimination is insulting, bullying, or any other behaviour aimed at causing pain in general based on gender. Racism takes race, belonging to a certain tribe, as the basis of insult, or, if necessary, forms of nationalism could also be introduced here, depending on the level of simplification. Radicalization represents the presentation of harsh/radical views that offend the views, faith, or ideological commitment of others. Other types of hate speech include insulting, for example, persons affected by illness or persons with some developmental disabilities, based on age, or some disorders. Insults,

© The Author(s), under exclusive license to Springer Nature Switzerland AG 2025
N. Filipović (Ed.): AAI 2024, LNNS 1446, pp. 107–114, 2025.
https://doi.org/10.1007/978-3-031-99201-8_11

cursing, profanity, blasphemy, vulgarities, and obscene language belong to this group. All the listed subtypes represent Offensive and Abusive language. Electronic violence is a special concept that, in addition to hate speech on networks, also introduces the exact existence of the target of hate speech and the category of insult. The insult is most often a simple swear word or any of the above-mentioned forms, while the target is most often the name of a specific person, that is, a person who is a victim of violence. Compared to General Hate Speech, electronic violence is more modern, although it includes all the elements of traditional insulting [1].

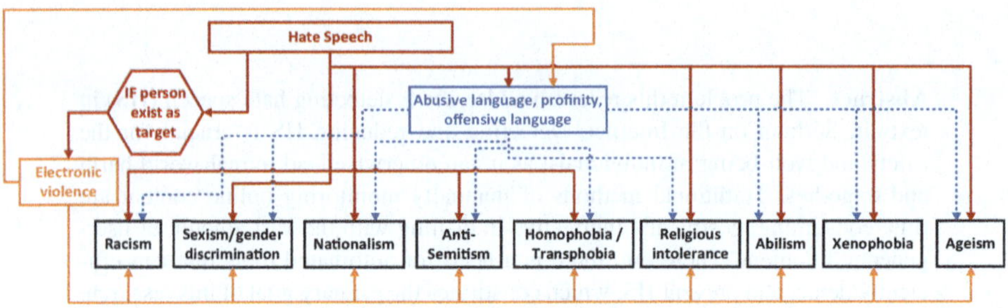

Fig. 1. Different types of hate speech.

The second chapter of this research paper describes the research problem and objectives. The third chapter describes the proposed model and key techniques that will be used to detect hate speech. The fourth chapter presents the data collection and processing in this research. The fifth chapter presents related work, and the last chapter describes further research guidelines.

2 Problem Description

The largest global IT companies, Facebook, Amazon, and Twitter are developing their hate speech detection methods using various algorithms and tools. Also, they have the largest databases of hate speech, but mostly for the English language as the most popular. The algorithms perform worse to detect hate speech in other languages with low resources, such as the Serbian language. Machine learning methods proved to be the most adequate method for solving the problem of detecting hate speech on the Internet. Figure 2 shows the increased number of papers on hate speech in the previous 10 years.

Unfortunately, it is increasingly difficult to protect users on social networks because the companies and individuals who own those networks are trying to leave as much freedom of speech as possible. It is often the case that the younger population post comments on social networks, which offend other people on various grounds and are not automatically detected by algorithms. The increased aggression of young people in web content and electronic violence often leads to great tragedies, as was the case recently in a Belgrade elementary school. On web portals, hate speech detection is mainly done manually, where an administrator approves or rejects comments that offend others.

In response to the alarming trends of growing xenophobia, racism and intolerance, misogyny, anti-Semitism and Islamophobia, around the world, on June 18, 2019, the

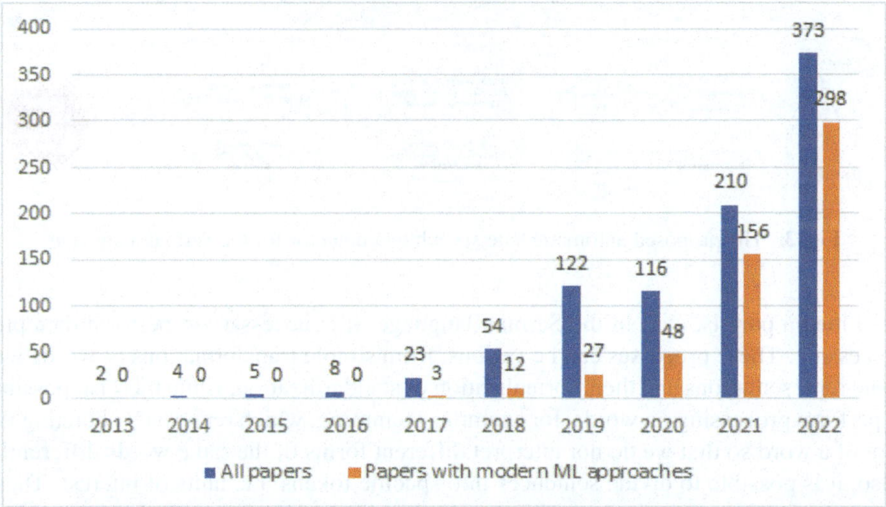

Fig. 2. Number of scientific papers on hate speech in the last decade [source: Google Scholar]

United Nations launched the Strategy and Action Plan against hate speech (HS). HS is recognized and sanctioned in Serbia by Article 11 of the Law on Prohibition of Discrimination, which prohibits the expression of ideas, information, and opinions that incite discrimination, hatred, or violence against a person or group of people because of their characteristics, in public media and other publications, on gatherings and places accessible to the public, by writing and displaying messages or symbols and in different ways.

One of the key problems of today's Internet is that hate speech needs to be regulated in a comprehensive manner, in a way that incorporates all principles, standards and determines the responsibilities of authors of hate speech, internet service providers, web forums, and hosts, social media platforms, online intermediaries, moderators of blogs and others performing similar roles. To solve this increasingly massive problem, it is desirable to automate the detection and classification of such written hate speech texts.

3 Methodology and Problem Solution

Artificial intelligence is increasingly used in the development of automated intelligent systems [2]. Machine learning refers to the ability of an intelligent software system to generalize based on previous experience (a set of data) and use the created generalizations to provide answers to questions about entities or phenomena it has not previously encountered. Machine learning can be applied in different fields and on different interest data to get the desired results [3]. In this case, the result of applying the machine learning model will be detecting hate speech and its classification. Figure 2 shows the diagram of the automatic hate speech detector, which we propose in this research.

The concept of the proposed solution uses the datasets that will be formed. These are raw data from various sources (comments from social networks, texts, and comments

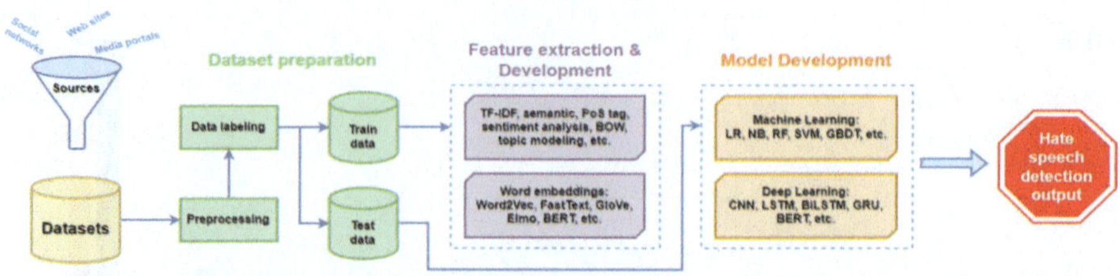

Fig. 3. The proposed automatic hate speech text detector for the Serbian language

from media portals, etc.) in the Serbian language. It is necessary to perform data pre-processing. These processes can be various, from simple transformations of words into some representations and their normalization or standardization, while it is also possible to perform processing on words, for example, stemming, which represents obtaining the root of a word so that we do not interpret different forms of the same words differently. Also, it is possible to divide sentences into specific tokens, i.e. units of interest. These are n-grams that represent a set of n consecutive words that go into further processing. There are also other pre-processing features in NLP, such as removing punctuation and stop signs or reducing lowercase letters (Fig. 3).

As our models are trained on the part of this data, it needs to be labeled; that is, part of the obtained data must be manually classified so that our model can learn the dependencies of our classes of interest. Then our labeled data is split into a training and a test set, where the training set is passed on. Based on the input data and their labels, in the process of supervised learning, through a certain number of training cycles or epochs, efforts are made to make the predictions of the labels as close as possible to the correct ones. Without going into the details of the given procedure, something called a loss is often defined that depends on the input and output values of the model and that loss the model tries to minimize. The test data is finally used to evaluate the resulting model. Classification success can be defined by different metrics: Accuracy, Precision, Recall, F1 Score, etc.

The second part of the procedure will be oriented to transforming the data into feature vectors of such a representation that our model can be trained as efficiently as possible. Bag of Words (BoW) and Term Frequency Inverse Document Frequency (TF-IDF) techniques can obtain the simplest features. The third part of the procedure is developing a model that should classify the received text as "hate speech" or derive some classes. With traditional models, it is only necessary to model the class boundary for certain classifications. These models are implemented using classic machine learning algorithms: Logistic Regression, Naive Bayes, Support Vector Machine, and others. With more modern models, it is necessary to model the distribution of classes as faithfully as possible. Such models include convolutional neural networks, recurrent neural networks, and expansion of the RNN architecture in the form of Long-Short-Term-Memory, Gated Recurrent Units, but combinations of CNN and RNN are increasingly used [4]. Our methodology for choosing the best model will include analyses of other researchers and the different architectures used for hate speech detection, such as CNNs, Bidirectional RNNs, LSTM, GRU, and the modern BERT architecture.

4 Data Collection and Processing

The research aims to generate and collect textual data in the Serbian language from various online sources, including social media platforms, media/news portals, blogs, online forums, and websites that encourage user-generated content. These sources have been identified as rich repositories of user interactions and public discussions, making them valuable for capturing instances of hate speech. The textual data collected will form the foundation for training, optimizing, and evaluating the hate speech detection software system.

To ensure the effectiveness of the hate speech detection software system, significant datasets are needed for the solution implementation. These datasets include a labeled hate speech dataset, a non-hate speech dataset, and a validation/test dataset. The labeled hate speech dataset comprises text samples that have been manually annotated with hate speech labels. It is essential to have a diverse collection of hate speech instances, specifically in the Serbian language, to train a robust detection model. The size of this dataset should ideally be substantial, containing thousands or tens of thousands of samples. The non-hate speech dataset, on the other hand, consists of text samples that do not contain hate speech but cover a wide range of topics and language usage, similar to the hate speech dataset. It is important to match or exceed the size of the hate speech dataset in terms of volume to ensure balanced training. Lastly, the validation/test dataset consists of labeled text samples, including hate speech and non-hate speech instances. This dataset is crucial for evaluating the performance of the trained model. Its size should typically be in the range of thousands of samples to ensure statistical significance.

To obtain the required data, a two-step process will be implemented. Firstly, web scraping tools will be developed to automate the collection of textual data from the identified online sources. These tools will crawl the web, retrieve relevant content, and store it in a structured format. They will be designed to gather diverse content, capturing different types of discourse, user interactions, and linguistic patterns. The web scraping tools will be tailored to the specific requirements of hate speech detection in the Serbian language, ensuring efficient and targeted data collection. Secondly, the collected data will undergo manual labeling. Trained human annotators will review the collected text samples and assign appropriate labels, indicating whether they contain hate speech or not. This manual labeling process is essential to create a reliable ground truth for training the hate speech detection model. Annotators will follow predefined guidelines and criteria to ensure consistency and accuracy in the labeling process. In order to minimize bias and increase reliability, multiple annotators will be involved, and inter-annotator agreement measures will be taken into consideration.

The labeled data will be properly documented and stored, ensuring traceability and transparency throughout the process. It will be treated with utmost care to protect the privacy and confidentiality of individuals represented in the dataset. Personally identifiable information (PII) will be anonymized or pseudonymized to minimize the risk of data breaches and adhere to privacy regulations. By removing or obfuscating identifiable information such as names, addresses, or social media handles, the privacy of individuals will be preserved while maintaining the usability of the dataset for research purposes. The pre-existing web tool Commano (developed by our organization) will be utilized for the purpose of annotation [5]. The development of this application aimed to accelerate the laborious and error-prone process of manual classification. Commano displays the presently selected comment, allowing annotators to select one of the predefined categories from a group of radio buttons. The collected data will be securely stored in a reliable storage system, utilizing formats such as.XLSX or.JSON. These files will be stored on local servers within the organization, implementing robust physical and logical security measures to ensure data protection. To further ensure secure data storage and usage, regular backups of the data will be performed.

5 Related Work

One of the first papers in this area [6] analyses the detection of racism as a specific form of Hate Speech, using a traditional ML model with SVM. A set of 3 million words where documents were annotated as racist or non-racist was used as a dataset. It is important to emphasize that the dataset was collected so that the number of these annotations is fairly balanced. Bag of words (BOW) and bigram tokenization were used as feature extraction and preprocessing techniques. After that, a long period of time was quite inactive, so the next research papers appeared only in 2012. In the research [7] from that period, a database collected from the YouTube platform was used. Specifically text sequences of comments on the 18 most popular videos. Techniques used in preprocessing are BOW, n-grams of order 1–5 and certain methods of assessing the level of violence in sentences. Also, since it is about informal texts in the comments, before the entire processing, a spell checker was made, specifically in this case for the English language. These assessment methods in the paper are a naive assessment where the very proximity of the words in the sentence is used and with that the weight of the words on the Appraisal method and Lexical Syntactic Feature (LSF).

Another study [8] analyses raw posts over a period of about a year, where it is ensured that 4.58% of posts have at least one offensive word, which was defined by the previous creation of a dictionary of offensive words, around 338 of the most common. We emphasize that a lot of posts have been collected, cleaned and classified by tools such as Hadoop. The paper proposes the use of four models for classification, namely SVM, LR, decision trees and the Random Forest method, which are standard machine learning algorithms used in classification. The main specificity of this research is the extraction of features from Twitter posts. The method is called "topic modelling". As each post is actually a mixture of multiple topics and assuming that topics can be modelled by distributions of words, the aim is to teach our model those distributions. The main idea is that the distribution of offensive words will most likely have offensive topics, and a lexical indicator is also introduced, so if there is a word from the corpus of 338 words that we mentioned, then it is automatically classified as an offensive/attacking post. Specifically, topic allocation is learned using the probabilistic generative model Linear Dirichlet Allocation (LDA).

The biggest breakthrough in modern NLP and hate speech detection is happening with the emergence of the BERT model [9]. In the paper [10], a special dataset is given that illustrates hate speech well. The dataset, which was mostly obtained from social networks with about 20 thousand posts, was labelled with the following labels: offensive, hate speech and normal. It is important to emphasize that along with these main labels, clarifications are also added in the form of target types representing African, Islam, Jewish, LGBTQ, Women, Refugee, Arab, Caucasian, Hispanic, Asian, with the LIME approach. The LIME approach implies that, during annotation, annotators emphasize which words in the sentence carry meaning. In addition to modern machine learning approaches in hate speech detection, the manuscript [11] shows the far-reaching nature of these simpler methods. The name hybrid gives us an insight into the fact that it is a mixture of a large number of available means and this paper is a kind of overview of various hybrid approaches. The same test is used that was collected from the Twitter platform for research work [12].

6 Conclusion

The research aims to generate and collect textual data in the Serbian language from various online sources, including social media platforms, media/news portals, blogs, online forums, and websites that encourage user-generated content. These sources have been identified as rich repositories of user interactions and public discussions, making them valuable for capturing instances of hate speech. The textual data collected will form the foundation for training, optimising, and evaluating the hate speech detection software system, that will be developed within the project "Software for Text Offences Prevention in Serbian: Al-driven Hate Speech Detection" [13].

It can be concluded that the topic of this research is state-of-the-art at the moment. On the other hand, it can be noted that more than 96% of scientific research on hate speech and the application of artificial intelligence in detection is oriented towards English, followed by Chincsc and Arabic languages. The number of research for the detection of hate speech in the Serbian language is negligible and the last research is in the field of classification of sports text comments from social networks using deep learning techniques [14]. Sets of big data are very important for this kind of hate speech detection and classification research. In [15], more than 200 thousand top-level English and Turkish tweets published in 2020 and 2021 were retreived from five hate domains—religion, gender, racism, politics, and sports, where each tweet can be classified to a single domain.

Acknowledgements. This research was supported by the Science Fund of the Republic of Serbia, Grant No 11113, *Software for Text Offences Prevention in Serbian: Al-driven Hate Speech Detection* - STOP.

This work was financially supported by the Ministry of Science, Technological Development and Innovation of the Republic of Serbia under contract number: 451-03-65/2024-03/200103.

References

1. Fortuna, P., Nunes, S.: A survey on automatic detection of hate speech in text. ACM Comput. Surv. **51**(4), 30 (2018). https://doi.org/10.1145/3232676
2. Draskovic, D., Cvetanovic, M., Nikolic, B.: SAIL - software system for learning AI algorithms. Comput. Appl. Eng. Educ. **26**(5), 1195–1216 (2018)
3. Milaković, A., Drašković, D., Nikolić, B.: Visual simulator for mastering fundamental concepts of machine learning. Appl. Sci. **12**(24), 1–22 (2022)

4. Drašković, D., Mitrić, M., Cincović, J., Nikolić, B.: Software tool for sentiment analysis of reviews (in Serbian). In: Proceeding of 29th ICT conference YU INFO. Informaciono društvo Srbije, pp. 106–110, Kopaonik, Serbia (2023)
5. Kostić, M., Batanović, V., Nikolić, B.: Monolingual, multilingual and cross-lingual code comment classification. Eng. Appl. Artif. Intell. **124**, 1–17 (2023)
6. Greevy, E., Smeaton, A.: Classifying racist texts using a support vector machine. In: Proceedings of the 27th annual international ACM SIGIR conference on Research and development in information retrieval, pp. 468–469 (2004). https://doi.org/10.1145/1008992.1009074
7. Chen, Y., Zhou, Y., Zhu, S., Xu, H.: Detecting offensive language in social media to protect adolescent online safety. In: International Conference on Privacy, Security, Risk and Trust and 2012 International Conference on Social Computing, Amsterdam, Netherlands, pp. 71–80 (2012). https://doi.org/10.1109/SocialCom-PASSAT.2012.55
8. Xiang, G., Fan, B., Wang, L., Hong, J., Rose, C: Detecting offensive tweets via topical feature discovery over a large scale twitter corpus. In: Proceedings of the 21st ACM international conference on Information and knowledge management (CIKM '12). Association for Computing Machinery, New York, NY, USA, pp. 1980–1984 (2012), https://doi.org/10.1145/2396761.2398556
9. Devlin, J., Chang, M., Lee, K., Toutanova, K.: BERT: Pre-training of deep bidirectional transformers for language understanding. In: Proceedings of the 2019 Conference of the North American Chapter of the Association for Computational Linguistics: Human Language Technologies, Volume 1 (Long and Short Papers). Association for Computational Linguistics, Minneapolis, Minnesota. pp. 4171–4186 (2019)
10. Mathew, B., Saha, P., Yimam, S., Bienmann, C., Goyal, P., Mukherjee, A.: Hatexplain: A benchmark dataset for explainable hate speech detection. Proceedings of the AAAI Conference on Artificial Intelligence (2021). https://doi.org/10.48550/arXiv.2012.10289
11. Shruthi, P., Kumar, K.M.: Novel approach for generating hybrid features set to effectively identify hate speech. Intel. Artif. **23**(66), 97–111 (2020). https://doi.org/10.4114/intartif.vol23iss66pp97-111
12. Waseem, Z., Hovy, D.: Hateful symbols or hateful people? Predictive features for hate speech detection on Twitter. In: Proceedings of the NAACL Student Research Workshop. Association for Computational Linguistics, San Diego, California, 88–93 (2016)
13. Software for Text Offences Prevention in Serbian: AI-driven Hate Speech Detection - STOP project web site, available at: http://stop.etf.bg.ac.rs, last accessed 2024/09/16
14. Vujičić Stanković, S., Mladenović, M.: An approach to automatic classification of hate speech in sports domain on social media. J. Big Data **10** (109), (2023)
15. Toraman, C., Şahinuç, F., Yilmaz, EH.: Large-scale hate speech detection with cross-domain transfer. arXiv (2022). https://doi.org/10.48550/arXiv.2203.01111

Data Analysis Techniques and Detection of Propaganda in Serbian Online Media in 2023

Dražen Drašković(✉), Mihajlo Ogrizović, Matija Dodović, and Miloš Obradović

University of Belgrade - School of Electrical Engineering, Belgrade, Serbia
drazen.draskovic@etf.bg.ac.rs

Abstract. Propaganda is a technique by which people manipulate information to direct the behavior of an individual or group, with the aim of provoking a desired reaction. Propaganda is strongly expressed in the spread of religion, political parties, and autocratic governments. In this paper, the most common propaganda techniques are presented, which propaganda implementers can use in shaping and realizing their messages. A certain content of short texts published on the portals of media companies in Serbia was collected. Then, using machine learning techniques, a model was developed, which was trained to classify the text as propaganda or not propaganda. This research included texts that were published on the portals RTS, Politika, B92, Blic, N1, and Danas in 2023. The texts were collected in real-time, because some media portals do not have an archive.

Keywords: natural language processing · data collection · data annotation · propaganda techniques · serbian language

1 Introduction

The detection of propaganda, or propaganda techniques, in the field of natural language processing (NLP), is defined as the task of text classification [1]. NLP represents a narrower area of the branch of computer science called machine learning and deals with the study of methods and techniques for computer processing and interpretation of textual data written in one of the natural languages. Detection of propaganda in the context of text classification is a task characterized by a high degree of complexity, primarily due to the existence of a large number of different propaganda techniques and the variety of topics they cover, and it represents a challenge for building models for binary, and especially for multi-class classification. In the spread of faith, and then in all reform and political movements, propaganda was the main means of persuasion and struggle [2]. The First World War and the propaganda efforts made by all participants, for certain authors, represent a turning point in the treatment of propaganda as a designed and organized communication, which is given serious importance in the conquest of economic, political, and war goals [3]. Therefore, propaganda, in the broadest sense, can be defined as the spreading of ideas, opinions, and attitudes, often with the use of lies and distorted facts.

© The Author(s), under exclusive license to Springer Nature Switzerland AG 2025
N. Filipović (Ed.): AAI 2024, LNNS 1446, pp. 115–123, 2025.
https://doi.org/10.1007/978-3-031-99201-8_12

Using manipulation techniques, propaganda affects the emotions and beliefs of an individual or group, ensuring that the recipient unknowingly adopts the messages and perceives them as their own beliefs. Sometimes, the propagandistic content can be difficult to spot. That is the reason why the use of new technologies and large language models in the automated detection of propaganda is very important.

The structure of this paper includes 6 chapters. In the second chapter, the problem of propaganda detection is described, and the most important research papers that applied machine and deep learning techniques are highlighted. In the third chapter, the propaganda techniques identified in the open literature are presented. The fourth chapter describes the process of collecting texts from the media. The fifth chapter presents the binary models that are trained for the described classification problem. At the end of the paper, a conclusion is given, in which other models and techniques that can be used in the continuation of the research are described.

2 Problem Description and Related Work

With the advent of deep neural networks, such as convolutional (CNNs) or recurrent networks (RNNs), the number of studies dealing with the detection of fake news has increased. Propaganda detection can be considered a subtask of fake news detection. Other subtasks are the detection of satire and deception. In the paper [4], the researchers present the use of supervised learning and present a new set of features, measuring the predictive performance of current approaches and features for automatic fake news detection.

The paper [5] by Shu provides an extensive review of fake news detection approaches focusing on data mining techniques. It categorizes the problem of fake news detection into several sub-problems, such as feature extraction, model training, and evaluation. The authors highlight the role of social context (e.g., user behaviors, user relationships) and content (textual, visual) in the detection process. They also explore challenges such as data scarcity and adversarial fake news attacks.

In a review paper [6], the authors survey the use of deep learning techniques in the automatic detection of fake news. The authors provide an overview of various architectures, including Attention, Generative Adversarial Networks, and Bidirectional Encoder Representations for Transformers. They highlight how these models outperform conventional machine learning methods and offer insights into their application in large-scale real-world datasets. The paper also outlines the challenges in the domain, including model interpretability and computational efficiency.

The conclusion of the paper [7] highlights the magic of Graph Neural Networks (GNNs) in fake news detection, though the field is still in its early stages, with limited studies published so far. While GNNs have shown excellent initial results, several challenges need to be addressed for more comprehensive outcomes. Key challenges include the lack of benchmark datasets, scalability issues due to complex graph structures, and limitations in early fake news detection due to rapidly evolving information. The authors also emphasize the need for dynamic, heterogeneous, and multiplex GNNs to better handle real-time updates, diverse media formats, and the integration of propagation, content, and context features. Addressing these challenges is likely to improve GNN-based fake news detection in the future.

Contrary to models based on deep learning, logistic regression, support vector machine (SVM), and Multinomial Naïve Bayesian (MNB) were used in research works [8] and [9]. In research [9], fine-tuning of SVM is being done by taking kernel Linear, Poly, and RBF.

The authors of this paper decided to use traditional artificial intelligence models in the first phase of this research and devote themselves to the application of deep learning using CNNs, RNNs, and GNNs in the following phases.

3 Propaganda Techniques

In the analysis of the existing propaganda techniques, it was observed that the size of the set of defined techniques differs significantly depending on the observed source, i.e. the author. Sociologist Clyde Miller distinguishes 7 techniques [10], while researchers Da San Martino et al., in their work on the detection of propaganda at the fragment level in the English language, highlight 18 techniques [11]. Table 1 provides a summary of all techniques.

Table 1. List of propaganda techniques from open literature.

Miller C. et al. (1936)	Da San Martino et al. (2020)
Band Wagon	*Appeal to authority*
Card Stacking	*Appeal to fear/prejudice*
Glittering Generalities	Bandwagon
Name-calling	Black-and-white fallacy
Plain Folks	Casual oversimplification
Testimonial	Doubt
Transfer	*Exaggeration or minimization*
	Flag-waving
	Loaded language
	Name-calling or labeling
	Obfuscation, confusion
	Red Herring
	Reductio ad Hitlerum
	Repetition
	Slogans
	Straw man
	Thought-terminating cliché
	Whataboutism

For the purposes of our research, we selected 8 propaganda techniques (italics in Table 1) based on the collected articles and determined to which classes those text articles would correspond. Table 2 provides examples of texts for each of the predefined categories.

Appeal to authority is a technique in which an assertion is insisted to be true simply because an expert or person who purports to be an expert in the field under consideration says it is true, without providing adequate evidence for it.

Appeal to fear/prejudice is an effort to gain support for a particular propaganda message or idea by instilling fear in the target audience towards an alternative that is usually based on preconceived notions rather than evidence.

Exaggeration or minimization is a technique in which the subject of conversation or the situation is exaggerated (things are presented as bigger, better, worse) or minimized (things are presented as less significant than they really are).

Flag-waving is a technique in which actions or ideas are justified by appealing to feelings of pride and duty towards the state, nation, city, and the like.

Glittering Generalities is a technique where the subject of propaganda is associated with a positive meaning. By using various types of generally accepted claims, as well as nice words that evoke positive feelings, the target audience is led to accept and approve of the propaganda message, without prior checking and careful consideration of the reasons.

Loaded language is a technique based on the use of words and expressions that cause strong negative emotional reactions in the target audience, such as disgust, repulsion, contempt, and others.

The name-calling technique involves using insults, using derogatory words, and assigning labels to people or groups.

Plain folks are an effort to facilitate the acceptance of the propaganda message and create a positive attitude towards the speaker or the subject of the propaganda, by presenting the speaker himself as a common man from the people who sympathizes with and understands the target audience.

4 Data Collection

This research included texts that were published on the portals RTS, Politika, B92, Blic, N1, and Danas in 2023. The texts were collected in real-time, because some media portals do not have an archive. Texts from social networks Facebook, platform X, and YouTube comments were not taken for analysis because in the last few years a large number of fake accounts have appeared, generating either fake content or propaganda. Thus, the company Twitter, on its X platform, removed thousands of accounts from Serbia, Egypt, Honduras, Indonesia, and Saudi Arabia at the beginning of 2020 for acting according to government directives or promoting pro-government content. Among them were as many as 8,558 active fake accounts that promoted the ruling party in Serbia, the Serbian Progressive Party (SNS), and its leader Aleksandar Vučić.

The analysis included news from the politics category, due to assumptions that the largest number of propaganda articles are in that category. By analyzing the static content of internet pages, it was concluded that all newspaper portals that are the subject of analysis are suitable for parsing. The Scrapy and BeautifulSoup libraries were used as the main tools for this purpose. The collection procedure, which included web crawling and web scraping, was not complex but it was time-consuming, considering that periodic cron jobs were created and executed every day.

Each text article that was collected had the following information: unique identifier of the source (media portal), identifier number of the article within the source, name of the source, link to the website of that article, publication date of the article (format YYYY-MM-DD), title of the article, and text of the article.

Four topics were selected in the research: the president, the opposition, elections in Serbia, Kosovo and Metohija, and the dataset after filtering contained about 8 thousand newspaper articles. After sampling, the research and definition of propaganda techniques that are present in the Serbian electronic media were started, after which, with the obtained set of eight propaganda techniques and their associated instructions, following the procedure of data annotation.

The distribution of articles in the data set of text articles, which includes 4 selected news categories, is shown in Fig. 1. A dataset of 1200 articles was selected for training, with 46% of the texts that contain propaganda and 54% that do not. Each article was annotated by two trained student annotators, and one teacher was responsible for the final review. Table 3 contains an overview of the most frequently used propaganda techniques for each of the topics defined in Table 4. It can be noted that the Exaggeration or minimization technique is represented with a high percentage in 3 out of 4 selected topics. In articles related to the topic of the President, the use of propaganda associated

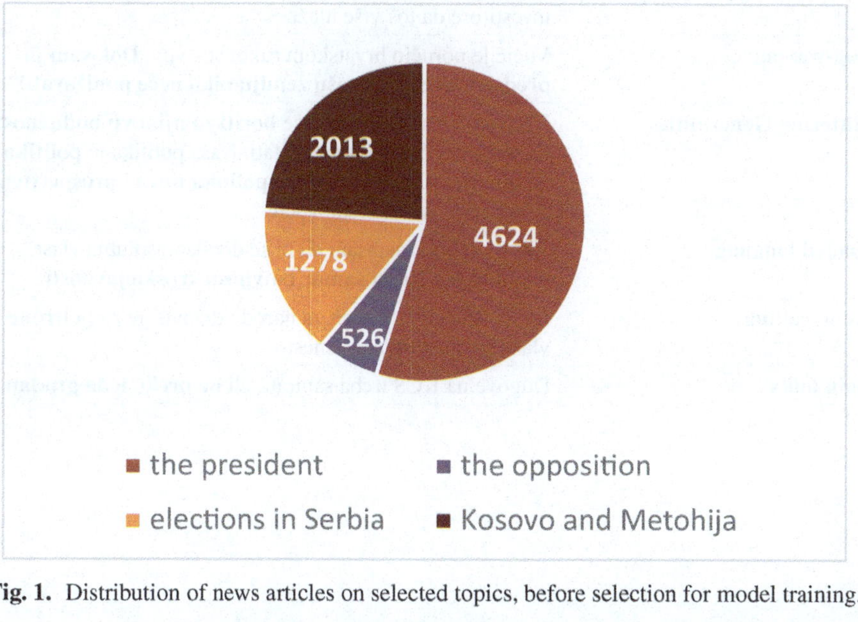

Fig. 1. Distribution of news articles on selected topics, before selection for model training.

with positive feelings and expressions of sympathy with citizens can be observed. Also, highlighting the importance of the nation (national pride) and the capital (Belgrade) is very common in articles about the president. In articles related to Kosovo and Metohija, in addition to feelings of national pride, expressions that instill fear in the target audience are often used. Articles that criticize the previous government and express negative feelings are most often associated with elections, with fears that life will be much worse in the future if the government changes. For the evaluation process, 300 newspaper articles were selected, 122 of which contained propaganda (40.67%), and 178 were not marked as propaganda (59.33%). The distribution of selected articles from the training set and the evaluation set by topic are shown in Table 4.

Table 2. Examples of collected articles that corresponded to selected techniques/classes.

Technique / Class	*Example (in Serbian)*
Appeal to authority	Pokret Dveri **pozdravlja stav Svetog arhijerejskog sabora SPC** da je očuvanje KiM u sastavu Srbije najvažniji nacionalni zadatak za sve. Smatramo da svako ko bi se oglušio o ovu poruku Svetog arhijerejskog sabora ne samo da bi se oglušio o SPC, već bi se odricanjem od KiM zarad puta Srbije u EU praktično ispisao iz srpske istorije, navodi se u saopštenju
Appeal to fear/prejudice	"Kurtiju ne vredi govoriti, ali ako se međunarodna zajednica ne dozove pameti, sve ovo **vodi u nove sukobe na Balkanu sa nesagledivim posledicama**" – izjavio je Ivica Dačić
Exaggeration or minimization	U tom smislu Aleksandar Vučić je ponovio da je **Srbija najbolje mesto za ulaganje na svetu** i pozvao francuske investitore da još više ulažu
Flag-waving	Vučić je poručio hrvatskom rukovodstvu: **„Dok sam ja predsednik Srbije našu zemlju niko neće ponižavati!"**
Glittering Generalities	Deca su bogatstvo i treba se **boriti za njihovu budućnost** Pa na njima će ovaj svet i ostati. Naša politika je **politika života, politika budućnosti, politika mira i prosperiteta**. – izjavio je Aleksandar Vučić
Loaded language	Ako su izbori raspisani da bi se dobila apsolutna vlast, bio bi to **još jedan šamar naivnosti srpske javnosti**
Name-calling	Dveri: Vučićev 'Pokret za narod i državu' je za **poltrone** vlasti, nama tamo nije mesto
Plain folks	Dugove na KCS treba sanirati, ali **ne preko leđa građana**

Table 3. Examples of collected articles that corresponded to selected techniques/classes.

Topic	*Propaganda techniques with the most examples*	*Representation in the subset for selected topic*
The president	Glittering Generalities	28%
	Flag-waving	22%
	Exaggeration or minimization	11%
The opposition	Exaggeration or minimization	26%
	Name-calling	17%
	Loaded language	15%
Elections in Serbia	Loaded language	19%
	Flag-waving	18%
	Appeal to authority	16%
Kosovo and Metohija	Exaggeration or minimization	24%
	Appeal to fear/prejudice	18%
	Glittering Generalities	15%

Table 4. Distribution of articles from the training set and the evaluation set by topic.

Topic	*Number of articles (%) in the training set*	*Number of articles (%) in the evaluation set*
The president	611 (50,92%)	127 (42,33%)
The opposition	157 (13,08%)	61 (20,33%)
Elections in Serbia	238 (19,83%)	78 (26%)
Kosovo and Metohija	194 (6,17%)	34 (11,33%)

5 Model Training and Evaluation

The main task in this research is the construction of a system for detecting propaganda, which in the context of NLP belongs to the problems of text classification. For these purposes, a linear classifier was used as the baseline model, which uses features obtained according to the bag-of-words (BOW) principle, that is, a bag of n-grams, where each text is represented in the form of an unordered set of words or n-grams in them, and each word or n-gram represents one feature. After that, linear models that use features obtained according to the bag-of-embeddings principle (BOE) were examined, where each text is represented as an average of the vector of word meanings in it, and each dimension of such an averaged vector is treated as a feature in the classification.

In our research, transfer learning was used. Transfer learning represents one of the approaches for improving the development of statistical solutions in the context of limited annotated resources. It includes a large number of methods, the basic idea of which is to train a model or some part of it on a problem or domain for which there is sufficient data, and then to use such a model to solve a specific problem of interest, for which there is a limited amount of available data.

In the processing of natural languages, the use of word meaning vectors is extremely widespread, as a form of transfer learning. They represent vectors of fixed length, which numerically represent the meaning of words, and are based on the distributional hypothesis that words with similar meanings are used in similar contexts. The vector representation of the meaning of words leads to the effect that semantically close words are located close to each other in the vector space and allows the meanings of words to be compared through the similarity of their vectors. Nowadays, the most widely used approach is the construction of word meaning vectors using neural networks, and these include the popular models word2vec, GloVe, and fastText, which were used when creating the solution in this paper.

In this section, the results of linear classifiers applied to features obtained by the bag of words principle will be presented. The following classifiers were used: logistic regression (LR) and Support Vector Machine (SVM). The following hyperparameters were taken: $C \in [0{,}01 - 1000]$ for LR and $C \in [0{,}1 - 100]$ for SVM.

Table 5 contains the results of bag-of-words classifiers trained for the binary propaganda detection problem. First, the results of the evaluation after the application of the basic text preprocessing methods, such as the removal of special characters, are presented. Then, the results of the application of morphological normalization (stemming and lemmatization) are presented, and finally weighting based on Term Frequency - Inverse Document Frequency (TF-IDF) values, as well as the addition of bigram features, are discussed. Based on the results shown in Table 5, it can be concluded that adding bigram features can positively affect the performance of both logistic regression and support vector methods. An SVM classifier (without kernel) was then applied to the features obtained using the Bag-of-embeddings model. For hyperparameter optimization, five-fold stratified cross-validation was used, with hyperparameters: $C \in [0{,}1 - 100]$ and $\gamma \in [0{,}001 - 1]$. In this case, the results obtained were worse than in the previous case with bag-of-words (F-measure = 0.299).

Table 5. Results of bag-of-words classifiers trained for the problem of binary detection of propaganda, expressed in the form of F-measure.

Topic	*LR*	*SVM*
Basic text preprocessing	0,258	0,271
Stemming	0,212	0,218
Lemmatization	0,333	0,292
TF-IDF weighting	0,269	0,275
Unigram + bigram features	0,367	0,325

6 Conclusion

This paper describes the procedure of collecting texts from media portals in Serbia, data preprocessing, propaganda techniques, and the development of a model that implements a binary classifier for automatic detection of propaganda in the Serbian language. The training set contained 1200 news articles, and the evaluation set contained 300 news articles.

The problem of classifying propaganda proved to be an extremely complex task during the annotation phase, and this complexity was also reflected during the training and evaluation of the developed models. Binary classification proved to be a good choice because it shows very well the ability to distinguish propagandistic from non-propagandistic content. The direction of further research can be the development of models with multi-class propaganda detection, as well as the application of other machine learning techniques, primarily deep learning techniques.

Acknowledgements. This research was supported by the Science Fund of the Republic of Serbia, Grant No 11113, *Software for Text Offences Prevention in Serbian: AI-driven Hate Speech Detection* - STOP.

This work was financially supported by the Ministry of Science, Technological Development and Innovation of the Republic of Serbia under contract number: 451-03-65/2024-03/200103.

References

1. Yoosuf, S., Yang, Y.: Fine-grained propaganda detection with fine-tuned BERT. In: Proceedings of the second workshop on natural language processing for internet freedom: censorship, disinformation, and propaganda, pp. 87–91 (2019)
2. Seidman, S.A.: Posters, Propaganda, and Persuasion in Election Campaigns Around the World and Through History. Peter Lang (2008)
3. Fox, J., and Welch, D.: Justifying war: Propaganda, politics and the modern age. In: Justifying War: Propaganda, Politics and the Modern Age, pp. 1–20. London: Palgrave Macmillan UK, 2012.Lasswell HD. Propaganda technique in the world war. Ravenio Books (2015)
4. Reis, J.C., Correia, A., Murai, F., Veloso, A., Benevenuto, F.: Supervised learning for fake news detection. IEEE Intell. Syst. **34**(2), 76–81 (2019)
5. Shu, K., Sliva, A., Wang, S., Tang, J., Liu, H.: Fake news detection on social media: a data mining perspective. ACM SIGKDD Explor. Newsl. **19**(1), 22–36 (2017)
6. Mridha, M.F., Keya, A.J., Hamid, M.A., Monowar, M.M., Rahman, M.S.: A comprehensive review on fake news detection with deep learning. IEEE access **9**, 156151–156170 (2021)
7. Phan, H.T., Nguyen, N.T., Hwang, D.: Fake news detection: a survey of graph neural network methods. Appl. Soft Comput. **139**, 110235 (2023)
8. Li, J., Ye, Z., Xiao, L.: Detection of propaganda using logistic regression. In: Proceedings of the Second Workshop on Natural Language Processing for Internet Freedom: Censorship, Disinformation, and Propaganda (pp. 119–124) (2019, November).
9. Khanday, A.M.U.D., Khan, Q.R., Rabani, S.T.: Detecting textual propaganda using machine learning techniques. Baghdad Sci. J. **18**(1), 0199 (2021)
10. Miller, C.R., Edwards, V.: The intelligent teacher's guide through campaign propaganda. Clear. House: J. Educ. Strategies Issues Ideas **11**(2), 69–73 (1936)
11. Martino, G.D.S., Shaar, S., Zhang, Y., Yu, S., Barrón-Cedeno, A., Nakov, P.: Prta: A system to support the analysis of propaganda techniques in the news. arXiv preprint arXiv:2005.05854 (2020).

AI-Enhanced Extended Reality in Medicine

Filip Filipović[1,2(✉)], Tijana Geroski[1,2], Aleksandra Vulović[1,2], and Nenad Filipović[1,2]

[1] Bioengineering Research and Development Center (BioIRC), Prvoslava Stojanovica 6, 34000 Kragujevac, Serbia
filipovicfilip1999@gmail.com, {tijanas,aleksandra.vulovic, fica}@kg.ac.rs

[2] Faculty of Engineering, University of Kragujevac, Sestre Janjic 6, 34000 Kragujevac, Serbia

Abstract. Artificial intelligence (AI) applied in Virtual Reality (VR) has emerged as a dominant interaction media in multiple application areas, thanks to its advantage of providing users with immersive experiences. Its use in rehabilitative medicine and psychiatry has made significant progress. This paper aims to make an extended reality application that can be used to help patients that have balance disorders. The idea behind the project was to make an application in which users try to catch falling apples controlling the movement of the basket by moving their head left and right with the Hololens device on their head. The application was made in Unity environment, using inputs for gaze, gesture, voice, spatial sound, spatial mapping. It was designed to be used on the Hololens device for the patients to improve their motor functions and reflexes, assisting them in managing balance disorders. The game was created in a 2D version featuring assets such as baskets, apples, rocks, and a background. The basket asset represents the tool that patients use to catch apples that are constantly falling from the rock asset that is moving continuously left and right. At the start of the game, a user has three baskets; when the user misses to catch an apple, the basket disappears, and the user continues to play until there are no more baskets left. In the bottom corners of the user's view, the current score representing the current number of apples the user has managed to catch is displayed, and there is also a high score field representing the highest number of apples the user has managed to catch. AI can enhance personalization in VR games by analyzing player behavior and preferences in real-time, tailoring immersive experiences to individual gaming styles and preferences. Future work can involve patients' improvement results while playing the game, using AI to develop more games to improve the effectiveness of the treatment.

Keywords: extended reality technologies · virtual reality · augmented reality · artificial intelligence · medicine

1 Introduction

Effective communication between healthcare professionals and patients has become a priority in medicine, as it has been shown to positively influence patients' well-being, adherence to treatment, recall and understanding of medical information, ability to cope

© The Author(s), under exclusive license to Springer Nature Switzerland AG 2025
N. Filipović (Ed.): AAI 2024, LNNS 1446, pp. 124–134, 2025.
https://doi.org/10.1007/978-3-031-99201-8_13

with disease, quality of life, and overall state of health 1, 2. AI can broadly be defined as the science and engineering of building intelligent machines 3. In medicine, AI is applied and used in a broad range of tools (e.g., genomics, some surgical robotics) and for a variety of purposes (e.g., precision diagnostics, precision therapeutics) 4. While the breadth of techniques that fall under this term is constantly evolving, branches such as machine learning—including neural networks, reinforcement learning and deep learning rule-based expert systems, and computer vision have shown great potential in healthcare 5, 6, 7. Meanwhile, VR has been described in the literature as encompassing simulators such as screen-based VR simulators (i.e., mixed or lightly augmented reality), screen-based virtual worlds (game-like environments), and immersive VR environments (often requiring a headset)—all of which intend to simulate natural settings 8. While formal and informal definitions of VR are evolving, we have opted to use the definitions and terms authors of peer-reviewed studies have used in this scoping review to help describe how VR is being both conceptualized and studied. Such an approach is also consistent with a broad approach to defining immersive technologies—one that is based on and accompanies technology, but is also conceptualized beyond it to include, for example, how effective a technology is in achieving the goals it was designed to perform 9. These simulation technologies have gained traction for technical skills training such as to create high-fidelity surgical simulators and planners 10, and for intraoperative support—such as applications using augmented reality to produce image guidance during a given surgical procedure 11. However, research regarding uses of VR for non-technical skills training has been limited 8. While the development and uptake of AI and VR technologies in medicine has been significant in areas including diagnostics 12, therapeutic selection 13, treatment delivery 14, patient risk stratification 15, and community-level healthcare 16, the implementation of such technologies to facilitate doctor-patient communication and ultimately empower patients in their own healthcare has lagged behind 17.

Starting in the mid-twenties, human body begins a progressive loss of lean body mass, skeletal muscles in particular. The muscle mass significantly decreases between the ages of 20 and 80. Lexell et al. [18] counted the number of fibres in autopsied vastus lateralis muscle and found an average 40% loss of muscles area and a similar loss of the number of fibres from the second to the eight decade. Normal ageing is associated with a decreased ability to maintain postural stability in standing and when responding to unexpected perturbations. Postural sway, which refers to the constant small deviations in the center of mass position, is a commonly used measure of standing balance. Control of postural sway involves continuous muscle activity in response to integrated sensory inputs [19].

Numerous studies have shown that well-prescribed exercise can improve physical abilities in older people. Strength can be improved by resistance training [20], fitness training can improve endurance [21, 22] and vestibular rehabilitation can lead to reduced dizziness [23–26]. Functional ability may also be improved by strength training, particularly among frailer older people [27]. Based on the health problems, it was noted that the designs aimed at prevention of falls and other conditions were evaluated in patients with Parkinson's disease. The study by Yen 28 resulted in improved sensory integration for postural control; however, the demand for attention was not altered after any Virtual Reality Therapy (VRT). Santana 29 and Loureiro 30, in experimental studies without a comparison group, concluded that training using the Nintendo Wii Fit Plus platform proved to be useful for the rehabilitation of degenerative diseases regarding motor performance, flexibility, lower limb joint stiffness and functional independence, in addition to improving the motor learning ability due to the cognitive stimuli provided by the videogame. The activation of neural circuits and structures by virtual games can play a key role in transferring the immediate effects of games to long-term effects. Lai 31 did not attribute maintenance of improvement to VRT; after the end of the interventions, the elderly improved their balance and felt confident to engage in physical activity, which justifies maintaining good results regarding balance.

The goal of this study is to characterize the use of AI and VR in assisting patients with balance disorders by creating practical exercises in VR for the patients and utilizing AI to help doctors analyze the results for further treatment. In this paper, a Virtual Reality application is developed with the focus on the physical rehabilitation of people with balance disorders, controlled by the movement of the head and controls of the virtual reality set. The game presents a simple task: a user catch as many falling apples as they can by moving their heads left and right to maneuver the basket. These are simple movements that do not require much strength, elasticity, or energy, making them suitable for patients with balance disorders.

2 Methods

The virtual environment is developed in the Unity engine version 2019.4.18f1, in which the patient interacts with the virtual elements using a headset. Two base stations cover the play area and pick up the signal from the headset according to the user's movements. In response to the received movements and actions, the virtual reality application updates the scene the user sees on the headset so that the users can control their movement in a virtual room and perform a series of simple actions to try to set the new high score.

The first game object called Rock is created and rock script is added in which the functions for the rock behavior are defined. The rock's speed is defined and the function that moves the rock to the left and right to some extension is added. In order to make that movement practical, a new game object called Bounds as well as a new component Edge Collider 2D are added, and the collider is placed to the edges of our scene covering sides and bottom that are needed for the game. At that point, some physics are added to the objects. To the object Rock, Box Collider 2D and Rigid Body 2D components are added. In the rock script function, OnCollisionEnter2D is added so that if the rock collides with the boundary, it changes direction. Now we want apple added to the scene, so the apple prefab is created. We need apples to drop so Rigid Body 2D is added. In the rock script, the apple prefab is added and the function for dropping apples every second from the rock object is created. Now we need the apple to disappear when it hits the ground so we do that by adding Circle Collider in the apple prefab and then we add a new script in the Bounds object with the function that destroys the apple when it hits the bottom boundary.

The basket object with the basket behavior script is created. We want our basket movement to be controlled with the mouse for now. We can achieve that by adding functions to the script that connects any movement of the mouse to the basket. Then we need to add Box Collider 2D to the basket so we can catch apples with the basket. In order to make the game interesting, we need to make a score that shows the user the number of apples that are caught. We do that by adding Increment Score function to the Boundary behavior script (Fig. 1) and call it in the basket behavior On Collision 2D function (Fig. 2).

```
public void IncrementScore()
{
    score += 1;

    audioSrcs[0].Play();
    if (score > highScore)
    {
        highScore = score;
        GameObject.Find("highScoreOutput").GetComponent<Text>().text = highScore.ToString();
        audioSrcs[2].Play();
    }
    GameObject.Find("scoreOutput").GetComponent<Text>().text = score.ToString();
}
private void RestartScene()
{
    PlayerPrefs.SetInt("HS", highScore);
    SceneManager.LoadScene("MainScene");
}
```

Fig. 1. Increment Score Boundary Behavior script

```
public class BasketBehavior : MonoBehaviour
{
    // Start is called before the first frame update
    void Start()
    {

    }

    // Update is called once per frame
    void Update()
    {
        Vector3 rawMouse = Input.mousePosition;
        Vector3 convertedMouse = Camera.main.ScreenToWorldPoint(rawMouse);
        Vector3 pos = transform.position;

        pos.x = convertedMouse.x;
        transform.position = pos;
    }
    private void OnCollisionEnter2D(Collision2D collision)
    {
        if (collision.gameObject.tag == "apple")
        {
            Destroy(collision.gameObject);
            GameObject.Find("Bounds").GetComponent<BoundaryBehavior>().IncrementScore();
        }
    }
}
```

Fig. 2. Basket Behavior script

We want to give the user three lives in the game so three baskets on top of each other are created. Therefore,when the game starts the user moves all three of them trying to catch the apples and when the user misses an apple one basket disappears and the user has two more tries to catch as many apples as possible (Fig. 3).

```
public class BoundaryBehavior : MonoBehaviour
{
    public int score;
    public int highScore;
    private List<GameObject> basketList;
    public GameObject basketPrefab;

    private AudioSource[] audioSrcs;
    // Start is called before the first frame update
    void Start()
    {
        audioSrcs = GetComponents<AudioSource>();
        highScore = PlayerPrefs.GetInt("HS", 10);
        GameObject.Find("highScoreOutput").GetComponent<Text>().text = highScore.ToString();

        basketList = new List<GameObject>();
        float bottomEdge = transform.position.y + 2.5f;
        for (int i = 0; i < 3; i++)
        {
            GameObject basket = Instantiate<GameObject>(basketPrefab);
            Vector3 pos = basket.transform.position;
            pos.y = bottomEdge + (i * .5f);
            basket.transform.position = pos;
            basketList.Add(basket);
        }
    }
```

Fig. 3. Creating and positioning of the baskets

This is done by creating a basket prefab in the Boundary Behavior script defining the number of baskets we want in our scene and defining their position. Now we need to destroy all apples when user misses to catch an apple. We do that in the Boundary Behavior script with On Collision Enter 2D function (Fig. 4).

```
private void OnCollisionEnter2D(Collision2D collision)
{

    if (collision.gameObject.tag == "apple")
    {

        audioSrcs[1].Play();
        int topBasket = basketList.Count - 1;
        if (topBasket > -1)
        {
            Destroy(basketList[topBasket]);
            basketList.RemoveAt(topBasket);
        }
        GameObject[] appleArray = GameObject.FindGameObjectsWithTag("apple");
        foreach (GameObject apple in appleArray)
        {
            Destroy(apple);
        }

        if (topBasket <= 0)
        {
            Destroy(GameObject.Find("Rock"));
            Invoke("RestartScene", .5f);
        }
    }

}
```

Fig. 4. Boundary Behavior on Collision Enter 2D

Sound effects for apples being dropped are added: when the user catches an apple, when the user misses an apple and when the basket gets destroyed. To the bottom of the screen, high score output is added that shows the highest number of apples the user has managed to catch.

3 Results and Discussion

In this application the assets for background, baskets, apples and the asset for the rock that is moving continuously left and right dropping apples are created (Fig. 5).

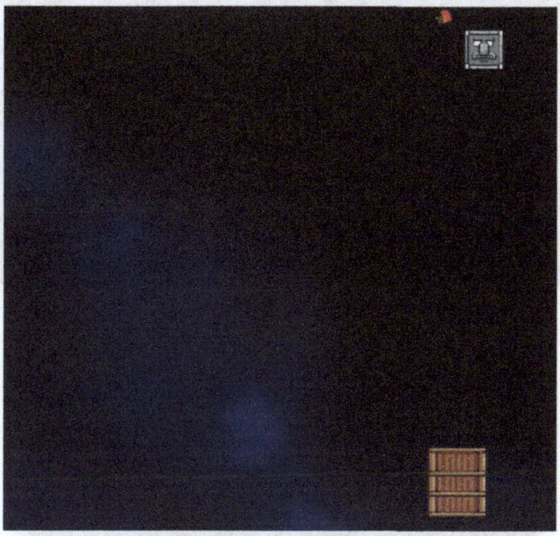

Fig. 5. Assets Background, Apple, Basket and Rock

At the bottom right corner, the current score is displayed that represents the number of apples that user has caught at a certain point of time, incrementing by one for every apple the user manages to catch. At the bottom left corner, the high score field is displayed representing the highest number of apples the user has managed to catch (Fig. 6).

A sound effect is added for each time an apple is created, for when the user catches an apple with the basket, and and for when the user fails to catch an apple and it hits the ground. From the start, the user has three baskets that are controlled by the movement of the user's head. When the user misses to catch an apple, one basket disappears and the game continues until all three baskets disappear, and then it restarts from the beginning (Fig. 7, Fig. 8, Fig. 9).

The user misses to catch an apple so one basket is lost and the game continues.

The user misses to catch another apple and only one basket is left to try to beat a high score. When the score obtained in this case is above 27 (Fig. 5), the user will set a new high score and can increase it until another apple is missed. After that, the game starts from the beginning.

We analyzed the paper in which Video game therapy (VGT) effects were tested and compared with a balance platform therapy (BPT) on balance and mobility in brain

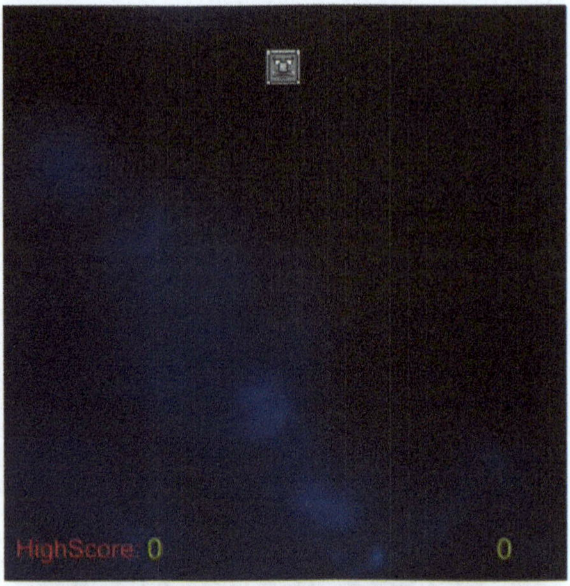

Fig. 6. Rock, High score, Current score

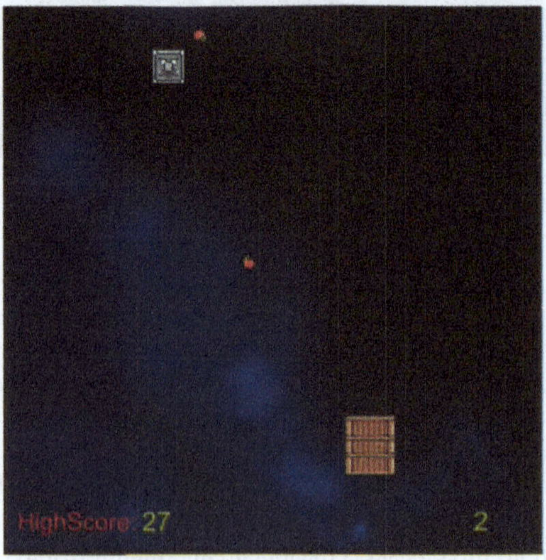

Fig. 7. Start of the game, three baskets

injury patients 32. The research enrolled 21 patients that randomly received VGT or BPT for 3 sessions per week during 6 weeks. Clinical outcome measures included the Community Balance & Mobility Scale (CB&M), the Unified Balance Scale (UBS), and the Timed Up and Go test (TUG). Video game therapy provided visual and augmented feedback. During each game, the patients exercised for 2–5 min under the supervision of a

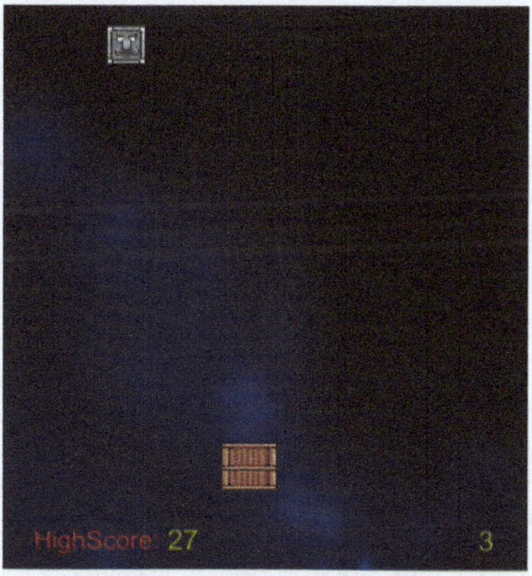

Fig. 8. Two baskets remaining

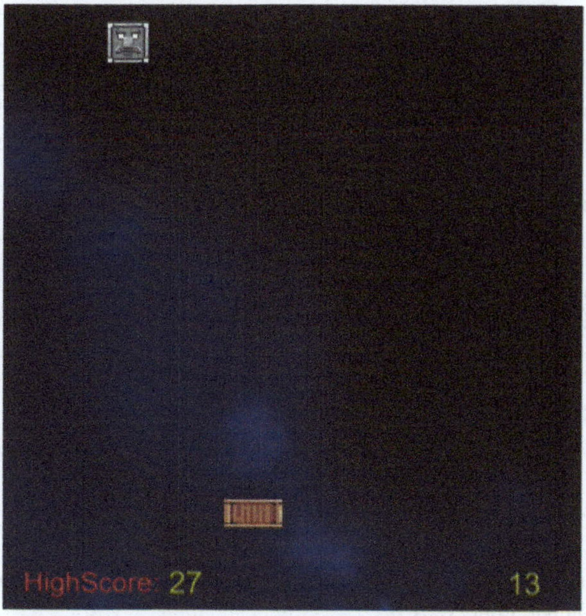

Fig. 9. One basket remaining

physiotherapist who monitored the safety of the patients and provided external feedback. The primary findings of this study indicate that dynamic balance and overall mobility improved after training, revealing significant cognitive engagement during VGT. This study, however, has several limitations: firstly, its findings cannot be generalized to the entire TBI (traumatic brain injury) population; secondly, the game was developed for a healthy population and is not adjustable for individuals with cognitive and balance disorders. Our game, on the other hand, was developed with the idea to be used by patients and specifically by patients with balance disorders. It is simple enough to be used daily during the rehabilitation process, while also providing doctors with information on the patient's progress and engagement level.

4 Conclusions

Approximately one in five elderly persons experiences annual problems with dizziness or balance. Given the significant prevalence and negative effect of balance problems on daily activities in the elderly, balance disorders merit special attention, especially considering an aging population. Balance is an important risk factor for falls and is affected by the progressive loss of sensorimotor functioning with increasing age. Deficits in proprioception, vision, vestibular sense, muscle function and reaction time contribute to balance disorders and the risk of falls in older people. Balance disorders manifest in poor performances in standing, leaning, stepping, and walking. The main goal of our work is to help doctors in analyzing patients' improvement in balance. By playing the game, which is simple enough for everyone, patients can find it engaging as they try to beat their high scores while simultaneously improving their disorders. Future work can focus on a more detailed analysis of patients' results by analyzing their reactions through different levels of the game, utilizing different scores achieved, and validating the status of disorder in order to obtain better outcomes in their treatment.

Acknowledgements. This research is funded by the European Union, under the Horizon Europe programme, GA number 101057747 (TeleRehaB project). Views and opinions expressed are however those of the author(s) only and do not necessarily reflect those of the European Union or the granting authority: European Health and Digital Executive Agency (HADEA). Neither the European Union nor the granting authority can be held responsible for them. Authors also acknowledges the funding by the Ministry of Science, Technological Development and Innovation of the Republic of Serbia, contract number [451–03-65/2024–03/200107 (Faculty of Engineering, University of Kragujevac)].

References

1. Wu, H.K., et al.: Current status, opportunities and challenges of augmented reality in education. Comput. Educ. (2013)
2. Eisenmenger, L.B. et al.: Application of 3-dimensional printing in a case of osteogenesis imperfecta for patient education, anatomic understanding, preoperative planning, and intraoperative evaluation. World Neurosurg. (2017)
3. Bragard, I., et al.: A virtual patient to improve doctor-patient communication: reality or fiction?. Rev. Med. Liege (2018)
4. Topol, E.J.: High-performance medicine: the convergence of human and artificial intelligence. Nat. Med. **25** (2019)

5. McCarthy, J.: What is artificial intelligence (2007)
6. Hamet, P., et al.: Artificial intelligence in medicine. Metabolism (2017)
7. Attaran, M., et al.: Machine learning: the new 'Big Thing' for competitive advantage. Int. J. Knowl. Eng. Data Min. (2018)
8. Davenport, T., et al.: The potential for artificial intelligence in healthcare. Future Health (2019)
9. Hashimoto, D.A., et al.: Artificial intelligence in surgery: promises and perils. Ann. Surg. **268** (2018)
10. Bracq, M.S., et al.: Virtual reality simulation in nontechnical skills training for healthcare professionals: a systematic review. Simul. Health (2019)
11. Cao, C., et al.: Virtual or augmented reality to enhance surgical education and surgical planning. Thorac. Surg. Clin. **29** (2019)
12. Rahman, R., et al.: Head-mounted display use in surgery: a systematic review. Surg. Innov. (2020)
13. Niazi, M.K.K., et al.: Digital pathology and artificial intelligence. Lancet Oncol. (2019)
14. Weiss, J., et al.: Machine learning for treatment assignment: improving individualized risk attribution. AMIA Annu. Symp. Proc. (2015)
15. Poplin, R., et al.: Prediction of cardiovascular risk factors from retinal fundus photographs via deep learning. Nat. Biomed. Eng. (2018)
16. Arksey, H., et al.: Scoping studies: towards a methodological framework. Int. J. Soc. Res. Methodol. (2005)
17. Perin, A., et al.: Informed consent through 3D virtual reality: a randomized clinical trial. Acta Neurochir. (2020)
18. Lexell, J., et al.: What is the cause of the ageing atrophy? total number, size and proportion of different fiber types studied in whole vastus lateralis muscle from 15 to 83-year-old men. J. Neurol. Sci. (1988)
19. Fitzpatrick, R., et al.: Proprioceptive, visual and vestibular thresholds for the perception of sway during standing in humans. J. Physiol. (Lond) (1994)
20. Baloh, R., et al.: Disorders of the vestibular system (1996)
21. Halmagyi, G.M., et al.: Assessment and treatment of dizziness. J. Neurol. Neurosurg. Psychiatry (2000)
22. Fife, T.D., et al.: Disequilibrium of unknown cause in older people. Ann. Neurol. (1993)
23. Herdman, S.J., et al.: Falls in patients with vestibular deficits. Am. J. Otol. (2000)
24. Katsarkas, A.: Dizziness in aging: a retrospective study of 1194 cases. Otolaryngol. Head Neck Surg. (1994)
25. Johnsson, L.G.: Degenerative changes and anomalies of the vestibular system in man. Laryngoscope (1971)
26. Richter, E.: Quantitative study of human Scarpa's ganglion and vestibular sensory epithelia. Acta Otolaryngol. (Stockh) (1980)
27. Tian, J.R., et al.: Effect of aging on the human initial interaural linear vestibulo-ocular reflex. Exp. Brain Res. (2002)
28. Yen, C.Y., Lin, K.H., Hu, M.H., Wu, R.M., Lu, T.W., Lin, C.H.: Effects of virtual reality–augmented balance training on sensory organization and attentional demand for postural control in people with Parkinson disease: a randomized controlled trial. Phys. Ther. **91**(6), 862–874 (2011)
29. Santana, C.M.F.D., et al.: Efeitos do tratamento com realidade virtual não imersiva na qualidade de vida de indivíduos com Parkinson. Rev. Bras. Geriatria Gerontologia **18**, 49–58 (2015)
30. Loureiro, A.P.C., Ribas, C.G., Zotz, T.G.G., Chen, R., Ribas, F.: Viabilidade da terapia virtual na reabilitação de pacientes com doença de Parkinson: estudo-piloto. Fisioter. mov 659–666 (2012)

31. Lai, C.H., Peng, C.W., Chen, Y.L., Huang, C.P., Hsiao, Y.L., Chen, S.C.: Effects of interactive video-game based system exercise on the balance of the elderly. Gait Posture **37**(4), 511–515 (2013)
32. Straudi, S., et al.: The effects of video game therapy on balance and attention in chronic ambulatory traumatic brain injury: an exploratory study. BMC Neurol. **17**, 1–7 (2017)

Temperature as a Factor Shaping Dissolved Oxygen in the Danube River

Irena Todorović[1(✉)], Branislav Kisačanin[2,3,4], and Milan Stojković[2]

[1] Department of Hydroecology and Water Protection, Institute for Biological Research "Siniša Stanković", National Institute of the Republic of Serbia, University of Belgrade, Bulevar despota Stefana 142, 11060 Belgrade, Serbia
irena.todorovic@ibiss.bg.ac.rs

[2] The Institute for Artificial Intelligence R&D of Serbia, Fruškogorska 1, 21000 Novi Sad, Serbia
{branislav.kisacanin,milan.stojkovic}@ivi.ac.rs

[3] Nvidia Corporation, Santa Clara, CA, USA

[4] Faculty of Technical Sciences, University of Novi Sad, 21000 Novi Sad, Serbia

Abstract. The health of terrestrial waters is of utmost importance for the sustenance of aquatic ecosystems, and dissolved oxygen (DO) represents one of the key indicators used to evaluate water health. In this study, the objective was to predict DO levels in the Danube river, using data on flow velocity and temperature, and to investigate intricate relationships among these variables. Firstly, a Random Forest regression algorithm was deployed to predict DO levels using flow velocity and temperature as predictors, and SHapley Additive exPlanations (SHAP) values were calculated. Then, an autocorrelation function was integrated into the Random Forest regression algorithm, incorporating DO data from previous time steps. Autocorrelation and partial autocorrelation functions, together with partial and cross-correlation analyses were also performed. Altogether, DO levels predicted, using flow velocity and temperature, resulted in model with R^2 equal to 0.88 (MSE = 0.59; RMSE = 0.77), and SHAP values highlighted that temperature exerted a more pronounced influence on DO levels than flow velocity. When DO data from previous time steps were used, this enhanced model achieved an R^2 equal to 0.97 (MSE = 0.11; RMSE = 0.33) and DO from preceding time steps had the most substantial influence on the predictive accuracy of this model. Autocorrelation and partial autocorrelation function plots for DO, underlined its strong correlation with adjacent time steps values. Partial and cross-correlation analyses revealed that temperature exhibited a strong negative correlation with DO. In conclusion, this research provides valuable insights into the dynamics governing terrestrial water health, emphasizing the indispensable role of temperature in modulating DO levels.

Keywords: water health · dissolved oxygen · Southeast Europe · hydrology · climate change

© The Author(s), under exclusive license to Springer Nature Switzerland AG 2025
N. Filipović (Ed.): AAI 2024, LNNS 1446, pp. 135–141, 2025.
https://doi.org/10.1007/978-3-031-99201-8_14

1 Introduction

With increasing industrialization and urbanization, human activities have led to numerous problems with climate (such as global warming and sea-level rise), as well as with terrestrial (such as deforestation and soil degradation) and aquatic ecosystems (such as eutrophication and hypoxia), which may have severe consequences on life on earth [1, 2]. The health of terrestrial waters is crucial when it comes to sustenance of aquatic ecosystems, and there are several ways to examine water health status, including the use of chemical (pH, chemical oxygen demand, biological oxygen demand, dissolved oxygen, etc.) and biological indicators (plants, animals and microorganisms) [3, 4]. Dissolved oxygen (DO, chemical symbol O_2) is one of the indicators frequently used to examine water quality, and it represents the amount of oxygen that is accessible to aquatic organisms [4, 5]. Each aerobic organism has its own lowest threshold level for DO that is needed for vital processes and, if DO levels are below this threshold, the consequences are detrimental [6]. Main sources of dissolved oxygen in the water are atmospheric exchange and photosynthesis by plants and algae, while respiration, decomposition and various chemical reactions consume oxygen [7]. Given the importance of DO in the aquatic ecosystems, the objective of this study was to predict DO levels, using data on flow velocity and temperature as predictors, and to assess the impact of flow velocity and temperature on DO in the Danube river.

2 Material and Methods

2.1 Data Description

In this study, we used daily data spanning from 2012 to 2021 on flow velocity (m^3/s) and temperature (°C) of the river Danube (measuring station Novi Sad; Fig. 1), obtained from the Republic Hydrometeorological Service of Serbia (RHMZ) website, and data on DO (mg/L), obtained from the Serbian Environmental Protection Agency (SEPA) website. All data were split into the training and testing datasets, the first containing 80% of the data, and the latter containing the remaining 20% of the data.

Fig. 1. Location of the study area: the Danube river, measuring station Novi Sad

2.2 Methods

All the analysis were performed in Python environment version 3.11.5. Random Forest (RF) regression algorithm, which is supervised algorithm that uses decision trees to predict target variables (Fig. 2), was deployed to predict DO levels using flow velocity and temperature as predictors. R-squared values (R^2), Mean Squared Error (MSE) and Root Mean Squared Errors (RMSE) were calculated. Subsequently, SHapley Additive exPlanations (SHAP) values were calculated to explain the output of the RF regression algorithm. Building upon this foundation, an autocorrelation function was integrated into the RF regression algorithm, incorporating DO data from previous time steps, and SHAP values were also calculated. Finally, autocorrelation and partial autocorrelation function plots for DO, together with partial and cross-correlation analyses of all the studied variables were performed.

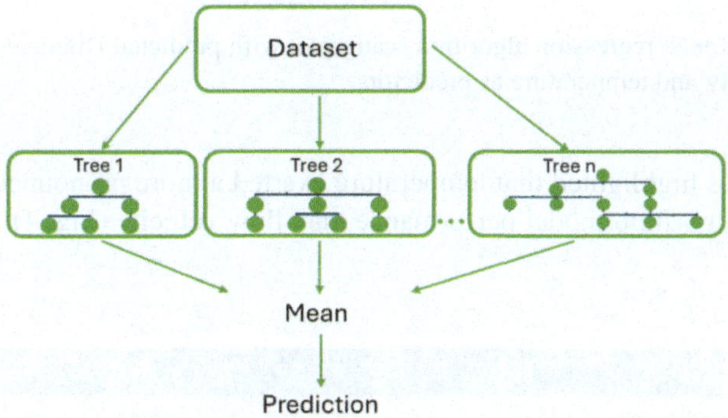

Fig. 2. Random Forest regression algorithm

3 Results

3.1 Prediction of Dissolved Oxygen Using Flow Velocity and Temperature as Predictors

When a RF regression algorithm was deployed to predict DO levels using flow velocity and temperature as predictors, the obtained model resulted in R^2 equal to 0.85, with MSE equal to 0.59, and RMSE equal to 0.77 (Fig. 3).

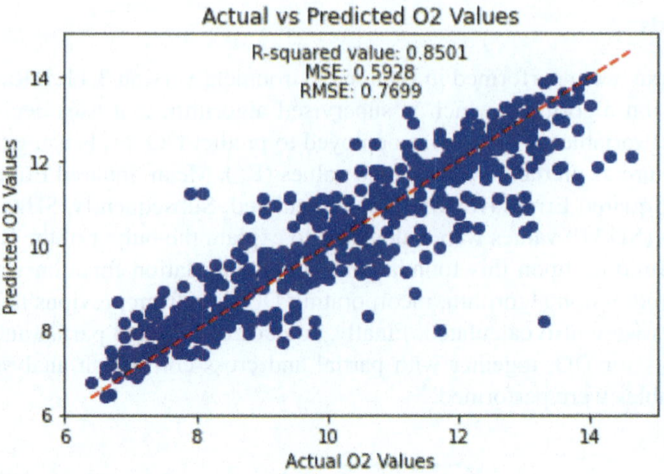

Fig. 3. Random forest regression algorithm scatterplot with predicted Dissolved oxygen values, using flow velocity and temperature as predictors

SHAP values highlighted that temperature exerted a more pronounced influence on RF regression algorithm model performance than flow velocity (Fig. 4).

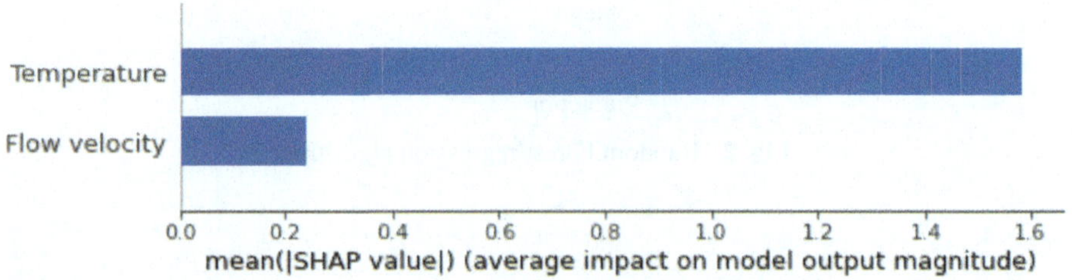

Fig. 4. SHAP values of flow velocity and temperature's influence on the RF regression algorithm model

3.2 Prediction of Dissolved Oxygen Using Dissolved Oxygen from the Preceding Time Steps, Flow Velocity and Temperature as Predictors

In the next step, an autocorrelation function was integrated into the Random Forest regression algorithm, incorporating DO data from previous time steps. This led to an enhanced model that achieved an R^2 equal to 0.97, with MSE equal to 0.11 and RMSE equal to 0.33 (Fig. 5).

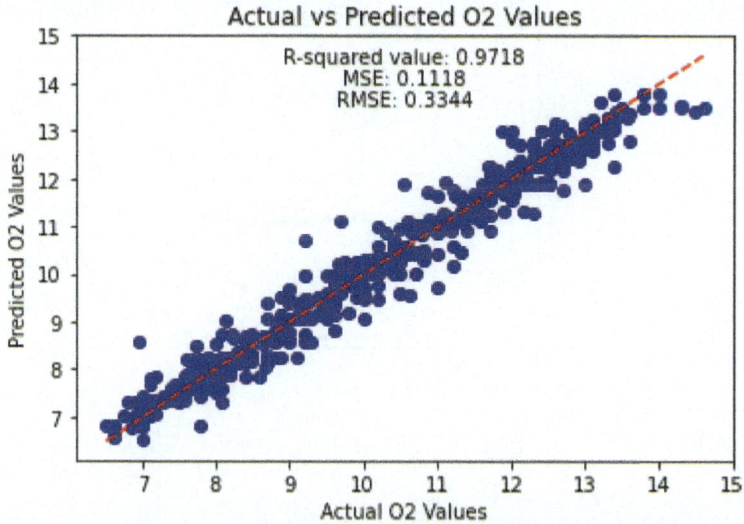

Fig. 5. Random forest regression algorithm scatterplot with predicted Dissolved oxygen using Dissolved oxygen from the preceding time steps, flow velocity and temperature as predictors

SHAP values further elucidated that DO from preceding time steps had the most substantial influence on the predictive accuracy of the model, followed by temperature and flow velocity (Fig. 6).

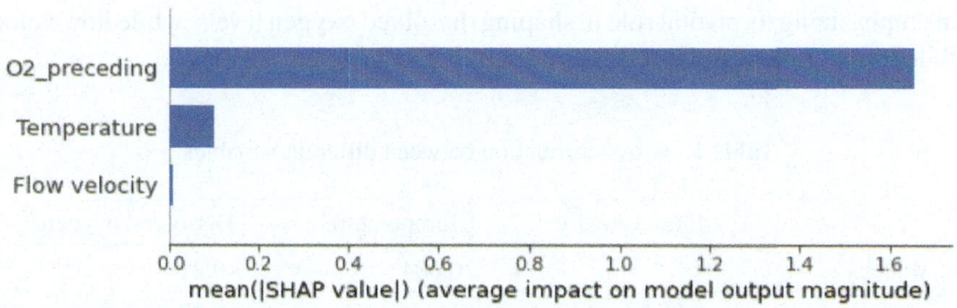

Fig. 6. SHAP values of Dissolved oxygen from the preceding time steps, flow velocity and temperature's influence on the model

3.3 Autocorrelation, Partial Autocorrelation, Partial Correlation and Cross-Correlation Analyses

Autocorrelation and partial autocorrelation function plots for DO, underlined its strong correlation with adjacent time steps (Fig. 7).

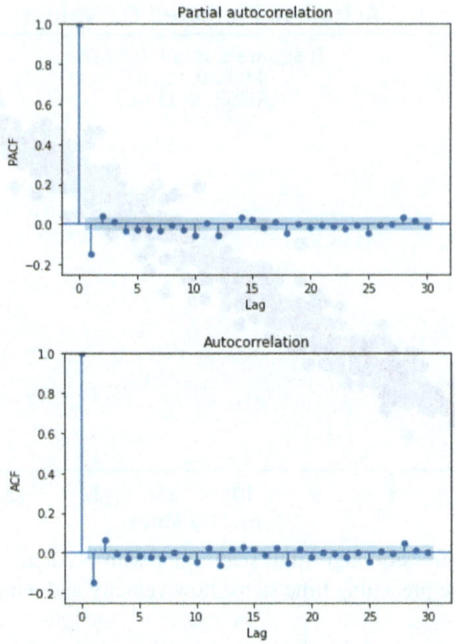

Fig. 7. Autocorrelation and partial autocorrelation function plots

Cross-correlation analyses of the studied variables revealed distinct relationships between them. Notably, temperature exhibited a strong negative correlation with DO, again emphasizing its pivotal role in shaping dissolved oxygen levels, while flow velocity exhibited moderate negative correlation with DO (Table 1).

Table 1. Cross-correlation between different variables

	Flow velocity	Temperature	Dissolved oxygen
Flow velocity	1	0.084	-0.21
Temperature	0.084	1	-0.89
Dissolved oxygen	-0.21	-0.89	1

Partial correlation between the different variables again confirmed a strong negative correlation between temperature and DO (Table 2).

Table 2. Partial correlation between different variables

	Flow velocity	Temperature
Flow velocity		−0.226
Dissolved oxygen	−0.295	−0.891

4 Conclusions

Altogether, this research outlined the importance of temperature and flow velocity's influence on DO levels in the water and showed that they can be used as predictors to anticipate DO levels. A model that can predict DO levels in the Danube river was trained, achieving an R^2 equal to 0.97. In conclusion, this research contributes valuable insights into the dynamics governing terrestrial water health, emphasizing the indispensable role of temperature in modulating DO levels. Understanding complex interrelationships between different environmental factors is crucial for effective water management and conservation strategies. In the face of escalating environmental challenges, as global temperatures continue to rise due to climate change, safeguarding the quality of terrestrial waters becomes increasingly imperative.

Acknowledgement. This paper results from project REmote WAter quality monitoRing anD IntelliGence – REWARDING [grant number 6707], supported by the Science Fund of the Republic of Serbia.

References

1. Webb, N.P., Marshall, N.A., Stringer, L.C., Reed, M.S., Chappell, A., Herrick, J.E.: Land degradation and climate change: building climate resilience in agriculture. Front. Ecol. Environ. **15**(8), 450–459 (2017)
2. Howarth, R., et al.: Coupled biogeochemical cycles: eutrophication and hypoxia in temperate estuaries and coastal marine ecosystems. Front. Ecol. Environ. **9**(1), 18–26 (2011)
3. Zaghloul, A., Saber, M., Gadow, S., Awad, F.: Biological indicators for pollution detection in terrestrial and aquatic ecosystems. Bull. Natl. Res. Cent. **44**(1), 1–1 (2020)
4. Zaghloul, A., Saber, M., El-Dewany, C.: Chemical indicators for pollution detection in terrestrial and aquatic ecosystems. Bull. Natl. Res. Cent. **43**, 1–7 (2019)
5. Bozorg-Haddad, O., Delpasand, M., Loáiciga, H.A.: Water quality, hygiene, and health. In: Economical, political, and social issues in water resources, Elsevier (2021)
6. He, H., Boehringer, T., Schäfer, B., Heppell, K., Beck, C.: Analyzing spatio-temporal dynamics of dissolved oxygen for the River Thames using superstatistical methods and machine learning. Sci. Rep. **14**(1), 21288 (2024)
7. Mader, M., Schmidt, C., van Geldern, R., Barth, J.A.: Dissolved oxygen in water and its stable isotope effects: a review. Chem. Geol. **473**, 10–21 (2017)

Recurrent Neural Networks for Energy Management Systems: A Case Study

Jelena Joksimović(✉), Jure Kos, Krištof Debeljak, and Janez Povh

Rudolfovo Science and Technology Center, Podbreznik 15, 8000 Novo Mesto, Slovenia
{jelena.joksimovic,jure.kos,kristof.debeljak,
janez.povh}@rudolfovo.eu

Abstract. Hybrid energy systems, which integrate diverse energy sources including solar power plants, supercapacitors, UPS batteries, generators, hydrogen cells, and the grid, represent sophisticated yet highly promising approaches to enhancing energy efficiency, reducing operational costs, and supporting renewable and grid-independent initiatives. The inherent complexity of these systems necessitates the energy management strategy (EMS) capable of judiciously allocating resources in line with demand forecasts. A critical component of devising an effective task scheduling system within this framework is the ability to generate precise forecasts of energy production from renewable sources, solar power in this case. This paper showcases the deployment and comparative evaluation of two advanced deep learning models, Long Short-term Memory Recurrent Neural Networks (LSTMs) and Bidirectional Long Short-term Memory Networks (BiLSTMs), and our proposed Ensemble model, which averages the forecasts from LSTM and BiLSTM models, developed at our Laboratory for Energy Management ({LabE). Our primary goal is to predict solar power output for three days at 15-min intervals. Incorporating thirteen weather features, our findings reveal that the proposed models perform well in predicting energy production data, with the Ensemble predictions showing the best performance for 15-min interval forecasts spanning three days.

Keywords: solar power · energy management · recurrent neural networks · LSTMs · BiLSTMs · deep learning

1 Introduction

Energy management (EM) is becoming a growing component of the business strategy of industrial companies [1], which are looking more closely at their energy profiles to identify opportunities for cost reduction, decarbonization, and resilience. They are increasingly adopting energy management technologies and considering onsite resources, such as renewable energy and energy storage, which require intelligent decision-making to optimize energy usage and reduce costs, this integration of renewable energy sources, energy storage systems, and smart grid technologies has significantly increased data volume and complexity in EMS, allowing for Machine learning (ML) and Artificial

© The Author(s), under exclusive license to Springer Nature Switzerland AG 2025
N. Filipović (Ed.): AAI 2024, LNNS 1446, pp. 142–157, 2025.
https://doi.org/10.1007/978-3-031-99201-8_15

Intelligence (AI) to be applied to the EM systems to address these challenges and to make them more efficient and reliable. ML algorithms can be used to analyze large amounts of data from smart meters, different energy sources, and consumer patterns, to identify trends, predict energy demand and supply, and identify areas of inefficiency that can be improved. Deep Learning (DL) techniques can be used to model complex relationships between variables and make accurate predictions. Neural networks can be used to predict energy consumption based on historical data and weather patterns [2, 3].

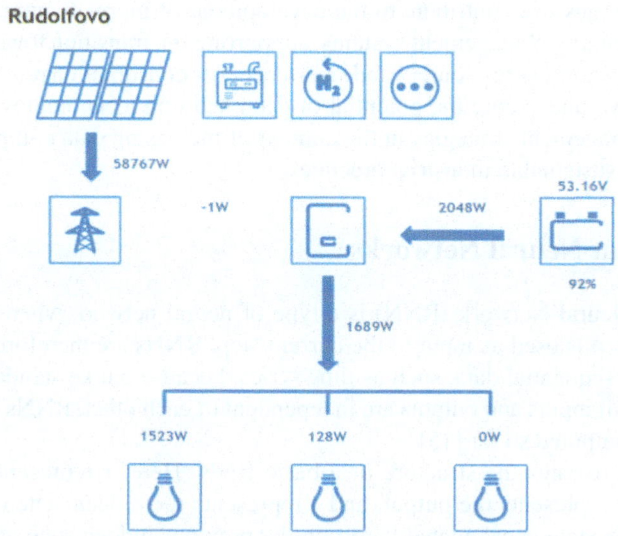

Fig. 1. {LabE energy flow diagram.

Recent recurrent neural network (RNN)-based models have demonstrated favorable performance in electric load forecasting [4].

LSTM is a popular improved model of artificial recurrent neural networks (RNN) that can model sequential or temporal aspects of data. Unlike the original RNN model, LSTM can solve the problem of gradient disappearance or explosion that often occurs when training continuous data [5–9].

LSTM networks-based models are proposed to capture periodicity in energy load consumption prediction. The method uses environmental and temporal variables related to industrial energy consumption data as model inputs for training and prediction, which had good generalization ability compared with traditional models such as ARMA and back propagation [10]. Bidirectional LSTMs (BiLSTMs) enable additional training by traversing the input data twice, left-to-right, and right-to-left. It was observed that BiLSTM models provide better predictions compared to ARIMA and LSTM models [11]. Within EMS, BiLSTM was, among others, proven useful in energy prediction and consumption optimization for individual home appliances [12], wind energy production prediction [13], and electric load forecasting in Energy Cell-Tissue Systems [14].

In Fig. 1 we see the integrated energy system in {LabE, showcasing the interaction between energy producers: solar power plant, grid, battery system, generator, hydrogen cells, and super-capacitor, and energy consumers (production line in the factory). This overview demonstrates the need for optimized energy management strategies employed to enhance efficiency within industrial operations.

In this paper, we advance the field of energy management by leveraging RNNs, specifically LSTM and BiLSTM models for precise solar power production forecasting of {LabE. Utilizing comprehensive sets of real data that encompass various environmental and temporal variables, we aim to enhance the accuracy and reliability of energy predictions within industrial operations. By integrating these advanced neural network techniques, we contribute to the development of more efficient, resilient, and cost-effective energy management systems, supporting the transition toward greater utilization of renewable energy sources and smarter energy consumption practices. Through our research, we underscore the potential of deep learning technologies to revolutionize energy management strategies in the context of increasing data complexity and the imperative for sustainable industrial practices.

2 Recurrent Neural Networks

A Recurrent Neural Network (RNN) is a type of neural network where the output of the previous step is used as input to the current step. RNNs are therefore very suitable for processing sequential data, such as time series, because unlike standard neural networks, where all inputs and outputs are independent of each other, RNNs can remember information computed so far [15].

Figure 2 illustrates the structure of a basic RNN. Here, x represents the input to the network, o represents the output, and s represents the hidden state of the network (memory). This state is calculated based on the previous hidden state and the input at the current step. The function used for this calculation is the activation function.

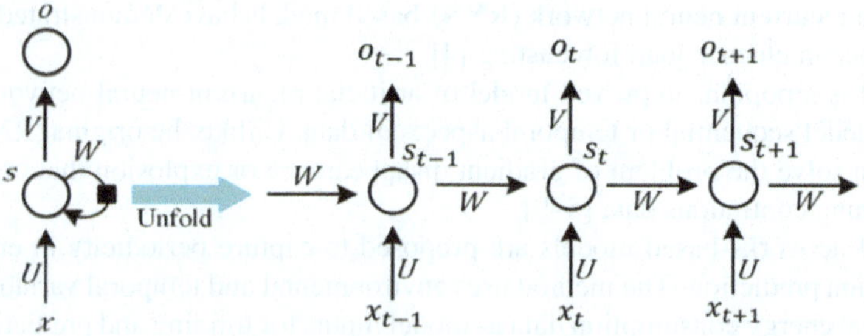

Fig. 2. Structure of a basic RNN.

More specifically,

- x_t = input at time t
- s_t = hidden state at time t; it is calculated based on the input vector and the previous hidden state:

$$s_t = f(U \cdot x_t + \mathrm{W} \cdot s_{t-1}) \quad (1)$$

where f is the activation function, with many alternatives such as *sigmoid*, *tanh* functions, and *ReLU* [16].

- o_t = output at time t :

$$o_t = f(V \cdot s_t) \tag{2}$$

- U, V, and W are the weights of the hidden layer, the output layer, and the hidden state, respectively.

RNN is one of the most successful and modern neural classes that can retain information. In this paper, we propose an RNN-based prediction method for the time series problem prediction of Solar Energy Production (kWh) within the 15-min granularity. However, classical RNNs suffer from the problem of vanishing gradients due to short memory [17]. Gradients are the values in neural networks which are used to update its weights. When RNN back propagates through time, the gradient shrinks, and if the value of the gradient becomes too small then it will not be able to contribute too much to the learning. LSTM RNNs come as an effective solution for combating vanishing gradients by using memory cells.

2.1 Long Short-Term Memory (LSTM)

LSTM is one of the very successful variants of state-of-art RNN and is widely used in the context of time series data. Even though RNNs provide good accuracy, they have been found to underperform for long-term memory as RNNs are unable to use information from the distant past. Also, LSTM can learn patterns with long dependencies when compared with traditional RNNs. A memory cell, which is the main reason LSTM is superior in these tasks, is composed of four units: an input gate, an output gate, a forget gate, and a self-recurrent neuron. The gates control the interactions be-tween neighboring memory cells and the memory cell itself, as illustrated in Fig. 3. Whether the input signal can alter the state of the memory cell is controlled by the input gate. On the other hand, the output gate can control the state of the memory cell on whether it can alter the state of the other memory cell. In addition, the forget gate can choose to remember or forget its previous state [5, 6].

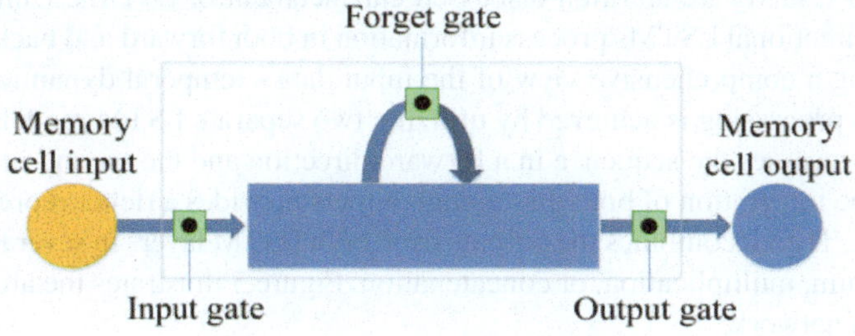

Fig. 3. LSTM memory cell [16].

Figure 4 shows an LSTM unit, which describes how the value of each gate is updated. More specifically:

- x_t is the input vector to the memory cell at time t
- W_i, W_f, $W_c$andW_o are weight matrices
- b_i, b_f, $b_c$andb_o are bias vectors
- h_t is the value of the memory cell at time t
- $i_t$and$\tilde{C}_t$ are values of the input gate and the candidate state of the memory cell at time t, respectively, which can be formulated as:

$$i_t = \sigma(W_i \cdot [h_{t-1}, x_t] + b_i) \tag{3}$$

$$\tilde{C}_t = tanh(W_c \cdot [h_{t-1}, x_t] + b_c) \tag{4}$$

- $f_t$andC_t are values of the forget gate and the state of the memory cell at time t, respectively, which can be formulated as:

$$f_t = \sigma(W_f \cdot [h_{t-1}, x_t] + b_f) \tag{5}$$

$$C_t = i_t * \tilde{C}_t + f_t * C_{t-1} \tag{6}$$

- o_t and h_t are values of the output gate and the value of the memory cell at time t, respectively, which can be formulated as:

$$o_t = \sigma(W_o \cdot [h_{t-1}, x_t] + b_o) \tag{7}$$

$$h_t = o_t * \tanh(C_t), \tag{8}$$

$*$ Where * denotes the Hadamard product (element-wise multiplication of vectors). The inclusion of additional training has resulted in the model extension of LSTM known as Bidirectional LSTM (BiLSTM). This model trains the input time series data twice through forward and backward directions [18].

2.2 Bidirectional Long Short-Term Memory (BiLSTM)

Bidirectional LSTMs are an often-discussed enhancement of LSTMs. Unlike standard LSTMs, Bidirectional LSTMs process information in both forward and backward directions, offering a comprehensive view of the input data′s temporal dynamics [19]. This bidirectional processing is achieved by utilizing two separate LSTM models, where the first model processes the sequence in a forward direction and the second in a backward direction. The integration of both directional outputs provides a richer representation of the sequence. Then it combines the outputs from both LSTM layers in several ways, such as average, sum, multiplication, or concatenation. Figure 5 illustrates the architecture of the BiLSTM network.

Finally, averaging received predictions from LSTM and BiLSTM models rained on the same parameters, we get a new *Ensemble* model, that shows better accuracy compared to the former two and the results are shown in section Results.

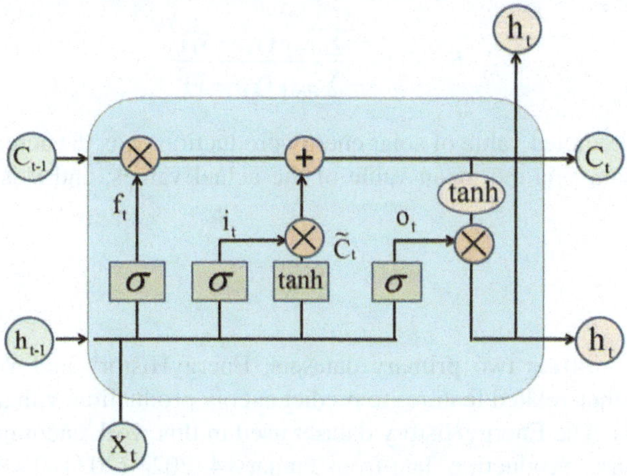

Fig. 4. The structure of LSTM units.

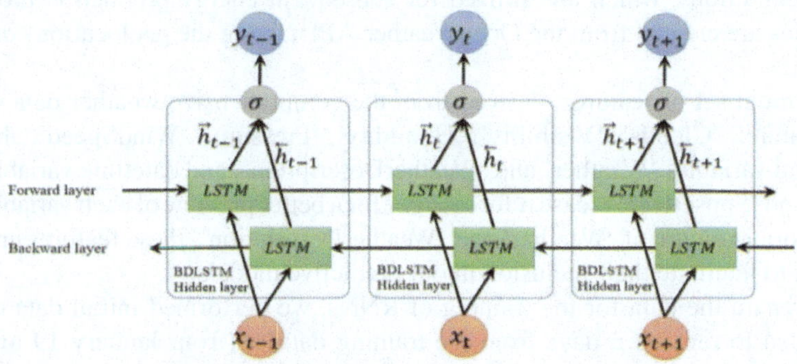

Fig. 5. Structure of BiLSTM neural network.

2.3 Evaluation Metrics

For model performance comparison, we use mean squared error (MSE), root mean square error (RMSE), and mean absolute error (MAE). Additionally, we evaluate the goodness-of-fit using the R^2 measure. These indicators are defined as:

$$MSE = \frac{1}{n} * \sum_{i=1}^{n} (\mathrm{y_i} - \widehat{y_i})^2 \tag{9}$$

$$RMSE = \sqrt{\frac{1}{n} * \sum_{i=1}^{n} (\mathrm{y_i} - \widehat{y_i})^2} \tag{10}$$

$$MAE = \frac{1}{n} * \sum_{i=1}^{n} |\mathrm{y_i} - \widehat{y_i}| \tag{11}$$

$$R^2 = 1 - \frac{\sum_{i=1}^{n} (y_i - \widehat{y_i})^2}{\sum_{i=1}^{n} (y_i - \overline{y})^2} \tag{12}$$

where $\widehat{y_i}$ is the predicted value of solar energy production, y_i is the actual value of solar energy production, $\overline{y}$ is the mean value of the actual values, and n is the number of samples.

3 Data

This case study utilizes two primary datasets: EnergyHistory and WeatherForecast, focusing on weather-related features to predict energy production with a temporal granularity of 15 min. The EnergyHistory dataset used in this work encompasses historical weather and energy production data from January 4, 2024 (2024–01-04 00:00:00), to March 31, 2024 (2024–03-31 23:45:00), totaling 8270 instances. This dataset serves as the basis for our training and evaluation. The WeatherForecast dataset provides future weather conditions, which are utilized for subsequent energy production predictions. Both tables are created from the OpenWeather API[1], using the geolocation component of {LabE.

The initial set of features, derived from the comprehensive weather data, includes 'Temperature', 'Clouds', 'Visibility', 'Humidity', 'Pressure', 'WindSpeed', along with categorical variables 'Weather' and 'WeatherDescription', and datetime variables 'SunriseDT' and 'SunsetDT'. Please refer to Fig. 7 for a better preview of each variable. Given the categorical nature of 'Weather' and 'WeatherDescription', these features underwent encoding to facilitate their inclusion in our predictive models.

To prepare the data for the training of RNNs, we performed initial data cleaning. We decided to remove 6 days from the training dataset (from January 19 to January 24). The reasons behind this decision can be seen in Fig. 6. It reveals a week without any energy production. Upon closer inspection, it was found that during these 6 days, the connection between the Solar Power plant and our battery storage system was off, introducing noise that is not indicative of regular occurrences. This figure also shows an increase in solar power production from January to February, with March exhibiting the most dark red color. This indicates that, with only this part of the year represented, we can′t expect the models to perform optimally. As different seasons are incorporated into our model learning, we anticipate more reliable predictions. Ideally, one would have at least a full year′s worth of temporal data.

3.1 Feature Selection

To enhance the model′s temporal understanding and prediction capabilities, we engineered additional time-based features: 'hour_sin', 'hour_cos', 'sin_day_of_month', 'cos_day_of_month', 'sin_day_of_year' and 'cos_day_of_year'. These features, derived from the cyclical nature of time (hours and days), enable the model to capture daily and seasonal patterns in energy consumption more effectively.

---

[1] https://openweathermap.org/api

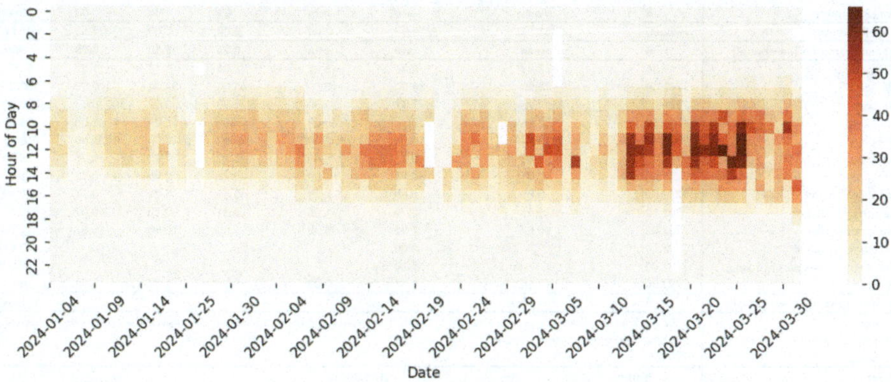

Fig. 6. Heatmap of Solar Energy Production (kWh). On the x-axis we have dates and on the y-axis, we have hours in a day.

To capture the cyclic nature of time within a day, we calculate the sine and cosine transformations of the hour of the day as follows:

$$hour\_sin = sin(\frac{2\pi \cdot hour}{24}) \tag{13}$$

$$hour\_cos = cos(\frac{2\pi \cdot hour}{24}) \tag{14}$$

where *hour* represents the hour component of the timestamp.

Similarly, to represent the cyclic behavior of days within a month, the sine and cosine transformations of the day of the month are calculated:

$$sin\_day\_of\_month = sin\left(\frac{2\pi \cdot day}{30}\right) \tag{15}$$

$$cos\_day\_of\_mont = cos\left(\frac{2\pi \cdot day}{30}\right) \tag{16}$$

where *day* denotes the day component of the timestamp.

Finally, sine and cosine transformations of the day of the year:

$$sin\_day\_of\_year = sin\left(\frac{2\pi \cdot day\_of\_year}{365}\right) \tag{17}$$

$$cos\_day\_of\_year = cos\left(\frac{2\pi \cdot day\_of\_year}{365}\right) \tag{18}$$

where *day_of_year* indicates the day's ordinal number within the year.

Furthermore, 'daylight_duration' was calculated to provide insight into the available daylight, a crucial factor for energy usage patterns.

In Fig. 7 we present the descriptive statistics of 16 features ('SunriseDT' and 'SunsetDT', the two datetime features, are removed, as they are used primarily for calculation of new features, such as 'daylight_duration', 'is_daylight' etc.). Throughout the

Target Variable	n	mean	std	min	Q1	Me	Q3	max
SolarPower (W)	7702.00	6451.37	1229.43	0.00	0.00	0.00	7502.50	73900.00
Feature variables (numerical)	**n**	**mean**	**std**	**min**	**0.25**	**0.50**	**0.75**	**max**
Temperature	7702.00	7.61	5.46	-8.01	4.04	8.18	11.27	24.37
Clouds	7702.00	75.50	34.03	0.00	55.00	97.00	100.00	100.00
Visibility	7702.00	9447.75	2097.22	0.00	10000.00	10000.00	10000.00	10000.00
Humidity	7702.00	82.36	12.27	35.00	76.00	85.00	92.00	100.00
Pressure	7702.00	1016.08	10.83	990.00	1008.00	1016.00	1025.00	1040.00
hour_sin	7702.00	0.00	0.71	-1.00	-0.71	0.00	0.71	1.00
hour_cos	7702.00	0.01	0.71	-1.00	-0.71	0.00	0.71	1.00
sin_day_of_month	7702.00	0.06	0.69	-0.99	-0.59	0.21	0.74	0.99
cos_day_of_month	7702.00	-0.01	0.72	-1.00	-0.81	-0.10	0.67	1.00
daylight_duration	7702.00	10.62	1.21	8.78	9.60	10.56	11.67	12.75
Weather_encoded	7702.00		0.00					3.00
WeatherDescription_encoded	7702.00		0.00					9.00
WindSpeed	7702.00	1.56	1.34	0.03	0.79	1.16	1.74	8.03
Feature variables (categorical)	**n**							**Values**
Weather	7702.00	'broken clouds', 'clear sky' ,'scattered clouds' ,'few clouds','light rain',' moderate rain' ,' light snow',' snow',' heavy intensity						
WeatherDescription	7702.00	'rain','clear' ,' snow',' clouds'						
is_daylight	7702.00	'0', '1'						
Index	**n**	**from**	**to**	**granularity**				
Time	7702.00	'2024-01-04 00:00:00'	2024-03-31 23:45:00'	15 minutes				

Fig. 7. Descriptive statistics of features in EnergyHistory data table.

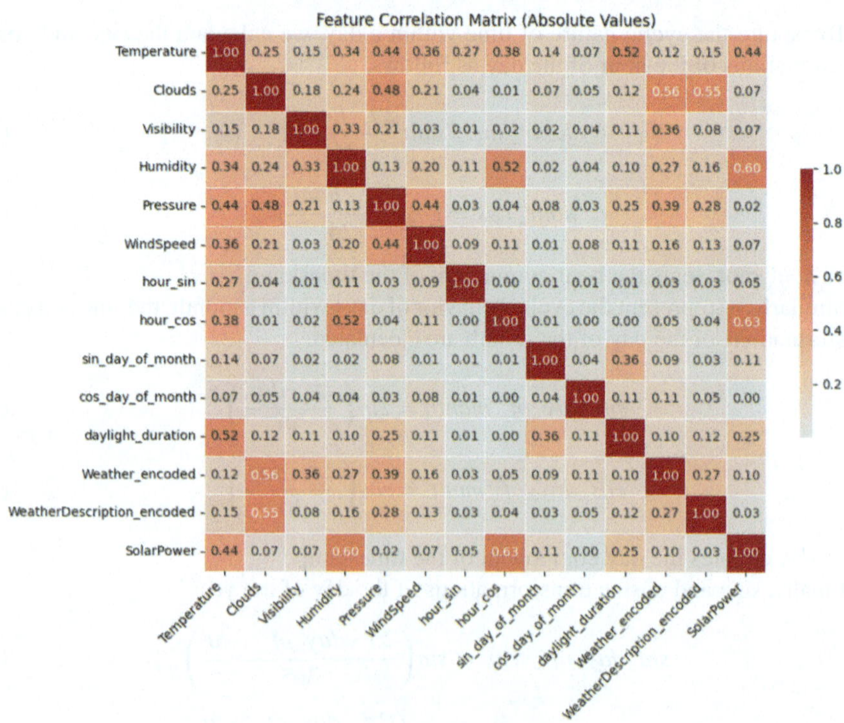

Fig. 8. Feature importance matrix. In boxes are shown absolute values of Pearson correlation coefficients between each feature, and also between them and the target variable (SolarPower). We seek features that don′t have a high correlation between themselves (higher than 0.8), as this can negatively impact the model.

feature importance analysis, we identified and also removed features that contributed minimally to the predictive performance of our models. We used the Pearson correlation coefficient to determine the linear correlation between features, and the correlation matrix is presented in Fig. 8. Additionally, we removed the features 'is_daylight' (0.83 correlation with 'cos_hour'), 'cos_day_of_year' (0.99 correlation with 'daylight_duration'), and 'sin_day_of_year' (0.97 correlation with 'cos_day_of_year'). The refined model utilizes a final set of 13 features, comprising 'Temperature', 'Clouds', 'Visibility', 'Humidity', 'Pressure', 'WindSpeed', 'hour_sin', 'hour_cos', 'sin_day_of_ month', 'cos_day_of_month', 'daylight_duration', 'Weather_encoded', and 'WeatherDescription_encoded'.

Our target variable is energy, which is initially given in Watts (W), but in our results, we will interpret it using kWh.

4 Results

4.1 Neural Architecture Search (NAS) and Models Evaluation

Now that we have decided on the feature space, we proceed to the training and evaluation of LSTM and BiLSTM models. Finding the optimal hyperparameter settings has a direct and significant impact on the model's performance [20] and for this purpose, we used the *HyperModel class* of the Keras Tuner API[2].

The LSTM model architecture, consisting of a single LSTM layer with 160 units, followed by a dense output layer, exhibited superior performance when evaluated against alternatives. Utilizing the 'ReLU' activation function and the 'adam' optimizer [21] set to a learning rate of 0.001, we trained the model for 10 epochs with a batch size of 32, incorporating an early stopping mechanism with a patience of 5, to achieve the best results.

Similarly, our exploration of neural architecture determined that a configuration of 60 units and a learning rate of 0.001 yielded optimal results for our BiLSTM model. The performance metrics, alongside the validation loss observed after each epoch, are detailed in Table 2 and depicted in Fig. 9. Given that our training dataset spans only three months, thus not encompassing all seasons, these outcomes affirm the robustness of our model's predictions.

As shown in Table 1, our chosen sequence length is 24, and the prediction time step is 1. This indicates that the model is designed to forecast energy production for the upcoming 15 min based on data from the preceding 6 h. This setup was selected after evaluating various options, proving to be the most effective for our current dataset.

However, future work should consider a broader exploration of these parameters, especially as we move towards datasets with larger volumes and 1-min granularity, where our current settings might become computationally less efficient. We adopted an 85/15 split for training and testing, and in Fig. 10, we showcase the cumulative solar energy predictions in kWh for one, two, and three days within our test set. It was observed that while the LSTM model tended to underestimate, the BiLSTM model generally overestimated, making the Ensemble approach, which averages these two

[2] https://keras.io/keras_tuner/

Table 1. The optimal parameter values of the LSTM and BiLSTM models. Sequence length and prediction time step are fixed prior to the search of the hyperparameters space.

Parameters	BiLSTM	LSTM
Sequence length	24	24
Prediction timestep	1	1
Number of units	60	160
Activation function	ReLU	ReLU
Learning rate	0.001	0.01
Batch size	32	32
Epochs	10	10

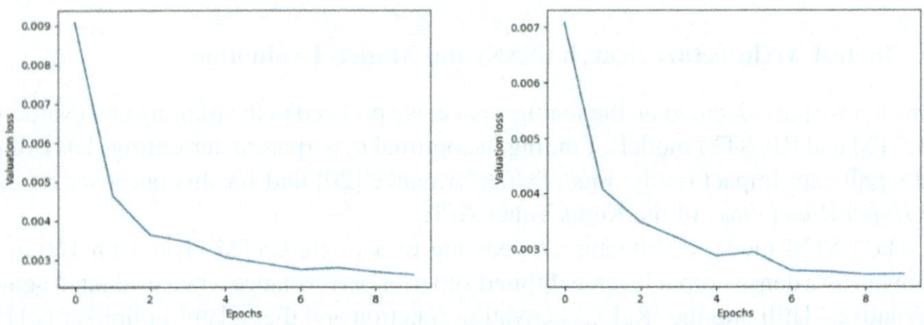

Fig. 9. Validation loss across 10 epochs for LSTM (left panel) and BiLSTM (right panel) models.

Table 2. The optimal parameter values of the LSTM and BiLSTM models. Sequence length and prediction time step are fixed prior to the search of the hyperparameters space.

	MSE	MAE	RMSE	R^2
LSTM	5.5	1.2	2.3	0.75
BiLSTM	6.5	1.4	2.5	0.70

models' predictions at 15-min intervals, the most accurate compared to actual data. As illustrated in the figure, the red and blue bars closely match and for commercial purposes, we are interested in seeing how many kWh we can expect to produce for the next day (or the next few days) so we can plan our business operations accordingly.

4.2 Forecasting the Solar Power Production on Unseen Data

Having built and assessed our models, we next apply them to the WeatherForecast data table, mentioned in section Data. This dataset constitutes new, unseen data for our models, spanning the initial three days following the final date in our EnergyHistory

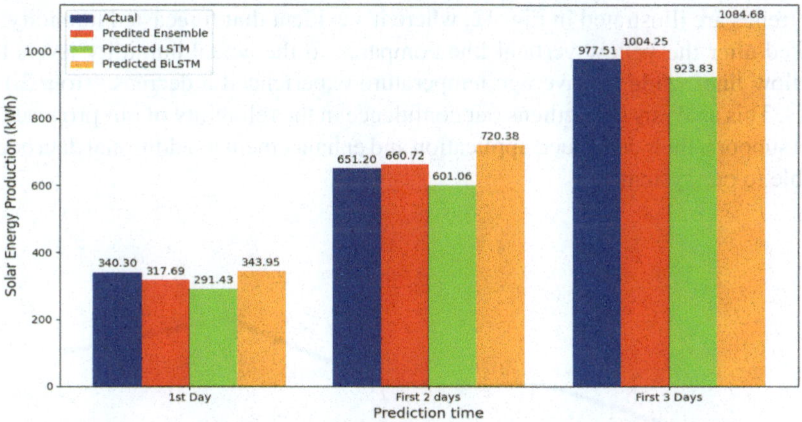

Fig. 10. Cumulative solar energy production error analysis predictions for one, two, and three days in advance.

dataset. A notable challenge with this dataset is its 3-h granularity, which we needed to adjust to the 15-min granularity required for our forecasting. To address this, we replicated the same weather forecasts across subsequent data points until the next forecast was available. For instance, if the weather is forecasted at 9 am and then at 12 pm, all intervening data points mirror the 9 am forecast until the 12 pm update. After aligning this dataset with our feature framework, we applied the same MinMax scalers used during the training phase to ensure consistency.

In Fig. 11, predictions for the period from April 1 to April 3, 2024, are displayed, showcasing forecasts from both our models and the ensemble predictions, which have been identified to yield the most precise outcomes. Observations reveal that the forecasted solar energy production was slightly below the actual figures in the preceding two days. This prompted an examination of the primary influencing features: 'Humidity', negatively correlated with our target variable (0.6), and 'Temperature', positively correlated with solar energy production (0.44). Please refer to Fig. 8 for these results.

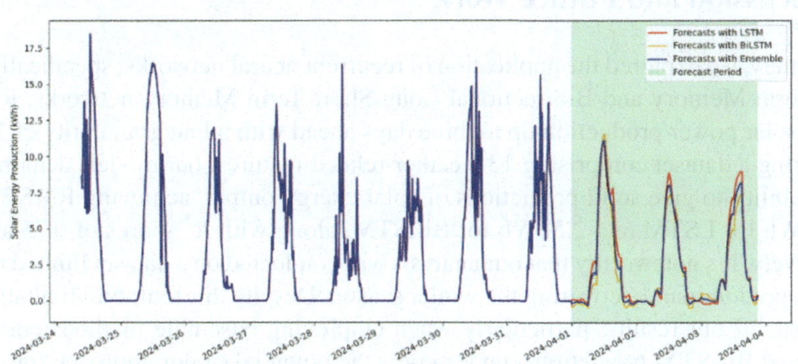

Fig. 11. Predictions on the serving dataset (unseen data).

These trends are illustrated in Fig. 12, where it′s evident that forecasted humidity levels increased after the yellow vertical line compared to the actual humidity levels before the yellow line, while the average temperature experienced a decrease from 20 to 13 degrees. This analysis strengthens our confidence in the reliability of our proposed models and supports their continued application and enhancement as additional data becomes available to our system.

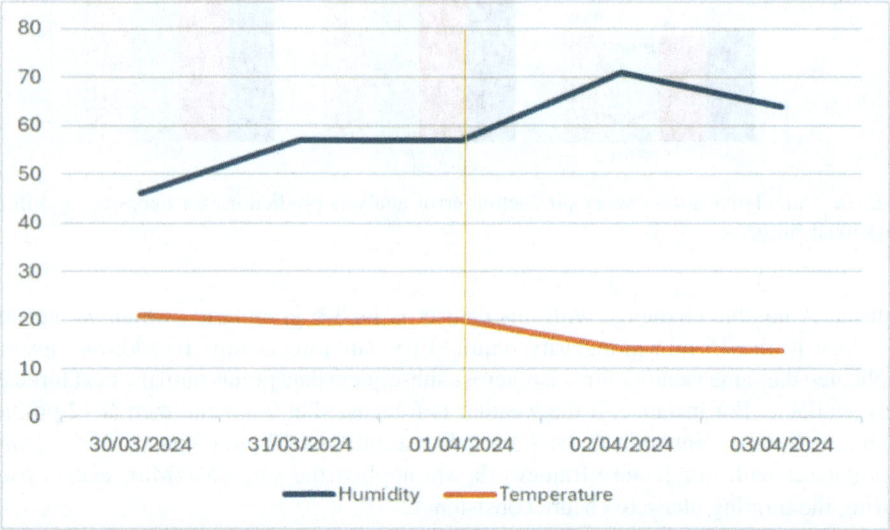

Fig. 12. Mean values of the 'Humidity' and 'Temperature' features recorded between 9 am and 3 pm, spanning from March 30th to April 3rd, 2024. The initial two days' data are derived from the training/testing dataset, while the data for the subsequent three days are sourced from the serving dataset.

5 Discussion and Future Work

In this study, we explored the application of recurrent neural networks, specifically Long Short-Term Memory and Bidirectional Long Short-Term Memory networks, for forecasting solar power production up to three days ahead with a fine granularity of 15 min. Leveraging a dataset comprising 13 weather-related features, our models demonstrated the capability to give solid predictions of solar energy output, achieving RMSE values of 2.3kWh for LSTM and 2.5kWh for BiLSTM, along with R^2 scores of 0.75 and 0.7, respectively. It′s noteworthy that our analysis was conducted on a dataset limited to three months, predominantly covering the winter season. Despite this temporal limitation, the robustness of our results, particularly when employing ensemble methods combining LSTM and BiLSTM predictions, underscores the potential of our approach for reliable solar power production forecasting. Another limitation of our analysis is the absence of the solar irradiance feature, which is one of the biggest influencers on solar output, besides temperature, humidity, and hour of the day [22, 23]. The reason for this omission is that this feature is a paid option from the OpenWeather data provider, but we have started obtaining it recently, for our future work.

Looking forward, we aim to integrate our predictive models with energy consumption forecasts and real-time electricity market pricing data for our {LabE. The objective is to enhance our capability to predict peak demand periods with high accuracy and devise optimal operational strategies for the lab. This includes the scheduling of upcoming tasks and efficient management of our energy storage systems to minimize costs and reduce reliance on grid electricity during peak times. To achieve these advanced modeling goals, we plan to utilize data with a higher resolution of 1-min granularity, drawing from historical production and consumption records. This approach promises to refine our predictive accuracy and operational efficiency, further contributing to sustainable and cost-effective energy management practices.

6 Reproducibility

The data and the code used for the analysis presented in this paper are available on *GitHub* page, URL: https://github.com/jelenajoksimovic/Energy_production_prediction.

Acknowledgements. The authors' research was co-funded by the Republic of Slovenia, the Ministry of HE, Science and Innovation, and the Slovenian Research and Innovation Agency of the European Union - NextGenerationEU through the DIGITOP project and by the Slovenian Research and Innovation Agency (ARIS) through the annual work program of Rudolfovo. The work is also funded through the business experiment Advanced Multi-objective Optimization of Energy Management - AIMED-HPC, which received funding through the FFplus project, which is financed by the European High-Performance Computing Joint Undertaking (JU) under the funding agreement No. 101163317. The JU receives support from the Horizon Europe program of the European Union.

References

1. Deloitte: Smart energy management for industrials. Deloitte Insights (2021). https://www2.deloitte.com/us/en/insights/industry/power-and-utilities/smart-energy-management.html. Accessed 25 Mar 2024
2. Jeong, H.: Predicting the output of solar photovoltaic panels in the absence of weather data using only the power output of the neighbouring sites. Sensors **23**(7), 3399 (2023). https://www.mdpi.com/1424-8220/23/7/3399
3. Jha, A.V., Appasani, B., Gupta, D.K., Ramavath, S., Khan, M.S.: Machine learning and deep learning approaches for energy management in smart grid 3.0. In: Appasani, B., Bizon, N. (eds.) Smart Grid 3.0, pp. 121–151. Springer, Cham (2023). https://doi.org/10.1007/978-3-031-38506-3_6
4. Vidyakant, J., Appasani, B., Gupta, D., Ramavath, S., Khan, M.S.: Machine learning and deep learning approaches for energy management in smart grid 3.0. Smart Grid 3.0, Springer, Cham, pp. 121–151 (2023)
5. Hochreiter, S., Schmidhuber, J.: Long Short-Term Memory. Neural Comput. **9**(8), 1735–1780 (1997). https://doi.org/10.1162/neco.1997.9.8.1735

6. Almalaq, A., Zhang, J.J.: Evolutionary deep learning-based energy consumption prediction for buildings. IEEE Access **7**, 1520–1531 (2019). https://doi.org/10.1109/ACCESS.2018.2887023
7. Kim, J., Moon, J., Hwang, E., Kang, P.: Recurrent inception convolution neural network for multi short-term load forecasting. Energy Build. **194**, 328–341 (2019). https://doi.org/10.1016/j.enbuild.2019.04.034
8. Lu, Y., Sheng, B., Fu, G., Luo, R., Chen, G., Huang, Y.: Prophet-EEMD-LSTM based method for predicting energy consumption in the paint workshop. Appl. Soft Comput. **143**, 110447 (2023). https://doi.org/10.1016/j.asoc.2023.110447
9. Cegovnik, T., Dobrovoljc, A., Povh, J., Rogar, M., Tomsic, P.: Electricity consumption prediction using artificial intelligence. CEJOR **31** (2023). https://doi.org/10.1007/s10100-023-00844-6
10. Wang, J.Q., Du, Y., Wang, J.: LSTM based long-term energy consumption prediction with periodicity. Energy **197** (2020). https://ideas.repec.org/a/eee/energy/v197y2020ics0360544220303042.html
11. Siami-Namini, S., Tavakoli, N., Namin, A.S.: The performance of LSTM and BiLSTM in forecasting time series. In: 2019 IEEE International Conference on Big Data (Big Data), pp. 3285–3292 (2019). https://doi.org/10.1109/BigData47090.2019.9005997
12. Kaur, S., Bala, A., Parashar, A.: GA-BiLSTM: an intelligent energy prediction and optimization approach for individual home appliances. Evol. Syst. (2023). https://doi.org/10.1007/s12530-023-09529-6
13. Song, D., Yu, M., Wang, Z., Wang, X.: Wind and wave energy prediction using an AT-BiLSTM model. Ocean Eng. **281**, 115008 (2023). https://doi.org/10.1016/j.oceaneng.2023.115008
14. Li, Z., Yu, F., Guan, Q.: BiLSTM network-based approach for electric load forecasting in energy cell-tissue systems. In: 2022 First International Conference on Cyber-Energy Systems and Intelligent Energy (ICCSIE), pp. 1–5 (2023). https://doi.org/10.1109/ICCSIE55183.2023.10175285
15. Zorko, L.: Uporaba rekurentnih nevronskih mrež za generiranje umetnih člankov. Diplomsko delo, University of Ljubljana (2021). https://repozitorij.uni-lj.si/IzpisGradiva.php?lang=slv&id=130488. Accessed 20 Mar 2024
16. Bao, W., Yue, J., Rao, Y.: A deep learning framework for financial time series using stacked autoencoders and long-short term memory. PLoS ONE **12** (2017). https://doi.org/10.1371/journal.pone.0180944
17. Palangi, H., et al.: Deep sentence embedding using long short-term memory networks: analysis and application to information retrieval. IEEE/ACM Trans. Audio Speech Lang. Process. **24**(4), 694–707 (2016). https://doi.org/10.1109/TASLP.2016.2520371
18. Abduljabbar, R.L., Dia, H., Tsai, P.W.: Development and evaluation of bidirectional LSTM freeway traffic forecasting models using simulation data. Sci. Rep. **11**(1), 23899 (2021). https://doi.org/10.1038/s41598-021-02949-6
19. Jamshidzadeh, Z., Ehteram, M., Shabanian, H.: Bidirectional long short-term memory (BILSTM) support vector machine: a new machine learning model for predicting water quality parameters. Ain Shams Eng. J. **15**(3), 102510 (2024). https://doi.org/10.1016/j.asej.2023.102510
20. Vincent, A.M., Jidesh, P.: An improved hyperparameter optimization framework for AutoML systems using evolutionary algorithms. Sci. Rep. **13**(1), 4737 (2023). https://doi.org/10.1038/s41598-023-32027-3
21. Dubey, S.R., Singh, S.K., Chaudhuri, B.B.: Activation functions in deep learning: a comprehensive survey and benchmark. Neurocomputing **503**, 92–108 (2022). https://doi.org/10.1016/j.neucom.2022.06.111

22. Kumari, P., Toshniwal, D.: Deep learning models for solar irradiance forecasting: a comprehensive review. J. Clean. Prod. **318**, 128566 (2021). https://doi.org/10.1016/j.jclepro.2021.128566
23. Sarmas, E., Dimitropoulos, N., Marinakis, V., Mylona, Z., Doukas, H.: Transfer learning strategies for solar power forecasting under data scarcity. Sci. Rep. **12**(1), 14643 (2022). https://doi.org/10.1038/s41598-022-18516-x

Impact of Cylindrical Object Diameter on Dielectric Constant in GPR Recordings

Daniel Štifanić[1], Jelena Štifanić[1](✉), and Zlatan Car[2]

[1] Faculty of Engineering, University of Rijeka, Vukovarska 58, 51000 Rijeka, Croatia
{daniel.stifanic,jelena.stifanic}@uniri.hr

[2] Faculty of Engineering, Catholic University of Croatia, Ilica 244, 10000 Zagreb, Croatia
zlatan.car@unicath.hr

Abstract. Often, the ground penetrating radar (GPR) is not considered a precise and rigorous measurement tool, but rather a tool for sub-surface exploration. One way of getting precise information about sub-surface objects and their surrounding media is excavation, which is usually demanding, time-consuming, and sometimes unfeasible. For that reason, the GPR has been widely used in numerous fields. However, the correct interpretation of acquired recordings tends to be challenging especially when the subsurface dielectric constant is unknown. To avoid wrong interpretations, the precise estimation of the dielectric constant is necessary. Underground cylindric objects appear within the B-scans as hyperbola-shaped features with different characteristics. A widely used approach for dielectric constant estimation is the hyperbola fitting method. Still, numerous factors can have an impact on the features and thus reduce the effectiveness of such a method. Accordingly, with the assistance of the genetic programming technique, the dielectric constant estimations of the hyperbola fitting method can be improved. The performance of the methods is evaluated by utilizing 5-fold cross-validation along with MAE, RMSE, and R^2 measures. Results reveal that cylindrical object diameter has a significant impact on the dielectric constant estimation of the hyperbola fitting method. Therefore, the proposed integration of the hyperbola fitting method with genetic programming successfully captures the impact of the diameter and improves dielectric constant estimation in GPR B-scans.

Keywords: ground penetrating radar · B-scan · hyperbola fitting · dielectric constant · genetic programming

1 Introduction

Ground Penetrating Radar (GPR) is a non-destructive electromagnetic geophysical tool used in various engineering applications to detect subsurface abnormalities and estimate the physical properties of medium [1]. GPR fault interpretation and underground object detection have typically relied on human-computer interaction. Due to the intricate nature of underground environments, inexperienced operators are likely to make mistakes. For that reason, one way of getting accurate information about sub-surface objects and their surrounding media is excavation, which is usually demanding, time-consuming, and

© The Author(s), under exclusive license to Springer Nature Switzerland AG 2025
N. Filipović (Ed.): AAI 2024, LNNS 1446, pp. 158–165, 2025.
https://doi.org/10.1007/978-3-031-99201-8_16

sometimes unfeasible [2]. Considering this, there has been a lot of interest in developing algorithms for automatic detection and interpretation of GPR data in recent years. However, the correct interpretation of acquired recordings tends to be challenging especially when the subsurface dielectric constant is unknown. To avoid wrong interpretations, precise dielectric constant estimation is necessary [3]. Underground cylindrical objects appear within the B-scans as hyperbola-shaped features with different characteristics. A widely used approach for dielectric constant estimation is the hyperbola fitting (HF) method. Still, numerous factors can have an impact on the features and thus reduce the effectiveness of such a method. Accordingly, with the assistance of the genetic programming (GP) technique, the electric constant estimations of the hyperbola method can be improved [4].

According to the literature review, only a few researchers have solved the challenges of estimating the dielectric constant, cylindrical object diameter, and burial depth. Ristic et al. (2009) [5] developed a method for determining both cylindrical object radius and electromagnetic wave propagation velocity or dielectric constant using GPR data. The proposed method has the advantage of avoiding velocity measurements and their impact on radius estimation quality. An accurate estimate of velocity leads to better soil characterization. This method surpasses existing methods in terms of accuracy and robustness, especially when dealing with noise and raw data [5]. Based on the hyperbolic patterns found in GPR data, He et al. (2024) provided a novel method for predicting the wave velocity, orientation, and burial depth of pipes. This method is a significant advancement since it allows for the simultaneous estimation of burial depth, dielectric constant, and pipe orientation from hyperbolic fitting [6].

The main contributions of this research are the following:

- Investigating the impact of cylindrical object diameter on dielectric constant estimation.
- Finding underlying mathematical expression that describes correlations between cylindrical object diameter and parameter *b* of hyperbola expression which resulted in correction of dielectric constant.

2 Materials and Methods

This section provides a description of the data used in this research as well as the genetic programming symbolic regression method utilized in order to identify the underlying mathematical expression. In the last subsection, the evaluation criteria used to estimate the performance of the model are briefly described.

2.1 Dataset Description

For the purposes of this research, synthetic B-scans are generated using GprMax simulation software [7]. Each B-scan consists of one hyperbola-shaped feature which represents the cylindrical object of defined parameters. Since the purpose of the research is to investigate the impact of different cylindrical object diameters on the dielectric constant estimation of the hyperbola fitting method, different values of parameters are used to generate the synthetic B-scans such as dielectric constant (εr), object depth (meters), and cylindrical object diameter COD (millimeter). Possible values of the aforementioned parameters are shown in Table 1.

Table 1. Parameters with corresponding values used for generating synthetic B-scans utilizing GprMax simulation software

Parameter	Possible values
Dielectric constant (ε_r)	3, 5, 7, 9, 11, 13, 17, 21
Object depth center (meter)	0.6, 0.8, 1.0, 1.25, 1.5, 1.75, 2.0
Cylindrical object diameter COD (millimeter)	100, 150, 200, 350, 500, 650, 800, 950

By randomly choosing parameter values from Table 1, a total of 89 B-scans are generated. In the case of this research, the recording trajectory is perpendicular to the cylindrical object. Examples of two B-scans generated with a dielectric constant of 21 and object depth of 1.175 m with different values of cylindrical object diameter are shown in Fig. 1.

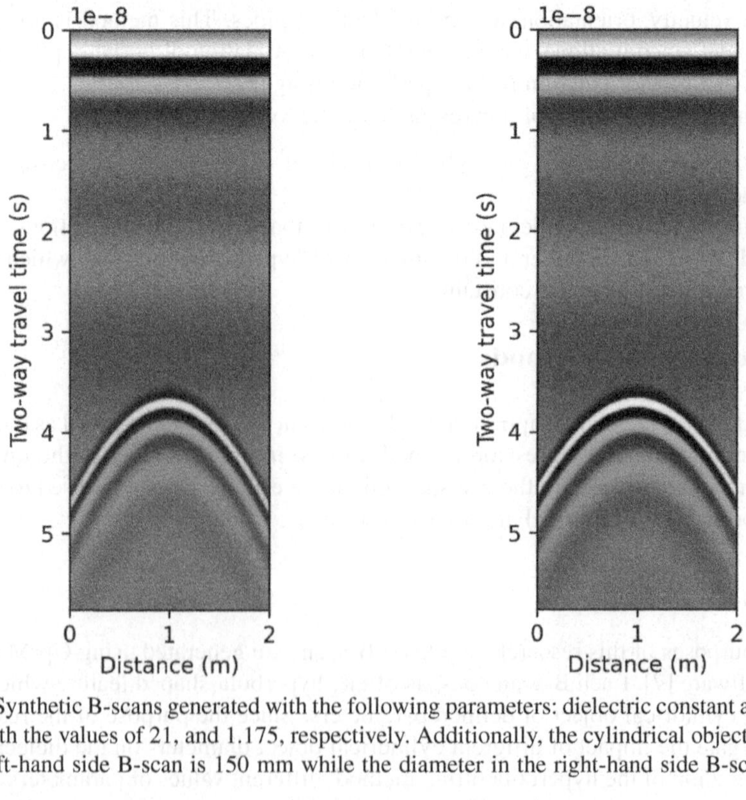

Fig. 1. Synthetic B-scans generated with the following parameters: dielectric constant and object depth with the values of 21, and 1.175, respectively. Additionally, the cylindrical object diameter in the left-hand side B-scan is 150 mm while the diameter in the right-hand side B-scan is 650 mm.

The x-axis of the B-scan directly corresponds to the distance GPR traveled, while the y-axis corresponds to the electromagnetic wave propagation into the ground. From Fig. 1 it can be seen that the hyperbola-shaped feature within the B-scan with smaller diameter is steeper than the feature found within the B-scan with bigger diameter.

2.2 Dielectric Constant Estimation – Hyperbola Fitting Method

Within synthetically generated B-scans, features that represent cylindrical objects are hyperbola-shaped, therefore, by fitting the features with the corresponding curve, certain parameters can be extracted [8]. The mathematical expression of hyperbola utilized for the purpose of fitting the hyperbolic reflections within the B-scan can be defined as:

$$\frac{y^2}{a^2} - \frac{(x - x_{pos})^2}{b^2} = 1, \tag{1}$$

where:

a – distance between the center and vertices along the y-axis,
b – distance between the center and vertices along the x-axis, and.
x_{pos} – displacement along the x-axis.

Carefully selecting the aforementioned parameters, the high-performance hyperbola fittings can be achieved. After the hyperbola-shaped features are fitted, a dielectric constant can be calculated using the following expression [9]:

$$\varepsilon_r = \left(\frac{ca}{2b}\right)^2, \tag{2}$$

where c is the speed of light in vacuum (c = 0.29979245 m ns-1).

2.3 Genetic Programming Symbolic Regression

One type of AI that draws inspiration from biological evolution and builds upon evolutionary algorithms (EA) is genetic programming. Koza first presented the methodology of GP as a domain-independent problem-solving method in which computer programs formed of functions and terminals are evolved to solve, or roughly solve, issues by creating a structured representation of the data [10]. The structural representation mimics naturally occurring genetic processes and the biological evolution of living things. The primary benefit of GP over ANNs is its capacity to present the relationship between a set of inputs and the related outputs in an easily understood mathematical form for users [11].

In GP, symbolic regression (GP – SR) is a thoroughly researched method for generating mathematical models in free form from observed data [12]. In traditional regression methods, the algorithm's starting point is a specific model, for which parameters must be optimized. For example, a linear regression model is based on the assumption that there is a linear relationship between the regression and the dependent variables [13]. However, no such a priori assumptions about the particular form of the function are needed in SR. Instead, one gives a mathematical expression space that contains candidate function building blocks, such as state variables, mathematical operators, constants, and analytic functions. Then, to find the best solution, symbolic regression is used to search through the space. In other words, symbolic regression optimizes both model structure and model parameters [13].

Due to the ability of GP to provide mathematical expression to represent the relationship between input and output variables, it is used in this research to investigate and mathematically describe the impact of cylindrical object diameter on dielectric constant estimations of the traditional HF method. Therefore, parameters such as a, b, and COD are used as input parameters to fit the GP – SR.

2.4 Evaluation Criteria

In order to evaluate the performance of the GP – SR model, mean absolute error (MAE), root mean squared error (RMSE), and coefficient of determination (R2) are utilized. The performance measures can be calculated as follows [14]:

$$MAE = \frac{1}{n}\sum_{i=1}^{n}|y_i - \hat{y}_i|, \tag{3}$$

$$RMSE = \sqrt{\frac{1}{n}\sum_{i=1}^{n}(y_i - \hat{y}_i)^2}, \tag{4}$$

and

$$R^2 = 1 - \frac{S_{res}}{S_{tot}} = 1 - \frac{\sum_i (y_i - \hat{y}_i)^2}{\sum_i (y_i - \overline{y})^2}, \tag{5}$$

where y_i is true data, $\hat{y}_i$ is predicted data, and $\overline{y}$ represents the mean of all observed true data. Due to the mathematical properties of Eqs. (3) and (4), the lower values of measures directly correspond to better model performance. On the other hand, the R2 measure is defined in the range between 0.0 and 1.0, where the value of 1.0 is assumed to be a perfect fit of the model and vice-versa.

3 Results and Discussion

After the hyperbola-shaped features within the B-scans are fitted with hyperbolas, and the parameters are acquired, the dielectric constants can be calculated by utilizing the expression (2). HF method resulted in MAE, RMSE, and R2 values of 1.65147 ± 0.30432, 2.16763 ± 0.34087, and 0.82420 ± 0.04037, respectively. Due to the mathematical properties of the HF method, the COD parameter is ignored during the calculations which leads to the lower performance values. In order to investigate the impact of cylindrical object diameter on dielectric constant estimation, the same parameter is included in the input data which is used to train the GP – SR model. Based on a, b, and COD parameters, the GP – SR model is used to estimate the b_{corr} value by which, a corrected ε_r can be calculated utilizing the expression (2).

Table 2. GP – SR best performing hyperparameter combination

Hyperparameter	Value
Constant range	-1, 1
Function set	add, sub, mul, div, sqrt, sin, cos, tan
Generations	100
Initial tree depth	(2, 6)
Initial method	half and half
Crossover prob	0.85
Hoist mutation prob	0.03
Point mutation prob	0.01
Point replace prob	0.05
Subtree mutation prob	0.1
Parsimony coefficient	9e5
Population size	10000
Tournament size	20

In order to find optimal hyperparameters of the GP – SR model, a random search is performed. Using the hyperparameters shown in Table 2, the highest values of performance measures are achieved.

After the hyperparameters of the best performing GP – SR model are found, the performance can be estimated using a 5-fold cross-validation. Performances of HF and HF + GP – SR methods in terms of 5-fold cross-validation with corresponding standard deviations are shown in Table 3.

Table 3. Performances of the HF method along with the improved HF + GP – SR method in terms of 5-fold cross-validation with corresponding standard deviation values

Method	MAE ($\pm\sigma$)	RMSE ($\pm\sigma$)	R^2 ($\pm\sigma$)
HF	1.65147 ± 0.30432	2.16763 ± 0.34087	0.82420 ± 0.04037
HF + GP – SR	0.13682 ± 0.03044	0.18201 ± 0.03762	0.99864 ± 0.00078

The HF + GP – SR method resulted in MAE, RMSE, and R2 performance values of 0.13682 ± 0.03044, 0.18201 ± 0.03762, and 0.99864 ± 0.00078, respectively. From the results it can be seen that the cylindrical object radius parameter has a significant impact on the parameter b and thereby affects the dielectric constant value. Analyzing the parameter's impact on the estimations, it can be concluded that the HF method produces satisfactory estimations only when the cylindrical object diameter is between 100 and 200 mm. Further increases in diameter value lead to larger errors in dielectric constant estimation. However, the HF + GP – SR method successfully captures the impact of cylindrical object diameter and corrects the estimation significantly.

The best performing mathematical expression produced by the GP – SR method that describes the correlation between parameter b and b_{corr} can be written as:

$$b_{corr} = b + \gamma COD\left(\frac{1}{a} + \frac{1}{a\sqrt{b}} - 10^9\right), \tag{6}$$

where γ is a constant value of 0.12984e-9.

The presented results indicate that by including the cylindrical object diameter parameter in the mathematical model can significantly improve performances in terms of estimating dielectric constant from the GPR B-scans. Since this research only focuses on hyperbolic reflections that are perpendicular to the cylindrical object, future work will introduce the approaching angle as an input parameter in order to examine the impact on the dielectric constant estimation.

4 Conclusions

According to the presented results, it can be concluded that cylindrical object diameter has a significant impact on the dielectric constant estimation if the hyperbola fitting method is utilized. In such a method, by increasing the cylindrical object diameter, the estimation error also increases. However, with the assistance of GP – SR, the cylindrical object diameter can be included as an input parameter which leads to significant improvements in dielectric constant estimation. Additionally, the mathematical expression that describes the aforementioned correlation is also proposed. The limitation of this research was data availability; therefore, future work should use more synthetic data with different variations to increase the robustness of the proposed method.

Acknowledgements. This research was (partly) supported by the CEEPUS network CIII-HR-0108, the European Regional Development Fund under the grant KK.01.1.1.01.0009 (DATACROSS), the project CEKOM under Grant KK.01.2.2.03.0004, the Erasmus+ project WICT, under the grant 2021-1-HR01-KA220-HED-000031177, and AISE project, under the grant KA220-SCH, and the University of Rijeka Scientific Grant uniri-tehnic-18–275-1447.

References:

1. Apaydın, O., İşseven, T., Çıtır, Y., Paker, S., Erer, I., Aydın, N.G.: Extracting tomographic images of interior structures of cylindrical objects and trees using ground penetrating radar method. Results Eng. **14**, 100410 (2022)
2. Catapano, I., Gennarelli, G., Ludeno, G., Noviello, C., Esposito, G., Soldovieri, F.: Contactless ground penetrating radar imaging: state of the art, challenges, and microwave tomography-based data processing. IEEE Geosci. Remote Sens. Mag. **10**(1), 251–273 (2021)
3. Elkarmoty, M., Colla, C., Gabrielli, E., Papeschi, P., Bonduà, S., Bruno, R.: In-situ GPR test for three-dimensional mapping of the dielectric constant in a rock mass. J. Appl. Geophys. **146**, 1–15 (2017)

4. Ponti, F., et al.: GPR radargrams analysis through machine learning approach. J. Electromagn. Waves Appl. **35**(12), 1678–1686 (2021)
5. Ristic, A.V., Petrovacki, D., Govedarica, M.: A new method to simultaneously estimate the radius of a cylindrical object and the wave propagation velocity from GPR data. Comput. Geosci. **35**(8), 1620–1630 (2009)
6. He, W., Lai, W.W.L.: Angle-corrected GPR hyperbolic fitting models for improved parameter estimation. Tunn. Undergr. Space Technol. **147**, 105741 (2024)
7. Warren, C., Giannopoulos, A., Giannakis, I.: GprMax: Open source software to simulate electromagnetic wave propagation for ground penetrating radar. Comput. Phys. Commun. **209**, 163–170 (2016)
8. Lei, W., et al.: Automatic hyperbola detection and fitting in GPR B-scan image. Autom. Constr. **106**, 102839 (2019)
9. Poluha, B., Porsani, J.L., Almeida, E.R., dos Santos, V.R.N., Allen, S.J.: Depth estimates of buried utility systems using the GPR method: studies at the IAG/USP geophysics test site. Int. J. Geosci. **8**(5), 726–742 (2017)
10. Koza, J.R.: Genetic programming as a means for programming computers by natural selection. Stat. Comput. **4**, 87–112 (1994)
11. Fatehnia, M., Amirinia, G.: A review of genetic programming and artificial neural network applications in pile foundations. Int. J. Geo-eng. **9**(1), 2 (2018)
12. Icke, I., Bongard, J.C.: Improving genetic programming based symbolic regression using deterministic machine learning. In: 2013 IEEE Congress on Evolutionary Computation, pp. 1763–1770. IEEE (2013)
13. Wang, Y., Wagner, N., Rondinelli, J.M.: Symbolic regression in materials science. MRS Commun. **9**(3), 793–805 (2019)
14. Štifanić, D., Musulin, J., Miočević, A., Baressi Šegota, S., Šubić, R., Car, Z.: Impact of COVID-19 on forecasting stock prices: an integration of stationary wavelet transform and bidirectional long short-term memory. Complexity **2020**(1), 1846926 (2020)

Development of Convolutional Neural Network for Classification of Heart Sounds Utilizing Mel-Frequency Cepstral Coefficients

Lazar Dašić[1,2](✉), Tijana Geroski[2,3], Ognjen Pavić[1,2], and Nenad Filipović[2,3]

[1] Institute for Information Technologies Kragujevac, University of Kragujevac, Jovana Cvijica Bb, 34000 Kragujevac, Serbia
lazar.dasic@kg.ac.rs

[2] Bioengineering Research and Development Center (BioIRC), Prvoslava Stojanovica 6, 34000 Kragujevac, Serbia

[3] Faculty of Engineering, University of Kragujevac, 6 Sestre Janjica, 34000 Kragujevac, Serbia

Abstract. Cardiovascular diseases are the leading cause of death worldwide, accounting for more than 30% of all global deaths. It is estimated that more than 17 million people lose their lives each year due to the heart attacks and strokes. In daily clinical practice, listening to and analyzing a patient's heart sounds via auscultation is essential in detecting any irregularities. This non-invasive initial screening procedure allows doctors to gather information about the heart's condition without any invasive procedures, while being quick and cost-effective. The high prevalence and mortality rate of heart diseases have led to an increased need for automatic and efficient diagnostic tools that could assist medical professionals in the analysis of heart sounds. This paper introduces a convolutional neural network (CNN) model that is designed to identify and categorize various irregularities in heart sounds from stethoscope audio recordings. In order to train the proposed CNN model, a publicly available dataset containing hundreds of heartbeat audio recordings was utilized. Since sound signals can be quite complex, it was necessary to extract the main characteristics of the sound and format them in a way suitable for training of the proposed CNN model. Mel-Frequency Cepstral Coefficients (MFCCs) are designed to capture key spectral properties of a sound signal automatically and format them as small sets of numbers that are easily understandable by a deep learning model. Our proposed model achieved promising results, not only in detection of irregular heart sounds, but also in a classification between different types of heart sound abnormalities.

Keywords: convolutional neural network · deep learning · heart disease · Mel-frequency cepstral coefficients · voice processing

1 Introduction

Cardiovascular diseases (CVDs) have been the leading cause of death globally for years, taking millions of lives annually [1]. Many of these cardiovascular disease-related deaths could be prevented by early detection and treatment of affected patients. Over the years,

© The Author(s), under exclusive license to Springer Nature Switzerland AG 2025
N. Filipović (Ed.): AAI 2024, LNNS 1446, pp. 166–174, 2025.
https://doi.org/10.1007/978-3-031-99201-8_17

numerous diagnostic procedures and tools have been developed, but still one of the first steps in CVD diagnosis is the physical examination of the patient's heart through the process of auscultation. Cardiac auscultation is the process of using a stethoscope to listen to the sounds of the heart and identify any abnormalities in its normal functionality [2]. These sounds are generated due to the mechanical activity of the heart and blood turbulence resulting from the opening and closing of the heart valves, as well as rapid accelerations and decelerations of blood flow in the heart chambers [3]. A normal heart sound consists of two clearly audible sounds forming a distinct "lub-dub" pattern:

- The first heart sound S1 ("lub") is generated at the beginning of systole by the closure of the mitral and tricuspid valves.
- The second heart sound S2 ("dub") is generated at the beginning of diastole by the closure of aortic and pulmonary valves.

Besides the standard S1 and S2 sounds, additional sounds can be generated by turbulent blood flow and heart muscle movements. These noisy signals often referred to as murmurs. Although some murmurs can be innocent (usually present in infants and children), abnormal murmurs may indicate underlying heart conditions that require further investigation.

Cardiac auscultation provides a cheap, fast and non-invasive approach for screening the condition of the human heart. However, auscultation is a difficult skill to master, requiring extensive training and years of clinical experience. Additionally, the evaluation of heart sounds is highly subjective. All of the aforementioned has resulted in a decline in cardiac auscultation proficiency, especially among young doctors [4]. The creation of a computer-aided decision system could provide an automatic way to analyze heartbeat sounds and assist in the diagnostic procedure. Our goal was to develop a system based on artificial intelligence (AI), capable of not only detecting abnormal heart sounds, but also distinguishing between them.

1.1 Related Works

The classification of heart sounds has been the topic of interest in research community for years. Before the advent of deep learning for classifications, traditional machine learning methods were primarily utilized.

In one of the earliest research papers, Vepa [5] tested three different classifiers: k-nearest neighbor (kNN) classifier, multilayer perceptron (MLP) neural networks and support vector machines (SVM) for the classification of heart sounds into normal, systolic murmurs and diastolic murmurs based on extracted features. Loukis et al. [6] utilized the Random Forest algorithm for the automatic identification of murmurs in heart sound signals, as well as for classifying them as systolic or diastolic.

With the recent advancements in the field of artificial intelligence, deep learning with its neural networks has been heavily utilized for the classification of heart sound recordings. Li et al. [7] extracted features from eight domains and applied convolutional neural networks (CNNs) to achieve high accuracy in detecting abnormal heart sounds. Norman et al. [8] designed a 1D-CNN that directly learns features from raw heart-sound signals, and a 2D-CNN that takes inputs of two-dimensional time-frequency feature maps based on Mel-frequency cepstral coefficients. Sudarsanan et al. [9] also based their methodology on CNNs, but they showed that incorporating background noise removal using smoothing filters and wavelet denoising positively affected the model's capability to detect abnormal heart sounds.

Based on the literature review,it is evident that most of the research has focused on binary classification, with the primary objective to distinguish between normal and abnormal heart sounds. However, this approach does not provide a clear picture of what might be wrong with the recorded patient's heart signals. This is the reason why we have shifted our focus to multiclass classification, which should provide more detailed information about abnormalities detected within the recorded signal.

2 Materials and Methods

We use a publicly available dataset containing recordings of hundreds of heart sounds. The important information from these audio recordings is represented as Mel-frequency cepstral coefficients (MFCCs), which are further used as input data for a CNN model that classifies heart sounds into one of four categories:

- **Normal** – the heartbeat sound of a healthy patient. A normal heart sound has a clear "lub-dub, lub-dub" pattern, with the time from "lub" to "dub" shorter than the time from "dub" to the next "lub".
- **Artifact** - noise in the heartbeat recording, with sounds coming from various sources (movement, breathing, poor contact, etc.) and interfering with the clarity of the heart sounds
- **Murmur** - heart murmurs sound as though there is a "whooshing, roaring, rumbling, or turbulent fluid" noise in one of two temporal locations: (1) between "lub" and "dub", or (2) between "dub" and "lub"
- **Extra heart sounds** - heart sound that is out of rhythm involving extra or skipped heartbeats, e.g. a "lub-lub dub" or a "lub dub-dub".

Differences between these classes of audio recordings of heart sounds are shown in Fig. 1.

2.1 Dataset

To develop a capable AI model that classifies heart sounds with a high degree of accuracy, a large amount of data is necessary. The proposed CNN model was trained on the PASCAL CHS dataset [10], which comprises samples from two different sources:

- Dataset A was collected on an unreported size population, using the iStethoscope Pro mobile app. It contains 176 samples, out of which 52 are not annotated
- Dataset B was gathered using a digital stethoscope system deployed in the Maternal and Fetal Cardiology Unit at Real Hospital Português (RHP) in Recife, Brazil. It contains 656 samples, out of which 196 are not annotated.

A combined dataset consists of 351 normal samples, 129 murmur samples, 65 extra heart sounds samples and 40 artifacts samples. Figure 2 shows the distribution of dataset samples.

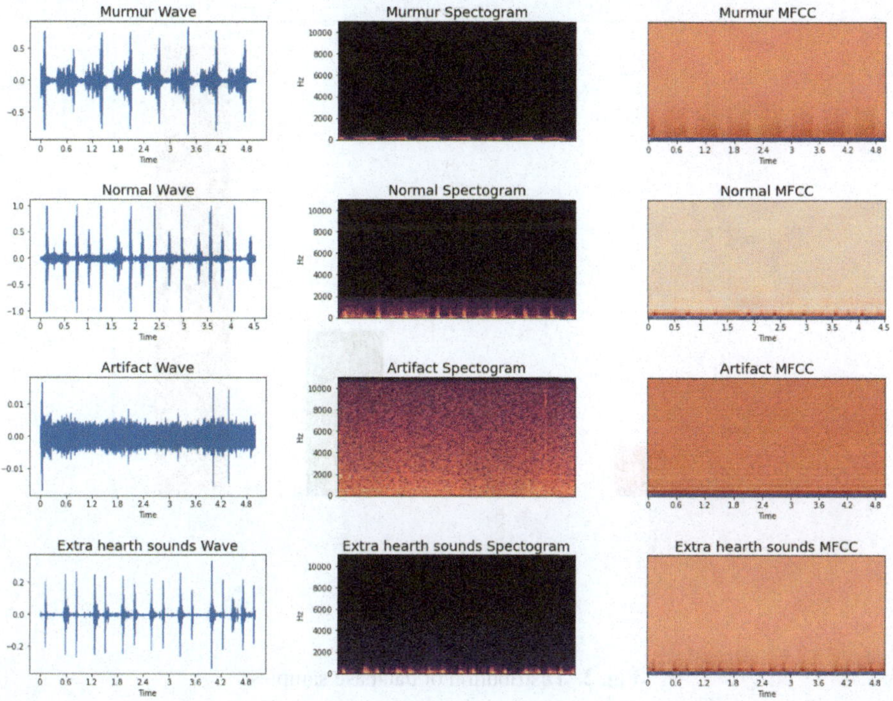

Fig. 1. Differences between different classes of audio recordings in visualized as phonocardiograms, spectrograms and MFCC

2.2 Data Preprocessing

The analysis of the used dataset revealed that not all heart sound audio recordings are of the same length. This variation does not pose a problem when training the model, but the inherent nature of short recordings presents an issue. Such recordings do not last long enough to contain and extract useful information, so it was decided to remove recordings shorter than 4 s from the dataset.

The remaining audio recordings were split into training and test datasets in an 80:20 ratio, resulting in a training dataset of 323 samples and a test dataset of 81 samples.

While it is possible to train artificial intelligence models directly on audio files, this is often not an optimal solution due to the dimensionality of audio files, large amount of information they contain, insufficient signal-to-noise ratio, etc. Therefore, during the preprocessing step, key features are often extracted from audio files that sufficiently describe the given audio recording, with the advantage of having smaller dimensions and containing less noise and unnecessary information. There are various ways to extract features from audio files, ranging from simple manual methods where each feature (e.g., timbre, frequency, rhythm, amplitude, etc.) is considered to more advanced algorithms that automatically create compact and high-quality representations of audio signals. One of these representations is the Mel-frequency cepstral coefficients. These coefficients were created using the Python programming language with its audio signal processing

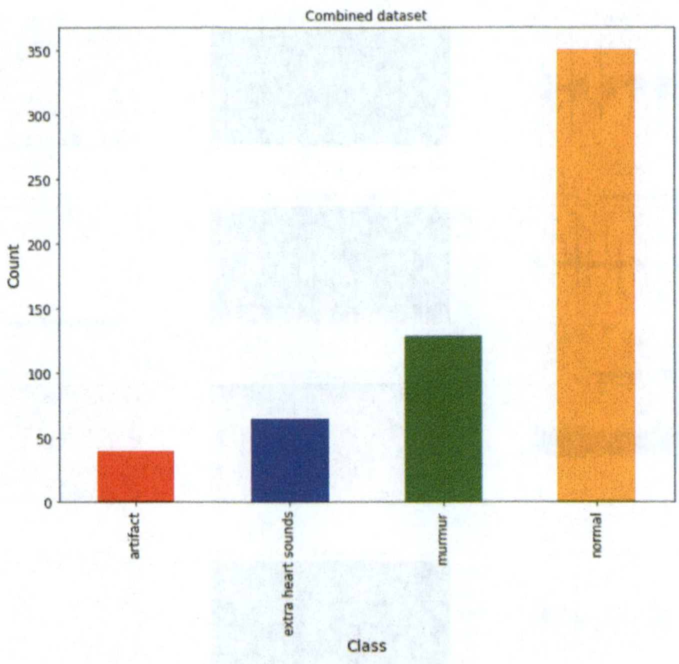

Fig. 2. Distribution of database samples

library called Librosa. As a result of this process, a representation of the dataset was created, with each audio signal represented by 128 components, i.e., coefficients. MFCCs are extracted from an audio signal by first framing the signal with a Hamming window and transforming it into the frequency domain using the Fast Fourier Transform. The resulting spectrum is then processed through a Mel-filter bank, followed by taking the logarithm of the filtered energies and applying the Discrete Cosine Transform to produce the MFCC vector, which captures essential perceptual features of the audio.

2.3 Creation and Training of a Model

Extracted MFCCs are used as input data for a CNN model. CNNs excel at capturing spatial patterns in structured data such as MFCCs, providing translation invariance and reduced computational complexity. The proposed CNN model consists of four convolutional blocks, each comprising a 2D Convolutional layer, followed by a 2D Max Pooling layer and a Dropout layer. The model uses convolutional filters of size 2×2, with the number of filters in each block being 16, 32, 64, and 128, respectively. The activation function chosen for the convolution layers is ReLU.

To address the issue of an unbalanced dataset, a weighted categorical cross-entropy loss function was utilized. The model was trained for 300 epochs, but early stopping was employed to prevent eventual overfitting, resulting in 143 epochs of training. The optimizer of choice was Adam. Figure 3 shows the loss function and accuracy metrics during the training procedure. The model creation and training were done using the Python programming language and TensorFlow library.

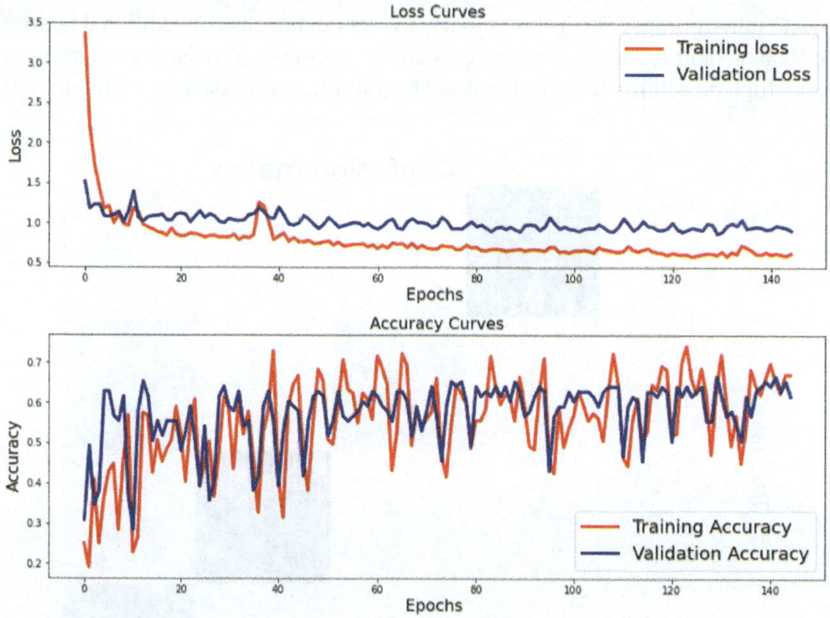

Fig. 3. Loss function and accuracy metrics during the training procedure

3 Results and Discussion

Accuracy is used as the primary metric for model evaluation in the heart sound classification task. The created CNN model achieves an accuracy of 68%, while other classification metrics for each class are presented in Table 1. The table shows that the model achieves the best results for classifying MFCC recordings representing heart tones with artifacts present. Additionally, there is a noticeable low sensitivity for classifying samples that belong to the class of extra heart tones, indicating that the model often misclassifies samples that actually belong to the class of extra heart sounds.

Table 1. Results of model evaluation

Class	Precision	Sensitivity	F1-score
Artifact	0.78	0.78	0.78
Extra heart sounds	0.70	0.39	0.50
Murmur	0.58	0.70	0.64
Normal	0.58	0.65	0.61

To gain insight into the model's performance for each class, a confusion matrix was created and is shown in Fig. 4. The confusion matrix indicates that the model often predicts that an audio recording of a heart belongs to the normal class, while it actually belongs to other groups. This is particularly evident in the case of additional heart tones, where 50% of samples are incorrectly classified as belonging to the normal tones group. There is a high possibility that this is caused by a significant imbalance within the dataset.

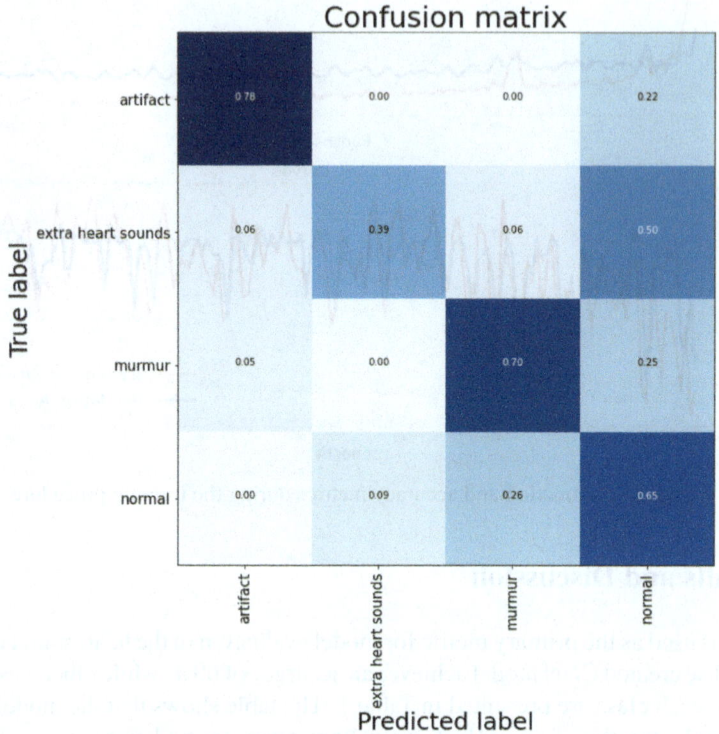

Fig. 4. Confusion matrix

Given the low quantity and imbalance of the used dataset, the model was able to differentiate between different classes and was especially good at the classification of murmurs and artifacts within the audio recordings. The model had difficulties distinguishing between normal recordings and recordings with extra heart sounds, which makes sense when looking at the similarities between the two as represented in Fig. 1.

Different, more complex CNN architectures have been tested, but these more complicated structures did not improve performance. Moreover, in some cases, they performed worse than the aforementioned architecture.

The proposed model would undoubtedly benefit from additional data samples, especially for the classes that are underrepresented in the dataset (extra heart sounds, murmur and artifacts). This would allow the model to learn how to better differentiate normal heart sounds recordings from the rest.

4 Conclusion

This paper introduces a deep learning method for the classification of heart sounds into the following four categories: normal, artifact, murmur and extra heart sounds. In order to extract useful information from audio recordings, Mel-frequency cepstral coefficients are utilized. These structured data representations of sounds are used as input data for a convolutional neural network model. The innovation lies in the multiclassification aspect of the model, where not only the detection of murmurs was involved, but also classification of other audio recordings.

The created CNN model achieved an accuracy of 68%, with high degrees of accuracy for classification of murmurs and artifacts within the heart recordings. On the other hand, distinguishing between normal heart recordings and recordings containing extra heart sounds proved to be challenging due to the similarities between the two classes.

While the model demonstrates promising results, the study showed some limitations. Besides the aforementioned difficulties in distinguishing between some classes, the model's overall accuracy score could be further improved. The model should also be evaluated on a much larger sample size, as the current size of the test dataset of 81 samples is not sufficient to accurately and objectively assess performance and generalizability of the model.

Future research will focus on addressing the aforementioned limitations and improving the methodology. The primary focus will be on the acquisition of new dataset samples, expanding the current training and test data size. Additional data would alleviate many issues related to model classification performance. Although MFCCs provide a lot of useful information, changes in data preprocessing could introduce new features (Discrete Wavelets Transform features, zero-crossing rate, spectral centroid, chroma features, etc.) that would contain information not present in the MFCCs representation.

Acknowledgements. This research is funded by the project that has received funding from the European Union's Horizon Europe research and innovation programme under grant agreement No 101080905 (STRATIFYHF). This research was also supported by the Ministry of Science, Technological Development and Innovation of the Republic of Serbia, contract numbers [451–03-65/2024–03/200107 (Faculty of Engineering, University of Kragujevac)] and [451–03-66/2024–03/200378 (Institute for Information Technologies)]. Authors T.G. and N.F. acknowledge to the Science Fund of the Republic of Serbia, as the research was supported under Grant Agreement No 7580, "Artificial INTELligence-based Decision Support System for Early and Accurate Diagnosis of HEART Failure – INTELHEART". The authors T.G. and N.F. are solely responsible for the content of the publication and the content does not reflect the views of the Science Fund of the Republic of Serbia.

References

1. Timmis, P., et al.: European society of cardiology: cardiovascular disease statistics 2021. Eur. Heart J. **43**(8), 716–799 (2022)
2. Montinari, M.R., Minelli, S.: The first 200 years of cardiac auscultation and future perspectives. J. Multidisciplinary Healthcare 183–189 (2019)

3. Randhawa, S.K., Singh, M.: Classification of heart sound signals using multi-modal features. Procedia Comput. Sci. **58**, 165–171 (2015)
4. Alam, U., Asghar, O., Khan, S.Q., Hayat, S., Malik, R.A.: Cardiac auscultation: an essential clinical skill in decline. Br. J. Cardiol. **17**(1), 8 (2010)
5. Vepa, J.: Classification of heart murmurs using cepstral features and support vector machines. In: 2009 Annual International Conference of the IEEE Engineering in Medicine and Biology Society (2009)
6. Loukis, E., Maragoudakis, M.: Heart murmurs identification using random forests in assistive environments. In: 3rd International Conference on Pervasive Technologies Related to Assistive Environments (2010)
7. Li, F., Tang, H., Shang, S., Mathiak, K., Cong, F.: Classification of heart sounds using convolutional neural network. Appl. Sci. **10**(11), 3956 (2020)
8. Noman, F., et al.: Short-segment heart sound classification using an ensemble of deep convolutional neural networks. In: ICASSP 2019–2019 IEEE International Conference on Acoustics, Speech and Signal Processing (ICASSP) (2019)
9. Sudarsanan, S., Aravinth, J.: Classification of heart murmur using CNN. In: 5th International Conference on Communication and Electronics Systems (ICCES) (2020)
10. Bentley, P.J., Nordehn, G., Coimbra, M., Mannor, S.: The PASCAL classifying heart sounds challenge 2011 (CHSC2011) Results. www.peterjbentley.com/heartchallenge/

Parameter Sensitivity Analysis in Multiscale Agent-Based Modeling of Atherosclerotic Plaque Progression

Lemana Spahić[1,2](✉), Leo Benolić[1,2], Safi Ur-Rehman Qamar[1,2], and Nenad Filipović[1,2]

[1] Bioengineering Research and Development Center (BioIRC), Prvoslava Stojanovica 6, 34000 Kragujevac, Serbia
lemana@kg.ac.rs

[2] Faculty of Engineering, University of Kragujevac, 6 Sestre Janjic, 34000 Kragujevac, Serbia

Abstract. Enhancement of precision and reliability in simulation models requires the exploration of advanced sampling techniques and parameter optimization strategies. This study integrates Latin Hypercube Sampling (LHS) with parameter adjustment methodology to explore the variability of simulation outputs, focusing on an Agent-Based Model (ABM) for atherosclerotic plaque progression sensitive to parameter variations. The research emphasizes improving model's generalization capabilities by systematically varying parameters within defined ranges, including constant and dynamic parameters, to evaluate their impact on the simulation results.

LHS was used to generate a comprehensive parameter space, consisting of 7 distinct parameters, with two set as constants and the remaining five subject to variation within specified ranges. This approach facilitated the generation of diverse, yet evenly distributed, parameter sets to feed into the ABM simulations.

The application of LHS in conjunction with the parameter adjustment strategy revealed significant insights into the model's sensitivity to various parameters. The findings indicate a pronounced impact of certain parameters on the simulation outcomes, specifically the parameter driving SMC proliferation in intima, increased probability of lipid infiltration and increased ECM degradation highlighting their potential as critical levers impacting model's behavior, suggesting areas for further investigation and optimization.

By systematically exploring the parameter space and identifying key parameters with pronounced impacts on model outcomes, this research lays the groundwork for further refinement of simulation practices. Future work will delve deeper into the mechanisms underlying the observed sensitivities and extend the approach to encompass a wider array of models and scenarios.

Keywords: agent-based model · Latin hypercube sampling · parameter optimization · atherosclerosis.0

© The Author(s), under exclusive license to Springer Nature Switzerland AG 2025
N. Filipović (Ed.): AAI 2024, LNNS 1446, pp. 175–184, 2025.
https://doi.org/10.1007/978-3-031-99201-8_18

1 Introduction

Finite element modeling of atherosclerosis plays a crucial role in understanding the biomechanical behavior of arterial walls under pathological conditions [1]. Atherosclerosis, a progressive disease characterized by the buildup of plaques within arterial walls, can lead to significant clinical complications such as myocardial infarction and stroke [2]. Computational models based on finite element analysis provide a powerful tool to simulate and analyze the complex mechanical interactions that occur within these diseased arteries [3]. At its core, finite element modeling of atherosclerosis involves discretizing the arterial wall into small geometric elements, each represented by a set of mathematical equations that describe its mechanical behavior [4–7]. These elements are interconnected at nodes, allowing researchers to simulate the distribution of stresses and strains throughout the arterial wall under various physiological conditions. Key factors influencing the mechanical behavior of atherosclerotic plaques include plaque composition (e.g. lipid core, fibrous cap), degree of calcification, and the overall geometry of the vessel [8]. By incorporating these factors into finite element models, researchers can predict stress concentrations within the plaque, assess the risk of plaque rupture, and evaluate the effectiveness of different therapeutic interventions. Moreover, finite element models enable researchers to explore how changes in blood flow patterns, such as those caused by stenosis (narrowing of the artery), influence plaque development and progression [4–7, 9, 10]. By integrating fluid-structure interaction simulations, these models can provide insights into the hemodynamic forces acting on the arterial wall and their role in plaque formation. Recent advancements in computational techniques, coupled with improvements in imaging modalities like MRI and CT angiography, have enhanced the accuracy and predictive capabilities of finite element models in studying atherosclerosis. These models not only contribute to our fundamental understanding of disease mechanisms, but also hold promise for personalized medicine by guiding clinicians in making informed decisions regarding patient-specific treatment strategies [11–13].

Given the inherent variability in biological tissues and the complexities of modeling pathological conditions, sensitivity analysis has emerged as a crucial methodology in computational biomechanics [14, 15]. Numerous studies have highlighted the importance of understanding how input parameters affect model outcomes, particularly in finite element models (FEMs) applied to medical and biological systems. Biological tissues, including arteries, exhibit complex, non-linear behaviors that are difficult to capture accurately in computational models due to their anisotropic, heterogeneous, and viscoelastic properties [16]. Furthermore, the variability in material properties such as elasticity, stiffness, and compliance introduces significant uncertainty into FEM simulations, especially when modeling patient-specific cases [4].

In the context of cardiovascular diseases such as peripheral artery disease (PAD), finite element modeling is frequently employed to simulate arterial wall mechanics, blood flow dynamics, and plaque progression [5]. However, the predictive accuracy of these models is often compromised by uncertainties in the input parameters, including plaque composition, arterial geometry, and boundary conditions [6, 13]. This uncertainty is compounded by the pathological changes in the arterial wall caused by plaque accumulation, which alters the mechanical properties of the tissue over time. Therefore, a comprehensive sensitivity analysis is necessary to identify the parameters that most significantly influence model predictions and guide model calibration.

Previous research has demonstrated that sensitivity analysis can reveal critical parameters that drive model behavior, enabling researchers to refine FEMs for improved accuracy. For instance, Suh et al. performed a sensitivity analysis on FEMs of arterial walls and found that parameters such as arterial wall stiffness and pressure significantly influenced the mechanical response of the model [17]. Similarly, in models of atherosclerosis, sensitivity to varying the intrinsic parameters of plaque progression ABM models has been analysed in the paper by Corti et al. where latin hypercube sampling was conducted to generate a variety of parameter combinations, followed by simulations to assess the impact of the varying parameters on simulation outcomes [18].

The insights gained from sensitivity analysis are critical for translating FEMs into clinical practice. Optimized models, calibrated based on sensitivity findings, can contribute to the development of more personalized treatment plans, where individual patient data is used to tailor interventions. The application of FEMs in personalized medicine has gained traction in recent years, particularly in cardiovascular research, where models have been employed to simulate stent deployment, plaque rupture risk, and post-surgical outcomes. Sensitivity analysis, therefore, serves as a foundational step toward achieving these personalized applications by ensuring that FEMs are robust, reliable, and capable of delivering clinically relevant predictions.

In light of these findings, this study applies sensitivity analysis to FEMs that simulate arterial occlusion and plaque progression in PAD. By leveraging a set of patient-specific geometries and parameter variations generated through Latin Hypercube Sampling (LHS), we aim to systematically evaluate the impact of various parameters—such as arterial wall stiffness, plaque material properties, and boundary conditions—on the outcomes of the model. Identifying the most sensitive parameters will not only help in improving the accuracy of the FEMs but will also enhance their applicability in clinical decision-making, particularly for predicting disease progression and assessing treatment efficacy.

2 Materials and Methods

LHS was used to generate a comprehensive parameter space, consisting of 7 distinct parameters, with two set as constants and the remaining five subject to variation within specified ranges.

This approach facilitated the generation of diverse, yet evenly distributed, parameter sets to feed into the ABM simulations. The process iteratively adjusted one of the varying parameters to its default value in each simulation run, ensuring a thorough exploration of the parameter space.

2.1 ABM Framework

Starting from an initial homeostatic state, the 3D agent-based model (ABM) simulates physiological or pathological arterial wall remodeling, depending on the wall shear stress (WSS) profile, by locally modeling processes such as cell division and apoptosis, extracellular matrix (ECM) production and degradation, and lipid infiltration into the intima. The model operates under the assumption that disease-promoting risk factors are already present. Consequently, plaque formation in a specific area is driven solely by hemodynamic factors, particularly the WSS profile.

The ABM was integrated with an initial Wall Shear Stress (WSS) profile, which induces pathological vascular remodeling by disrupting baseline cellular activity and promoting lipid infiltration and accumulation within the arterial wall. A finite element mesh was created for both fluid and solid domains, with governing equations and numerical methods applied to the fluid domain. Fluid-structure interaction and mesh motion were implemented, with all algorithms incorporated into the PAK - Athero program [19].

At each step, the finite element method (FEM) was used to calculate the velocity distribution, LDL transport, pressure distribution, and shear stress within the lumen area. The ABM utilizes the shear stress and initial LDL distribution from the lumen to begin iterative calculations within the wall for lipid infiltration and accumulation, using a random number generator for each time step. After the ABM iterations, both the lipid distribution and wall geometry are updated, which directly affects the artery's geometry. This geometry, modeled using finite elements that include ABM elements, is then recalculated by running the fluid-structure solver for the lumen domain [13].

A basic solution was developed by carefully adjusting the agent dynamics to stabilize the system near an equilibrium point. This state represented the homeostasis of a healthy artery, which was then disturbed to simulate the formation of atherosclerotic plaque. Before constructing the fully integrated framework, the ABM behavior was tested under both normal and atherogenic conditions.

2.2 Architecture Description

The DenseNet121 architecture was used as the basis of all models created within the study. DenseNet121 is a deep convolutional neural network with 121 layers involved in the network training process along with an input layer, a data embedding layer, an output layer and 6 auxiliary data vectorization layers. The trainable layers are organized into blocks of two convolutional neuron layers for learning progressively larger image features followed by one max-pooling layer each. The convolutional layers of the DenseNet121 architecture perform convolution in 3 channels. Max-pooling layers use a 2x2 kernel which reduces the image size based on the maximum value under the kernel during each stride.

DenseNet121 is suitable for training in cases of multi-label classification with a large number of labels present simultaneously, as well as multi-class classification with a large number of possible output classes, due to its depth and the ability to learn specific information. Shallow networks are rarely suitable for solving problems of the nature explored within this study because a large number of activation functions are required to fully separate the data set into classes in an n-dimensional space. However, although the depth of the network in this situation approaches the higher ceiling of the performance that can be achieved, it is accompanied by an extremely large increase in the time required for training as well as for the final decision making.

Other than DenseNet121, there exist two other similar architectures which could theoretically increase the prediction capabilities of the final decision support system, these being DenseNet169 and DenseNet201. These architectures provide an even greater number of activation functions for distributing the n-dimensional space described by the dataset into targeted classes. However, these architectures did not have any suitable pretrained weights for the application of transfer learning. Because of this, these networks would have to be trained from random starting weights in their deeper neuron layers and were not used in the scope of this study.

2.3 Parameter Sensitivity Analysis

The framework of sensitivity analysis conducted in this paper is presented in Fig. 1.

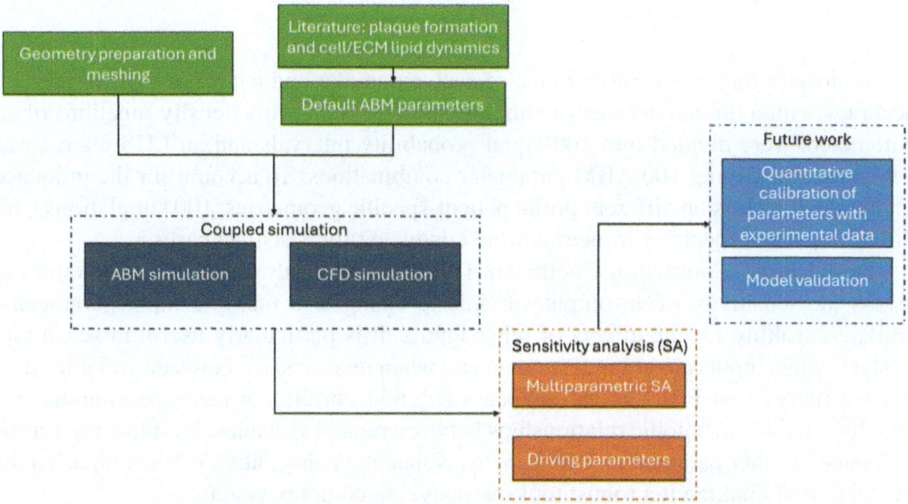

Fig. 1. Multiparametric sensitivity analysis of PAKAthero ABM

The intrinsic ABM parameters driving the simulation were initially defined by Garbey et al. [20] and the constant parameters that drive cellular events are:

- Probability of mitosis and apoptosis
- Smooth muscle cell (SMC) division in the intimal layer
- Extracellular matrix (ECM) deposition in intimal layer
- ECM deposition in medial layer
- SMC division in medial layer
- Outward remodeling driven by shear forces
- Outward remodeling driven by tensile forces

All of the aforementioned parameters have been defined in literature and their respective physiological ranges determined (Table 1.).

Multi-parametric sensitivity analysis was conducted using Latin hypercube sampling (LHS) to randomly sample the triangular probability density function of each parameter and define the parameter set for the ABM simulations. This method allowed for the

Table 1. ABM parameters and ranges

Parameter name	Parameter description	Parameter type	Range
α_1	Probability of mitosis and apoptosis	Constant	0.05
α_2	Probability of SMC proliferation in tunica media	Variable	2–17
α_3	Probability of SMC proliferation in intima	Variable	0–0.5
α_4	Probability of ECM degradation	Constant	0.008
α_5	Probability of lipid infiltration	Variable	0–0.106
α_6	Outward remodeling driven by shear forces	Variable	0–24.46
α_7	Outward remodeling driven by tensile forces	Variable	1.84–100

exploratory testing of the entire range of each parameter and is proven to achieve good accuracy with a limited number of simulations. The probability density functions of all parameters were divided into 100 equal probability intervals and an LHS matrix was generated identifying 100 ABM parameter combinations. To account for the influence of these parameters on different initial patient-specific geometries, 100 simulations with different patient-specific geometries were conducted for 3 distinct cases.

Partial Rank Correlation Coefficient (PRCC) was used as a statistical method to assess the sensitivity of an output variable to changes in multiple input parameters, while controlling for the effects of other inputs. It is particularly useful in sensitivity analysis when inputs are not independent and when relationships between variables may not be strictly linear. PRCC relies on Spearman's rank correlation, measuring the strength and direction of monotonic relationships between ranked variables. By adjusting for the influence of other parameters, PRCC helps isolate the contribution of each input on the model output, making it a robust tool for analyzing complex systems.

3 Results and Discussion

Figures 2, 3, 4, and 5 show the PRCCs between the variable model inputs ($\alpha2$, $\alpha3$, $\alpha5$, $\alpha6$ and $\alpha7$) and target model outputs such as arterial wall, arterial and final content of fibrous plaque type and calcified plaque type, respectively.

As it can be seen from Fig. 2, $\alpha2$, $\alpha3$, $\alpha5$ and $\alpha6$ exhibit relatively low but statistically significant PRCC. This implies that the input parameters have a meaningful influence on the output, even if the relationship is not extremely strong. This indicates increasing coefficients of SMC proliferation, lipid infiltration and outward remodeling driven by shear forces leads to a slight but consistent reduction in a physiological outcome, which was expected as all of these parameters should stimulate increased plaque growth thus constricting the arterial lumen.

There are no statistically significant PRCC scores for the influence of variable simulation parameters on remodeling of the arterial wall. The arterial wall's response might be driven by a combination of factors working together, rather than any single parameter exerting a dominant influence.

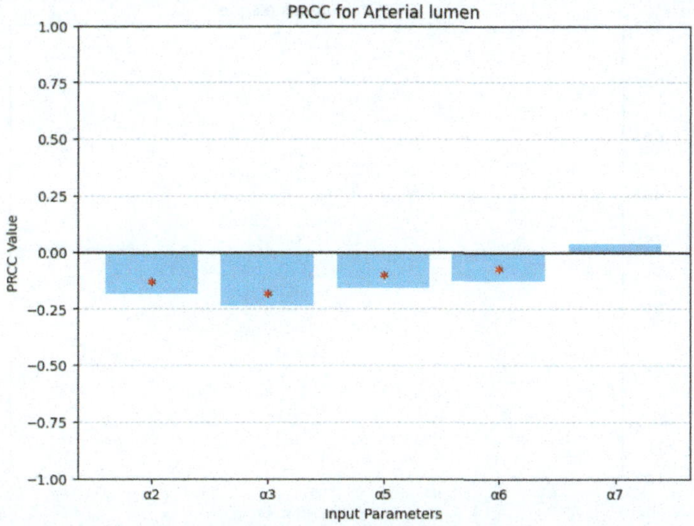

Fig. 2. PRCC for arterial lumen

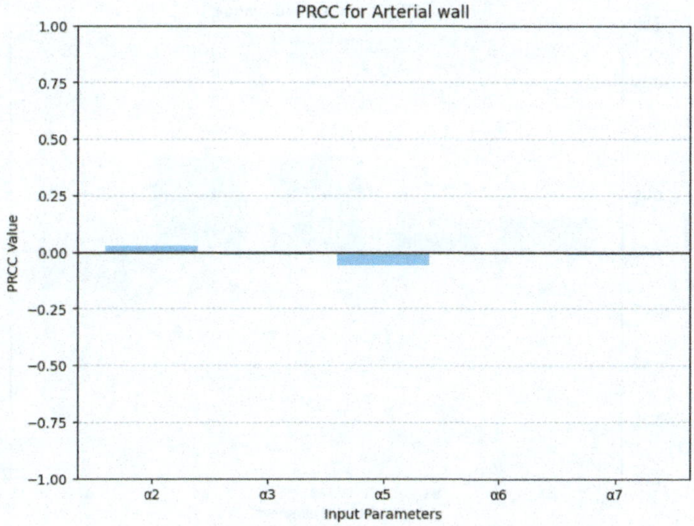

Fig. 3. PRCC for arterial wall

The only parameter shown to be statistically significant influence on fibrous plaque decrease is the probability of lipid infiltration. From a biological point of view, lipid infiltration leads to plaque progression towards transition to calcified plaque.

Variable parameters α2, α4, α5, α6 show a statistically significant positive PRCC while α7 shows a relatively low, but statistically significant, PRCC. A positive PRCC for α2 suggests that increased smooth muscle cell activity contributes to plaque growth,

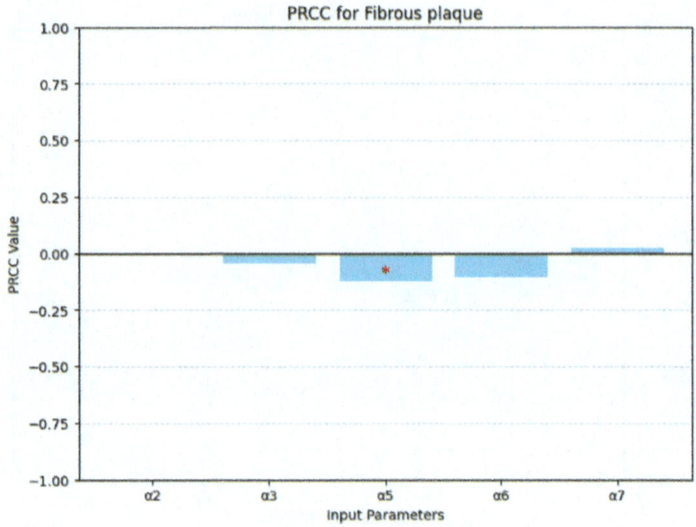

Fig. 4. PRCC for fibrous plaque

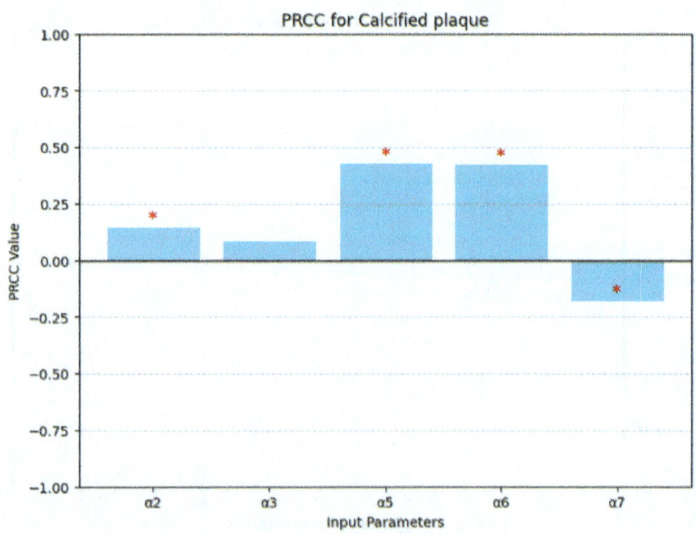

Fig. 5. PRCC for calcified plaque

while α4 indicates that weakening of the extracellular matrix exacerbates arterial occlusion. Similarly, α5 plays a crucial role in plaque formation by increasing lipid accumulation within the artery and α6 positively affects the lumen, suggesting that hemodynamic forces help maintain or expand the arterial diameter, while α7 shows a weaker, but still significant, influence on outward remodeling. These findings underscore the complex interplay between cellular proliferation, lipid dynamics, and mechanical forces in the progression of atherosclerosis.

4 Conclusion

The sensitivity analysis conducted on the ABM provided valuable insights into the model's mechanisms by identifying key influencing parameters, interactions among agent dynamics, and the impact of SMC, ECM, and lipid dynamics on changes in the lumen area. For the arterial wall output, several parameters, including α2 (SMC proliferation), α3 (SMC proliferation in the intima), α5 (lipid infiltration), and α6 (outward remodeling driven by shear forces), exhibit relatively low, but statistically significant, PRCCs. This suggests that, while these factors influence arterial changes, the relationships are not strongly pronounced. The lack of statistically significant PRCCs for the arterial wall remodeling suggests that its response is likely influenced by multiple factors in combination, rather than any single parameter dominating the process. In contrast, the fibrous plaque output shows that lipid infiltration plays a significant role in plaque progression, transitioning fibrous plaques into more advanced stages, such as calcified plaques. For the calcified plaque output, α2, α4, α5, and α6 demonstrate a strong positive influence, highlighting the roles of SMC proliferation, ECM degradation, and lipid infiltration in driving plaque calcification, while α7 also plays a role, albeit with a weaker influence. These findings emphasize the importance of cellular dynamics, lipid infiltration, and mechanical forces in the development and progression of atherosclerosis.

Acknowledgements. This paper is supported by the DECODE project (www.decodeitn.eu) that has received funding from the European Union's Horizon 2020 research and innovation programme under grant agreement No 956470. This article reflects only the authors' view. The Commission is not responsible for any use that may be made of the information it contains.

References

1. Kwak, B., et al.: Biomechanical factors in atherosclerosis: mechanisms and clinical implications. Eur. Heart J. **35**(43), 3013–3020 (2014)
2. Björkegren, J.L., Lusis, A.J.: Atherosclerosis: recent developments. Cell **185**(10), 1630–1645 (2022)
3. Jakka, V.V.S.V., Burša, J.: Finite element simulations of mechanical behaviour of endothelial cells. BioMed Res. Int. (2021)
4. Djorovic, S., Šaveljić, I., Filipovic, N.: Computational simulation of carotid artery: from patient-specific images to finite element analysis. J. Serbian Soc. Comput. Mech. (2019)
5. Filipovic, N.: Computational modeling of atherosclerosis, pp. 1–39 (2020)
6. Filipovic, N., Smiljana, T., Andjela, B., Branko, A., Miloš, A., Djukić, T.: Modeling of plaque progression in the carotid artery using coupled agent based with finite element method. IEEE-EMBS Int. Conf. Biomed. Health Inform. (BHI) **2022**, 1–4 (2022)
7. Nenad, D.F., Smiljana, T., Branko, A., Miloš, A., Djukić, T., Končar, I.: Agent based with finite element method for plaque progression in the carotid artery. In: 2023 24th International Conference on Digital Signal Processing (DSP), pp. 1–5 (2023)
8. Karimi, A., Navidbakhsh, M., Faghihi, S., Shojaei, A., Hassani, K.: A finite element investigation on plaque vulnerability in realistic healthy and atherosclerotic human coronary arteries. Proc. Inst. Mech. Eng. [H] **227**, 148–161 (2013)

9. Milos, J., Milos, D.R., Milosevic, Z., Filipovic, N., Obradovic, Z.: Machine learning approach for predicting wall shear distribution for abdominal aortic aneurysm and carotid bifurcation models. IEEE J. Biomed. Health Inform. **22**, 537–544 (2018)
10. Spahić, L., et al.: 'Prediction of coronary plaque progression using data mining and artificial neural networks. In: Book Prediction of Coronary Plaque Progression Using Data Mining and Artificial Neural Networks', pp. 3–13. Springer (2023)
11. Arsic, B., Tomasevic, S., Anic, M., Gakovic, B., Koncar, I., Filipovic, N.: Atherosclerotic plaque characterization using deep learning methods. In: AORTA, p. A085 (2022)
12. Avril, S., Mousavi, S.J.: Computational modeling of aneurysm growth in mechanobiology: Biomechanics of the Aorta, pp. 243–264. Elsevier (2024)
13. Filipovic, N., Tomasevic, S., Arsic, B., Anic, M., Djukic, T., Koncar, I.: Agent based with finite element method for plaque progression in the carotid artery. In: Book Agent Based with Finite Element Method for Plaque Progression in the Carotid Artery, pp. 1–5. IEEE (2023)
14. Mester, R., Alfonso, L., Chris, R., Lange, K.: 'Differential methods for assessing sensitivity in biological models. PLoS Comput. Biol. 18 (2021)
15. Pjotr, L.J.H., Quicken, S., Frans, N.v.d.V., Huberts, W.: Efficient sensitivity analysis for biomechanical models with correlated inputs. Int. J. Numer. Methods Biomed. Eng. (2023)
16. Holzapfel, G., Ogden, R.: Constitutive modelling of arteries. Proc. Roy. Soc. A: Math. Phys. Eng. Sci. **466**, 1551–1597 (2010)
17. Suh, G.-Y., et al.: Quantification of particle residence time in abdominal aortic aneurysms using magnetic resonance imaging and computational fluid dynamics. Ann. Biomed. Eng. **39**, 864–883 (2011)
18. Corti, A., Chiastra, C., Colombo, M., Garbey, M., Migliavacca, F., Casarin, S.: A fully coupled computational fluid dynamics–agent-based model of atherosclerotic plaque development: multiscale modeling framework and parameter sensitivity analysis. Comput. Biol. Med. **118**, 103623 (2020)
19. Filipovic, N., et al.: PAK Athero, Finite element program for atherosclerosis. In: Book PAK Athero, Finite element program for atherosclerosis' BioIRC (2023)
20. Garbey, M., Casarin, S., Berceli, S.A.: Vascular adaptation: pattern formation and cross validation between an agent based model and a dynamical system. J. Theor. Biol. **429**, 149–163 (2017)

Folder Design Optimization with Genetic Algorithm for Drug Coated Balloon Folding

Leo Benolić[1,2(✉)], Efstathios Stratakos[3], Elisabetta Stretti[4], Giancarlo Pennati[3], Spyridon Psarras[4], Vassilis Kostopoulos[4], and Nenad Filipović[1,2]

[1] Bioengineering Research and Development Centre (BioIRC), Prvoslava Stojanovica 6, 34000 Kragujevac, Serbia
leo.benolic@kg.ac.rs

[2] Faculty of Engineering, University of Kragujevac, Sestre Janjica 6, 34000 Kragujevac, Serbia

[3] Laboratory of Biological Structure Mechanics, Department of Chemistry, Materials and Chemical Engineering "Giulio Natta", Politecnico di Milano, Milan, Italy

[4] Laboratory of Applied Mechanics, Mechanical Engineering and Aeronautics Department, University of Patras, Patra, Greece

Abstract. Peripheral Artery Disease (PAD) significantly impacts patient life quality by reducing blood flow to the extremities, leading to severe complications. The treatment protocol often includes the use of drug-coated balloons (DCB) during percutaneous transluminal angioplasty, which can reduce the recurrence of arterial narrowing. However, the effectiveness of DCBs is highly dependent on the precision of the balloon folding process prior to deployment. In this paper, we introduce an innovative application of artificial intelligence (AI) to optimize the folding geometry of DCBs. Utilizing a dataset from Boston Scientific, we employ a genetic algorithm (GA) to explore and optimize folder shapes, aiming to enhance the effectiveness and consistency of balloon folding. Our methodology includes defining fitness functions that evaluate the geometrical configuration and operational functionality, ensuring that the folding process minimizes potential damage to the drug coating and maximizes therapeutic outcomes. Results indicate that AI-driven optimization can significantly refine the folding process, offering potential improvements in the clinical application of DCBs. This study does not only advance the engineering of medical devices, but it also illustrates the potential for AI to enhance therapeutic efficacy in the treatment of PAD.

Keywords: peripheral artery disease · DCB · design optimization · genetic algorithms · AI optimization

1 Introduction

Peripheral Artery Disease (PAD) is a prevalent disorder that impacts the arteries beyond heart and brain, most notably those in legs. This condition arises from the accumulation of plaque within the arteries, leading to diminished blood flow to the extremities. This reduced circulation can cause symptoms such as pain, cramps, and numbness. In its more severe forms, PAD can result in tissue damage, infections, or even necessitate amputation.

© The Author(s), under exclusive license to Springer Nature Switzerland AG 2025
N. Filipović (Ed.): AAI 2024, LNNS 1446, pp. 185–191, 2025.
https://doi.org/10.1007/978-3-031-99201-8_19

Prompt recognition and management of PAD are essential to avert these serious outcomes and to enhance life quality, particularly among the elderly and individuals with risk factors like smoking, diabetes, and hypertension.

Percutaneous transluminal angioplasty (PTA) with a drug-coated balloon (DCB) is a refined medical approach for managing peripheral arterial disease (PAD), particularly in the lower extremities. This technique utilizes a special type of angioplasty balloon that is coated with a medication designed to prevent restenosis (re-narrowing) of the artery after treatment. The DCB delivers the drug directly to the arterial wall during the angioplasty procedure, enhancing the therapeutic effect by inhibiting the proliferation of cells that can lead to restenosis. The procedure involves the insertion of a thin, flexible catheter through a small incision in the skin, usually at the groin or arm. The catheter is navigated through the vascular system to the site of the arterial blockage. Once in place, the drug-coated balloon at the tip of the catheter is inflated. This inflation compresses the plaque against the artery wall and simultaneously releases the medication, which is absorbed by the arterial tissue. The use of a DCB typically extends the procedure time slightly compared to standard PTA, due to the need for precise delivery of the drug. However, it can be performed on an outpatient basis under local anesthesia, with most patients able to return home the same day. The introduction of DCBs has improved the long-term outcomes of percutaneous interventions by reducing the incidence of restenosis, thus potentially decreasing the need for repeat procedures. Despite these advancements, it is important to note that while DCBs offer significant benefits over traditional balloons, they do not eliminate the risk of restenosis entirely. Continuous follow-up is crucial to monitor and address any recurrent narrowing of the treated arteries [1].

Previous studies on balloon folding modeling in the context of stents have utilized advanced numerical methods like finite element (FE) analysis to simulate preoperative processes such as folding, pleating, and stent crimping, including detailed analyses of residual stresses and their impact on the accuracy of stent expansion models. Other research has explored optimal stent design strategies through the development and comparison of different expansion modeling approaches, emphasizing the importance of realistic balloon geometry and detailed examination of the interactions between stent material and balloon during expansion [2, 3].

In this paper, we employ AI techniques to optimize the folder shape for the folding process of drug coated balloons. Through the utilization of AI algorithms, we aim to develop an efficient methodology for determining the ideal folder configuration that maximizes the effectiveness of the folding process. This optimization process will involve the exploration of various folder shapes and their impact on the folding outcome. The proposed approach is expected to enhance the efficiency and accuracy of balloon folding procedures, ultimately leading to improved product quality and performance in medical applications.

2 Material and Method

In this study, we utilized a dataset provided by the manufacturer Boston Scientific, which includes the diameters for various sizes of the drug-coated balloon (DCB) and the number of folds. This dataset was essential for designing the initial geometry [4] and establishing limits on the size and positioning within the geometry, enabling our optimization algorithm to focus on a specific design and enhancing its ability to converge towards a solution [5].

For optimization, we employed a genetic algorithm (GA), inspired by the processes of natural selection and genetics. The choice of a GA was driven by its ability to efficiently search through large and complex solution spaces. The algorithm began with an initial population of 1,000 chromosomes, each representing a potential solution. A specific fitness function [1] was defined to measure the quality of each chromosome, essentially assessing how well the solution meets the established criteria, which is a crucial aspect of this study for determining the superior design.

$$\mathrm{FitnessF}() = \textit{evaluation of parametric model for the chromosome} \qquad (1)$$

The selection process for chromosomes for reproduction was based on roulette wheel selection, and crossover was executed using a single-point crossover with a probability of 0.7. Mutation was applied to 1% of the population at a mutation probability of 0.01 per chromosome. The iterative process of selection, crossover, and mutation continued through 1000 generations or until a satisfactory level of convergence towards the optimal solution was achieved. Parameters such as population size, crossover probability, mutation probability, and the number of generations were carefully selected based on prior experiments and literature recommendations tailored to the type of problem at hand [6].

This methodological approach enabled efficient exploration of the solution space and provided valuable insights into potential optimal strategies for addressing the problem. We ran the genetic algorithm multiple times to identify the best result, characterized by the lowest value of the fitness function for this task. The implementation was carried out in Python, and the use of parallel computing through the multiprocessing library maximized the utilization of computer resources, enabling the GA to explore a broader range of possible solutions more rapidly.

3 Results and Discussion

The results of the study were derived using a folder geometry similar to that depicted in Fig. 1. The folders utilized were from an MSI FFS1075S machine. In this geometry, variables such as 'R satellite'—the pivot distance from the center—and 'satellite angle' were defined. The design included only one folder, with the other N_satellite folders were being created by rotating this primary folder. The folder's shape could be selected from either a 3-point arc or a spline, with the points being variable.

All these variables were outputs from the genetic algorithm (GA), and the feedback was provided via a fitness function. This fitness function is the most crucial part of the study, divided into three components: (1) collision, where negative points are assigned for any collisions between folders, (2) negative points for translational movement during balloon contact, and (3) the direction of the tangent to the shape of the balloon fold. The methods of scoring were illustrated in Fig. 2, which shows the Finite Element Method (FEM) simulation of the folding process of a 4mm balloon with 5 folds, and in Fig. 3, which depicts the shape of the balloon after pleating.

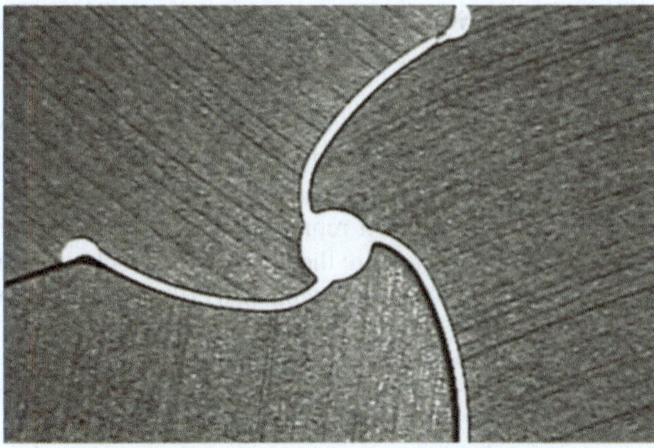

Fig. 1. Folders from MSI FFS1075S machine [5]

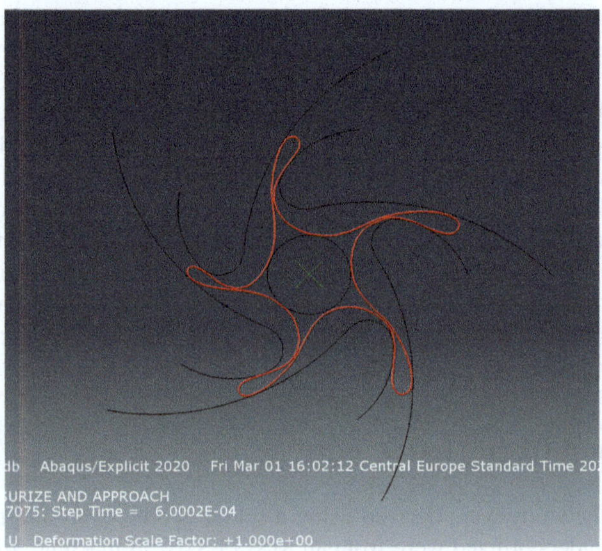

Fig. 2. Finite Element Method (FEM) Simulation of the Folding Process of a 4 mm Balloon with 5 Folds

One of the key design considerations was to avoid collisions between the folders; if one folder came too close to another—less than the thickness of two balloon walls—the part of the fitness function [1] that calculated their overlap would assign a negative score with a factor of 50. Another part of the function penalized large translational movements of the balloon during the final stages of the folding process, which could lead to damage to the coating, with a negative factor of 20. The third component ensured that the final shape of the balloon was not symmetrical, but spiraled in a clockwise direction (tangent

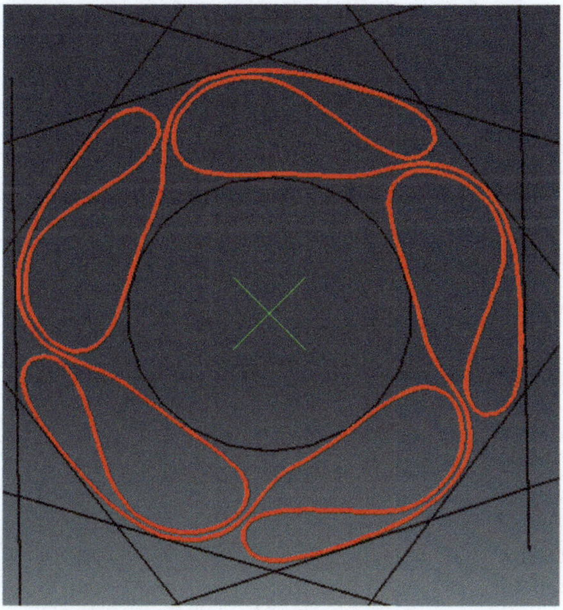

Fig. 3. Shape of the Balloon after Pleating

intersects the y-axis on the positive side), facilitating the pleating process.

$$\text{FitnessF}() = -50 * \textit{overlap} - 20 * \textit{Ltranslation} + \textit{Spiral shape evaluation} \quad (2)$$

By adjusting these factors, various designs could be achieved, each with its advantages depending on what was more critical for the user. One potential solution is shown in Fig. 4 – an example of possible folders design in the final compressed position of the balloon, while intermediate positions are displayed in Fig. 5 – an example of possible folders design, intermediate position. This approach allowed for a nuanced understanding and optimization of the balloon folding design, catering to specific operational needs and enhancing the overall functionality of the DCB.

Previous research has not been extensively focused on balloon design as the emphasis has been on stents rather than DCB, which have distinct requirements [2, 3]. Therefore, we believe this approach is beneficial. This model can provide not just one, but multiple solutions, depending on which factors are considered most critical, or which properties are given greater importance or weight in the success function. On the other hand, this approach will not always work if the set success function does not have a pronounced minimum, i.e., the best solution. In such cases, the genetic algorithm may struggle to converge to a solution or remain in local minima and fail to find the best solution. As simple as the use of this algorithm may be, it is crucial to properly evaluate each solution, i.e., to create a robust success function.

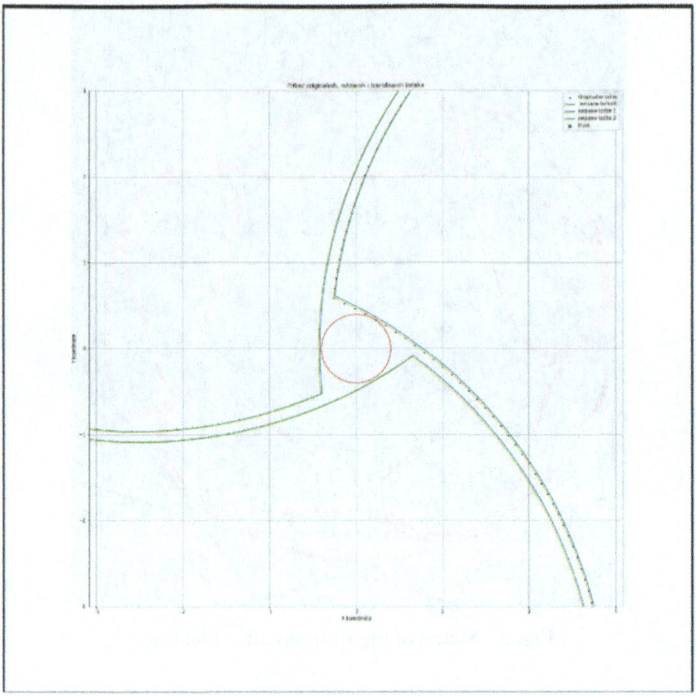

Fig. 4. Example of possible folders design, final position

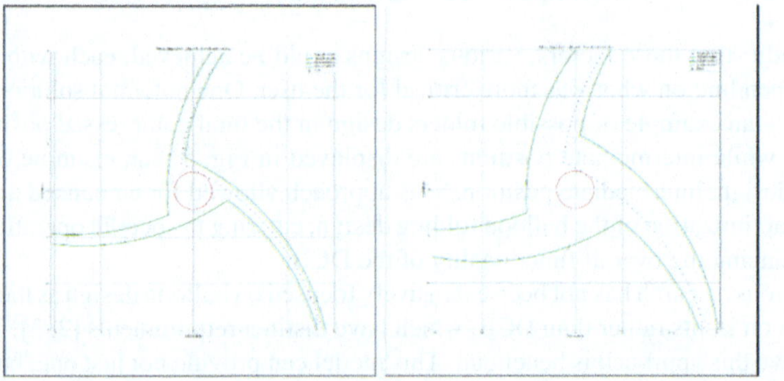

Fig. 5. Example of possible Folders design, intermediate position

4 Conclusion

In conclusion, this study effectively employs advanced artificial intelligence techniques, particularly a genetic algorithm (GA), to optimize the folding process of drug-coated balloons used in percutaneous transluminal angioplasty for treating Peripheral Artery Disease (PAD). By tailoring the folding geometry through computational optimization, we mitigated potential complications such as balloon damage during deployment, critical for ensuring the efficacy of the treatment. The GA facilitated a systematic exploration of design variables, enhancing deployment effectiveness and clinical outcomes by ensuring consistent drug delivery and reducing the risk of restenosis. The findings underscore the importance of precision in medical device engineering and suggest that further refinements in genetic algorithms could lead to more personalized and effective treatments. This integration of genetic algorithms into the design process exemplifies the potential of combining engineering with biomedicine, promising a future of increasingly precise, personalized, and effective medical treatments.

Acknowledgment. This research is supported by the DECODE project that has received funding from the European Union's Horizon 2020 research and innovation programme under the Marie Skłodowska-Curie grant agreement No 956470. This article reflects only the author's view. The Commission is not responsible for any use that may be made of the information it contains.

References

1. Tatò, F., Heilmeier, B.: MMW - Fortschritte der Medizin **160**(2), 54–62 (2018). https://doi.org/10.1007/s15006-018-0002-6
2. Geith, M., Swidergal, K., Hochholdinger, B., Schratzenstaller, T., Wagner, M., Holzapfel, G.: On the importance of modeling balloon folding, pleating, and stent crimping: a FE study comparing experimental inflation tests. Int. J. Numer. Meth. Biomed. Eng. **35**, e3249 (2019). https://doi.org/10.1002/cnm.3249
3. De Beule, M., et al.: Realistic finite element-based stent design: the impact of balloon folding. J. Biomech. **41,** 383–389 (2008). https://doi.org/10.1016/j.jbiomech.2007.08.014
4. MSI FFS1075S drug-eluting balloon pleating and folding equipment. https://msi.equipment/product/ffs975s/. Accessed 20 Mar 2024
5. https://www.bostonscientific.com/en-EU/products/balloons--drug-coated/ranger.html. Accessed 1 Apr 2024
6. https://towardsdatascience.com/genetic-algorithm-an-optimization-approach-dd60e23261e6. Accessed 24 Mar 2024

Utilizing Fuzzy Logic for Enhanced Characterization of Ovarian Tumors

Marija Branković[1](✉), Ana Mirić[1], Tijana Geroski[2,3], Marko Živanović[1], and Nenad Filipović[2,3,4]

[1] Institute for Information Technologies Kragujevac, University of Kragujevac, Jovana Cvijica Bb, 34000 Kragujevac, Serbia
marija.brankovic@uni.kg.ac.rs

[2] Faculty of Engineering, University of Kragujevac, Sestre Janjic 6, 34000 Kragujevac, Serbia

[3] 3 BioIRC – Bioengineering Research and Development Center , University of Kragujevac, Prvoslava Stojanovića 6, Kragujevac, Serbia

[4] Centre for Interdisciplinary and Multidisciplinary Studies, University of Montenegro, Podgorica, Montenegro

Abstract. Over time, ovarian tumors have remained a prominent cause of mortality among women worldwide. Given the complex and uncertain origins of the disease, various sophisticated diagnostic models have emerged to differentiate between benign and malignant tumors. This present study aims to employ fuzzy logic for tumor characterization, focusing on the expression of apoptotic genes *Bax*, *Bcl-2*, and *Cas3*. Using MATLAB software, we devised a fuzzy system to classify ovarian tumors based on qualitative polymerase chain reaction (qPCR) data from eleven samples. The system outputs three potential diagnoses: benign, borderline, and malignant tumors, with assessment scores ranging from 0 to 1. Scores below 0.5 suggest benign tumors, scores above 0.5 indicate invasive potential, and those near 0.5 indicate borderline cases. The system consistently produced results aligned with known sample diagnoses upon inputting relative gene expression values. This method shows promise in advancing ovarian tumor characterization, potentially improving treatment outcomes. Nonetheless, further comprehensive research and an extensive literature review are essential to validate these initial findings.

Keywords: ovarian tumors · apoptosis · qualitative polymerase chain reaction · gene expression · Bax · Bcl-2 · Cas3 · fuzzy logic

1 Introduction

Over the years, ovarian tumors have been one of the leading causes of death in women [1]. Although early diagnosis can be crucial for the progression of the disease, the outcome is highly dependent on the applied therapy. The etiology of ovarian tumors is not entirely understood [2], making it challenging to distinguish benign and malignant lesions and ensure the appropriate treatment. Considering that, this endeavor aims to examine the expression of apoptotic genes and employ AI to establish a system that will permit a more precise characterization of ovarian tumors.

© The Author(s), under exclusive license to Springer Nature Switzerland AG 2025
N. Filipović (Ed.): AAI 2024, LNNS 1446, pp. 192–202, 2025.
https://doi.org/10.1007/978-3-031-99201-8_20

Benign tumors do not invade nearby or distant tissues; instead, they stay in the primary site of their origin. They have boundaries that are easily defined and grow more slowly. On the other hand, malignant tumor cells grow and divide uncontrollably, spreading through the lymphatic or circulatory systems to different parts of the body [3]. Although all women face some risk of illness, the human body has natural defenses that identify and eliminate abnormal cells. Programmed cell death, or apoptosis, is one such defense mechanism. The expression of different apoptotic factors (genes) controls this process by initiating a set of morphological and biochemical alterations [4–6]. In this context, the cornerstone of clinical oncology is the invention of treatments that encourage the death of tumor cells through apoptosis [7, 8]. Research indicates that a higher presence of apoptotic cells suggests a lower risk of aggressive tumor forms. Conversely, a deficiency in this mechanism may elevate the risk of developing malignant formations [9] (Table 1).

Table 1. Differences between benign and malignant tumors.

Benign tumors	Malignant tumors
Do not invade nearby or distant tissues	Do invade nearby or distant tissues
Remain at the primary site of their origin	Move from one area to another through the lymphatic or circulatory systems
Have boundaries that are easily defined	Have boundaries that are not easily defined
Grow more slowly	Grow rapidly and divide uncontrollably
Contain more apoptotic cells	Contain fewer apoptotic cells

Identifying benign from malignant tumors at an early stage is one of the most challenging problems in modern gynecology. It is a notably difficult task for inexperienced gynecologists and small medical centers without the resources to perform specialized medical exams [10, 11]. For this reason, oncologists and gynecologists have developed a wide range of diagnostic models, most of which rely on scoring systems and logistic regression [12–15].

Biological systems are non-linear and time-varying [16]. Additionally, they are frequently characterized qualitatively, with terms such as weakly expressed or moderately active. Despite these limitations, fuzzy logic can predict outcomes in biomolecular networks and cellular processes [17–19]. Multiple studies utilize fuzzy logic to aid in the early and more precise diagnosis of ovarian tumors [11, 20–26]. This study focuses on characterizing ovarian tumors according to the apoptotic gene expression of Bax, Bcl-2, and Cas3.

2 Materials and Methods

2.1 Gene Expression Values

Gene expression involves converting genetic information into functional gene products, such as proteins. Researchers can identify which genes are active or expressed within a specific cell, tissue, or organism under particular conditions by measuring gene expression [27].

This study analyzed gene expression values (Table 2) sourced from the research conducted by Perić et al. The biopsied tumors included 7 benign, 1 borderline, and 3 malignant samples. The relative expression values were obtained through RNA isolation and reverse transcription, followed by qualitative polymerase chain reaction (qPCR) [9].

Table 2. Relative gene expression values for each sample.

Sample	*Bax*	*Bcl-2*	*Cas3*	Tumor type
O1T	0.687771	1.04681	0.126701	Borderline
O2T	0.047268	2.63536	0.179555	Malignant
O3T	2.015308	1.873843	3.032484	Benign
O4T	2.550889	1.484524	1.533683	Malignant
O5T	0.700764	1.331913	0.547906	Benign
O6T	0.427057	0.882703	0.426022	Benign
O7T	0.298644	1.6449	0.422493	Benign
O8T	6.081807	1.667284	0.804687	Benign
O9T	5.212589	3.00945	1.153885	Benign
O10T	4.384766	2.252458	0.941805	Malignant
O11T	1.750783	1.718917	1.096192	Benign

Expression levels were evaluated relative to the reference β-actin gene, which is constant and integral to basic cellular processes [28]. Accordingly, the standard (control) expression level is 1 (expression of the reference gene). Levels below 1 indicate low expression and levels above 1 indicate high expression (Table 3) [29].

Table 3. Expression levels defined relative to the reference β-actin gene.

Low expression	Normal (control) expression	High expression
< 1	1	> 1

2.2 Fuzzy Logic Implementation

Input data included relative gene expressions (Bax, Bcl-2, Cas3), and the output (diagnosis) resulted in three possible outcomes: benign, borderline, or malignant tumor (Fig. 1). We have introduced subcategories for high and low expression ranges to improve accuracy. The breadth of expression dictates the number of subcategories utilized. This variability is particularly noticeable in the case of the Bax gene. Conversely, the expression range for the Bcl-2 and Cas3 genes is narrower, resulting in fewer subcategories (Table 4).

Table 4. Subcategory classification based on the relative expression levels of *Bax*, *Bcl-2*, and *Cas3* genes compared to *β-actin*.

	Very low	Low	**Normal** (*β-actin*)	Medium	High	Very high
Bax (0.04 –6.08)	0 – 0.5	0.51 – 0.99	1	1.01 – 2	2.01 – 4.5	4.51 – 6.1
Bcl-2 (0.8 – 3.009)	-	0 – 0.99	1	1.01 – 2	2.01 – 3.1	-
Cas3 (0.12 – 3.03)	-	0 – 0.99	1	1.01 – 2	2.01 – 3.1	-

These subcategories are based on the expression of the reference β-actin gene and the values detailed in Table 2. Membership functions used Gaussian normal distributions with standard deviations from 0.1 to 0.5. Finally, data from Table 5 defined 11 rules (one for each sample) (Fig. 2).

The initial step in creating rules using a fuzzy logic rule editor is identifying the variables that define the system. In these variables, terms or phrases are used instead of numerical values. IF-THEN relationships connect these two types of variables. Utilizing logical operators like AND, OR, or NOT allows for deriving conclusions based on precise conditions. Fuzzy rules have the following basic form [30]:

$$\text{IF } x \text{ is A THEN } y \text{ is B}$$

where x and y are linguistic variables.

In our scenario, the subcategories for apoptotic gene expression determine the linguistic input, and the linguistic output is the tumor diagnosis. Specific conclusions were derived using the AND logical operator.

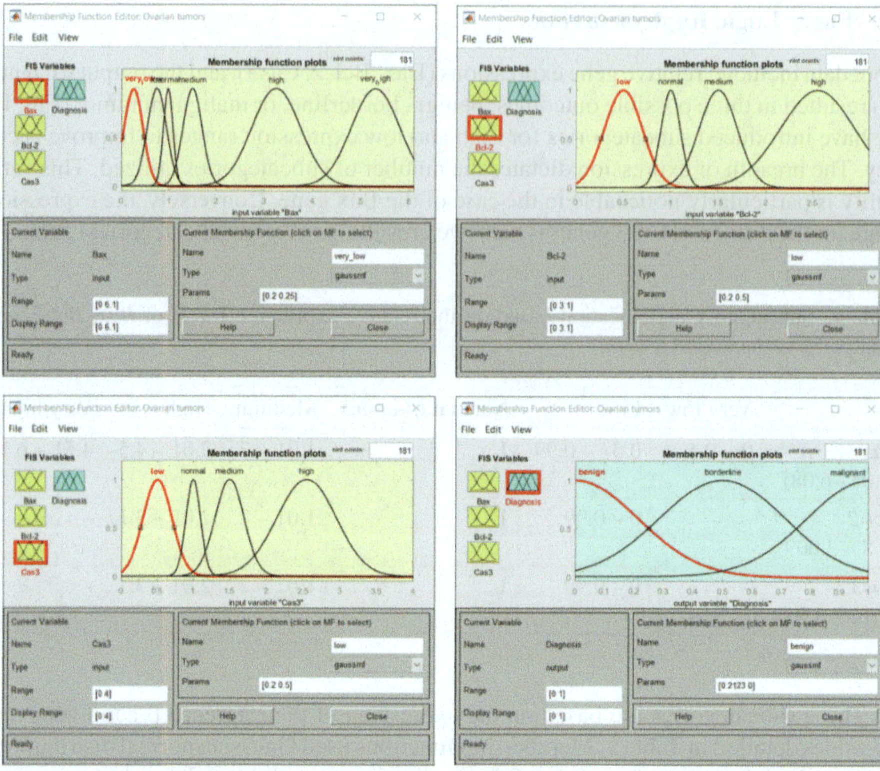

Fig. 1. Input (*Bax*, *Bcl-2*, *Cas3*) and output (diagnosis) variables. The *Bax* gene expression values are divided into six subcategories: very low, low, normal, medium, high, and very high. Additionally, the *Bcl-2* and *Cas3* gene expression values are categorized as low, normal, medium, and high. Specific criteria outlined in Table 4 define the boundaries for each subcategory. Tumor diagnosis (output), based on a scale from 0 to 1, categorizes tumors as benign, borderline, or malignant. A score below 0.5 indicates benign tumors, while scores above 0.5 indicate malignant tumors. Results near 0.5 indicate borderline tumors.

3 Results and Discussion

The fuzzy system was used to input expression values from the referenced samples, generating diagnostic scores aligned with the tumor types listed in Table 2 (Figs. 3, 4, 5, Table 6). As previously noted, the output data present three potential outcomes: benign, borderline, and malignant tumors. The assessment score ranges from 0 to 1. Scores below 0.5 suggest benign tumors, while scores above 0.5 suggest potential malignancy. This boundary distinguishes benign and malignant diagnoses, identifying borderline tumors near the 0.5 threshold.

Table 5. Input subcategories and tumor types (known diagnosis) for each sample.

Sample	*Bax*	*Bcl-2*	*Cas3*	Tumor type (*Perić et al.*)
O1T	Low	Medium	Low	Borderline
O2T	Very low	High	Low	Malignant
O3T	High	Medium	High	Benign
O4T	High	Medium	Medium	Malignant
O5T	Low	Medium	Low	Benign
O6T	Very low	Low	Low	Benign
O7T	Very low	Medium	Low	Benign
O8T	Very high	Medium	Low	Benign
O9T	Very high	High	Medium	Benign
O10T	High	High	Low	Malignant
O11T	Medium	Medium	Medium	Benign

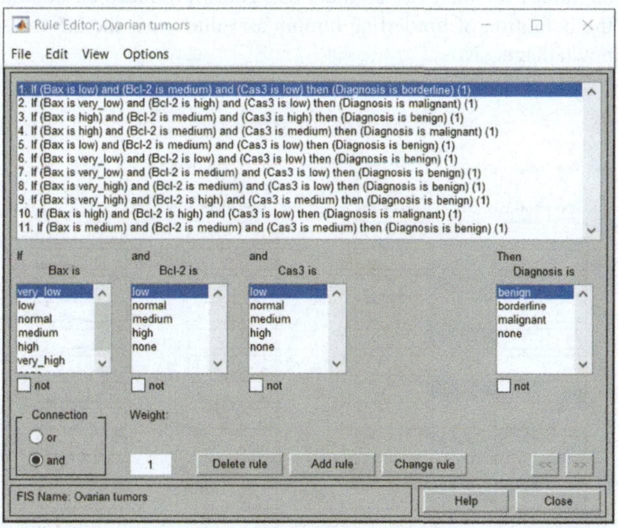

Fig. 2. Fuzzy logic rule editor. The rules are created using the expression levels and diagnostic status (tumor type) provided in the study by *Perić et al.* For instance, in the sample O1T, the *Bax* gene expression of 0.68 is categorized as "low expression" according to Table 2 and Table 4. The same categorization process is used for the *Bcl-2* and *Cas3* genes, with their expressions categorized as medium and low, respectively. Based on the borderline nature of the tumor, the corresponding diagnosis is applied to this sample in the rules editor. In this manner, rules were established for each sample.

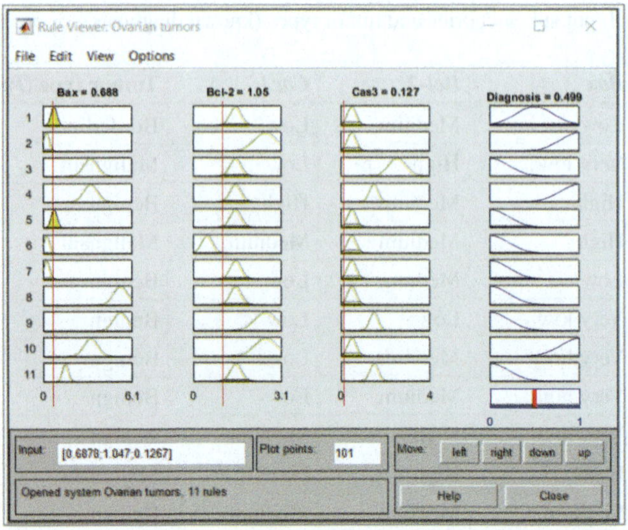

Fig. 3. Result of the created fuzzy system for the sample O1T (borderline tumor). The input of relative expression values for *Bax*, *Bcl-2*, and *Cas3* genes produced an output value of 0.499. Consistent with the definition of borderline tumors as values near the 0.5 threshold, this result aligns with the known diagnosis.

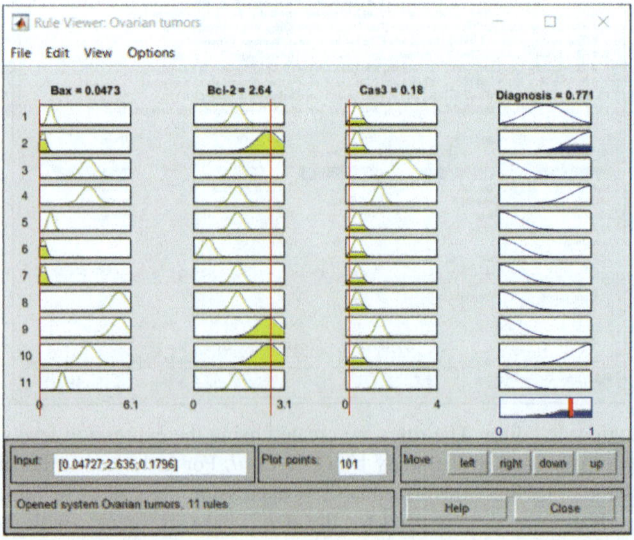

Fig. 4. Result of the created fuzzy system for the sample O2T (malignant tumor). The input of relative expression values for *Bax*, *Bcl-2*, and *Cas3* genes produced an output value of 0.771. Consistent with the definition of malignant tumors as values above the 0.5 threshold, this result aligns with the known diagnosis.

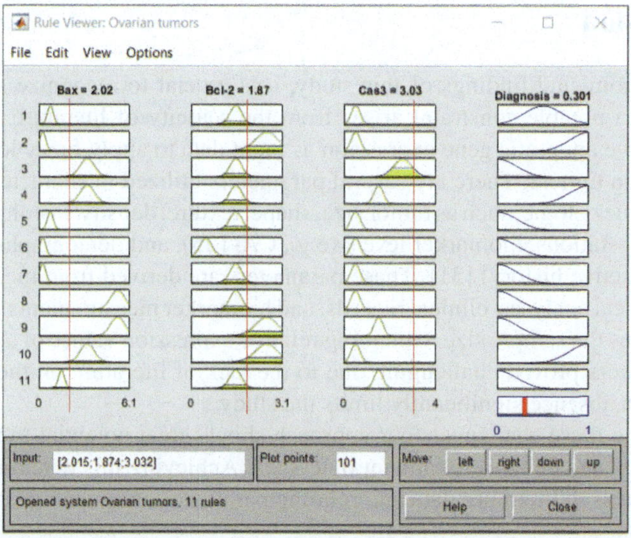

Fig. 5. Result of the created fuzzy system for sample O3T (benign tumor). The input of relative expression values for *Bax*, *Bcl-2*, and *Cas3* genes produced an output value of 0.301. Consistent with the definition of benign tumors as values below the 0.5 threshold, this result aligns with the known diagnosis.

Table 6. Tabular representation of data for each sample. The results are consistent with the known diagnoses.

Sample	Known diagnosis	Input data			Output data	0.5	Resulting diagnosis
		Bax	*Bcl-2*	*Cas3*			
O1T	Borderline	0.687771	1.04681	0.126701	0.499	≈ 0.5	Borderline
O2T	Malignant	0.047268	2.63536	0.179555	0.771	> 0.5	Malignant
O3T	Benign	2.015308	1.873843	3.032484	0.302	< 0.5	Benign
O4T	Malignant	2.550889	1.484524	1.533683	0.733	> 0.5	Malignant
O5T	Benign	0.700764	1.331913	0.547906	0.420	< 0.5	Benign
O6T	Benign	0.427057	0.882703	0.426022	0.270	< 0.5	Benign
O7T	Benign	0.298644	1.6449	0.422493	0.251	< 0.5	Benign
O8T	Benign	6.081807	1.667284	0.804687	0.226	< 0.5	Benign
O9T	Benign	5.212589	3.00945	1.153885	0.239	< 0.5	Benign
O10T	Malignant	4.384766	2.252458	0.941805	0.642	> 0.5	Malignant
O11T	Benign	1.750783	1.718917	1.096192	0.290	< 0.5	Benign

4 Conclusion

Despite the promising findings of this study, it is crucial to recognize several limitations. Firstly, a notable constraint arises from the scarcity of literature. No study has used the relative apoptotic gene expression as input data to apply fuzzy logic to characterizing ovarian tumors. There are several parameters utilized in characterizing ovarian tumors using fuzzy logic, such as tumor size, shape, texture, density, histological features, clinical manifestations, biomarker levels (e.g., CA-125), and demographic factors (e.g. age or reproductive history) [31]. These parameters are derived from medical imaging, histopathological analyses, clinical records, and biomarker measurements. A further limitation concerns the sample size. Obtaining relative expression values of apoptotic genes in ovarian tumors proved challenging due to the lack of literature on their expression. Thus, this data absence significantly limits the study.

Yet, despite these challenges, our approach shows great potential to improve diagnostic accuracy and enhance treatment outcomes. Achieving this potential will require increasing access to literature data and refining parameter optimization strategies.

Funding. The research was funded by the Ministry of Science, Technological Development and Innovation of the Republic of Serbia, contract numbers [451–03-66/2024–03/200378 (Institute for Information Technologies Kragujevac, University of Kragujevac) and 451–03-65/2024–03/200107 (Faculty of Engineering, University of Kragujevac)].

References

1. Bray, F., Ferlay, J., Soerjomataram, I., Siegel, R.L., Torre, L.A., Jemal, A.: Global cancer statistics 2018: GLOBOCAN estimates of incidence and mortality worldwide for 36 cancers in 185 countries. CA Cancer J. Clin. **68**(6), 394–424 (2018). https://doi.org/10.3322/caac.21492
2. Brett, M.R., et al.: Epidemiology of ovarian cancer: a review. Cancer Biol. Med. **14**(1), 9–32 (2017). https://doi.org/10.20892/j.issn.2095-3941.2016.0084
3. Patel, A.: Benign vs malignant tumors. JAMA Oncol. **6**(9), 1488 (2020). https://doi.org/10.1001/jamaoncol.2020.2592
4. Kiraz, Y., et al.: Major apoptotic mechanisms and genes involved in apoptosis. Tumor Biol. **37**(7), 8471–8486 (2016). https://doi.org/10.1007/s13277-016-5035-9
5. Nagata, S.: Apoptosis and clearance of apoptotic cells. Annu. Rev. Immunol. **36**(1), 489–517 (2018). https://doi.org/10.1146/annurev-immunol-042617-053010
6. Koff, J., Ramachandiran, S., Bernal-Mizrachi, L.: A time to kill: targeting apoptosis in cancer. Int. J. Mol. Sci. **16**(2), 2942–2955 (2015). https://doi.org/10.3390/ijms16022942
7. Carneiro, B.A., El-Deiry, W.S.: Targeting apoptosis in cancer therapy. Nat. Rev. Clin. Oncol. **17**(7), 395–417 (Jul.2020). https://doi.org/10.1038/s41571-020-0341-y
8. Pfeffer, C., Singh, A.: Apoptosis: a target for anticancer therapy. Int. J. Mol. Sci. **19**(2), 448 (Feb.2018). https://doi.org/10.3390/ijms19020448
9. Perić, A., et al.: Expression of proapoptotic-antiapoptotic genes in malignant, borderline and benign ovarian tumors. Kragujev. J. Sci. **45**, 121–128 (2023). https://doi.org/10.5937/KgJSci2345121P
10. Du Bois, A., Rochon, J., Pfisterer, J., Hoskins, W.J.: Variations in institutional infrastructure, physician specialization and experience, and outcome in ovarian cancer: a systematic review. Gynecol. Oncol. **112**(2), 422–436 (2009). https://doi.org/10.1016/j.ygyno.2008.09.036

11. Żywica, P., Dyczkowski, K., Wójtowicz, A., Stachowiak, A., Szubert, S., Moszyński, R.: Development of a fuzzy-driven system for ovarian tumor diagnosis. Biocybern. Biomed. Eng. **36**(4), 632–643 (2016). https://doi.org/10.1016/j.bbe.2016.08.003
12. Alcázar, J.L., Mercé, L.T., Laparte, C., Jurado, M., López-García, G.: A new scoring system to differentiate benign from malignant adnexal masses. Am. J. Obstet. Gynecol. **188**(3), 685–692 (2003). https://doi.org/10.1067/mob.2003.176
13. Szpurek, D., Moszynski, R., Zietkowiak, W., Spaczynski, M., Sajdak, S.: An ultrasonographic morphological index for prediction of ovarian tumor malignancy. Eur. J. Gynaecol. Oncol. **26**(1), 51–54 (2005)
14. Timmerman, D., et al.: Logistic regression model to distinguish between the benign and malignant adnexal mass before surgery: a multicenter study by the international ovarian tumor analysis group. J. Clin. Oncol. **23**(34), 8794–8801 (2005). https://doi.org/10.1200/JCO.2005.01.7632
15. Jacobs, I., Oram, D., Fairbanks, J., Turner, J., Frost, C., Grudzinskas, J.G.: A risk of malignancy index incorporating CA 125, ultrasound and menopausal status for the accurate preoperative diagnosis of ovarian cancer. BJOG Int. J. Obstet. Gynaecol. **97**(10), 922–929 (1990). https://doi.org/10.1111/j.1471-0528.1990.tb02448.x
16. Makkar, R.: Application of fuzzy logic: a literature review. Int. J. Stat. Appl. Math. **3**(1), 357–359 (2018)
17. Nobile, M.S., et al.: Fuzzy modeling and global optimization to predict novel therapeutic targets in cancer cells. Bioinformatics **36**(7), 2181–2188 (Apr.2020). https://doi.org/10.1093/bioinformatics/btz868
18. Morris, M.K., Saez-Rodriguez, J., Sorger, P.K., Lauffenburger, D.A.: Logic-based models for the analysis of cell signaling networks. Biochemistry **49**(15), 3216–3224 (2010). https://doi.org/10.1021/bi902202q
19. Samaga, R., Klamt, S.: Modeling approaches for qualitative and semi-quantitative analysis of cellular signaling networks. Cell Commun. Signal **11**(1), 43 (2013). https://doi.org/10.1186/1478-811X-11-43
20. Tan, T.Z., Quek, C., Ng, G.S., Razvi, K.: Ovarian cancer diagnosis with complementary learning fuzzy neural network. Artif. Intell. Med. **43**(3), 207–222 (2008). https://doi.org/10.1016/j.artmed.2008.04.003
21. Consiglio, A., et al.: Explaining ovarian cancer gene expression profiles with fuzzy rules and genetic algorithms. Electronics **10**(4), 375 (2021). https://doi.org/10.3390/electronics10040375
22. Wang, D., Quek, C., See Ng, G.: Ovarian cancer diagnosis using a hybrid intelligent system with simple yet convincing rules. Appl. Soft Comput. **20**, 25–39 (2014). https://doi.org/10.1016/j.asoc.2013.12.018
23. Kalaiyarasi, M., Rajaguru, H.: Performance analysis of ovarian cancer detection and classification for microarray gene data. BioMed Res. Int. **2022**, 1–16 (2022). https://doi.org/10.1155/2022/6750457
24. Acharya, U.R., et al.: Use of nonlinear features for automated characterization of suspicious ovarian tumors using ultrasound images in fuzzy forest framework. Int. J. Fuzzy Syst. **20**(4), 1385–1402 (2018). https://doi.org/10.1007/s40815-018-0456-9
25. İPek, S.L., Göktürk, D.: Evaluation of artificial neural network and adaptive-network-based fuzzy inference system for ovarian and lung cancer prediction. J. Health Sci. Med. **7**(1), 80–88 (2024). https://doi.org/10.32322/jhsm.1360782
26. Zhou, J., Zeng, Z.Y., Li, L.: Progress of artificial intelligence in gynecological malignant tumors. Cancer Manag. Res. **12**, 12823–12840 (2020). https://doi.org/10.2147/CMAR.S279990
27. Brazma, A., Vilo, J.: Gene expression data analysis. FEBS Lett. **480**(1), 17–24 (2000). https://doi.org/10.1016/S0014-5793(00)01772-5

28. Joshi, C.J., Ke, W., Drangowska-Way, A., O'Rourke, E.J., Lewis, N.E.: What are housekeeping genes? PLOS Comput. Biol. **18**(7), e1010295 (2022). https://doi.org/10.1371/journal.pcbi.1010295
29. Norouzi-Barough, L., et al.: CRISPR/Cas9, a new approach to successful knockdown of ABCB1/P-glycoprotein and reversal of chemosensitivity in human epithelial ovarian cancer cell line. Iran. J. Basic Med. Sci. (2017). https://doi.org/10.22038/ijbms.2017.25145.6230
30. Guo, Z.X., Wong, W.K.: Fundamentals of artificial intelligence techniques for apparel management applications. In: Optimizing Decision Making in the Apparel Supply Chain Using Artificial Intelligence (AI), pp. 13–40. Elsevier (2013). https://doi.org/10.1533/9780857097842.13
31. Madu, E.O., et al.: Preoperative ovarian cancer diagnosis using neuro-fuzzy approach. In: Proceedings of the European Conference on Emergent Aspects in Clinical Data Analysis, Italy, pp. 1–8 (2005)

Analysis of Association Between Social Media Signals and Stock Return on Asian Markets

Milan Zdravković[1,2](✉) and Pavel Dudko[2]

[1] Faculty of Mechanical Engineering, University of Niš, Al. Medvedeva 14, 18000 Niš, Serbia
milan.zdravkovic@masfak.ni.ac.rs
[2] ZenPulsar LTD, 20-22 Wenlock Road, London N1 7GU, UK

Abstract. Research behind this paper explores the relationship between social media sentiment signals and market prices, and the predictive strength of these signals for selected assets traded on Taiwanese and Hong Kong markets. The research utilizes data sourced from PTT Bulletin Board System (BBS) and ET Net forum, analyzing sentiment signals and their association with Return on Investment (ROI) through various analytical techniques including signal association analysis, causality analysis, and model back-testing. The findings demonstrate a significant association between high sparsity sentiment signals and ROI, with machine learning models showing improved performance over traditional Buy and Hold scenarios in both markets. Out of 14 assets represented on ET Net forum, machine learning models outperform the Buy and Hold scenarios in 9 cases. Out of 10 assets with data from PTT BBS, models trained with data have shown improved results for 6 assets. ML approach outperforms Buy and Hold scenarios in both markets, with more significant gains (20% in absolute ROI) on Taiwanese market, where also all benchmarks are exceeded.

Keywords: financial market · market forecasting · machine learning · advanced analytics

1 Introduction

In recent years, the development of AI has created new roadmaps for innovative approaches to predicting stock market movements. Social media evolves as the venue for sharing market information, becoming as important as the traditional financial media outlets.

One particularly promising area of research focuses on leveraging the vast data generated by social media platforms to forecast stock returns. In fact, sentiments expressed in social media posts, comments, and discussions encapsulate the collective mood and perspectives of the investing public (especially on markets with significant uptake of individual investors, such as Chinese markets), potentially offering valuable insights into future market trends. By employing advanced analytics and machine learning algorithms to interpret these sentiments, researchers and investors gain the capability to explore new frontiers in financial prediction. This approach not only challenges traditional models

© The Author(s), under exclusive license to Springer Nature Switzerland AG 2025
N. Filipović (Ed.): AAI 2024, LNNS 1446, pp. 203–213, 2025.
https://doi.org/10.1007/978-3-031-99201-8_22

based on historical price data (technical analysis) and fundamental analysis but also highlights the evolving relationship between public sentiment and market dynamics in the digital age.

There are numerous works in scientific literature addressing the above challenges. Extracting the sentiment from the news articles is still the primary approach by the investors [1]. Recently, the research started to turn to the social media outlets [2–5]. However, most of the work suffers from a few key weaknesses.

First, social media is used as a source for market decisions, mostly by individual investors. This may restrict its potential impact. In US, individual investors account for about 6% of the market trading value only while institutional investors account for more than 93%, while in China, this proportion is 90:10 [6].

Second, most of the published research works use data from X (formerly Twitter) social media. X is a general media, covering a wide range of topics without providing an explicit structure making it very difficult to extract relevant data; it is prone to manipulations, commonly tried by the bots which identification is usually difficult; finally, it is used worldwide, which introduces geographical and cultural sampling challenges when sentiment is extracted [7].

Finally, most of the research works performs experiments using a few popular US stocks or cryptocurrencies, especially so-called "meme" stocks. The reason is straightforward – acquiring data about popular stocks at scale is easier.

In our previous work [8], we have demonstrated the relevance and credibility of Reddit as a source for social media sentiment having a predictive strength in forecasting cryptocurrencies market dynamics, including an effectiveness of using the original approach combining topic modeling and Large Language Models for social media data extraction and sentiment classification. In addressing the above-mentioned challenges, we turn now to the Asian markets and specialized social media channels.

The objective of research behind this paper was to demonstrate the clear evidence of association between social media sentiment signals and market prices and predictive strength of the former for selected assets traded on Taiwanese and Hong Kong markets. In Sect. 2 of this paper, we introduce the detailed methodology to make this demonstration. Then, in Sect. 3, results of selected experiments and discussion are presented. Finally, in Sect. 4, we introduce the key takeaways and important conclusions.

2 Methodology

Social media data was sourced from PTT BBS and ET Net forum. PTT (PTT Bulletin Board System, or PTT BBS) is an online bulletin board system, one of the largest and most influential web forums in Taiwan, featuring a wide variety of topics ranging from news, politics, to technology and finance. PTT BBS hosts numerous discussion boards where users share information, insights, and opinions on stocks, investments, and economic trends specific to Taiwan's financial markets. The ET Net forum is part of ET Net (etnet.com.hk), a website that provides a wide range of financial information, including real-time stock quotes, market analysis, financial news, and investment strategies focusing on the Hong Kong stock market.

The analysis was carried out for every asset with available data and per specific source. ET Net data included assets traded on Hong Kong market (HK), while PTT BBS considered assets traded on Taiwanese market (TW). After merging with financial data, the overall size of the dataset was 3178 datapoints for ET Net source and 2430 datapoints for PTT BBS source. There were 14 available assets for analysis traded on the HK market and 10 available assets traded on TW market. HK market assets were: 0728.HK, 0939.HK, 1024.HK, 1088.HK, 1398.HK, 1658.HK, 1810.HK, 2318.HK, 2333.HK, 3328.HK, 3690.HK, 3968.HK, 3988.HK, 9999.HK. TW market was: 1303.TW, 2002.TW, 2317.TW, 2357.TW, 2454.TW, 2603.TW, 2609.TW, 2881.TW, 2882.TW, 2891.TW.

Methodology included data pre-processing and feature engineering, hourly series signal association analysis and causality analysis and model back-testing.

2.1 Sentiment Analysis

Our methodology for sentiment analysis incorporates a comprehensive approach to data collection, labeling, and model fine-tuning, with a focus on both English and Mandarin financial texts. Data for sentiment classifier was collected from diverse platforms including Hugging Face, Kaggle, and our proprietary databases. The objective was to amass a large collection of financial-related content, with a particular emphasis on posts from financial boards and Mandarin social media platforms.

Given the presence of unlabeled datasets, advanced labeling techniques such as weak supervision, zero-shot learning, and pseudo-labeling were used to assign sentiment categories. Additionally, we undertook language translation tasks, converting a subset of Traditional Chinese posts to Simplified Chinese to ensure linguistic diversity and balance. The approach leveraged transformer-based models renowned for their efficacy in NLP tasks, specifically: FinBert, CryptoBert, and CryptoRoberta. These models were fine-tuned on curated datasets. The selection of the Roberta model with Whole Word Masking was predicated on comparative performance analyses, ensuring the chosen model was best suited for the task.

Recognizing the challenges posed by the Mandarin language, particularly due to the scarcity of pre-trained models in the financial domain, we collected financial posts from platforms like PTT BBS and ET Net forum. These sources provided a rich foundation for model training, albeit with the hurdle of limited labeled data. Our methodology adapted to these challenges, applying similar data augmentation and labeling strategies to develop a model capable of analyzing sentiments within Mandarin financial texts effectively.

2.2 Data Pre-Processing and Feature Engineering

After extracting the raw sentiment data (bullish/bearish/neutral) by using NLP pipeline, sentiment and volume signals were engineered by using the approach as follows.

Volume corresponds to the number of posts/comments/likes. Two different ways to calculate sentiment were used. Sentiment delta corresponds to the difference between the number of bullish and bearish posts/comments/likes. Sentiment ratio (Sentiment_r) is calculated as the natural logarithm of ratio of number of bullish (N_{bu}) and bearish (N_{be}) posts.

$$Sentiment_r = \ln(\frac{1 + N_{bu}}{1 + N_{be}})$$

The latter is important because natural logarithms amplify the effect of bearish posts which are normally less produced in social media when compared with bearish ones. The social media "optimism" is a well-known and common occurrence across all social media channels discussing market dynamics.

Weighted rolling sum was implemented for all sentiment time series. Implementing rolling sum embodies the assumption that past sentiment data also has some effect on the return variable. N past points were considered with decaying effect, which is implemented by using weighted rolling sum where weights (0.1–1.0) have quadratic distribution. Statistical analysis which is using the complete dataset adopts assumed N = 10, while all other analyses consider the optimization of the period based on Kendall correlation coefficient.

2.3 Signal Association Analysis on Hourly Series

Association between all volume and sentiment signals and Return of Investment (ROI), namely the difference between a market price in two consecutive timepoints on different lags [0,7] was calculated. Three different metrics were used for assessment of association of two signals - Spearman and Kendall correlation coefficient and mutual information. Although it is industry standard, Pearson was not used due to its high sensitivity to outliers, which are common in sentiment signals.

Granger causality is calculated between each of those signals and ROI.

Rolling Spearman cross-correlation (sentiment and volume signals vs. ROI) test was implemented. The goal was to enable exploration of the strength of causation in the specific periods in the overall dataset. The rolling window for ET Net and PTT BBS sources contains 140 timepoints, which corresponds to approx. 1 month of data. Cross-correlation is calculated by sentiment and volume signals shifted one step in the past. Local maxima and minima were found on each of the smoothed signals, by using argrelextrema Python function. The goal was to highlight the change points in sentiment and volume signals, for better illustration of the social media sentiment trend (see Fig. 1).

Another analysis was carried out to assess the cross-correlation coefficients in the dominantly bullish and bearish areas. To highlight the bullish and bearish areas, or trend change points, local extrema points were calculated on exponential moving average of the market close price signal. For each bullish and bearish interval data, Spearman cross-correlation coefficients between different sentiment and volume signals and ROI were calculated and highlighted on the plots.

Spearman cross-correlation for bullish or bearish areas (C_s) (see Fig. 2) was calculated as weighted average of coefficients in the respective areas (S_i), where weights correspond to the size of the areas (N_i) divided with the overall number of datapoints (N) in all areas.

$$C_s = \sum_{i=0}^{n} \frac{N_i S_i}{N}$$

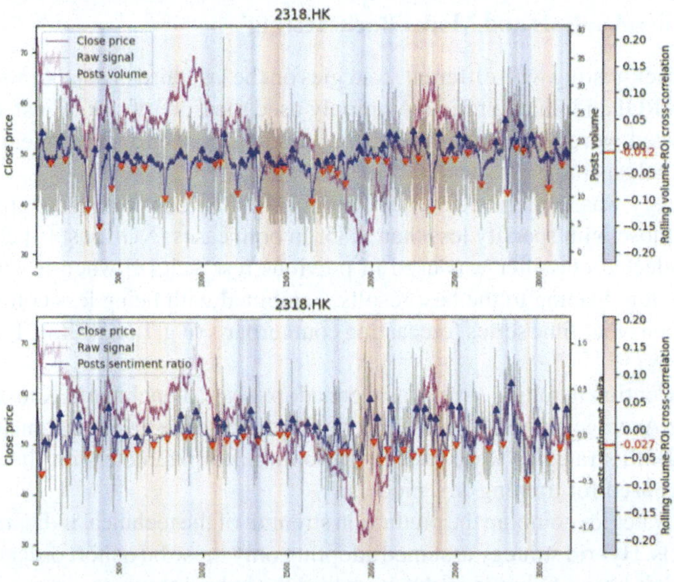

Fig. 1. Examples of cross-correlation heatmaps.

The objective of rolling cross-correlation analyses was to understand how social media users respond to the changes driving the market dynamics for assets and the dominating sentiment. This information can be used in hybrid strategies to strengthen the conviction of decisions suggested, for example, by Machine Learning models.

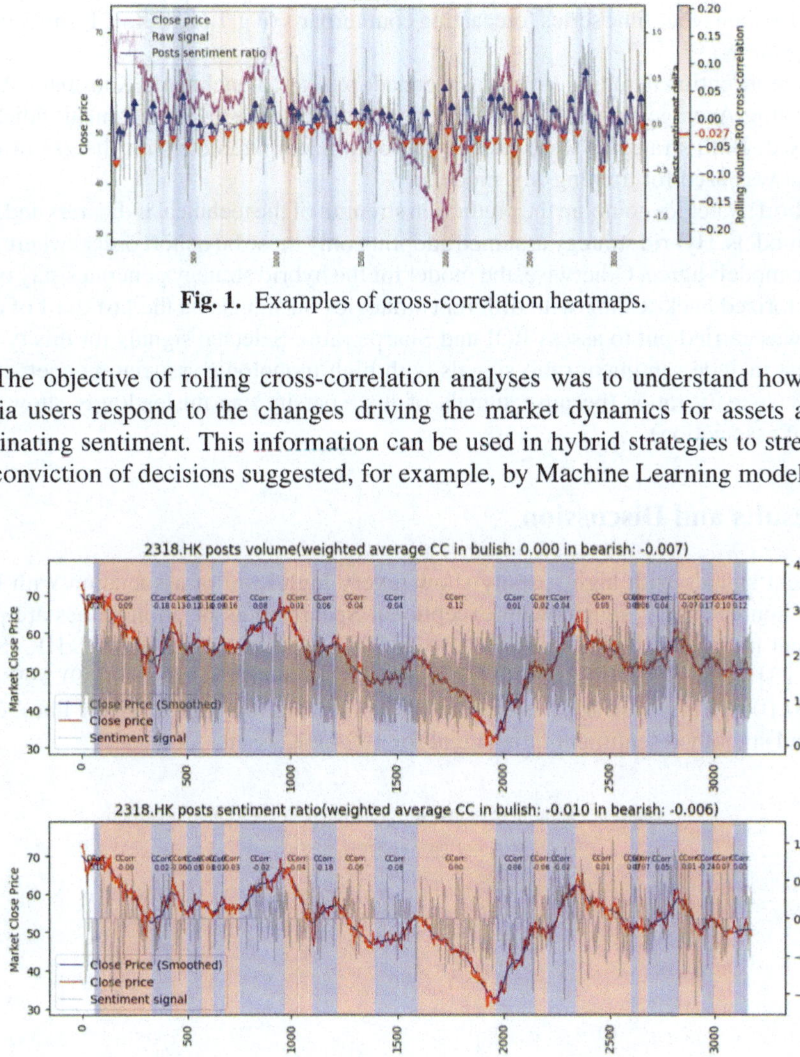

Fig. 2. Examples of cross-correlation analysis in bullish and bearish periods.

2.4 Causality Analysis and Model Back-Testing

Vectorized back-testing with different strategies on the last third of datapoints was carried out to assess ROI and Sharpe ratio (commonly used measure of risk-adjusted ROI). The strategies tested were ML driven forecasting models, conventional strategies based on Technical Analysis indicators and hybrid strategies.

ML strategy was implemented in two approaches: by considering all signals and by taking only those with sparsity less than 0.85. In both cases, XGBoost [9] classifier was used. The selected classifier was used in previous research [8] when it was proven as the best solution, leading to the best results, combined with being less computationally demanded then other time series forecasting counterparts (e.g. Long-Short Term Memory configurations).

Two conventional trading strategies were adopted as a benchmark. Kaufmann Adaptive Moving Average (KAMA) and Relative Strength Index (RSI) technical indicators were used. Both strategies were tested with optimized parameters on the 2/3 of data, data that was used for training ML models.

Hybrid strategies combine the prediction strength of the technical indicators and ML-driven models. Hybrid strategy assumed adopting only those buy/short orders where both of those models agree. Otherwise, the model for the hybrid strategy generates stay order.

Vectorized back-testing with different strategies on signals on the last third of datapoints was carried out to assess ROI and Sharpe ratio. Selected signals for this type of back-test include sentiment ratio signals with high monotonic association coefficients and low sparsity ratios (because signals of low sparsity can misleadingly show high levels of association).

3 Results and Discussion

Sentiment signals with high sparsity show a very high level of association with ROI. ET Net source assets' signals with exceptional Spearman association values are posts sentiment ratios for 1024.HK (Sp = 0.163), 1088.HK (Sp = 0.084), 1810.HK (Sp = 0.173), 2318.HK (Sp = 0.083), 3690.HK (Sp = 0.131), 3988.HK (Sp = 0.069), 9999.HK (Sp = 0.106). This is clearly visible in Fig. 3, showing raw signal (silver line), close price and smoothed sentiment ratio signal for 1024.HK asset.

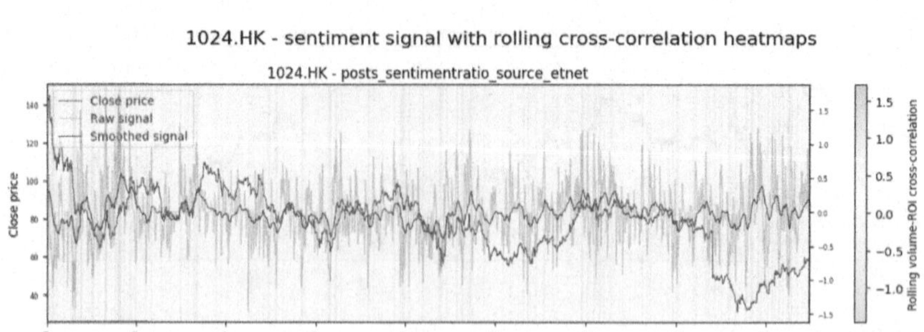

Fig. 3. Illustration of raw and smoothed sentiment ratio signal for 1024.HK.

Predictive strength of the signals is reflected in high values of cross-correlation coefficients between sentiment at lag = 1 (previous time period) and current ROI, such as 1088.HK (Sp = 0.043), 1024.HK (Sp = 0.028), 1810.HK (Sp = 0.04) and 3988.HK (Sp = 0.024). Among those assets represented on PTT BBS source (TW stocks), 2603.TW stands out with a high level of association (Sp = 0.039–0.045) as well as cross-correlation (0.02–0.024).

In general, a signal that is considered to be relevant for trading strategies is sentiment signal. In some cases, volume signals are characterized by the high level of positive correlation with ROI. However, there are also volume signals with high negative correlation.

3.1 Testing on All Signals with ML

Machine Learning models are normally not well accepted by the investors as tools for using social media data for market forecasting, due to lack of interpretability of local forecasts. Still, they are valuable for demonstrating the causality, namely the impact of possible use of social media signals in trading strategies (see Fig. 4). Out of 14 assets represented on ET Net source, ML models beat the market in 9 cases. Out of 10 assets with data from PTT BBS, models trained with data have shown improved results for 6 assets.

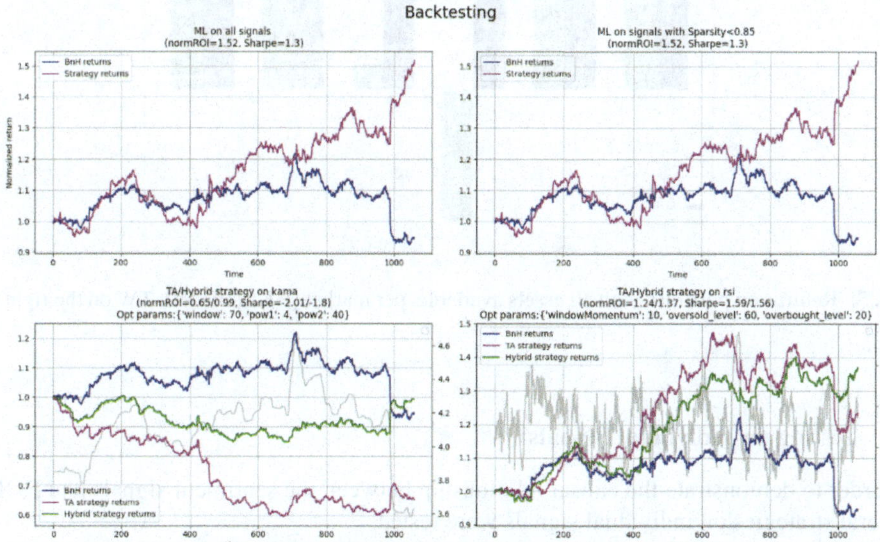

Fig. 4. Back-testing results for 1398.HK.

Figure 5 shows the back-testing results for both markets on an equally weighted portfolio level, where all assets with available data are considered. It is very important to highlight that the portfolio includes all selected assets, satisfying the sparsity level criteria; no exclusion of worse performing assets has been made. ML approach beats both HK and TW markets, with more significant gains (20% in absolute ROI) on TW, where also all benchmarks are beaten.

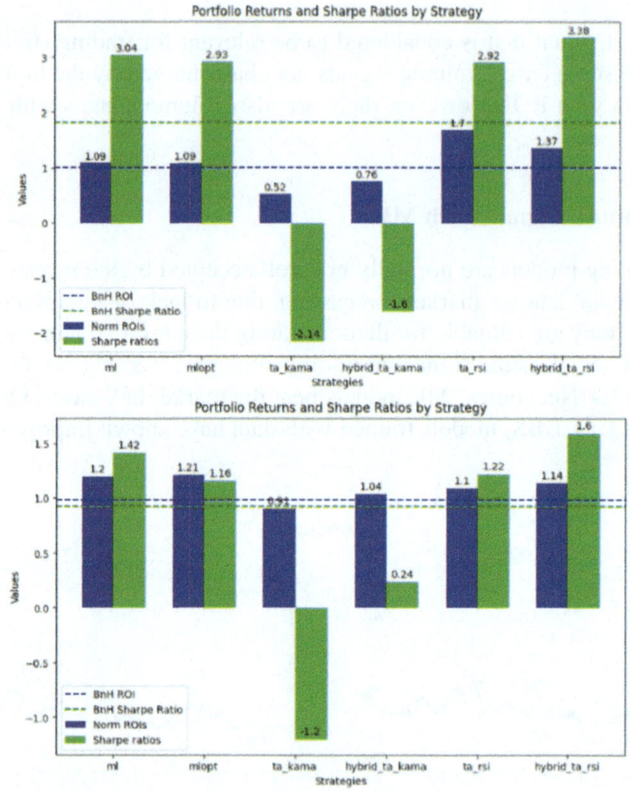

Fig. 5. Returns on portfolio with all assets available, per market (HK at the left, TW on the right).

3.2 Testing on Individual Signals

In order to demonstrate the causal relationship between the sentiment signals and ROI, several strategies on individual signals were tested.

Strategies used were:

- Trading on local extrema corresponding to the peaks and valleys of the smoothed sentiment signal. Different online change point detection approaches were tried. Adopted strategy was using online local extremes in the past. This signal is very conservative as it involves double delay: on smoothing (exponential smoothing with lag of 30 and) and extrema detection (additional lag of 20, in which period local extremes are looked for). It is presented only for illustrative purposes.

- Naïve momentum strategy. Long order is generated on occurrence of p consecutive timepoints of increased sentiment in the past. Short order is generated on occurrence of q consecutive timepoints of decreased sentiment in the past. Results with default $p = 7$, $q = 7$ (one day of trading) are recorded.
- Naïve momentum boosted strategy. The orders acquired by the Naïve momentum strategy are validated by the cross-correlation signal. Only orders produced at the timepoint of Spearman cross-correlation > 0.02 are adopted. Other orders are replaced with "stay" orders.
- Percentile rank oscillator. Percentile strategy is implemented by applying percentile rank transformation of sentiment signals. Long order is generated when sentiment reaches the highest percentile. Short order is generated when sentiment touches the lowest percentile. The floors and ceilings of the signals were the parameters, default being floor = 0.8 and ceil = 0.2 (quintiles).

Figure 6 illustrates the performance of the above methods for one asset from the HK market.

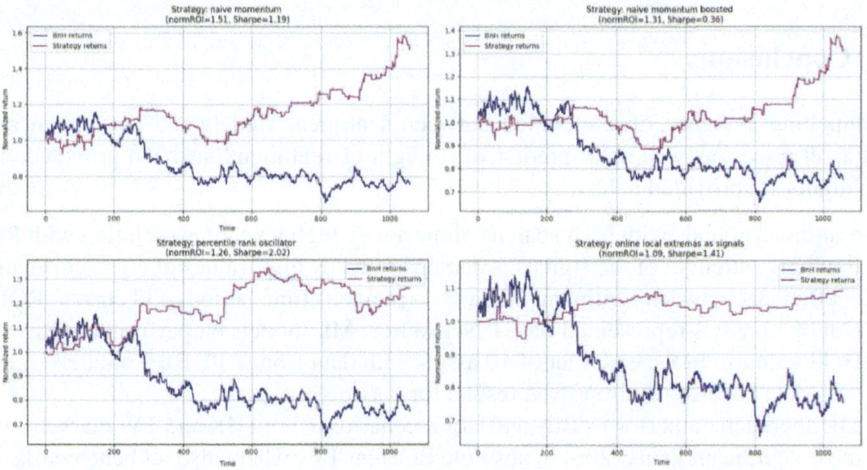

Fig. 6. Examples of financial features and trading strategies for 3690.HK.

The above tests have been carried out only on signals from ET Net social media (HK market) due to lower sparsity in case of PTT BBS (TW market). In fact, only assets with signals with sparsity ratio lower than 0.6 were included in the sample. As also illustrated in Fig. 7, the naïve momentum strategy has been confirmed as the one bringing the highest ROI.

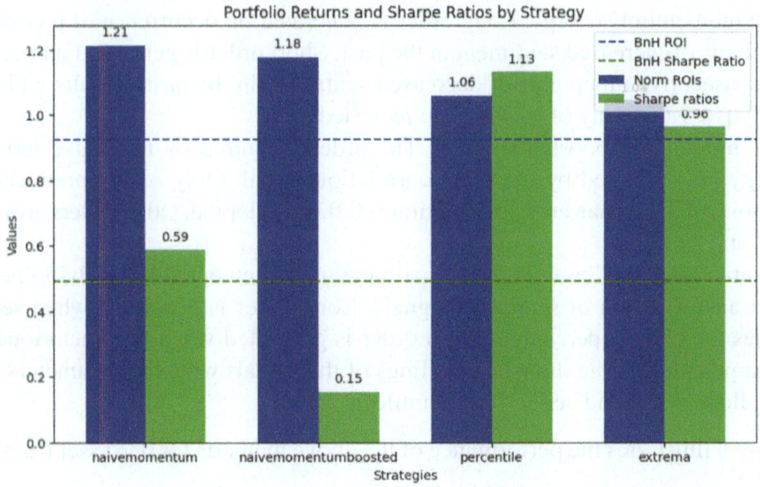

Fig. 7. Returns on portfolio with all assets available in HK market (ET Net source).

4 Conclusions

Compelling evidence of association between sentiment signals and two Asian market asset prices, as well as of predictive strength of sentiment has been provided. The highlights are provided below:

- Sentiment signals with high sparsity show a very high level of association with ROI. Predictive strength of the signals is demonstrated by high values of cross-correlation coefficients between sentiment at lag $= 1$ (previous time period) and current ROI.
- Out of 14 assets represented on ET Net source, ML models outperform the Buy and Hold scenario in 9 cases. Out of 10 assets with data from PTT BBS, models trained with data have shown improved results for 6 assets.
- ML approach outperforms Buy and Hold scenarios in both HK and TW markets, with more significant gains (20% in absolute ROI) on TW, where also all benchmarks are exceeded.
- Strategies driven by the individual signals (Naïve momentum strategy being the best performer) bring ROI of 21% in the test period at the portfolio level.

There are also other important takeaways from this research. Social media sentiment volume signals were found to be not a reliable source for gaining the advantage over market in the strategies. Log transformation of sentiment ratio appeared as a very important factor for performance by amplifying the bearish signal, especially in conventional strategies on smoothed signals.

All the above highlights also confirm the appropriateness and quality of our sentiment classification approach [8], which is usually identified in the literature as the most important factor for strong performance [3, 10]. However, there is still room for improvement. Sentiment classification for PTT BBS source may be too conservative, as distribution of uptake of neutral posts/comments/likes in the overall number of posts/comments/likes across assets is very high, over 80% on average. Similar conclusion stands for data from ET Net source (however, with lower uptakes of neutral posts, from 50% up).

One of the most important findings as an important guideline for future work is that the different assets respond differently to social media. When building predicting models, dynamic modeling per market asset is a must. Dynamic modelling learns about the statistical properties and then uses this knowledge for configuration of the predictive models. Some of the parameters that are specific to the asset/source that should be considered in trading strategies are measures related to: relevance of the volume signal (bullish/bearish), strength of the social media signals (posts/comments/likes), cross-correlation on different lags, and others.

References

1. Case, J., Clements, A: the impact of sentiment in the news media on daily and monthly stock market returns. In: Xu, Y., Wang, R., Lord, A., Boo, Y.L., Nayak, R., Zhao, Y., Williams, G. (eds.), Data Mining Vol. 1504, pp. 180–195 (2021)
2. Kordonis, J., Symeonidis, S., Arampatzis, A.: Stock price forecasting via sentiment analysis on twitter. Proceedings of the 20th Pan-Hellenic Conference on Informatics, pp. 1–6 (2016)
3. Zimbra, D., Abbasi, A., Zeng, D., Chen, H.: The state-of-the-art in twitter sentiment analysis: a review and benchmark evaluation. ACM Trans. Manag. Inf. Syst. **9**(2), 1–29 (2018)
4. Khan, W., Ghazanfar, M.A., Azam, M.A., Karami, A., Alyoubi, K.H., Alfakeeh, A.S.: Stock market prediction using machine learning classifiers and social media, news. J. Ambient Intelligence and Humanized Computing (2020)
5. Tsapeli, F., Bezirgiannidis, N., Tino, P., Musolesi, M.: Linking Twitter Events With Stock Market Jitters. arXiv:1709.06519 (2017)
6. Sun, Y., Zeng, X., Zhou, S., Zhao, H., Thomas, P., Hu, H.: What investors say is what the market says: measuring China's real investor sentiment. Pers. Ubiquit. Comput. **25**(3), 587–599 (2021)
7. Bollen, J., Mao, H., Zeng, X.: Twitter mood predicts the stock market. J. Computational Science **2**(1), 1–8 (2011)
8. Zdravković, M., Dudko, P., Kucherov, M.: Can memes beat the market? forecasting financial asset returns by using social media data and machine learning. In: Trajanovic, M., Filipovic, N., Zdravkovic, M. (eds.), Disruptive Information Technologies for a Smart Society Vol. 872, pp. 423–434 (2024)
9. Chen, T., Guestrin, C.: XGBoost: a scalable tree boosting system. Proceedings of the 22nd ACM SIGKDD International Conference on Knowledge Discovery and Data Mining, pp. 785–794 (2016)
10. Nguyen, T.H., Shirai, K., Velcin, J.: Sentiment analysis on social media for stock movement prediction. Expert Syst. Appl. **42**(24), 9603–9611 (2015)

Deep Learning Segmentation of the Porcine Femoral Arteries IVOCT Images

Miloš Anić[1], Sotirios Nikopoulos[2], Konstantinos Siaravas[2], Christos S. Katsouras[2,3], Vassiliki Potsika[1], Nenad Filipović[4,5], and Dimitrios I. Fotiadis[1,6](✉)

[1] Department of Materials Science and Engineering, Unit of Medical Technology and Intelligent Information Systems, University of Ioannina, 45110 Ioannina, Greece
anic.milos@kg.ac.rs, fotiadis@uoi.gr
[2] Department of Cardiology, Medical School, University of Ioannina, 45110 Ioannina, Greece
cskats@yahoo.com
[3] Michaelidion Cardiac Center, University of Ioannina, 45110 Ioannina, Greece
[4] Faculty of Engineering, University of Kragujevac, 34000 Kragujevac, Serbia
[5] Bioengineering Research and Development Center (BioIRC), 34000 Kragujevac, Serbia
[6] With the Biomedical Research Institute FORTH, University, Campus of Ioannina, Ioannina, Greece

Abstract. Peripheral artery disease is an atherosclerotic disease that represents accumulation of cholesterol in peripheral arteries, causing thickening of the arterial walls making it difficult for the blood to flow to lower extremities. For these reasons, it is extremely important to detect the disease as soon as possible. In this paper we provide a deep learning approach for lumen and intima segmentation of porcine femoral arteries as an important pre-step for the use in humans. Proposed methods focus on different convolutional neural networks and highlights the difference in results and include modified U-Net, SegNet and PSPNet networks. Methods are evaluated on manually annotated OCT images in terms of Dice coefficient, Hausdorff distance and overall quality of the reproduced masks. Experimental results show that modified U-Net provides us with the best results for lumen and intima segmentation with Dice coefficient 0.985 and 0.98, and Hausdorff distance 0.194 and 0.101 mm, respectively.

Keywords: convolutional neural networks · segmentation · peripheral artery · optical coherence tomography

1 Introduction

Peripheral artery disease, commonly known as PAD, is a chronic disease that progresses slowly due to the accumulation of plaques caused by atherosclerosis in the peripheral arteries, leading to partial or complete occlusion. PAD

© The Author(s), under exclusive license to Springer Nature Switzerland AG 2025
N. Filipović (Ed.): AAI 2024, LNNS 1446, pp. 214–224, 2025.
https://doi.org/10.1007/978-3-031-99201-8_23

typically results in the stiffening and constriction of the abdominal aorta, iliac arteries, femoral arteries and, occasionally, the upper extremities [12]. It often involves different illnesses that impact arteries outside of the brain and heart [16]. Peripheral artery disease impacts around 200 million individuals globally and 8.5 million in the United States [2,13].

When it comes to the diagnoses of atherosclerosis, imaging is usually the way to go. While the type of imaging used depends on what is available and how likely we are to detect atherosclerosis, duplex ultrasound, CT angiography, and MRI angiography are all commonly used [24]. OCT (optical coherence tomography) is also a popular choice because it can take really detailed pictures of arteries. One of the advantages of OCT is that it gives you a comprehensive view of the arterial wall, including the lumen and intima [23]. This level of detail is important for understanding the composition of atherosclerotic plaques, how they form, and how severe they are. All of this is crucial for determining the progression of peripheral artery disease and the risk of future cardiovascular events [23].

The quantitative analysis of OCT images entails dividing the artery into the lumen and the arterial wall into its specific layers: intima, media, and adventitia. After precise segmentation, thickness of arterial wall can be measured to evaluate plaque build-up and pinpoint vulnerable plaques that may rupture. Nevertheless, the intricate nature of atherosclerotic lesions in PAD presents difficulties in precisely delineating the impacted arterial wall due to changes in the usual arterial structure. Moreover, the difficulty in segmenting atherosclerotic plaque growth is exacerbated by its constantly changing nature, including calcification, lipid accumulation, and neovascularization [5]. Despite these challenges, it is crucial to conduct accurate segmentation to properly evaluate the illness and enable the implementation of treatments such as angioplasty or stenting, which aim to restore normal blood flow in affected areas [5].

Traditional segmentation techniques have heavily relied on expert manual delineation, which is time-consuming and subject to inter-observer variability. Yet, progress in technology, particularly artificial intelligence and machine learning, has resulted in the creation of automated and semi-automated segmentation algorithms, showing potential in improving the precision and consistency of OCT image analysis [14].

Lately, there has been an increased focus on creating deep learning methods for segmenting OCT images in atherosclerosis diagnostics, showing better performance than conventional techniques. These methods depend on using extensive image datasets labeled by doctors to teach neural networks to identify subtle characteristics in the arterial wall that signal particular tissue types [9]. Abdolmanafi et al. proposed a method for coronary artery classification using a fully automated tissue classification approach [1]. They employed a pre-trained CNN as a feature extractor to eliminate the classification layers before training Random Forest and support vector machine algorithms using the activations of the last fully connected layer. The algorithms' sensitivities were 0.97, 0.95, and 0.87, in that order [1]. Lee and colleagues developed a convolutional neural network specifically for detecting microchannels automatically, and obtained a Dice coefficient of 0.92 [17]. Zahnd et al. created a CNN in their study for automatic segmentation using arterial wall layer contours, which achieved a Dice coefficient of 0.93 [25]. Cao et al. introduced a new method for segmenting lumens in IVOCT images using level set approach to address irregularities in the lumen. The results of the experiment indicated that the method suggested is resilient, achieving an average Dice score of 0.98 [8].

Current advanced techniques for arterial segmentation do not meet the needs for a complete tool for arterial segmentation. Due to the substantial clinical impact of PAD, the necessity of precise and prompt diagnosis, and the difficulties involved in manually measuring OCT images, it is essential to create a strong algorithm for thorough segmentation. In our latest work, we have shown that it is feasible to differentiate lumen and intima in OCT images of coronary and femoral arteries [3,4]. Several diverse convolutional neural networks were employed for the job. Here, we expand on the results providing the comparison of said networks with other deterministic methods and state-of-the-art CNN models. Evaluation of results is done based on Dice coefficient and Hausdorff distance.

The reminder of our paper encompasses five sections. In Sect. 2, the methodology for the segmentation of lumen and intima is proposed. Section 3 provides additional information about the dataset and describes the experimental setup. In Sect. 4 we present the achieved results and compare them in terms of Dice coefficient and Hausdorff distance. In Sect. 5 we discuss the results, their drawbacks and advantages of using CNN. Conclusion of our study and suggestions for the future work are presented in Sect. 6.

2 Methodology

Segmentation of images is crucial as a preliminary step for accurate 3D reconstruction of arteries and models in general [11]. In our previous research, we focused on the deep CNN approach for the task of binary segmentation from intravascular OCT images without relying on any form of preprocessing methods for the images [3,4]. Initially, traditional deterministic techniques such as thresholding and active contours were utilized for the binary segmentation of the lumen and intima region. Despite being outdated, Otsu threshold method was chosen for thresholding as it was a reliable technique for accurately extracting the necessary features. Although it has demonstrated high performance in terms of Dice coefficient, Hausdorff distance has exhibited significant discrepancies. The snake algorithm was chosen for active contours for its precise extraction of necessary feature borders. The primary issue with active contours is the requirement to choose the region of interest. Because the pixel spacing between the lumen and intima contours is minimal, it is crucial to select the regions accurately to avoid mistakenly identifying the intima layer instead of the lumen border. Selecting the regions is very challenging for us because the artery is not consistently positioned in the center of the image, resulting in a low algorithm performance. In order to compensate for this, a blend of thresholding and active contours was employed. Initially, we use a threshold to identify the lumen of the artery for the purpose of extracting the geometric center of the artery. This is carried out to identify the precise location of the artery in the image, with the center being utilized to choose a suitable region for active contours. Despite the superior performance of this method when compared to using the methods separately, especially in terms of Hausdorff distance, it still did not meet our needs.

In our earlier studies [3,4], we utilized modified U-Net models for binary segmentation of lumen and intima. Different from the original U-Net design by Ronneberger et al. (2015) [21], U-Net architectures in this case come with greater depth. Furthermore, the initial adaptation of U-Net features distinct convolutional layers. In contrast to [21], where the image shrinks by a factor of 2 and the initial depth ranges from 64 to 512, here the depth starts at 24 and reaches 768, resulting in improved performance [6,11]. Just like in Arsic et al. (2019) [6], the second iteration also adopted the methodology of Ronneberger et al. (2015) [21], with each convolutional layer including padding to maintain the initial size of the activation map. Hence, the resulting segmentation map will possess identical resolution to the initial image. Moreover, batch normalization is utilized following each convolutional layer and before the ReLU activation function, resulting in better performance than the original U-Net model as shown in [26]. In addition to the U-Net architectures, as in previous research [3,4], PSPNet and SegNet with ResNet 50 encoders were used for the comparison.

3 Data Acquisition and Experimental Setup

Collecting human data has always posed a challenge for conducting thorough research, often taking a considerable amount of time to gather. Consistent absence of data and ethical approvals for utilization have consistently posed a significant obstacle, particularly for deep learning due to the need for clinical experts to annotate the data. Public datasets are available in many instances, however, they may not be present in others. A significant obstacle we faced in advancing the research was the absence of human data. Using intravascular OCT, which is a fairly new method of imaging, faces the same issue. A significant amount of studies concentrate on animal trials and animal data sets, particularly using mice, rabbits, and pigs because of their resemblance to human anatomy and/or physiology [7,15,19,20]. Collecting data on pigs with peripheral artery disease (PAD) is challenging, despite the fact that they experience the condition similar to humans [10,18]. These reasons also make induced PAD on porcine specimens suitable for learning purposes, including AI scenarios. Eight porcine samples were scanned using intravascular OCT at various points for the study. The pork samples were separated into two parts. Each specimen showed a developed case of PAD. The primary distinction between the two groups was that 4 samples had a native artery, whereas the other 4 samples had a stented artery. In order to guarantee the quality of the specimens, a particular workflow had to be considered. On day 0, the pigs had a Boston Scientific Mustang over the wire balloon inflated in both left and right arteries as part of a procedure. For research, 4 specimens also received Boston Scientific Innova Vascular Nitinol self-expanding stent in their arteries and imaging was done at each stage. Following the successful introduction of PAD, arteries were imaged once more 30 d post the initial surgery. Later, group 1 underwent conventional balloon angioplasty (POBA) followed by drug-coated balloon angioplasty for both groups, with imaging of the arteries at each step. This led to more than 13000 images across 36 sequences.

As most plaques are found in the intima-media layer, the lumen and intima boundaries were identified as the key regions for annotation [22]. Clinical experts carefully analyzed all sequences for this research, resulting in the annotation of 400 images with lumen and intima layers. Following the exclusion of outliers, the dataset was shortened to 300 images displaying healthy and diseased arteries. The images had dimensions of 512×512 pixels, with each pixel equating to 0.017857×0.017857 mm, and a slice thickness of 0.2 mm.

All of the tests have been performed on NVIDIA GeForce GTX 1080 Ti with 11 GB of GDDR5 RAM memory. Python V3.12.1 was used as programming language and the codes were implemented in Keras framework which regards to Tensorflow framework as a backend and PyTorch.

4 Results

State-of-the-art deep CNN models are selected for binary segmentation of lumen and intima. The effectiveness of the original U-Net architecture with small datasets has been demonstrated, with depth-modified U-Nets showing better results than the original design. Hausdorff distance and Dice coefficient are employed for quantitative assessment of the models. The results of the models without any additional processing are presented in the Table 1.

As shown in the table, deterministic methods excelled with the Dice coefficient but performed poorly with the Hausdorff distance. Furthermore, the extensive processing time required for utilizing methods in mixed reality must also be taken into account. Among CNN models, the modified U-Net structures showed the highest performance, particularly when the depth was set to 768. The U-Net 1024 architecture outperformed all others for lumen segmentation, with performance only slightly behind that of U-Net 768. When considering both matrices, U-Net 1024 emerged as the obvious victor. Even though deterministic methods achieved high performances from the aspect of Dice coefficient for both lumen and intima, same can not be said for Hausdorff distance. Results of deterministic methods can be observed on the cases of healthy and diseased arteries on Fig. 1 and Fig. 2

As it can be observed from the images, threshold method lacks performance when trying to segment either lumen or intima region. Unlike it, a better performance can be observed for active contours and the combination of 2 methods. This can be observed even more in the case of diseased artery. The results of deep CNN models, in comparison to the deterministic methods, show relatively high precision and can be observed on Fig. 3 and Fig. 4.

Table 1. Overview of model details

Models	Dice coeff.	Hausdorff dist. [mm]
Lumen Otsu Threshold	0.914	0.701
Lumen Active Contour	0.894	0.583
Lumen Active Contour + Otsu Threshold	0.922	0.48
Lumen Extended U-Net (768)	0.985	0.194
Lumen Extended U-Net (1024)	0.986	0.119
Lumen SegNet with ResNet50	0.982	0.438
Lumen PSPNet with ResNet50	0.915	0.942
Intima Otsu Threshold	0.919	0.749
Intima Active Contour	0.937	0.623
Intima Active Contour + Otsu Threshold	0.938	0.588
Intima Extended U-Net (768)	0.98	0.101
Intima Extended U-Net (1024)	0.977	0.251
Intima SegNet with ResNet50	0.962	1.463
Intima PSPNet with ResNet50	0.969	0.622

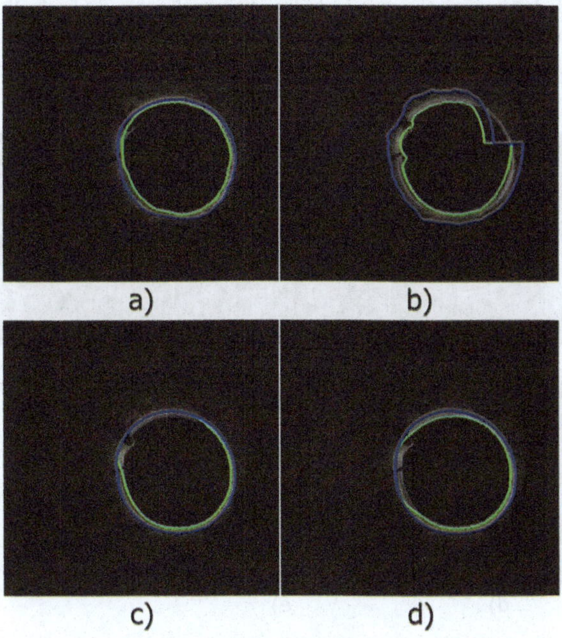

Fig. 1. Results of deterministic methods for healthy artery. Green line represents segmented lumen border while blue line represents intima border. a) manually annotated image; b) Otsu Threshold; c) Active contour; d) threshold & active contour.

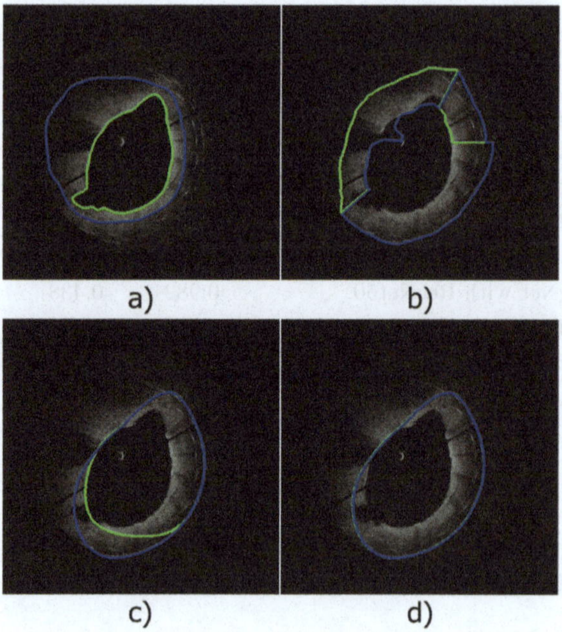

Fig. 2. Results of deterministic methods for diseased artery. Green line represents segmented lumen border while blue line represents intima border. a) manually annotated image; b) Otsu Threshold; c) Active contour; d) threshold & active contour.

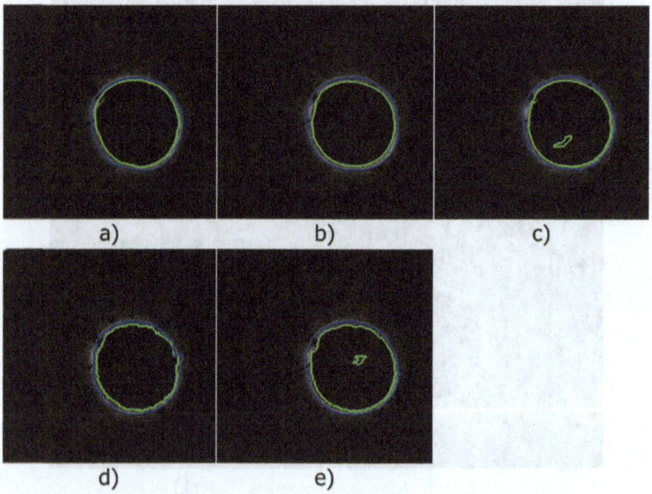

Fig. 3. Results of CNN models for healthy artery. Green line represents segmented lumen border while blue line represents intima border. a) manually annotated image; b) U-Net 768; c) U-Net 1024; d) PSPNet with ResNet50 encoder; e) SegNet with ResNet50 encoder.

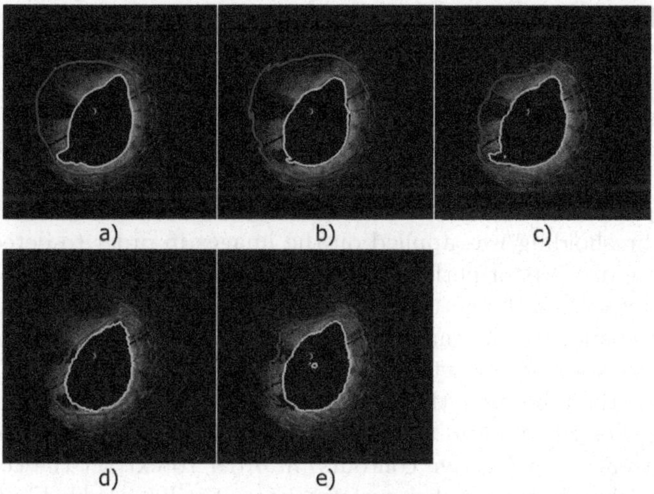

Fig. 4. Results of CNN models for diseased artery. Green line represents segmented lumen border while blue line represents intima border. a) manually annotated image; b) U-Net 768; c) U-Net 1024; d) PSPNet with ResNet50 encoder; e) SegNet with ResNet50 encoder.

5 Discussion

For the task of lumen and intima segmentation from porcine arteries IVOCT images, several different methods were employed. While deterministic methods show high performance from the aspect of Dice coefficient for both healthy and diseased arteries, same can not be said in terms of Hausdorff distance. Otsu threshold achieved Dice coefficient 0.914 and 0.919 and Hausdorff distance 0.701 mm and 0.749 mm for lumen and intima, respectively. Active contour method achieved Dice coefficient 0.894 and 0.937 and Hausdorff distance 0.583 mm and 0.623 mm, while the combination of two methods achieved Dice coefficient 0.922 and 0.938 and Hausdorff distance 0.48 mm and 0.588 mm for lumen and intima, respectively. As it can be observed the combination of two methods achieved best performance. The lower Dice coefficient for lumen segmentation using active contours can be explained by the fact that the region of interest, needed for accurate detection, can not be properly selected without understanding where the artery on the image is located which is solved by the combination of threshold and active contour. Unlike deterministic methods, which have shown poor performance from the aspect of Hausdorff distance, CNN models proved to be more efficient. U-Net 1024 achieved the best performance for lumen segmentation achieving Dice coefficient 0.986 and Hausdorff distance 0.119 mm. While it has shown great performance in intima segmentation as well, it was slightly outperformed by U-Net 768 which achieved Dice coefficient 0.98 and Hausdorff distance 0.101 mm. Both models outperformed other configurations which can also be observed on Fig. 3 and Fig. 4.

6 Conclusion

This study explores the application of Convolutional Neural Networks (CNNs) for the segmentation of lumen and intima contours from porcine femoral arteries imaged using IVOCT imaging. We explored different deterministic methods for image segmentation and compared them with manually annotated images from the aspect of Dice coefficient and 95th percentile Hausdorff distance. At first, Otsu thresholding was applied on the images in order to detect said contours. Because of it's poor performances, active contour method were employed. This method proved to be better then thresholding but still lacked in the terms of Hausdorff distance. The main problem of active contours is that region of interest has to be precisely set in order to precisely extract the contours. This proved to be tricky because the location of artery is not always in the center of the image. For this reason, a combination of thresholding, in order to detect where the artery is, and active contours, in order to extract the contours, was employed. While achieving higher performances, it still lagged behind CNN models, which were later deployed. In addition to the deterministic methods, deep CNN models were employed for said task and outperformed deterministic methods by significant margin, especially in the case of Hausdorff distance. Extended U-Net configurations (768 and 1024) proved to be the best while working with a relatively small dataset while outperforming other models and deterministic methods, achieving Dice coefficients Dice coefficients 0.985 and 0.986 and Hausdorff distance 0.194 mm and 0.119 mm for lumen, and Dice coefficients 0.98 and 0.977 and Hausdorff distance 0.101 mm and 0.251 mm for intima segmentation, respectively.

The identification of the lumen and intima borders is the first stage of the fully automatic 3D reconstruction of the femoral arteries from IVOCT images. In the future, a fully automatic 3D reconstruction will be performed which will be visualized employing mixed reality.

References

1. Abdolmanafi, A., Duong, L., Dahdah, N., Cheriet, F.: Deep feature learning for automatic tissue classification of coronary artery using optical coherence tomography. Biomed. Opt. Express **8**(2), 1203–1220 (2017)
2. Allison, M.A., et al.: Ethnic-specific prevalence of peripheral arterial disease in the united states. Am. J. Prev. Med. **32**(4), 328–333 (2007)
3. Anic, M., Fotiadis, D., Potsika, V.: Convolutional neural networks for the segmentation of coronary arteries. In: 2023 IEEE 23rd International Conference on Bioinformatics and Bioengineering (BIBE), pp. 308–314. IEEE (2023)
4. Anic, M., et al.: Convolutional neural networks for the segmentation of coronary arteries. In: 2024 3rd Serbian International Conference on Applied Artifical Intelligence (SICAAI) (2024)
5. Araki, M., et al.: Optical coherence tomography in coronary atherosclerosis assessment and intervention. Nat. Rev. Cardiol. **19**(10), 684–703 (2022)

6. Arsic, B., Obrenovic, M., Anic, M., Tsuda, A., Filipovic, N.: Image segmentation of the pulmonary acinus imaged by synchrotron x-ray tomography. In: 2019 IEEE 19th International Conference on Bioinformatics and Bioengineering (BIBE), pp. 525–531. IEEE (2019)
7. Bassols, A., Costa, C., Eckersall, P.D., Osada, J., Sabria, J., Tibau, J.: The pig as an animal model for human pathologies: a proteomics perspective. PROTEOMICS-Clinical Appli. **8**(9–10), 715–731 (2014)
8. Cao, Y., et al.: Automatic lumen segmentation in intravascular optical coherence tomography images using level set. Comput. Math. Methods Med. **2017**(1), 4710,305 (2017)
9. Carpenter, H.J., Ghayesh, M.H., Zander, A.C., Li, J., Di Giovanni, G., Psaltis, P.J.: Automated coronary optical coherence tomography feature extraction with application to three-dimensional reconstruction. Tomography **8**(3), 1307–1349 (2022)
10. Deppen, J.N., et al.: A swine hind limb ischemia model useful for testing peripheral artery disease therapeutics. J. Cardiovasc. Transl. Res. **14**(6), 1186–1197 (2021)
11. Djukic, T., Arsic, B., Djorovic, S., Filipovic, N., Koncar, I.: Validation of the machine learning approach for 3d reconstruction of carotid artery from ultrasound imaging. In: 2020 IEEE 20th International Conference on Bioinformatics and Bioengineering (BIBE), pp. 789–794. IEEE (2020)
12. by: the European Stroke Organisation (ESO), E., Members, A.F., Tendera, M., Aboyans, V., Bartelink, M.L., Baumgartner, I., Clément, D., Collet, J.P., Cremonesi, A., De Carlo, M., et al.: Esc guidelines on the diagnosis and treatment of peripheral artery diseases: document covering atherosclerotic disease of extracranial carotid and vertebral, mesenteric, renal, upper and lower extremity arteries the task force on the diagnosis and treatment of peripheral artery diseases of the european society of cardiology (esc). European Heart J. **32**(22), 2851–2906 (2011)
13. Fowkes, F., et al.: Comparison of global estimates of prevalence and risk factors for peripheral artery disease in 2000 and 2010: a systematic review and analysis. Lancet **382**(9901), 1329–1340 (2013)
14. Hormel, T.T., Hwang, T.S., Bailey, S.T., Wilson, D.J., Huang, D., Jia, Y.: Artificial intelligence in oct angiography. Prog. Retinal and Eye Res. **85**, 100,965 (2021)
15. Hou, N., Du, X., Wu, S.: Advances in pig models of human diseases. Animal Models Experim. Med. **5**(2), 141–152 (2022)
16. Kullo, I.J., Rooke, T.W.: Peripheral artery disease. N. Engl. J. Med. **374**(9), 861–871 (2016)
17. Lee, J., et al.: Automatic microchannel detection using deep learning in intravascular optical coherence tomography images. In: Medical Imaging 2022: Image-Guided Procedures, Robotic Interventions, and Modeling, vol. 12034, pp. 166–173. SPIE (2022)
18. Long, C.A., et al.: An endovascular model of ischemic myopathy from peripheral arterial disease. J. Vasc. Surg. **66**(3), 891–901 (2017)
19. Lunney, J.K., Van Goor, A., Walker, K.E., Hailstock, T., Franklin, J., Dai, C.: Importance of the pig as a human biomedical model. Science Trans. Med. **13**(621), eabd5758 (2021)
20. Meurens, F., Summerfield, A., Nauwynck, H., Saif, L., Gerdts, V.: The pig: a model for human infectious diseases. Trends Microbiol. **20**(1), 50–57 (2012)
21. Ronneberger, O., Fischer, P., Brox, T.: U-Net: convolutional networks for biomedical image segmentation. In: Navab, N., Hornegger, J., Wells, W.M., Frangi, A.F. (eds.) MICCAI 2015. LNCS, vol. 9351, pp. 234–241. Springer, Cham (2015). https://doi.org/10.1007/978-3-319-24574-4_28

22. Seekircher, L., et al.: Intima-media thickness at the near or far wall of the common carotid artery in cardiovascular risk assessment. Euro. Heart J. Open **3**(5), oead089 (2023)
23. Shlofmitz, E., Sosa, F.A., Ali, Z.A., Waksman, R., Jeremias, A., Shlofmitz, R.: Oct-guided treatment of calcified coronary artery disease: breaking the barrier to stent expansion. Current Cardiovascular Imaging Rep. **12**, 1–8 (2019)
24. Vlachopoulos, C., Georgakopoulos, C., Koutagiar, I., Tousoulis, D.: Diagnostic modalities in peripheral artery disease. Curr. Opin. Pharmacol. **39**, 68–76 (2018)
25. Zahnd, G., et al.: Contour segmentation of the intima, media, and adventitia layers in intracoronary oct images: application to fully automatic detection of healthy wall regions. Int. J. Comput. Assist. Radiol. Surg. **12**, 1923–1936 (2017)
26. Zhou, X., Yang, G.: Normalization in training u-net for 2-d biomedical semantic segmentation. IEEE Robot. Autom. Lett. **4**(2), 1792–1799 (2019). https://doi.org/10.1109/LRA.2019.2896518

Transforming Learning: Adapting to Generative AI Technologies in the Serbian Educational Paradigm

Nebojša Stanojević[1](✉), Ranka Popovac[1], Tijana Čomić[2], Uroš Milićević[3], and Dragan Vukmirović[1]

[1] Faculty of Organizational Science, University of Belgrade, 157 Jove Ilića Street, 11000 Belgrade, Serbia
ns20205003@student.fon.bg.ac.rs, dragan.vukmirovic@fon.bg.ac.rs
[2] UNICEF, 58 Svetozara Markovica Street, 11000 Belgrade, Serbia
[3] Ministry of Internal Affairs of the Republic of Serbia, 2 Mihajlo Pupin Boulevard, 11000 Belgrade, Serbia

Abstract. This paper explores the transformative potential of Generative Artificial Intelligence (GAI) technologies in revolutionizing the Serbian educational system. With the rapid development and widespread adoption of GAI tools like ChatGPT-4, Serbian education stands at the cusp of a significant paradigm shift. This document provides a comprehensive examination of the integration of GAI technologies, highlighting the challenges, opportunities, and the crucial need for strategic measures for successful implementation. It delves into critical areas such as infrastructure enhancement, ethical considerations, data protection, and the professional development of educators. By outlining a phased implementation strategy, the paper aims to equip policymakers, educators, and technologists with actionable insights for leveraging GAI to enhance educational equity, improve learning outcomes, and prepare students for a future dominated by digital literacy. Ultimately, this exploration contributes to the broader discourse on leveraging advanced technologies to foster an inclusive, innovative, and effective educational environment, emphasizing the importance of responsible planning and committed preparation in adopting new technologies.

Keywords: Generative Artificial Intelligence · Serbian educational system · digital literacy · ethical considerations · technological integration

1 Introduction

This position paper explores the strategic integration of GAI into the Serbian educational system, highlighting the transformative potential of these technologies in enhancing learning environments and academic outcomes. As digital technologies rapidly evolve, the integration of GAI presents a unique and urgent opportunity to significantly improve the quality and accessibility of education in Serbia [1]. This paper aims to provide

© The Author(s), under exclusive license to Springer Nature Switzerland AG 2025
N. Filipović (Ed.): AAI 2024, LNNS 1446, pp. 225–235, 2025.
https://doi.org/10.1007/978-3-031-99201-8_24

a comprehensive blueprint for policymakers, educators, and technologists to effectively harness GAI's benefits within the national education framework, emphasizing the importance of its benefits and the urgency of its implementation.

1.1 Serbian Educational Context

Serbia's educational system, like many systems worldwide, faces numerous challenges, including varying levels of digital literacy, inconsistent access to technology, and a need for modernization of teaching methodologies and content. According to recent studies, Serbia shows a promising increase in digital engagement, yet disparities in access and quality of digital resources remain across different regions and socioeconomic groups [2, 3]. The national strategy for education has increasingly emphasized the importance of digital literacy, viewing it as critical to preparing students for a digital economy [4]. The COVID-19 pandemic has further highlighted the essential role of digital technologies in education, accelerating the shift towards more digital and remote learning environments [5].

While improving, the current infrastructure requires significant enhancements to support deploying advanced technologies like GAI. Moreover, Serbian educational policies are undergoing revisions to incorporate digital tools and learning methodologies better, aiming to create an inclusive, flexible, and forward-thinking educational system [6]. In this context, integrating GAI technologies is feasible and critical to bridging the educational divide and enhancing the learning experience at all levels.

This backdrop of emerging digital literacy and infrastructural development makes integrating GAI particularly relevant and challenging. By addressing these challenges head-on and navigating the complexities of technological integration, this position paper seeks to lay down a strategic framework that can guide the successful adoption of GAI in Serbia's educational paradigms.

2 Pioneering Educational Transformation with Generative AI in Serbia

Adopting GAI within the Serbian educational system marks a cornerstone in the country's educational reform. GAI technologies are poised to be pivotal in reshaping teaching methodologies and enhancing learning experiences across various academic levels. This section delves into specific GAI tools and applications that could be integrated into Serbian schools and explores how these technologies can be tailored to meet the needs of primary, secondary, and higher education settings.

2.1 Specific GAI Tools and Applications

Several GAI tools and applications are particularly suited for integration into the Serbian educational landscape [7]:

- ChatGPT-4 for Language Learning: This advanced conversational AI can enhance language teaching and learning. By interacting with ChatGPT-4, students can improve their language skills through practice conversations, receive instant feedback, and engage in interactive, personalized learning sessions.

- AI-Driven Analytics Tools: Tools like IBM Watson's AI-powered analytics can monitor student performance, analyze learning patterns, and provide educators with insights that help tailor educational content to their students' needs. This can lead to a more customized learning experience and better academic outcomes.
- AI Content Generators: Tools such as Jasper (formerly known as Jarvis) can assist in creating educational content. They can generate readable texts, create engaging presentations, and even help develop educational games and simulations that cater to diverse learning styles.
- AI in Science Education: AI-powered simulations can provide students with virtual labs for chemistry and physics, where they can conduct experiments in a safe and controlled environment, enhancing their understanding and engagement.

Application Across Educational Levels

The integration of GAI across different educational levels can be approached by considering each group's specific needs and capabilities [8]:

- Primary Education: GAI tools can introduce basic literacy and numeracy concepts through interactive games and tailored learning activities at this level. For instance, AI-driven programs can adapt to each child's learning pace, providing a highly personalized education that fosters foundational skills.
- Secondary Education: In secondary schools, GAI can support more complex subjects such as mathematics, science, and languages through interactive tutorials and problem-solving activities. AI tools can also help prepare students for standardized tests by offering personalized practice sessions and identifying areas for improvement.
- Higher Education: At the university level, GAI can facilitate advanced research by quickly helping students and researchers sift through large volumes of data and literature. AI can also support the development of more complex interdisciplinary projects, integrating information from various fields into cohesive learning experiences. Additionally, AI can simulate real-world scenarios in engineering and medicine, providing students with practical, hands-on experience in virtual environments.

Potential Benefits and Challenges

While GAI's benefits in education include increased engagement, personalized learning, and efficient knowledge transfer, some challenges must be addressed. These include ensuring the ethical use of AI, protecting student data privacy, and mitigating any biases in AI algorithms. Additionally, the varying levels of digital infrastructure across different regions can pose a significant challenge in the equitable implementation of GAI tools.

By addressing these specific tools and applications and tailoring them to the needs of various educational levels, Serbia can harness GAI's full potential to enhance its educational system. This strategic integration aims to improve academic outcomes and prepare Serbian students to thrive in an increasingly digital world.

3 Preconditions for the Successful Implementation of GAI in the Serbian Educational System

Several preconditions must be met to integrate GAI technologies into the Serbian educational system. These pave the way for a smooth transition and ensure that adopting these technologies is beneficial and sustainable. This section discusses these preconditions in detail, focusing on practical steps for achievement and addressing specific ethical considerations.

Robust and Inclusive Infrastructure

A fundamental requirement for integrating GAI into education is the development of a robust and inclusive digital infrastructure. This includes high-speed internet connectivity, adequate hardware, and accessible software across all educational institutions, including those in remote and underserved areas [9, 10].

- Funding Sources: To finance these infrastructure upgrades, Serbia can explore a mix of government funding, international grants, and partnerships with private sector technology providers. Initiatives like the European Union's Digital Education Action Plan could be potential financial and technical support sources.
- Government Initiatives: National digital transformation initiatives should prioritize education. Policies to subsidize internet access and provide schools with necessary technological equipment will be crucial.
- International Partnerships: Collaborating with international bodies and other countries that have successfully implemented similar technologies can provide financial support and valuable insights.

Comprehensive Data Protection Framework

Implementing GAI requires handling large amounts of data, making a robust data protection framework that aligns with international standards such as the General Data Protection Regulation (GDPR) imperative.

- Legislation: Update or enact new legislation that specifically addresses the unique challenges posed by AI in education, focusing on student data protection, consent mechanisms, and data usage transparency.
- Standards and Protocols: Develop standards and protocols for data collection, storage, and processing to ensure they are secure and compliant with national and international laws.

Diverse and Representative Data Sets

To mitigate biases in GAI applications, it is crucial that the data used to train these models are diverse and representative of the Serbian student population.

- Monitoring and Evaluation: Establish a system for ongoing monitoring and evaluation of AI models to ensure they perform pretty and effectively across different demographics.

- Collaboration: Engage with local universities and research institutions to gather and analyze data that reflects the diversity of the Serbian population.

Ethical Guidelines and Policies
The ethical deployment of GAI in education requires clear guidelines that dictate how these technologies are used and the measures taken to safeguard users.

- Development of Ethical Frameworks: Collaborate with educational and technological experts to develop an ethical framework that addresses AI transparency, accountability, and the potential impact on students.
- Stakeholder Engagement: Include educators, parents, students, and policymakers in discussions about the ethical use of GAI to ensure that the policies developed are comprehensive and inclusive.

Economic and Collaborative Support Structures
An ecosystem supporting innovation and collaboration is needed for GAI to integrate effectively into the Serbian educational system.

- Partnerships with Tech Companies: Forge partnerships with technology companies to access advanced AI tools and platforms at reduced costs or through educational grants.
- Academic Collaboration: Facilitate collaborations between educational institutions and tech companies to foster innovation and ensure the relevance of educational technologies to Serbia's needs.

Educational Framework Adaptation
Incorporating digital literacy and AI education into the curriculum is essential for preparing students and educators to interact effectively with new technologies.

- Curriculum Updates: Integrate AI and digital literacy modules into the national curriculum at all educational levels.
- Professional Development: Provide ongoing training for teachers to become proficient in using GAI tools and understanding AI concepts.

Hybrid Learning Models and Human Oversight
To maintain a balanced educational approach, blended learning models that combine traditional teaching methods with AI-driven tools should be developed.

- **Implementation of Blended Learning**: Develop and implement hybrid models that utilize in-person teaching and AI enhancements.
- **Human Oversight Mechanisms**: Establish guidelines for human oversight to ensure that AI tools supplement rather than replace human teaching.

By establishing these preconditions, Serbia can ensure that the integration of GAI into its educational system is successful, ethical, and aligned with the country's academic goals and standards. These steps address immediate needs and lay a foundation for sustainable and responsible use of AI in education.

4 Research Results

While there's enthusiasm for AI's potential in education, specific policies, explicitly regarding GAI tools like ChatGPT, are still relatively rare (as critical policies and initiatives related to generative AI in education, drawing from UNESCO's global guidance and the OECD's examination [11, 12]. At this very moment, it is difficult to find many fully-formed policies specifically focused on the term "generative AI in education" on a national or governmental level worldwide.

The potential reasons and the critical considerations surrounding this could be:

- Novelty: GAI tools are very recent, particularly the scale and complexity of models like ChatGPT-4. Governments, educational institutions, and policymakers often need time to evaluate a technology's impact before creating dedicated policies.
- Complexity: GAI raises complex, multifaceted concerns. Issues like bias, plagiarism, student dependency, and assessment integrity lack simple solutions, requiring careful deliberation for effective policymaking.
- Proactive vs. Reactive: Some countries might take a wait-and-see approach, observing how other nations or institutions handle challenges before enacting their policies. Others might be more proactive, but internal policymaking processes can be slow.
- Focus on Broader AI: Many existing policies target AI in general. As the field becomes more precise and use cases more defined, specific applications like GAI in education might receive dedicated attention.

Developing policies for generative AI in education is crucial for Serbia for several potential reasons:

Harnessing the Benefits Responsibly

- Targeted potential: A clear policy framework will help Serbia channel GAI's benefits towards improving educational outcomes, addressing existing learning gaps, and preparing students for a digitally driven future.
- Managing risks: Proactive policies help mitigate risks such as bias, plagiarism, and over-dependence on technology, ensuring that GAI is a tool for empowerment, not a source of new problems.

Specific to Serbia:

- Digital divide: Serbia has varying levels of technology access and digital literacy, so policies can address equity concerns and ensure that all students have the potential to benefit from GAI.
- Language and culture: Policies should consider how to adapt GAI tools to the Serbian language and educational context, avoiding the dominance of models primarily trained on English or culturally distant content.
- Serbian context: Issues like assessment methodology, data privacy, and teacher training must be framed within the Serbian education system.

Policy Development Considerations

- Guidance, not restriction: Policies should create a supportive framework for educators, giving them guidelines and resources to integrate GAI effectively and ethically.
- Flexibility: As GAI rapidly evolves, Serbia's policies must be adaptable to stay relevant and practical.
- Collaboration: Developing these policies should involve educators, researchers, technologists, parents, and students. This ensures a nuanced and inclusive approach.

Serbia's Opportunity

- Regional leader: By creating well-crafted policies in this emerging area, Serbia can become a frontrunner in responsible and innovative GAI integration in education within the Balkans and the broader European context.
- Attracting talent and investment: Proactive policies demonstrate commitment to innovation,

Developing a policy on generative AI in education could be a strategic step for Serbia, which has been making strides in digital transformation and education reform. It would address the immediate challenges of integrating such technologies and align with broader educational and economic goals. As with any policy, the key will be balancing innovation and technological adoption with safeguards that protect and enhance the educational experience for all students [13].

5 Action Plan

A successful integration of GAI into Serbia's educational system requires a detailed and phased implementation strategy. This action plan outlines the steps to ensure GAI technologies are adopted effectively and responsibly across academic institutions. Each phase includes specific milestones and Key Performance Indicators (KPIs) to monitor progress and assess the implementation's impact.

5.1 Phase 1: Preparation and Planning

- Objective: Establish a foundational understanding and readiness for GAI integration.
- Actions:

  Form a cross-sector AI in Education task force that includes policymakers, educators, technologists, and community representatives.
  Conduct a comprehensive needs and infrastructure assessment to identify technological and educational gaps.
  Develop legal and ethical frameworks tailored to integrating GAI in education.

- Milestones:

  Taskforce established by Q? (for example, Q2 2025).
  Completion of needs assessment by Q? (for example, Q4 2025).
  Frameworks approved by Q1 202?. (for example, Q1 2026)

- KPIs:

  Number of task force meetings and resulting action items.
  Percentage of schools assessed for technological readiness.
  Legal and ethical guidelines ratified and published.

Phase 2: Building and Strengthening Infrastructure

- Objective: Enhance digital infrastructure to support GAI applications in education.
- Actions:

  Upgrade internet connectivity and provide necessary hardware and software across all educational levels.
  Develop and pilot GAI educational tools customized for the Serbian educational system.

- Milestones:

  Infrastructure upgrades completed in 50% of targeted schools by Q3 2026.
  Pilot tools developed and ready for testing by Q1 2027.

- KPIs:

  Number of schools with completed infrastructure upgrades.
  Feedback from initial testing of pilot tools in schools.

Phase 3: Capacity Building and Awareness

- Objective: Enhance the capabilities of educators and raise awareness among students and parents about the benefits and challenges of GAI.
- Actions:

  - Implement professional development programs for educators on the use of GAI tools.
  - Integrate AI education into the student curriculum and conduct community awareness sessions.

- Milestones:
  - Training programs launched by Q2 2027.
  - AI modules integrated into curricula by the start of the academic year 2027–2028.
- KPIs:
  - Number of educators trained in GAI tools.
  - Student and parent surveys measuring understanding and acceptance of GAI technologies.

Phase 4: Implementation and Evaluation

- Objective: Full-scale implementation of GAI tools in educational settings and continuous impact evaluation.
- Actions:
  - Begin broad implementation of GAI tools across schools.
  - Establish monitoring and evaluation mechanisms to continually assess the effectiveness and fairness of GAI tools.
- Milestones:
  - GAI tools will be implemented in over 80% of schools by the end of 2028.
  - Ongoing evaluation and adjustments implemented by mid-2029.
- KPIs:
  - Percentage of schools using GAI tools regularly.
  - Reports on student performance improvements and satisfaction ratings.

Following this action plan, Serbia can systematically integrate GAI into its educational system. This strategy addresses the immediate challenges and lays a foundation for the sustainable and ethical use of GAI in education, ensuring that the educational landscape in Serbia is transformed for the better. Each phase builds on the success of the previous one, allowing for adjustments based on real-world feedback and changing needs.

6 Conclusion

The integration of GAI into Serbia's educational system represents a transformative opportunity to enhance the quality and accessibility of education. This paper has outlined a strategic framework and action plan aimed at harnessing the potential of GAI technologies to revolutionize learning environments and educational outcomes across the country. The phased implementation strategy ensures that adopting these technologies is sustainable and aligned with the specific needs and challenges of the Serbian educational landscape.

Key Points Summary

- Educational Transformation: GAI offers unprecedented opportunities to personalize and enrich the learning experience, making education more engaging and effective.
- Infrastructure and Readiness: The successful deployment of GAI technologies requires robust digital infrastructure and comprehensive planning to address the current disparities in access and technology readiness.
- Ethical and Practical Considerations: Ethical guidelines, data protection, and bias mitigation are critical to ensure that GAI tools are used responsibly and beneficially in educational settings.
- Capacity Building: Educators must have the necessary skills and knowledge to utilize GAI tools effectively, requiring ongoing professional development and support.

Call to Action

A collaborative approach is essential to realize the vision of a digitally advanced and inclusive educational system. We call upon policymakers, technologists, educators, and the broader community to:

- Foster Collaborative Partnerships: Engage in partnerships that bridge the gap between technology development and educational needs, ensuring that GAI tools are tailored to the real-world requirements of Serbian education.
- Commit to Continuous Learning and Adaptation: Embrace a culture of continuous learning and flexibility, allowing educational practices to adapt to technological advancements.
- Ensure Inclusive and Ethical Practices: Prioritize the development of inclusive and ethically sound practices that respect the diversity and rights of all students, ensuring that no one is left behind in the digital transition.

Serbia can position itself at the forefront of educational innovation by taking these steps, leveraging GAI technologies to improve academic outcomes and prepare students for a future where digital literacy and technology fluency are paramount. The journey towards integrating GAI into Serbia's educational framework is both ambitious and necessary, marking a significant leap towards a future-ready education that is more accessible, engaging, and aligned with the demands of the 21st century.

References

1. Vukmirović, D., Comic, T., Stanojevic, N., Milicevic, U.: Emerging tech for future experts: unleashing the potential of generative ai in sustainable education. In [Editors' Names] (Eds.), Proceedings of the International Conference on Sustainable Energy Education - SEED 2024 (2024). In press

2. Djukic, G.P.: Digital inclusion for vulnerable groups and transformation: a comparative case study. In: Handbook of Research on Global Institutional Roles for Inclusive Development, pp. 19–38 (2022). IGI Global
3. Ožegović, J.: Report on digital inclusion in the Republic of Serbia from 2014 to 2018. Belgrade: Team for Social Inclusion and Poverty Reduction of the Government of the Republic of Serbia (TSUSS) (2019)
4. Ministry of Education, Science, and Technological Development, MPNTR (2020) Ex-post analysis of the Strategy for Education Development implementation in the Republic of Serbia (SROS) until 2020. SROS 2020 Realization analysis. http://www.mpn.gov.rs/wp-content/uploads/2021/02/3-Ex-post-analiza_0202.pdf
5. Kovács Cerović, T., Mićić, K., Vračar, S.: A leap to the digital era—what are lower and upper secondary school students' experiences of distance education during the COVID-19 pandemic in Serbia? Eur. J. Psychol. Educ. **37**(3), 745–764 (2022)
6. Kayyali, M.: Rebuilding higher education systems impacted by crises: navigating traumatic events, disasters, and more: navigating traumatic events, disasters, and more. IGI Global (2024)
7. Vukmirović, D.: Primena generativne VI u nastavi, Stručni skup-tribina "Integracija i implementacija generativne veštačke inteligencije (VI) u procesu obrazovanja", Regionalni centar za profesionalni razvoj zaposlenih u obrazovanju (2023). https://doi.org/10.13140/RG.2.2.36616.19202
8. Vukmirović, D.: Integracija i implementacija generativne veštačke inteligencije u proces obrazovanja, Stručni skup-tribina "Integracija i implementacija generativne veštačke inteligencije (VI) u procesu obrazovanja", Regionalni centar za profesionalni razvoj zaposlenih u obrazovanju (2023. https://doi.org/10.13140/RG.2.2.23194.41921
9. Ethical dimensions of information and knowledge societies. InWSIS 2023 High-Level Interactive Policy Sessions, pp. 1–6 (2023). ITU International Telecommunication Union
10. Mandic, D.: Report on Smart Education in the Republic of Serbia. InSmart Education in China and Central & Eastern European Countries, pp. 271–291 (2023). Springer Nature Singapore, Singapore
11. UNESCO, Guidance for generative AI in education and research, UNESCO (2023). https://www.unesco.org/en/articles/guidance-generative-ai-education-and-research
12. Miao, F., Holmes, W., Huang, R., Zhang, H.: AI, and Education: Guidance for Policymakers. UNESCO Publishing (2021)
13. Vukmirović, D., Popovac, R., Stanojević, N., Čomić, T.: Challenges of integration and implementation of generative artificial intelligence in the process of higher education. International Multidisciplinary Conference "Challenges of Contemporary Higher Education" - CCHE (2024)

Dataflow Hardware Advancements for Supporting Artificial Intelligence Algorithms

Nenad Korolija(✉)

School of Electrical Engineering, University of Belgrade, Belgrade, Serbia
nenadko@etf.rs

Abstract. The raising need for artificial intelligence applications poses new demands on computer hardware and algorithm development. Both faster processing and lower power consumption are needed. Dataflow hardware proves to be more energy-efficient due to lower frequencies and more dense computing compared to control-flow programming. However, the utilization of dataflow chip die surface is not uniform. The diversity is the result of dataflow hardware implementation of algorithms and diversity in algorithm execution. If an algorithm is presented as a dataflow graph, certain paths might be utilized more than others. However, each path requires a certain amount of FPGA resources comparable to the length of the path. This work aims to present a new model for controlling frequencies of dataflow kernels adjusted to the expected amount of processing of available kernels. Results indicate that trading between frequencies of certain kernels can improve the execution speed of the dataflow hardware, without jeopardizing the durability of the chip die due to the overall heating.

Keywords: dataflow hardware · dataflow kernels · hardware frequencies

1 Introduction

Remarkable increase in exploitation of artificial intelligence (AI) affects the production of computer hardware. Specialized AI hardware includes neural processing units (NPUs) [1], cloud tensor processing units (TPUs) [2], graphics processing units (GPUs) [3], and dataflow hardware [4], which are designed to accelerate the processing demands, including those of AI algorithms.

This work is based on accelerating algorithms using the dataflow hardware, as it offers the possibility to be reconfigured, enabling execution of multiple high-performance computing algorithms, but is known to be more energy efficient compared to the control-flow hardware [5, 6]. Various high-performance computing algorithms can be accelerated using the dataflow hardware [7–9].

The dataflow hardware is optimized for the repetitive execution of the same set of instructions. Most algorithms include instructions that are repeatedly executed, but also instructions needed for initialization of the data that needs to be processed, and for handling results. Utilization of dataflow hardware for those instructions that are rarely

© The Author(s), under exclusive license to Springer Nature Switzerland AG 2025
N. Filipović (Ed.): AAI 2024, LNNS 1446, pp. 236–247, 2025.
https://doi.org/10.1007/978-3-031-99201-8_25

executed is not justifiable. Therefore, dataflow hardware is almost always utilized along with control-flow hardware.

One of the research and industry problems of utilizing dataflow hardware is the ability to optimize the performance of the hardware in terms of utilization of the chip die and the frequencies of specific parts of the dataflow chip. This research aims to show the potentials of adjusting frequencies of portions of dataflow hardware separately from each otherto reduce the total algorithm execution time on dataflow hardware.

The following chapters present the problem statement, existing research, proposed model, results of the analysis, and conclusions. This paper is structured in accordance with what is proposed to be the best method for presenting research results [10].

2 Problem Statement

Algorithm execution on a dataflow hardware often includes multiple kernels that the dataflow hardware is configured to execute. Although they are available during the algorithm execution, they are often not utilized equally. While a kernel can reside on a critical path of a dataflow execution graph, another kernel can be responsible solely for handling edge cases. Therefore, the total time of utilization of each kernel can vary drastically.

The research question this work poses is whether there is a possibility to manipulate frequencies of specific dataflow kernels in order to reduce the total high-performance computing algorithm execution time. Another issue this research considers is a power consumption. By increasing frequency of a portion of dataflow hardware by a given factor, the power dissipation increases approximately by the factor equal to the square of the previous one. Therefore, increasing the frequency can lead to faster processing, but more power consumption at the same time. On the other hand, it is possible to limit the total power consumption of dataflow hardware and manipulate frequencies of portions to achieve faster processing. In this scenario, it is required that faster processing kernels process enough elements to justify this approach.

3 Existing Research

Although dataflow computing has existed for more than half a century, control-flow computing was predominantly in use both in personal computers and high-performance computing architectures. However, due to advantages that dataflow hardware brings, both the industry and research community have recognized the possibilities of combining dataflow hardware with control-flow hardware. As a result, various types of hybrid architectures were introduced to increase the speed of communication between control-flow and dataflow hardware [11–17].

Some of the hybrid computer architectures also include internet of things support [18, 19]. The work presented in this paper has no connections with internet of things, but most of the concepts introduced in these hybrid computers also apply for this research.

The work [20] proposes power minimization for real-time dataflow applications. The authors consider the dataflow modeled real-time applications that can be mapped onto multiple hardware resources. Their model supports frequencies of tasks to be constrained, as well as frequencies of processors.

Researchers have tried to improve the efficiency of dataflow hardware by scheduling task so that the total execution of all tasks is minimized [21].

Availability of the dataflow hardware often presents a problem for a programmer to start working with a dataflow hardware. The web-based integrated development environment has been implemented to support developing dataflow hardware applications without having the dataflow hardware available on the local personal computer [22].

One way of tackling the problem of applying spatial computing instead of temporal computing (used for control-flow architectures) is working on improving cache performance [23–26]. This clearly presents a way to improve the total performance of the system, but the control-flow parts of the architecture still present the bottleneck.

Working with dataflow hardware is considered to require more effort compared to control-flow programming [27, 28]. However, due to the applicability of dataflow hardware, various tools have been introduced to help with transition of control-flow algorithms into the dataflow representation [29, 30]. Researchers and industry professionals have recognized that providing open-source dataflow applications can help bridge the gap between programming control-flow and dataflow applications [31, 32].

Researchers have been working on partial configuration of FPGA-based dataflow hardware for decades. Runtime partial reconfiguration is known to be capable of reducing the number of FPGA-based devices or the size of a single device, thereby reducing the size of the system and total power consumption [33, 34]. Partial configurations can be produced by assembling components from a previously created library, thus enabling the embedded system development faster and with less effort compared to the conventional design flow [35, 36].

FPGA embedded processors based on run-time partial reconfiguration introduce another layer of complexity, thereby affecting fault-tolerance. However, precompiled configurations with spare resources can be generated for each duplicated module. They can further be used to repair defective modules at runtime [37].

The dynamic reconfiguration ability of clock managers of FPGAs is known to be capable of changing the operating frequency of FPGAs at runtime [38]. However, while partial and dynamic reconfiguration at run-time enable sharing the hardware between multiple independent functions from different applications, the resource allocation tends to fragment the FPGA logic space. In order to get enough contiguous space to efficiently implement incoming functions, a defragmentation needs to be performed after certain number of reconfiguring the FPGA resources [39, 40].

Another aspect of the FPGAs is the clock-frequency that is fixed for most FPGA designs. It is usually determined through static analysis in FPGA vendor tools, limiting the exploitation of advantages of the full run-time potentials that applications can achieve by running on FPGAs. Several methods have been developed to overcome this limitation by using dynamic clock-frequencies [41, 42].

4 Proposed Model

It is known that dataflow hardware based on FPGA is capable of accelerating algorithms that include uniform processing of large amounts of data. These algorithms include AI algorithms. The trend is that more and more processing power is needed for processing AI algorithms. Moreover, the percentage of computation that is related to these algorithms keep growing. However, not all algorithms are equally suitable for dataflow hardware. The fact that once configured hardware based on FPGA cannot be utilized for other algorithms before reconfiguring the hardware makes it suitable for algorithms with predictable type of operations needed to be computed. One of the key elements that enable using dataflow hardware for multi layered neural networks are multiplexers. These elements are used for controlling which input will be provided to the processing element. For example, a multiplexer can be used to determine whether the dataflow kernel will start processing by reading the input data from the control-flow processor, or from a local memory of the dataflow card. Another example would be to utilize it to read appropriate weights as inputs to neural network nodes. This way, a multi layered neural network can be directly modeled into the dataflow hardware based on FPGA. This implementation enables parallel processing in all nodes, producing the output in the number of clock cycles that is proportional to the number of layers. This can be further optimized if the algorithm is used on multiple instances of data simultaneously, enabling processing multiple data in parallel. For example, while the last layer is processing the data from the first instance of data, the previous layer can process data from the second instance of data, etc. This way, the dataflow hardware can produce the output faster, with the rate equal to the number of layers. However, changes to the number of layers or nodes within a layer require reconfiguring the dataflow hardware.. While the limitation on the number of layers and nodes might seem too rigorous, it should be noted that the time needed for reconfiguring the dataflow hardware is neglectable comparing to the processing time for many AI algorithms. Besides that, the dataflow hardware configured for a certain algorithm can be utilized repeatedly. An important aspect of this solution is that the dataflow hardware for different neural network configurations can be implemented on separate dataflow hardware elements within a computer cloud, offering faster processing without the need for reconfiguration of the dataflow hardware before executing an algorithm.

Other solutions to the previously mentioned limitation can be found in streaming data through dataflow kernel responsible for a general node, so that the length of the data that is streamed through the kernel directs the number of nodes of the neural network. Of course, there could be multiple instances of such kernels, enabling parallel processing of multiple nodes within a layer simultaneously, or even processing multiple instances of data simultaneously. This solution is left out of the scope, as it requires more complex control of data flowing into the kernel and complex control for storing results.

This chapter describes possibilities to accelerate the execution of dataflow kernels by manipulating frequencies on dataflow hardware, while the total power consumption can be lowered as well. In order to demonstrate the proposed method, we could consider an algorithm that processes element by element of a given matrix. The processing differs for the elements on edges of the matrix and the rest of elements belonging to the middle part of the matrix. One way of processing a matrix is to have a single kernel with various conditions that direct the processing depending on the position of an element in the matrix. The other approach is to have separate kernels for processing edge elements, keeping kernels simple and the processing uniform. The analysis performed in this research is based on the assumption that there are five kernels involved in the matrix processing, four for edge elements and one for the middle elements. Those kernels can run in parallel, but the processing time of them vary. Another assumption is that the complexity of kernels is similar. It is known that processing edge elements requires somewhat less processing (e.g., 25%), but in some cases, there could be separate kernels for corner cases, increasing the complexity and time needed for managing streams to all the kernels.

The total time needed for processing the matrix is equal to the time needed for processing the middle elements. Without jeopardizing the generality, we can assume that the square matrix needs to be processed. The dimensions of matrix may vary depending on a problem. We can also assume that separate frequencies can be applied to kernels. The existing approach is to have the same frequency for all kernels. In this research, the frequency of the middle kernel is increased, while the frequency of other kernels is decreased. The total power consumption should not be changed to justify the proposed approach.

A simple implementation that would enable two clock cycles per a single FPGA image (dataflow hardware) would be to have portions of the total FPGA area that would pass the operating frequency to the following one, or from the previous one, depending on a single multiplexer. This implementation does not add a remarkable burden in terms of required transistor count, but it imposes certain limitations: there could be only two operating frequencies, and each functionality has to be restricted to a number of portions of FPGA. Sharing of the same portion by two kernels running at different frequencies would not be possible. The first limitation can be avoided if each portion would have a clock assigned to it, thereby increasing the complexity of the system. The fragmentation imposed by the second one can be lowered by increasing the number of portions. For the purpose of this research, only two operating frequencies of the FPGA at the same time are enough.

Without jeopardizing the generality, we can assume the default dataflow hardware frequency, without having multiple kernels operating at different frequencies. We can also assume that the kernels processing elements on matrix edges are working on the same frequency, while the kernel processing middle elements can operate at higher frequency. Let us denote the optimal frequency of the faster processing kernel with x. In order to be able to lower the frequency of slower processing kernels (denoted as slowFreq), the processing time of edge elements should not be higher than the processing time of middle elements. If nEdges represents the number of elements on edges of the matrix, and nMiddle represents the number of elements in the middle, the following equations can be applied for calculating maximal optimal frequency of the faster processing kernel. Equations 1 and 2 define the ratio between faster and slower frequencies into achieve the same processing time of all kernels. Factor ratio in Eq. 2 represents the ratio between nMiddle and nEdges.

$$\text{slowFreq} * \text{nEdges} = x * \text{nMiddle} \tag{1}$$

$$\text{slowFreq} = \text{x} * \text{ratio} \tag{2}$$

It is clear that the ratio between the frequencies equal to the ratio between the number of elements in the middle and those on edges of a matrix is not always possible to achieve, nor is it reasonable to lower the slower processing kernels frequency so much that it presents a problem for streaming data, but the theoretical value is used for presenting the proposed approach. Since the power consumption is almost proportional to the square of the frequency, frequencies couple of order of magnitude lower than the frequency of faster processing kernel led to neglect-able power consumption of slower processing kernels.

Equation 3 defines the equality of the power consumption of faster and slower processing kernels as if all kernels would be operating at 300 MHz. The left-hand side of the equation defines the power consumption of the dataflow hardware if the frequency of 300 MHz is applied to the whole dataflow hardware. This power consumption should be equal to the power consumption of the dataflow hardware whose 80% is operating at slower frequency defined in Eq. 2 and 20% is operating at faster frequency x. As it can be seen, the optimal frequency of the faster processing kernel depends on the ratio between the number of middle elements and the number edge elements. Values 4/5 and 1/5 denote the effect of four out of five kernels on the power consumption, and the effect of the faster processing kernel on the power consumption.

$$300^2 = (4/5) * (\text{x}/\text{ratio})^2 + (1/5) * \text{x}^2 \tag{3}$$

Equations 4 and 5 define the x as a function of ratio and percentages of dataflow hardware operating at lower and faster frequencies. As it can be seen from the formula, the percentage of slower hardware is divided by the square of ratio between middle and edge elements. This ratio further rapidly increases with increasing dimensions of a square matrix. Therefore, this portion of power consumption becomes neglectable even for relatively small matrix dimensions.

$$300^2 = \text{x}^2 * \left((4/5)/\text{ratio}^2 + (1/5)\right) \tag{4}$$

$$\text{x} = 300/\text{sqrt}\left((4/5)/\text{ratio}^2 + (1/5)\right) \tag{5}$$

Table 1 presents matrix dimensions for validating results. The column ratio increases with the square of matrix height. As it can be seen from Eq. 5, the squared ratio divides the effect of four kernels on the heating. Therefore, the square root from Eq. 5 approaches the value of the square root from 1/5, denoting that almost all the power consumption is related to the faster processing kernel.

Equation 6 defines the power factor (denoted as PowerFactor) that represents the power reduction of the dataflow hardware without the proposed approach and the dataflow hardware with the proposed approach but limiting the faster processing kernel frequency to 300 MHz. This approach is based on the assumption that the frequency of the dataflow hardware is limited and that even with relatively low power consumption, it cannot be increased further. As a result, the faster processing kernel will consume

Table 1. Dimensions of matrices.

Matrix dimensions	Edge elements	Middle elements	Ratio
10 x 10	36	81	2.25
50 x 50	196	2401	12.25
100 x 100	396	9801	24.75
200 x 200	796	39601	49.75
500 x 500	1996	249001	124.75
1000 x 1000	3996	998001	249.75
10000 x 10000	39996	99980001	2499.75
100000 x 100000	399996	9999800001	24999.75
1000000 x 1000000	3999996	999998000001	249999.75

the same amount of power as if the whole dataflow hardware would be working on the same frequency. However, the slower processing kernels will have lowered operating frequency, reducing their power consumption compared to the case where the whole dataflow hardware operates on the same frequency. Even in this scenario, the example with five kernels processing the matrix justifies the proposed frequency manipulation, as the total power consumption will be reduced with raising matrix dimension, approaching the power consumption of the faster processing kernel.

$$\text{PowerFactor} = 300^2/\left(300^2 * 1/5 + (300/\text{ratio})^2 * 4/5\right) \tag{6}$$

As compilers cannot estimate the execution time of slow and fast segments of code, they can be augmented by artificial intelligence to automatically detect edge kernels and middle elements kernels. Edge elements have minimal and maximal indices, while middle have ranges in-between.

5 Results and Analysis

In this section, results of the analysis of the proposed solution are examined. First, the acceleration is calculated under the condition that the total power consumption of the dataflow card remains the same as if there was a single frequency applied for all dataflow kernels.

In order to evaluate the acceleration possibility of the proposed approach, frequencies of dataflow kernels are manipulated so that the total power consumption of the dataflow hardware remains unchanged. Without jeopardizing the generality, we can assume that the dataflow hardware operates at a frequency of 300 MHz. Table 2 shows the acceleration possibilities of the proposed approach, where Fast frequency denotes the frequency of the kernel processing middle elements of a matrix, Slow frequency denotes the frequency of the kernels processing edge elements, and Frequency factor show how much is the frequency of the kernel processing middle elements higher than the default frequency of 300 MHz. The value of Frequency factor represents the acceleration of the algorithm.

Table 2. Frequencies of slow and fast kernels.

Matrix dimensions	Fast frequency	Slow frequency	Frequency factor
10 x 10	501.38	222.83	1.67
50 x 50	662.05	54.05	2.21
100 x 100	668.64	27.02	2.23
200 x 200	670.28	13.47	2.23
500 x 500	670.73	5.38	2.24
1000 x 1000	670.80	2.69	2.24
10000 x 10000	670.82	0.27	2.24
100000 x 100000	670.82	0.03	2.24
1000000 x 1000000	670.82	0.00	2.24

Table 3 presents the frequency factor in the case that the fast kernel has the same frequency that the dataflow hardware would have without the proposed method. As it can be seen from Table 3, power factor relatively quickly becomes around five, which is to be expected, as the power of the dataflow hardware almost reaches the power consumption of the faster processing kernel, because of the noticeable lower frequencies of four kernels.

Table 3. Reducing power consumption without increasing the processing time.

Matrix dimensions	Fast frequency	Slow frequency	Power factor
10 x 10	300.00	133.33	2.79
50 x 50	300.00	24.49	4.87
100 x 100	300.00	12.12	4.97
200 x 200	300.00	6.03	4.99
500 x 500	300.00	2.40	5.00
1000 x 1000	300.00	1.20	5.00
10000 x 10000	300.00	0.12	5.00
100000 x 100000	300.00	0.01	5.00
1000000 x 1000000	300.00	0.00	5.00

The fact that the power consumption is almost proportional to the squared frequency limits the acceleration possibilities. Even with 80% of the hardware surface being disabled, the remaining 20% of the hardware would be accelerated only by the factor around 2.24. On the other side, the increasing dimension of square matrix rapidly changes the ratio between middle and edge elements, justifying the proposed approach.

The analysis was demonstrated on a single problem of processing algorithms whose main computing problem is performing operations on matrices. The same type of analysis can be applied to dataflow kernels that process other algorithms. For example, if an algorithm implementation on the dataflow hardware consists of two kernels that are similar in size measured in required resources on the dataflow hardware, while one of them is dominant, the acceleration factor would tend to approach the value of the square root of number two.

The proposed solutions are general enough to be applicable to the dataflow implementation of any existing algorithms within the dataflow paradigm. The achievable acceleration is directed almost solely on the ratio between the amount of resources on the dataflow hardware that are occupied by the faster processing kernels and the amount of resources on the dataflow hardware that are occupied by the slower processing kernels. The ratio of processing tasks between faster and slower kernels is not as much important as the power consumption of a portion of dataflow hardware that is almost proportional to the squared frequency. This can be seen from Tables 2 and 3. The analysis is performed based on the assumption that the problem is known in a compile time. This would require prior knowledge of the amount of processing for all problems that will be executed using the dataflow hardware. Since the acceleration using the proposed algorithm does not depend drastically on the problem size (see Table 2), this requirement can be eliminated, and the dataflow hardware can be optimized for relatively small problem sizes. In this manner, the acceleration factor will be almost the same for any problem size larger than the problem size for which the dataflow hardware is configured.

One of the drawbacks of the proposed solution is that additional limitations must be included in mapping kernels to transistors, as unequal power distribution could damage the chip faster than the uniform power distribution for the same power consumption. However, this can be automated, as the software for mapping a program written in a high-level programming language into the FPGA has existed for more than a decade. This automation could start by converting high-level programming constructs into an intermediary form that can be implemented in dataflow hardware. In a high-level programming language, programmers often write a code that is easier to work with compared to hardware description languages. The intermediary form would capture the essentials of the algorithm that should be converted into the dataflow hardware, while abstracting away specific syntax and constructs of the original high-level programming language. Further, the intermediary form can be processed by a synthesis tool, so that it is converted into a dataflow hardware representation. The most important task of a synthesis tool is to create a dataflow graph and determine the possibility of executing operations in parallel. The synthesized representation can be further mapped to the dataflow hardware. This process is needed for organizing logic blocks, look-up tables, and other dataflow hardware elements. After mapping, the design undergoes placement. The most important aspect of placement is in deciding where each element will go on the dataflow hardware and routing. One of the most complicated tasks that this process includes is optimizing the design to fit within the dataflow hardware, while the algorithm should be executed as fast as possible using the hardware. The final step would involve generating a bitstream file, containing the configuration data that will be loaded onto the FPGA.

This data defines how it will operate based on the original high-level programming language code. The bitstream is typically directly uploaded to the FPGA. Several tools and frameworks are developed to help previously explained possible translation of the high-level programming language code into the dataflow hardware, including high-level synthesis tools by Xilinx and Intel. Some tools even enable the translation of CUDA code into the dataflow hardware. However, in some cases, the translation process can be relatively complex, requiring deep knowledge of both the high-level programming language and the dataflow hardware. Furthermore, achieving the best performance often requires additional tuning and optimizations. With even more increased complexity of the system by having faster and slower processing kernels that should be spread over the dataflow hardware surface so that the heating is distributed over the hardware, the required skills may even be increased, unless the process is completely automated.

6 Conclusions

One of the key aspects in enabling fast processing that artificial intelligence applications require is the computer hardware. This research is based on dataflow hardware augmented by control-flow hardware. This kind of architecture is proven to be able to accelerate processing and lower power consumption at the same time. This comes from the fact that the dataflow hardware is configured solely for what it needs to perform, while the frequencies are lower. However, due to the nature of high-performance computing algorithms, the utilization of dataflow chip die surface is not uniform. The goal of this research was to check whether it is possible to manipulate freequencies of dataflow hardware per kernel, not the whole hardware, in order to achieve faster algorithm execution and lower power consumption. The problem that is considered is based on processing matrix elements, where different kind of processing is needed for elements on edges and those in the middle. This is often implemented on the dataflow hardware by creating multiple kernels, where the one processing the middle of the matrix would be utilized noticeably more than other kernels. Results indicate that increasing the frequency of the middle kernel while lowering the frequency of other kernels can lead to accelerating algorithm execution and lowering power consumption, without changing the total heating of the dataflow hardware.

References

1. Fiack, L., Rodriguez, L., Miramond, B.: Hardware design of a neural processing unit for bio-inspired computing. In: 2015 IEEE 13th International New Circuits and Systems Conference (NEWCAS), IEEE, 1–4 (2015)
2. Jouppi, et al.: In-datacenter performance analysis of a tensor processing unit. In: Proceedings of the 44th Annual International Symposium on Computer Architecture, pp. 1–12 (2017)
3. Brodtkorb, A.R., Hagen, T.R., Sætra, M.L.: Graphics processing unit (GPU) programming strategies and trends in GPU computing. J. Parallel and Distributed Computing **73**(1), 4–13 (2013)
4. Milutinovic, V., Salom, J., Veljovic, D., Korolija, N., Markovic, D., Petrovic, L.: Maxeler AppGallery Revisited. DataFlow Supercomputing Essentials: Research, Development and Education, pp. 3–18 (2017)

5. Babović, Z., et al.: Research in computing-intensive simulations for nature-oriented civil-engineering and related scientific fields, using machine learning and big data: an overview of open problems. Journal of Big Data **10**(1), 1–21 (2023)
6. Milutinović, V., Furht, B., Obradović, Z., Korolija, N.: Advances in high performance computing and related issues. Mathematical Problems in Engineering (2016)
7. Flynn, M.J., et al.: Moving from petaflops to petadata. Commun. ACM **56**(5), 39–42 (2013)
8. Babović, Z., et al.: Teaching computing for complex problems in civil engineering and geo-sciences using big data and machine learning: synergizing four different computing paradigms and four different management domains. Journal of Big Data **10**(1), 89 (2023)
9. Korolija, N., Djukic, T., Milutinovic, V., Filipovic, N.: Accelerating lattice-boltzman method using maxeler dataflow approach. Transactions on Internet Research **9**(2), 5–10 (2013)
10. Milutinovic, V.: The best method for presentation of research results. IEEE TCCA Newsletter, pp. 1–6 (1996)
11. Popović, J., Jelisavčić, V., Korolija, N.: Hybrid supercomputing architectures for artificial intelligence: analysis of potentials. In: 1st Serbian International Conference on Applied Artificial Intelligence (SICAAI), Kragujevac, Serbia (2022)
12. Miladinović, D., Bojović, M., Jelisavčić, V., Korolija, N.: Hybrid manycore dataflow processor. Proceedings, IX International Conference IcETRAN, Novi Pazar, Serbia (2022)
13. Milutinović, V., Trifunović, N., Korolija, N., Popović, J., Bojić, D.: Accelerating program execution using hybrid control flow and dataflow architectures. In: 2017 25th Telecommunication Forum (TELFOR), 1–4, IEEE (2017)
14. Korolija, N., Milfeld, K.: Towards hybrid supercomputing architectures. J. Computer and Forensic Sciences **1**(1), 47–54 (2022)
15. Popović, M., Korolija, N., Štrbac-Savić, S.: Hybrid control-flow and dataflow processor: algorithm granularity analysis. In Zbornik 29. konferencije YUINFO (2023)
16. Liu, S., Zhu, Y., Yu, B., Gaudiot, J.L., Gao, G.: Dataflow accelerator architecture for autonomous machine computing. In: Proceedings of the 43rd IEEE/ACM International Conference on Computer-Aided Design, pp. 1–7 (October 2024)
17. Djukic, T., Filipovic, N.: Computational Medicine in Data Mining and Modeling (Vol. 626). In: Rakocevic, G., Milutinović, V., (eds.) Springer, New York (2013))
18. Milutinović, V., et al.: The ultimate dataflow for ultimate supercomputers-on-a-chip, for scientific computing, geo physics, complex mathematics, and information processing. 10th Mediterranean Conference on Embedded Computing, IEEE, pp. 1–6 (2021)
19. Milutinović, V., et al.: The ultimate data flow for ultimate super computers-on-a-chip. Handbook of Research on Methodologies and Applications of Supercomputing, IGI Global, pp. 312–318 (2021)
20. Nelson, A., et al.: Power minimisation for real-time dataflow applications. In: 2011 14th Euromicro Conference on Digital System Design, pp. 117–124, IEEE (2011)
21. Korolija, N., Bojić, D., Hurson, A.R., Milutinovic, V.: A runtime job scheduling algorithm for cluster architectures with dataflow accelerators. Advances in computers, Elsevier **126** (2022)
22. Korolija, N., Zamuda, A.: On cloud-supported web-based integrated development environment for programming dataflow architectures. In: Exploring the DataFlow Supercomputing Paradigm, New York: Springer Cham, pp. 41–51 (2019)
23. Milutinovic, V., Tomasevic, M., Markovi, B., Tremblay, M.: A new cache architecture concept: the split temporal/spatial cache. Proceedings of 8th Mediterranean Electrotechnical Conference on Industrial Applications in Power Systems, Computer Science and Telecommunications (MELECON 96). Vol. 2. IEEE (1996)
24. Milutinovic, V., Markovic, B., Tomasevic, M., Tremblay, M.: The split temporal/spatial cache: a complexity analysis. In Proceedings of the SCIzzL **6**, 89–96 (1996)
25. Gabbay, F.: Microarchitecture. In: Handbook of Computer Architecture, pp. 3--45. Springer Nature Singapore, Singapore

26. Popović, J., Korolija, N., Marković, Ž., Bojić, D.: Developing algorithmic skills of pupils in Serbian schools using code. org materials. In: 2017 25th Telecommunication Forum (TELFOR), pp. 1--4. IEEE, November 2017
27. Milutinović, V., Salom, J., Trifunović, N., Giorgi, R.: Guide to Dataflow Supercomputing. Springer Nature **10**, 978–3 (2015)
28. Popovic, J., Bojic, D., Korolija, N.: Analysis of task effort estimation accuracy based on use case point size. IET Software **9**(6), 166–173 (2015)
29. Korolija, N., Popović, J., Cvetanović, M., Bojović, M.: Dataflow-based parallelization of control-flow algorithms. Advances in Computers **104**, pp. 73–124, Elsevier (2017)
30. Milutinovic, V., et al.: Transforming applications from the control flow to the dataflow paradigm. Dataflow Supercomputing Essentials, Springer Cham, New York, pp. 107–129 (2017)
31. Trifunovic, N., Milutinovic, V., Korolija, N., Gaydadjiev, G.: An AppGallery for dataflow computing. Journal of Big Data **3**(1), 1–30 (2016)
32. Trifunovic, N., Perovic, B., Trifunovic, P., Babovic, Z., Hurson, A.R.: A novel infrastructure for synergistic dataflow research, development, education, and deployment: the Maxeler AppGallery project. Advances in Computers, Elsevier **106**, 167–213 (2017)
33. McDonald, E.J.: Runtime FPGA partial reconfiguration. In 2008 IEEE Aerospace Conference, pp. 1–7 (2008), IEEE
34. Raha, A., Mathaikutty, D.A., Ghosh, S.K., Kundu, S.: FlexNN: a dataflow-aware flexible deep learning accelerator for energy-efficient edge devices. arXiv preprint arXiv:2403.09026 (2024)
35. Silva, M.L., Ferreira, J.C.: Support for partial run-time reconfiguration of platform FPGAs. J. Syst. Architect. **52**(12), 709–726 (2006)
36. Schweitzer, O., Weiser, U., Gabbay, F.: Enhancing DNN Computational Efficiency via Decomposition and Approximation. IEEE Access (2024)
37. Vavousis, A., Apostolakis, A., Psarakis, M.: A fault tolerant approach for FPGA embedded processors based on runtime partial reconfiguration. J. Electron. Test. **29**, 805–823 (2013)
38. Jayasinghe, D., Ignjatovic, A., Parameswaran, S.: RFTC: Runtime frequency tuning countermeasure using FPGA dynamic reconfiguration to mitigate power analysis attacks. In Proceedings of the 56th Annual Design Automation Conference 2019, pp. 1–6 (2019)
39. Gericota, M.G., Alves, G.R., Silva, M.L., Ferreira, J.M.: On-line defragmentation for run-time partially reconfigurable FPGAs. In: International Conference on Field Programmable Logic and Applications, pp. 302–311 (2002). Berlin, Heidelberg: Springer Berlin Heidelberg
40. Korolija, N.: Fine grain algorithm parallelization on a hybrid control-flow and dataflow processor. J. Big Data **12**(1), 42 (2025)
41. Bower, J.A., Luk, W., Mencer, O., Flynn, M.J., Morf, M.: Dynamic clock-frequencies for FPGAs. Microprocess. Microsyst.Microsyst. **30**(6), 388–397 (2006)
42. Korolija, N., Salom, J., Milutinovic, V.: Recent advancements in dataFlow computing. IPSI Trans. Adv. Res. **20**(1), 1–5 (2024)

Multilabel Classification Process Optimization Through the Utilization of Transfer Learning Approaches Supported by Decision Postprocessing Techniques

Ognjen Pavić[1,2](✉), Tijana Geroski[2,3], Lazar Dašić[1,2], and Nenad Filipović[2,3]

[1] Institute for Information Technologies Kragujevac, University of Kragujevac, Jovana Cvijica bb, 34000 Kragujevac, Serbia
opavic@kg.ac.rs

[2] Bioengineering Research and Development Center (BioIRC), Prvoslava Stojanovica 6, 34000 Kragujevac, Serbia

[3] Faculty of Engineering, University of Kragujevac, 6 Sestre Janjica, 34000 Kragujevac, Serbia

Abstract. Transfer learning, as a machine learning technique, is based on transferring previously acquired knowledge from one classification task to another. The application of transfer learning in the creation of deep neural networks allows for the newly created model to start with predefined weights instead of randomly assigned weight values, and thus achieved better classification achievements have reduced training time and even in some situations required less training data. In this concrete study focused on a multi-label classification problem of uncovering one or more respiratory diseases or radiological findings in chest X-ray images, a comparison was made between models trained with and without the utilization of transfer learning. Both approaches were supported by a result postprocessing system which was made to rectify possible mistakes which can arise as a byproduct of the nature of multi-label classification. Comparison was made between these models using two modified classification accuracy metrics. Neural networks trained with randomized starting weight values achieved 61% and 94% patient-based and disease-based accuracy, respectively. In contrast, the model which utilized transfer learning by using weights from a network which solved a similar respiratory disease classification problem achieved 64% and 96% patient-based and disease-based classification accuracy, respectively. While the increase of 3% over the stricter accuracy metric does not seem like a large improvement, the amount of mistakes made on a disease to disease basis was reduced by one third, since the number of incorrectly classified instances was reduced from 6% down to 4%. In addition, the models trained from transferred weights required significantly less training time to reach results that are up to standard, requiring 5 h for model training, in contrast to 16 h required for the other approach.

Keywords: deep learning · multi-label classification · transfer learning · postprocessing

© The Author(s), under exclusive license to Springer Nature Switzerland AG 2025
N. Filipović (Ed.): AAI 2024, LNNS 1446, pp. 248–255, 2025.
https://doi.org/10.1007/978-3-031-99201-8_26

1 Introduction

Lung diseases are among the most prominent and commonly diagnosed diseases worldwide. A statistical analysis conducted by the National Institutes of Health (NIH) shows that nearly 545 million people around the world live with chronic respiratory disease [1]. This number makes up more than 7% of the world's population. It is also important to note that these statistics do not take into account non-chronic lung diseases, which include diseases and conditions caused by bacterial and viral infections and physical factors such as permanent physical trauma in the chest area. These figures have proven that lung diseases are a huge burden on medical systems worldwide, with many patients seeking diagnosis, hospitalization and treatment [2]. In order to reduce the burden on medical centers, reduce the mortality rate caused by diseases of the respiratory system and improve the overall quality of life of patients and medical professionals, it is of great importance that new and better automatic diagnosis and decision support systems are developed to speed up the process and increase the accuracy of diagnosis and treatment, therapies and rehabilitation processes.

Since the advent of artificial intelligence (AI), attempts have begun to be made to create automated decision support systems based on raw data and statistical connections and correlations to improve medical centers around the world in many areas of medical expertise [3]. The use of neural networks has opened up the possibility of analyzing medical images and even uncovering important information that is not visible to the human eye [4].

A large number of studies has been conducted over the years regarding the use of various AI approaches in the analysis of medical images for the diagnosis of respiratory diseases. One of the most important studies is that of Rajpurkar et al. [5] within which, CheXNeXt, a deep neural network was created for the classification of 14 different respiratory diseases and radiological findings. The network was created using the transfer learning approach. Network weight and bias coefficients were transferred from ImageNet onto the DenseNet121 standard architecture. A study by Ge et al. [6] relies on exploiting correlations between X-ray data and their labels. In order to reveal hidden data contained in X-ray images, two loss functions were created that better manage error correction in case of multi-class classification. These loss functions include a Softmax loss function adapted to work with multiple simultaneous tags and a correlation loss function. In the field of medical image processing, as well as x-ray image processing specifically, one of the frequently used methods is attention-based residual neural networks. A study by Guan and Huang [7] points to the possibility of improving the predictive ability of artificial intelligence models using special embedded feature modules and attention region learning modules. Using the mentioned techniques, a better distinction is made between the physical characteristics of different respiratory diseases. However, it is also more difficult to distinguish between diseases and radiological findings that have similar physical properties, such as nodules and masses. Alauzi et al. [8] use the DenseNet121 network architecture as a primary disease classifier chained with an n-th number of binary classifiers specialized on separate diseases that make a decision for each detected disease, whether it is actually present or not. In this way, the training pressure of the master classifier is reduced and the risk of predicting a positive class for a disease that is absent in reality is eliminated. In addition to sequential chaining of classifiers, parallel systems of multiple neural networks have also been investigated in this area. A study by Chen et al. [9] uses two collaborative neural networks to simultaneously perform lung region-of-interest segmentation and multi-class classification of respiratory diseases based on the extracted region and the original X-ray image.Although this study achieves a high level of prediction accuracy and shows excellent results in terms of sensitivity to key diagnoses, the results can be further improved in the future by using knowledge transfer from sources closer to the nature of the classification problem of specific respiratory diseases.

Transfer learning is a machine learning technique used to transfer knowledge learned on one problem to solving another problem of a similar nature [10]. In deep learning and neural network training, transfer learning usually refers to using the weight coefficients learned by one model as a starting point for training a different model [11]. In this way, the initial weights are not randomly chosen and require less tuning during the training process, require fewer epochs to achieve stable results, and can be paired with lower learning rate values for more stable gradient descent of the fit function. Transfer learning also provides a way to continuously improve predictive accuracy, which is critical in medical fields, especially when it comes to diagnosing life-threatening diseases.

Decision post-processing is another technique that can be used to further improve results. Unlike transfer learning, post-processing of the results does not come into play during the creation of the deep learning model, but instead is used to process the results after the neural network has already reached a consensus on the decision to be displayed in the output of the decision support system. The post-processing modules are, therefore, built separately from the network and do not interact with it during the training process. These modules enforce widely agreed upon medical knowledge and logical conclusions to further increase the accuracy of the final report provided to the end user of the decision support system. The use of post-processing results reduces the strain on the neural network during the learning process.

2 Materials and Methods

In this study, we have created a convolutional neural network capable of classifying 17 different lung diseases and radiological findings by analyzing chest X-ray images. The classification was implemented as a multi-label classification with a single model capable of predicting multiple output classes at the same time, without the need to train multiple models to distinguish similar diseases.

2.1 Dataset

The dataset used in the conducted study was compiled from several publicly available databases of lung X-rays of common respiratory diseases and radiological findings. The publicly available databases used in compiling the final training and testing set of the proposed deep learning models are:

- Chest X-Ray Images (Pneumonia) [12]
- COVID-19 image data collection [13]
- ChestX-Ray 14 Dataset [14]

- COVID-QU-Ex Dataset [15]
- VinBigData Chest X-Ray Abnormalities Detection Dataset [16]
- Tuberculosis (TB) Chest X-Ray Database [17]
- SIIM-ACR Pneumothorax Segmentation dataset [18]
- Curated Datasets for COVID-19 Posterior-Anterior Chest Radiography Images[19]

The dataset collected from publicly available sources was composed of 191660 frontal chest radiographs and contained information on atelectasis, cardiomegaly, consolidation, pulmonary oedema, effusion, emphysema, fibrosis, hernia, parenchymal infiltration, parenchymal thickening, non-viral pneumonia, pneumothorax, viral pneumonia (covid-19), tuberculosis as well as tumors in the form of masses and nodules.

2.2 Architecture Description

The DenseNet121 architecture was used as the basis of all models created within the study. DenseNet121 is a deep convolutional neural network with 121 layers involved in the network training process along with an input layer, a data embedding layer, an output layer and 6 auxiliary data vectorization layers. The trainable layers are organized into blocks of two convolutional neuron layers for learning progressively larger image features followed by one max-pooling layer each. The convolutional layers of the DenseNet121 architecture perform convolution in 3 channels. Max-pooling layers use a 2x2 kernel which reduces the image size based on the maximum value under the kernel during each stride.

DenseNet121 is suitable for training in cases of multi-label classification with a large number of labels present simultaneously, as well as multi-class classification with a large number of possible output classes, due to its depth and the ability to learn specific information. Shallow networks are rarely suitable for solving problems of the nature explored within this study because a large number of activation functions are required to fully separate the data set into classes in an n-dimensional space. However, although the depth of the network in this situation approaches the higher ceiling of the performance that can be achieved, it is accompanied by an extremely large increase in the time required for training as well as for the final decision making.

Other than DenseNet121, there exist two other similar architectures which could theoretically increase the prediction capabilities of the final decision support system, these being DenseNet169 and DenseNet201. These architectures provide an even greater number of activation functions for distributing the n-dimensional space described by the dataset into targeted classes. However, these architectures did not have any suitable pretrained weights for the application of transfer learning. Because of this, these networks would have to be trained from random starting weights in their deeper neuron layers and were not used in the scope of this study.

2.3 Model Creation

As part of the study, 3 deep learning models were created. The first model did not use transfer learning and post-processing of results. The second model used post-processing of the results, also without transfer learning. The third model used both techniques applied with the aim of improving the final performance of the model.

The models were evaluated using specially developed accuracy metrics named lenient accuracy and strict accuracy. Strict accuracy was used as a measure of the reliability of neural networks in predicting all present and absent diseases and radiological findings at the patient level, while lenient accuracy was used as a measure of prediction at the level of individual markers and served as an instrument to discover which diseases and radiological findings cause the greatest problems during classification. The strict accuracy metric is calculated at the patient level and its value is increased only if all predicted classes for a given patient are predicted correctly. Lenient accuracy was calculated at the level of each output class separately for each patient and its value was increased every time the given disease was correctly predicted.

Transfer learning was incorporated into the developed models by transferring the knowledge learned within the existing CheXNeXt network, with the addition that the network architecture had to be expanded with 4 additional convolutional layers for retraining the network on diseases that were not present when training the original model and new extended output layer with the aim of increasing the number of output neurons according to the number of classes present in the data set.

The technique of result post-processing consisted of several steps of logical error correction based on widely accepted knowledge of rules and laws that govern the observed system. The first step of result post-processing was to raise the confidence threshold required to classify the disease or radiological finding as present to 0.8. The neural network itself had a default threshold of 0.5 and, in these cases, made a large number of errors by selecting a large number of diseases as present under the influence of their high representation in the initial data set. The post-processing rules included logical rules that made a decision about whether the patient was healthy in the event of the absence of all diseases, as well as a rule that made a decision about whether the tumor appeared as a nodule or a mass on the X-ray. Both of the previously mentioned rules were based on the certainty of the model for each class when making the decision. The post-processing of the results did not affect the network training process.

3 Results and Discussion

The original neural network, trained from randomly selected initial weight coefficients without result post-processing had major problems with disease imbalance in the initial data set. Diseases that appeared most often in the set, as well as diseases that appeared a very small number of times under the influence of weight equalization, were predicted too many times. This model achieved modified accuracy metric values of 23% and 51% for strict accuracy and lenient accuracy, respectively.

Applying result post-processing to a model trained without using transfer learning drastically raised the accuracy metrics to 61% and 94% for strict and lenient accuracy, respectively. The application of the transfer learning technique did not have a large impact on the absolute values of the accuracy metrics increasing performance to 64% and 96% strict and lenient accuracy, respectively. The performance comparison in terms of modified accuracy metrics among the created deep learning models is shown in Table 1.

Table 1. Performance of modified accuracy metrics

Model baseline	Strict accuracy	Lenient accuracy
Base model	23%	51%
Application of post-processing	61%	94%
Application of transfer learning	64%	96%

The use of result post-processing significantly limits the amount strain on the final decision that the equalization of class weights due to data imbalance puts on the classification model. Setting the decision threshold to 0.8 sets a tight constraint that a disease or radiological finding can only be considered present when the model is almost certain of its decision. Setting a decision threshold also gives the model a great advantage in the case of using the model in the role of a decision support system, by avoiding labeling the normal lung variations of different patients on radiographs as if a single or multiple diseases or anomalies have been detected. This is especially useful in the case of physical deformities and their labeling with classes of pneumothorax, masses and nodules. Setting additional distinction rules between masses and nodules is also of great importance for a decision support system as it provides medical professionals with additional information about the nature and size of the tumor. Cases in which the model predicts that a person is healthy with the presence of any other class does not affect the final accuracy metrics to a large extent, but avoids the occurrence of an out of the question situation that would be impossible to delineate without additional tests during preparation in daily medical practice.

Applying knowledge transfer from a pretrained model does not bring a large absolute increase in custom accuracy metrics compared to applying result post-processing. However, relative changes in performance are noticeable. The increase in lenient accuracy from 94% to 96% eliminates a third of incorrectly classified labels across the entire patient set. This shift is extremely sensitive when using the model on a large number of different patients over a long period of time.

4 Conclusion

Within the study, three deep learning models were created based on the same initial model architecture. The models were trained on a created dataset consisting of frontal chest x-rays. The aim of the classification was to detect respiratory diseases and radiological findings in lung images. The data set was created using publicly available databases on the Internet.

The first of the three models was created without using transfer learning and result post-processing. The second model used post-processing of the results to improve the overall model performance, while the third model was created using both transfer of learning and post-processing of the results. The post-processing of the results achieved a drastic improvement in the prediction capabilities of the model. Transfer learning had less impact on the final performance of the model. However, it still reduced the number of errors occurring based on individual output labels.

The conclusion is that deep learning models can be significantly improved using external knowledge both during the training and after network training. Transfer learning provides opportunities to reduce the amount of data required for training and the amount of time required reaching a steady state. Result post-processing provides the ability to use logical reasoning and established expertise to modify results and avoid states that are physically impossible to reach.

Future research will include attempts to improve the model by increasing the knowledge base of diseases and radiological findings that are rarely represented in the currently collected data set. Also, the use of new architectures provides the possibility of additional performance improvement. In addition, reducing the problem of multi-label classification with multiple simultaneously present outputs to binary classification using a large number of classification models can help determine more precisely which diseases are present and which are absent.

Acknowledgements. This research is funded by the project that has received funding from the European Union's Horizon 2020 research and innovation programme under grant agreement No 101016834 (HOSMARTAI). This research was also supported by the Ministry of Science, Technological Development and Innovation of the Republic of Serbia, contract numbers [451–03-65/2024–03/200107 (Faculty of Engineering, University of Kragujevac)] and [451–03-66/2024–03/200378 (Institute for Information Technologies Kragujevac)].

References

1. GBD Chronic Respiratory Disease Collaborators. Prevalence and attributable health burden of chronic respiratory diseases, 1990–2017: a systematic analysis for the Global Burden of Disease Study 2017. Lancet Respir Med., **8**(6), pp. 585–596 (2020)
2. World Health Organisation. WHO COVID-19 dashboard. https://data.who.int/dashboards/covid19/cases
3. Jean, F., et al.: Clinical artificial intelligence quality improvement: towards continual monitoring and updating of AI algorithms in healthcare. NPJ digital medicine **5** (2022)
4. Chan, H.-P., Samala, R.K., Hadjiiski, L.M., Zhou, C.: Deep learning in medical image analysis. In: Advances in Experimental Medicine and Biology.: Springer, Cham., pp. 3–21 (2020)
5. Rajpurkar, P., Irvin, J., Ball, R.L., et al.: Deep learning for chest radiograph diagnosis: a retrospective comparison of the CheXNeXt algorithm to practicing radiologists. PLOS MEDICINE (2018)
6. Ge, Z., et al.: Improving multi-label chest X-ray disease diagnosis by exploiting disease and health labels dependencies. Multimedia Tools and Applications **79**, 14889–14902 (2019)
7. Guan, Q., Huang, Y.: Multi-label chest X-ray image classification via category-wise residual attention learning. Pattern Recogn. Lett. **130**, 259–266 (2020)
8. Allaouzi, I., Ahmed, M.B.: A novel approach for multi-label chest X-ray classification of common thorax diseases. IEEE Access **7**, pp. 64279–64288 (2019)
9. Chen, B., Zhang, Z., Lin, J., Chen, Y., Guangming, L.: Two-stream collaborative network for multi-label chest X-ray Image classification with lung segmentation. Pattern Recogn. Lett. **135**, 221–227 (2020)
10. Weiss, K., Khoshgoftaar, T.M., Wang, D.: A survey of transfer learning. Journal of Big Data **3**(9) (2016)

11. Zhuang, F., et al.: A comprehensive survey on transfer learning. Proc. IEEE **109**(1), 43–76 (2021)
12. Kermany, D., Zhang, K., Goldbaum, M.: Labeled optical coherence tomography (OCT) and chest x-ray images for classification. Mendeley Data **2** (2018)
13. Cohen, J.P., et al.: COVID-19 Image Data Collection: Prospective Predictions Are the Future. arXiv:2006.11988 (2020)
14. Wang, X., et al.: ChestX-Ray8: hospital-scale chest X-Ray database and benchmarks on weakly-supervised classification and localization of common thorax diseases. IEEE Conference on Computer Vision and Pattern Recognition (CVPR) (2017)
15. Chowdhury, M., et al.: COVID-QU-Ex Dataset (2021). https://www.kaggle.com/datasets/anasmohammedtahir/covidqu
16. VinGroup Big Data Intitute. VinBigData Chest X-ray Abnormalities Detection (2021). https://www.kaggle.com/c/vinbigdata-chest-xray-abnormalities-detection
17. Rahman, T., et al.: Reliable tuberculosis detection using chest x-ray with deep learning, segmentation and visualization. IEEE Access **8**, 191586–191601 (2020)
18. Zawacki, A., Wu, C., Shih, G., et al.: SIIM-ACR Pneumothorax Segmentation 2019 Dataset (2019). https://datasetninja.com/siim-acr-pneumothorax-segmentation
19. Sarkar, O., et al.: An empirical model of classifying lung affected diseases to detect COVID-19 using chest x-ray employing convolutional neural architecture. In: 2022 IEEE International Conference on Artificial Intelligence in Engineering and Technology (IICAIET), Kota Kinabalu, Malaysia, pp. 1–6 (2022)

Scientific Production and Collaboration Patterns of Medical Researchers: A Case Study in Epidemiology and Infectious Diseases

Iva Potkonjak, Predrag Obradović(✉), and Marko Mišić

School of Electrical Engineering, University of Belgrade, 73 Kralja Aleksandra Boulevard, 11000 Belgrade, Serbia

pi233099m@student.etf.rs, {predrag.obradovic, marko.misic}@etf.bg.ac.rs

Abstract. This study employs social networks analysis to investigate scientific collaboration within the field of immunology and infectiology at the University of Belgrade, Faculty of Medicine. A primary dataset for this analysis is obtained using publicly available data about employees and their publications registered in the Scopus database between 1982 and 2020. A co-authorship network is constructed and analyzed using both integer and fractional counts for co-authorship. Basic network metrics reveal the network's classification as a small-world network with medium density and high clustering. Further analysis focuses on the organizational division of authors into four departments, unveiling significant collaboration between authors across departments. Community detection techniques are employed to reveal finer community structures within the network. Assortativity analysis, calculation of the rich-club coefficient, and exploration of power-law properties indicate a healthy research environment with minimal centralization of node influence. Centrality and brokerage analyses distinguish researchers with numerous co-authors from those with fewer co-authors, validating the utility of both weighting schemes, while ranking the distinguished researchers by each of these criteria, thus giving further insight into the collaboration dynamics of the environment.

Keywords: social networks analysis · co-authorship networks · community detection · rich-club phenomenon · small world networks

1 Introduction

World scientific production grows at a tremendous pace each year, especially in the fields such as biology, medicine, physics, and their interdisciplinary applications. Recent analyses have shown robust growth in scientific production in the period from 1996 onwards when the production of articles per year tripled to around 3 million papers per year in 2020 [1, 2]. Countries such as China, India, South Korea, Brazil, Turkey, and Iran largely increased their portion in the world scientific production, outside the Euro-American block [1, 3]. Such a tremendous volume of data constantly imposes challenges and requires new bibliometric and scientometric methods and tools [4] to analyze them, often including machine learning and artificial intelligence-based approaches [5, 6].

© The Author(s), under exclusive license to Springer Nature Switzerland AG 2025
N. Filipović (Ed.): AAI 2024, LNNS 1446, pp. 256–271, 2025.
https://doi.org/10.1007/978-3-031-99201-8_27

Social networks analysis has been successfully applied to different aspects of scientific production analysis. Collaboration patterns are one of the most widely studied areas, both on the individual [7], institutional, and country level [3]. Different collaboration patterns have been observed in mathematics, physics and biology using social network analysis methods [8]. Also, several studies [3, 9] suggest increase in international collaboration, which can be seen in a wider context of globalization of science.

In this paper we perform the bibliometric and scientometric analysis of scientific production and collaboration at the University of Belgrade, Faculty of Medicine, with the goal of exploring collaboration patterns and unraveling the social structure in the network of academic workers and medical professionals working in the field of epidemiology and infectious diseases. In order to conduct this case study, we use complex network theory methods such as network modeling, network metrics computation and visualization. We construct two networks of authors, which we name Authors Network (ANet) and Authors Network with Fractional Counting (FCANet). Analyzing standard network metrics such as node degree and its distribution, various centrality measures and clustering coefficients, we give insights into the topological structures of the modelled networks.

The paper is divided into several sections. The second section describes the constructed data set, its refined form and furnishes a comprehensive outline of the employed methodology for network modeling. The third section presents the results of the research, with appropriate quantitative and qualitative analyses of the data, as well as the generated visualizations. Finally, the concluding section offers recommendations for prospective research endeavors, coupled with a succinct summary of the study's findings.

2 Dataset and Methodology

This section of the paper starts with illustrating the primary and secondary datasets used in our research, as well as documenting the data cleaning and refinement procedures used to obtain the secondary dataset from the primary dataset. Next, the process of modeling the constructed networks is described. Finally, this section contains a short overview of the software stack used to perform the analysis.

2.1 Primary Dataset

In this research the scientific production in the field of epidemiology and infectious diseases from four departments of the University of Belgrade, Faculty of Medicine, is analyzed: the Department of Immunology, the Department of Epidemiology, the Department of Infectious Diseases, and the Department of Microbiology. The analysis focuses on scientific publications in journals with an impact factor, based on queries in the Scopus database. Data on scientific publications were collected using information about authors affiliated with academic positions at the aforementioned departments as of October 2020, obtained from the websites of these faculties. Other authors of scientific publications appearing in the primary dataset are not relevant for the scope of this research.

2.2 Data Cleaning and Refinement

The primary dataset comprises five tables derived from the Scopus database. Four tables contain information about papers authored by scientists from respective departments, while the fifth table focuses on author details. Data cleansing primarily involved the following steps:

- Rows lacking author information were removed;
- Special characters in Serbian Latin script were standardized;
- Journal titles were uniformly formatted to lowercase;
- Irrelevant columns (volume, issue, article number, page range, number of pages, and source) were discarded;
- Empty fields, notably in the citation count column, were filled with zeros;
- Non-essential document types were excluded, retaining Article, Article in Press, Review, Book Chapter, Letter, and Note;
- Unique identifiers were assigned to papers, based on their respective links to Scopus webpages;
- A comprehensive analysis was conducted to resolve instances of scientists with two surnames, resulting in the addition of 152 paper entries while maintaining data integrity.

The author table remained largely unchanged, containing complete department affiliations. Minor modifications included merging researchers' names and surnames for streamlined analysis. Additionally, presumed signatures were established for each author to facilitate further investigation of paper authorship. Thus, the secondary dataset was obtained.

2.3 Network Modeling of Author Networks ANet and FCAnet

In order to model the scientific collaboration between the researchers in the Faculty of Medicine, we have constructed two complex networks. Due to the symmetric property of the coauthorship relation, both formed networks are represented with weighted undirected graphs. Both constructed networks, Authors Network (ANet) and Authors Network with Fractional Counting (FCANet) comprise 59 nodes and 305 edges. The number of nodes is directly determined by the number of significant researchers, while the edges are constructed to connect nodes that have collaborated at least once. The weight of an edge in ANet indicates the number of times the authors collaborated, regardless of the number of authors on the paper. In order to account for papers with a large number of authors, fractional counting is often used in scientometric analysis, where the weight of the edge is normalized by the number of authors of the paper [10]. This is how FCANet is obtained.

2.4 Software Stack

Data cleaning and transformation, as well as elementary statistical analysis, were conducted using Python 3 and the Pandas module. A module for complex networks modelling and analysis NetworkX [11] was used for the construction of required networks and network metrics computations. Network model fitting was accomplished using the powerlaw package [12]. Visualizations were performed using NetworkX [11] and Gephi [13].

3 Results

This section presents the results of the conducted analysis. Firstly, basic network properties of ANet and FCANet are discussed. Secondly, the decomposition of the network of authors with regards to the departments the authors belong to is analyzed. Afterwards, a community detection procedure is performed using three clustering techniques and the results are compared between the techniques. Finally, collaboration patterns and nodes that individually stand out are further explored and displayed.

3.1 Network Metrics and Properties of ANet and FCANet

Commonly used network metrics for the author collaboration networks ANet and FCANet were calculated. As the topological structure of ANet and FCANet is identical and they only differ in edge weights, most network metrics and properties of ANet and FCANet are shared. Therefore, we present them together in Table 1.

The density of the networks is 0.18, indicating that 18% of possible edges are formed. The networks consist of two connected components, with 58 and 1 node, respectively. This comprises one large connected component and one isolated node representing a scientist with no recorded collaborations with colleagues during this period. The geodesic diameter of the large connected component of the network, disregarding edge weights, is 4, with an average distance of 2.2. Considering that only one node is not in the connected component and the average distances between nodes are small, it can be inferred that the network is well-connected. Further analyses will be conducted primarily on the giant component of the network. The density of the giant component is slightly higher, at 18.45%.

Network centralization was analyzed by calculating the centrality by degree, betweenness, and closeness [14]. The obtained values are 0.18, 0.08, and 0.23, respectively. Since all three values are low, it can be concluded that the network is not centralized, which means that the research work done on the Faculty of Medicine is well distributed.

The global clustering coefficient of both networks is 0.51, while the average clustering coefficient is 0.61. To ascertain the strength of clustering, the network will be compared to Erdős-Rényi and Barabási-Albert networks of comparable size. The corresponding Erdős-Rényi network has a global and average clustering coefficient of 0.18. The Barabási-Albert network has a global clustering coefficient of 0.5, while the average clustering coefficient of the Barabási-Albert network is 0.52. It can be observed that the co-authorship network exhibits significantly higher clustering coefficients compared to the randomly generated Erdős-Rényi network. Conversely, the coefficients are slightly higher than those in the Barabási-Albert network. Histograms of the distribution of local clustering coefficients for each network can be seen in Fig. 1.

Table 1. Network metrics of author collaboration networks ANet and FCANet.

Number of nodes	59
Number of edges	305
Network density	0.18
Geodesic diameter	4
Average node distance	2.2
Global Clustering Coefficient	0.51
Average Local Clustering Coefficient	0.61
Network degree centralization	0.18
Network closeness centralization	0.08
Network betweenness centralization	0.23

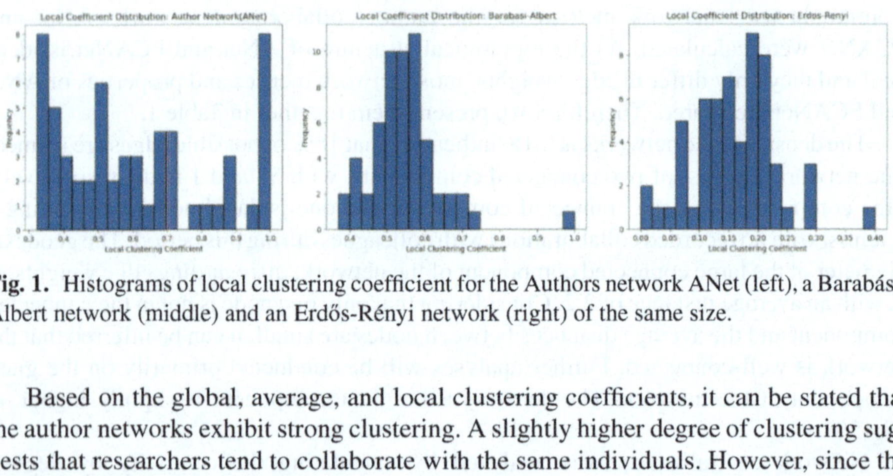

Fig. 1. Histograms of local clustering coefficient for the Authors network ANet (left), a Barabási-Albert network (middle) and an Erdős-Rényi network (right) of the same size.

Based on the global, average, and local clustering coefficients, it can be stated that the author networks exhibit strong clustering. A slightly higher degree of clustering suggests that researchers tend to collaborate with the same individuals. However, since the network is well-connected and many nodes have moderate values of the local clustering coefficient, authors likely have a larger number of different coauthor groups with whom they collaborate.

A small-world network [15] is characterized by a short average path length between nodes, combined with a high level of local clustering. ANet and FCANet satisfy the small-world property because they demonstrate both a short average path length (geodesic diameter of 4) and a relatively high clustering coefficient, indicating that nodes tend to form clusters of interconnected neighbors despite the overall network being well-connected. Additionally, to confirm this assumption, the omega coefficient of the small world network [16] is computed using the NetworkX omega function, which returns a value close to zero if the network is a small-world network. The obtained value is -0.1, confirming the observation.

3.2 Partitioning of ANet and FCANet into Departments

Information about departments and their employees is available in the primary dataset. Using this information as a metadata label for nodes in Anet and FCANet a visual representation of the decomposition can be obtained, as shown in Figs. 2 and 3, where Fig. 2 shows the whole network, while Fig. 3 focuses on the individual departments, in order to allow more detailed inspection. In these figures, green, blue, orange and purple nodes correspond to the researchers from the departments of Epidemiology, Immunology, Infectiology and Microbiology, respectively. There is ample collaboration between members of the same departments, but also some co-authorship between authors from different departments, which should suggest that the decomposition of the networks according to the departments is not the only logical decomposition and other node clusterings should be explored. That is done in detail in Sect. 3.3.

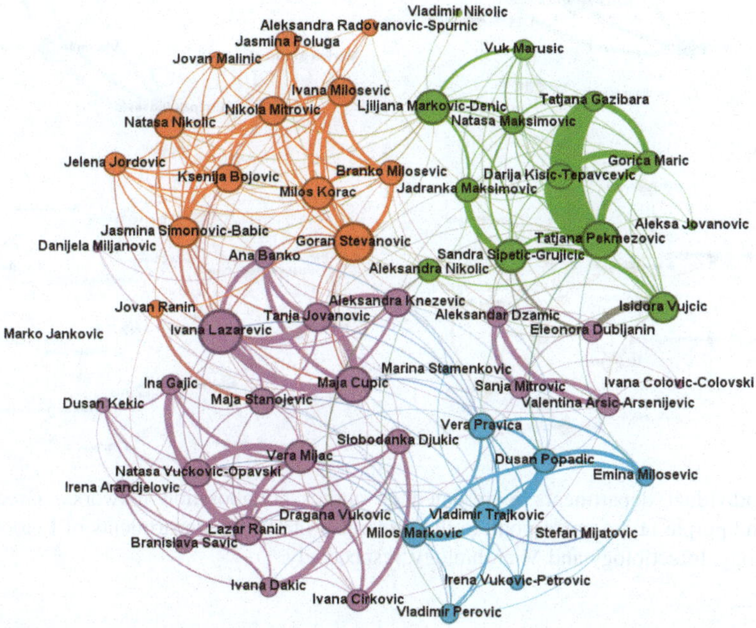

Fig. 2. A visual representation of the decomposition of ANet and FCANet into departments. Green, blue, orange and purple nodes correspond to the researchers from the departments of Epidemiology, Immunology, Infectiology and Microbiology, respectively.

To gain a more nuanced understanding of author interactions within departments, we will now delve into specific characteristics observed within the subgraphs of each department. Initially, the Microbiology network emerges with the lowest network density among the four, measuring at 0.29. Subsequent analysis will reveal that this department exhibits the highest degree of fragmentation in community partitioning. Moving to the network of Epidemiologists, which possesses the second lowest density at 0.68, we note the presence of significantly elevated edge weights, indicative of extensive collaborations among select members. Within the Immunology department, characterized by a density of 0.71, robust connectivity is observed, particularly evidenced by a central node serving as a broker to multiple members. Finally, the Infectiology department boasts the highest density at 0.72, reflecting strong interconnections. Notably, the most influential actor within the subgraph is identified as Miloš Korać; however, this dynamic shifts when interactions with actors from other departments are considered, underscoring a propensity for interdisciplinary collaboration.

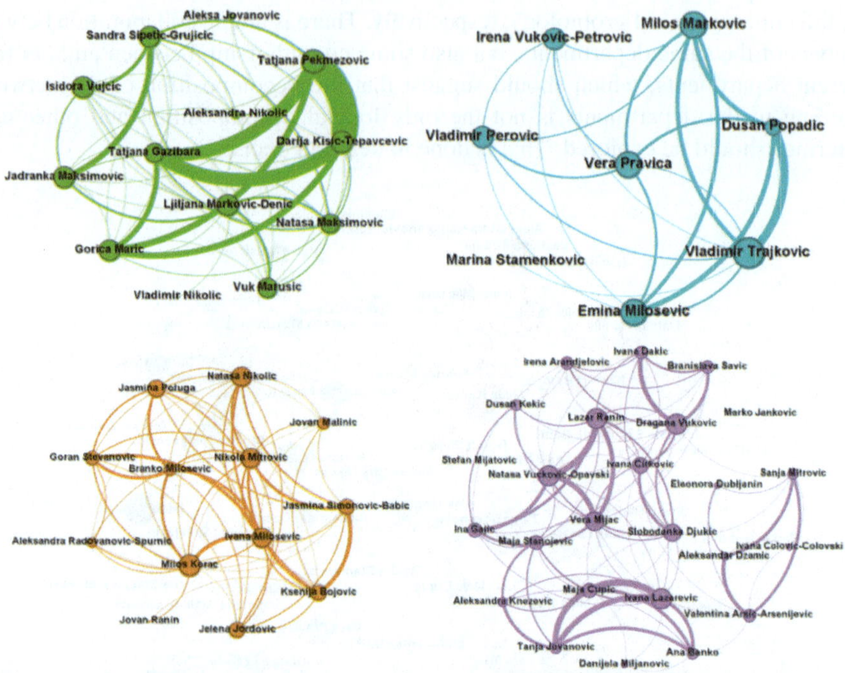

Fig. 3. Individual departments represented as author collaboration networks. Green, blue, orange and purple nodes correspond to the researchers from the departments of Epidemiology, Immunology, Infectiology and Microbiology, respectively.

In order to further assess the collaboration between departments, assortative mixing was examined through the co-authorship matrix of authors from different departments and is shown in Table 2. It can be observed that, as expected, departments collaborate most within themselves. However, it is interesting to note that the collaboration values between Immunology and Microbiology are measurably higher. Additionally, it is observed that the Departments for Immunology and Infectious Diseases have no recorded collaborations.

3.3 Community Detection and Analysis

Even though the information about the partitioning of researchers into departments is available in the dataset, it is possible that researchers from the same department

Table 2. Assortative mixing between the departments.

	Epidemiology	Immunology	Infectiology	Microbiology
Epidemiology	0.174	0.006	0.021	0.026
Immunology	0.006	0.065	0	0.046
Infectiology	0.021	0	0.184	0.048
Microbiology	0.026	0.046	0.048	0.282

do not collaborate. Conversely, researchers from different departments do collaborate, each providing their expertise, as shown in Fig. 2 and Table 2. Therefore, in order to justly analyze the collaboration patterns, we perform community detection in the collaboration network, showcasing different approaches to partitioning the collaboration network into distinct communities or clusters based on the available data about co-authorship. We employ three community detection algorithms to identify groups within the network: the Louvain method [17], Girvan–Newman algorithm [18] and the scikit-learn implementation of the spectral clustering algorithm, based on the works of Shi and Malik [19], Yu and Shi [20], Knyazev [21] and Damle et al. [22]. As each method uses a different approach to split the network into clusters, each algorithm provides a different insight into the interaction among actors. It is important to note that the Louvain method was applied to the whole network, while the Girvan-Newman algorithm and spectral clustering were applied to the giant component of the network, containing 58 nodes.

Louvain Clustering.
Given the significance of the resolution parameter for the quality of the network partition, we conducted an optimization process to select the resolution value that maximizes the modularity of the split. This optimization involved running the Louvain method with a range of resolution parameters, allowing us to evaluate the resulting modularity scores. We determined that the optimal resolution value for our analysis was 1.35, with a modularity score of 0.68. The network itself was split into 7 communities and the produced visualization can be seen in Fig. 4.

In this case, the network is divided into 7 clusters, one of which is always trivial, as it consists of an isolated node. We observe that the cluster of infectiologists contains nearly all nodes from this department, except for Jovan Ranin, who is attached to the central cluster of microbiologists. Microbiology is a department whose members are largely distributed across various clusters in this partition. Specifically, members are found in the central purple cluster, pink cluster, and one group is attached to the turquoise cluster. Additionally, Stefan Mijatović is attached to the cluster of immunologists, who all remain in the same community. The Department of Epidemiology is thus divided into two clusters: a smaller green one consisting of authors with stronger connections, and the turquoise cluster.

Girvan-Newman Clustering.
When applying the Girvan-Newman algorithm [18], an iterative process is used to split

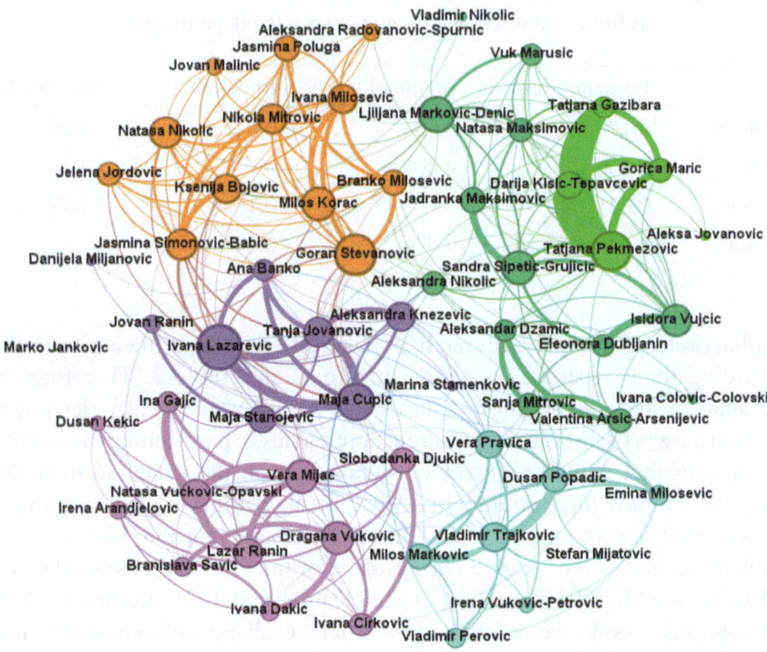

Fig. 4. Louvain clustering of the authors network using the optimal resolution value of 1.35.

the network into communities by removing one branch at a time, based on the betweenness centrality of the branch. This method produces several alternative partitions for the same network. We selected the partition with the highest modularity value for further analysis. The partition in question has a modularity value of 0.58 and 11 communities are detected, where singular nodes represent 5 of those communities, meaning there are 6 nontrivial clusters, exactly like when the Louvain clustering was employed. The constructed clusters largely align with the clusters obtained by the Louvain clustering and spectral clustering, with a notable 4-member cluster of microbiologists that is also obtained by the spectral clustering, but omitted by the Louvain method. A dendrogram of this decomposition is shown in Fig. 5.

Spectral Clustering.
In order to detect communities using spectral clustering, it is necessary to specify the number of groups that need to be detected. This is done with the use of graph Laplacian whose eigenvectors and eigenvalues provide the suggestion for the optimal number of communities. We found that the best partitions were made for 7 and 8 clusters respectively and the visualizations induced by these partitions are shown in Fig. 6.

Instead of optimizing modularity, which would force the obtained clusters to be largely of the same size, spectral clustering optimized the normalized cut metric.

By exhibiting enhanced sensitivity to clusters of varying sizes, spectral clustering provides the most detailed partitioning compared to previous methods. Specifically, upon examination of the partition with 8 clusters, it reveals a finer granularity within the Microbiology department, discerning four distinct groups. Furthermore, while spectral

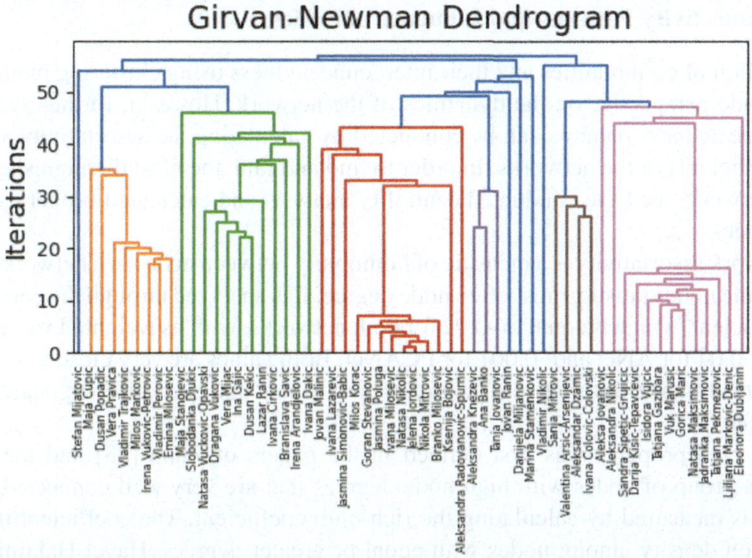

Fig. 5. A dendrogram constructed using the Girvan-Newman algorithm, decomposing the authors graph into 11 clusters. Out of those, 5 clusters are trivial, but the remaining 6 largely overlap with the clusters constructed by the other two employed clustering algorithms.

clustering successfully delineates the partition within the Microbiology department into separate clusters, it achieves this without sacrificing the granularity observed within the Epidemiology department, unlike the Girvan-Newman algorithm.

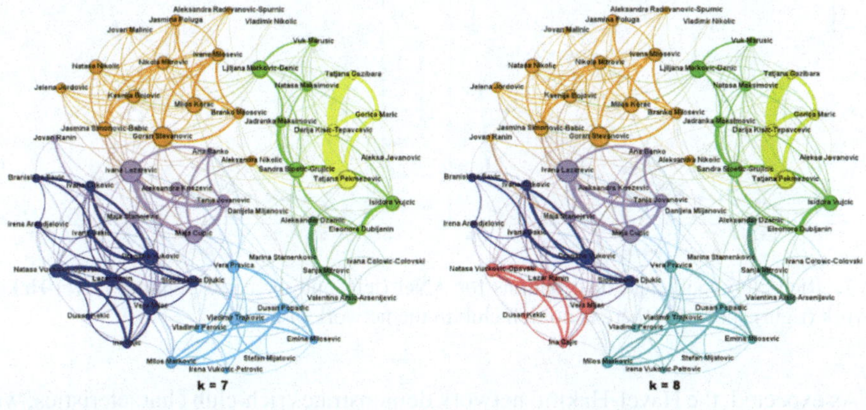

Fig. 6. Communities in the authors network obtained by spectral clustering for 7 and 8 clusters. When increasing the cluster number from 7 to 8, a small group of microbiologists that collaborate closely becomes its own cluster, decomposing the blue cluster hierarchically.

3.4 Connectivity Patterns and Influence of Nodes

Observation of communities and their interconnectedness using clustering methods can give a wide perspective on the dynamics of the network. However, further analysis of interconnectedness patterns can be conducted by calculating the assortativity and rich-club coefficients of the networks. In order to find and rank the most distinguished nodes in the network, we have conducted centrality analysis and calculated brokerage scores of the nodes.

Network assortativity is a measure of homophily between nodes in a network according to some node statistic, most often node degree. It is analyzed through the corresponding coefficient, which is equal to -0.02 for both networks, with its weighted variant being equal to -0.04 for ANet and -0.001 for FCANet. Both values are very close to zero, indicating a neutral non-assortative structure, telling us that nodes of differing degrees often collaborate.

Rich-club property was first defined in the papers of Zhou [23] and means that there is a group of nodes with high node degrees that are very well connected to each other. It is measured by calculating the rich-club coefficient. The coefficient indicates connection density among nodes with equal or greater degrees. Havel-Hakimi graphs [24] are constructed from a node degree sequence using the Havel-Hakimi algorithm and exhibit the rich club property. Therefore, it is possible to check if a network contains a rich club by comparing the rich-club coefficient distribution of the network to its distribution in the Havel-Hakimi graph with the same node degree sequence, as shown in Fig. 7.

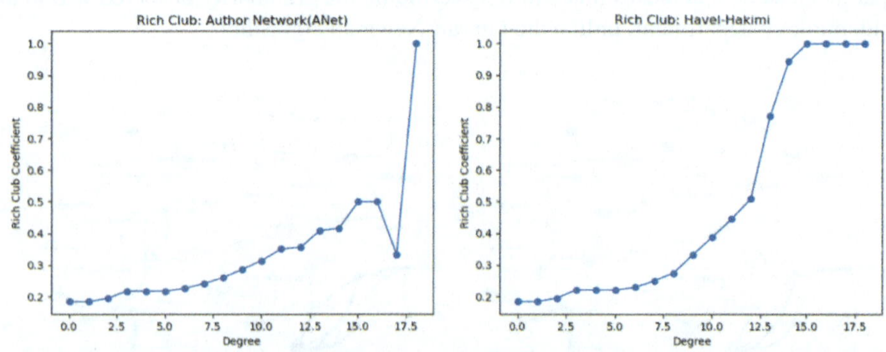

Fig. 7. Rich-club coefficient distributions for ANet (left) and the corresponding Havel-Hakimi network (right) show that there is no rich club in the network of authors.

As expected, the Havel-Hakimi network demonstrates rich-club characteristics, with a sharp increase in density as node degrees rise. Coefficient values for nodes with high degrees approach one, signifying strong node interconnectivity. Conversely, the co-authorship network exhibits marginal coefficient increments. At the terminal point, a rich-club coefficient of value 1 is achieved by only two nodes. A notable decline before the terminal point is observed. Upon manual inspection of the network, this decline stems from the fact that there are 4 nodes with a degree greater or equal to 16, 3 of them being strongly connected and forming a triad. However, one of the nodes has a degree of 16, while the remaining 2 have greater node degrees. Therefore, a triad's disintegration happens upon raising the degree threshold to 17, leading to a decrease in the rich-club coefficient. In summary, the network of authors does not reflect the typical traits of a rich-club network due to the lack of a significant number of dense interconnections among high-degree nodes.

Given the small size of the network of authors, attempting to fit its distribution to a power law seems impractical and unjustifiable. The observed distribution exhibits an irregular shape, diverging significantly from both power law distribution and those encountered in random networks. This conclusion is further validated through analysis using the powerlaw package [12], which returns probability values of 0.14 and 0.08 for unweighted and weighted degree variants respectively, indicating poor alignment with the expected distribution. This shows that the network's characteristics do not conform to a traditional power law distribution model, therefore, it seems that the dynamic of scientific collaboration among the researchers from the Faculty of Medicine does not express preferential attachment, leading to less centralization and a more robust and uniformly distributed influence. This finding is further reinforced by the analysis of centrality measures and their distribution in the authors networks.

When talking about individual nodes that stand out, one could consider a multitude of criteria by which to rank the nodes, but the centrality measures and brokerage scores are the most commonly used. The concept of brokers was formally introduced by Gould and Fernandez [25, 26] and corresponds to the nodes that connect parts of the network that would otherwise become poorly connected. That means the brokers are key for the spread of information in the network and, in the context of co-authorship, enhance scientific collaboration between groups of researchers. To identify the brokers in the network, we used a combination of betweenness centrality, local node constraint [27] in the network, and the brokerage metric calculated according to the Gould-Fernandez framework. This combined brokerage metric is computed for each node in the unweighted representation of ANet and FCANet, as well as for the weighted versions of these networks. The nodes are ranked and the top three in each network according to the combined brokerage metric are shown in Table 3. Researcher Goran Stevanović stands out, being among the top three in all three network representations.

Table 3. Top three researchers by combined brokerage metric in the unweighted representation of the network of authors, as well as weighted networks of authors ANet and FCANet.

	Rank	Unweighted networks	ANet weighted	FCANet weighted
Combined brokerage score	1	Tatjana Pekmezović	Maja Čupić	Tatjana Pekmezović
	2	Ivana Lazarević	Goran Stevanović	Goran Stevanović
	3	Goran Stevanović	Ljiljana M. Denić	Ljiljana M. Denić

To assess the influence of actors based on centrality measures, we will present the top three ranked actors according to degree, betweenness, closeness, and eigenvector centrality metrics. This analysis will be conducted on both the unweighted structures, which are equivalent in ANet and FCANet, as well as their weighted variants. Subsequently, the findings will be synthesized into two composite metrics to discover the most influential actors within the network. Additive composite rank orders the researchers by

Table 4. Top three researchers by degree, betweenness, closeness and eigenvector centrality, as well as by the additive and multiplicative composite rank.

	Rank	Unweighted networks	ANet weighted	FCANet weighted
Degree centrality	1	Ivana Lazarević	Tatjana Pekmezović	Tatjana Pekmezović
	2	Goran Stevanović	Darija K. Tepavčević	Darija K. Tepavčević
	3	Tatjana Pekmezović	Ivana Lazarević	Tatjana Gazibara
Betweenness centrality	1	Tatjana Pekmezović	Tatjana Pekmezović	Tatjana Pekmezović
	2	Ivana Lazarević	Maja Čupić	Miloš Korać
	3	Goran Stevanović	Ljiljana M. Denić	Ljiljana M. Denić
Closeness centrality	1	Goran Stevanović	Ivana Lazarević	Ivana Lazarević
	2	Ivana Lazarević	Jasmina S. Babić	Maja Čupić
	3	Maja Čupić	Maja Čupić	Tanja Jovanović
Eigenvector centrality	1	Ivana Lazarević	Darija K. Tepavčević	Darija K. Tepavčević
	2	Goran Stevanović	Tatjana Pekmezović	Tatjana Pekmezović
	3	Nataša Nikolić	Tatjana Gazibara	Tatjana Gazibara
Additive composite rank	1	Ivana Lazarević	Tatjana Pekmezović	Tatjana Pekmezović
	2	Tatjana Pekmezović	Darija K. Tepavčević	Darija K. Tepavčević
	3	Goran Stevanović	Maja Čupić	Tatjana Gazibara
Multiplicative composite rank	1	Ivana Lazarević	Tatjana Pekmezović	Tatjana Pekmezović
	2	Goran Stevanović	Darija K. Tepavčević	Darija K. Tepavčević
	3	Tatjana Pekmezović	Tatjana Gazibara	Tatjana Gazibara

the weighted sum of their centralities, while the multiplicative composite rank does the same, but multiplies the centralities instead of adding them up. The selection of weights was determined empirically to capture the significance of each centrality measure, considering its variance within the dataset. For instance, centrality measures with lower variability, such as closeness centrality, are assigned lower weights in this scheme shown in Table 4.

We observe that the inclusion of edge weights accentuates the prominence of nodes involved in edges with higher weights, exemplified by Darija Kisić Tepavčević and Tatjana Gazibara. These nodes demonstrate a pattern of more frequent collaboration with the same counterparts, at the expense of greater variance in collaborators. Furthermore, they represent a densely interconnected cluster that is itself well-connected to the rest of the graph. Conversely, researchers such as Ivana Lazarević and Goran Stevanović exhibit heightened prominence in the absence of weights, indicative of numerous connections with lower weights. Additionally, the node represented by Maja Čupić gains visibility when weights are taken into consideration, however, her influence lowers when fractional counting is employed, indicating a tendency towards collaboration within larger cohorts.

4 Conclusion

In this study, we addressed the analysis of scientific collaboration within the domain of immunology and infectiology at the University of Belgrade, Faculty of Medicine, employing techniques of social network analysis. A co-authorship network was constructed, considering both integer and fractional counts for co-authorship. Through examination of basic network metrics and properties, the network was identified as a small-world network characterized by medium density and high clustering. Subsequently, considering the organizational division of authors into four departments, smaller collaboration networks were constructed and analyzed. Assortative mixing patterns between departments indicated significant collaboration between authors from different departments, necessitating further investigation and justifying the application of community detection techniques. Utilizing three distinct clustering algorithms (Louvain, Girvan-Newman, and spectral clustering), finer community structures within the co-authorship network were revealed, previously obscured without the application of social network analysis.

Additionally, through assortativity analysis, calculation of the rich-club coefficient, and exploration of power-law properties, it was demonstrated that the collaboration network lacked a rich club, preferential attachment, or discrimination with regards to node degree, indicating a healthy research environment with minimal centralization of node influence. To further analyze key nodes in the collaboration network, centrality and brokerage analyses were conducted using both integer and fractional weights, successfully distinguishing researchers with numerous co-authors from those with fewer co-authors, thereby validating the utility of both weighting schemes.

Future directions for research could involve expanding the network to encompass all researchers at the Faculty of Medicine, or even all medical professionals in Serbia. Additionally, an analysis of journals where observed researchers publish could be pursued using similar methods, facilitating the identification of journal groups for potential subscription by the Faculty of Medicine. Finally, the constructed dataset could be leveraged to predict future co-authorships among researchers using Granovetter's theory of weak and strong ties [28], facilitating potential collaborations on scientific projects.

Acknowledgement. This paper was financially supported by the Ministry of Science, Technological Development and Innovation of the Republic of Serbia under contract number: 451–03-47/2023–01/200103.

References

1. White, K.: Publications Output: US Trends and International Comparisons. Science & Engineering Indicators. NSB-2021–4. National Science Board, National Science Foundation (2021)
2. Oliveira, E.A., Oliveira, M.C.L., Colosimo, E.A., Martelli, D.B., Silva, L.R., Silva, A.C.S.E., et al.: Global scientific production in the pre-Covid-19 Era: an analysis of 53 countries for 22 years. An Acad. Bras. Cienc. **94**, e20201428 (2022)
3. Marginson, S.: Global science and national comparisons: beyond bibliometrics and scientometrics. Comp. Educ. **58**(2), 125–146 (2022)
4. Moral-Muñoz, J.A., Herrera-Viedma, E., Santisteban-Espejo, A., Cobo, M.J.: Software tools for conducting bibliometric analysis in science: an up-to-date review. Prof Inf. **29**(1), 4 (2020)
5. Safder, I., Hassan, S.U.: Bibliometric-enhanced information retrieval: a novel deep feature engineering approach for algorithm searching from full-text publications. Scientometrics **119**, 257–277 (2019)
6. Correia, A., Jameel, S., Schneider, D., Paredes, H., Fonseca, B.: A workflow-based methodological framework for hybrid human-AI enabled scientometrics. In: 2020 IEEE International Conference on Big Data (Big Data). IEEE, pp. 2876–2883 (2020)
7. Glänzel, W.: Analysis of co-authorship patterns at the individual level. Transinformação. **26**, 229–238 (2014)
8. Newman, M.E.: Coauthorship networks and patterns of scientific collaboration. Proc Natl Acad Sci. **101**(suppl_1), 5200–5 (2004)
9. Mitrović, I., Mišić, M., Protić, J.: Exploring high scientific productivity in international co-authorship of a small developing country based on collaboration patterns. J Big Data. **10**, 64 (2023)
10. Perianes-Rodriguez, A., Waltman, L., Van Eck, N.J.: Constructing bibliometric networks: a comparison between full and fractional counting. J Informetrics. **10**(4), 1178–1195 (2016)
11. Passos, P., Davids, K., Araújo, D., Paz, N., Minguéns, J., Mendes, J.: Networks as a novel tool for studying team ball sports as complex social systems. J. Sci. Med. Sport **14**(2), 170–176 (2011)
12. Alstott, J., Bullmore, E., Plenz, D.: powerlaw: a python package for analysis of heavy-tailed distributions. PLoS One. **9**(4) (2014)
13. Bastian, M., Heymann, S., Jacomy, M.: Gephi: an open source software for exploring and manipulating networks. In: International AAAI Conference on Weblogs and Social Media (2009). http://www.aaai.org/ocs/index.php/ICWSM/09/paper/view/154
14. Hanneman, R.A., Riddle, M.: Introduction to social network methods. Riverside, CA: University of California, Riverside (2005). http://faculty.ucr.edu/~hanneman/
15. Watts, D.J., Strogatz, S.H.: Collective dynamics of 'small-world' networks. Nature **393**(6684), 440–442 (1998)

16. Telesford, J.Q., Hayasaka, S., Burdette, J.H., Laurienti, P.J.: The ubiquity of small-world networks. Brain Connectivity. **1**(0038), 367–375 (2011). https://doi.org/10.1089/brain.2011.0038
17. Blondel, V.D., Guillaume, J.L., Lambiotte, R., Lefebvre, E.: Fast unfolding of communities in large networks. J Stat Mech. **2008**(10), P10008 (2008)
18. Girvan, M., Newman, M.E.: Community structure in social and biological networks. Proc. Natl. Acad. Sci. **99**(12), 7821–7826 (2002)
19. Shi, J., Malik, J.: Normalized cuts and image segmentation. IEEE Trans. Pattern Anal. Mach. Intell. **22**(8), 888–905 (2000)
20. Shi, J.: Multiclass spectral clustering. In: Proceedings ninth IEEE international conference on computer vision IEEE, pp. 313–319 (2003)
21. Knyazev, A.V.: Toward the optimal preconditioned eigensolver: locally optimal block preconditioned conjugate gradient method. SIAM J. Sci. Comput. **23**(2), 517–541 (2001)
22. Damle, A., Minden, V., Ying, L.: Simple, direct and efficient multi-way spectral clustering. Inf Inference. **8**(1), 181–203 (2019)
23. Zhou, S., Mondragón, R.J.: The rich-club phenomenon in the internet topology. IEEE Commun. Lett. **8**(3), 180–182 (2004)
24. Hakimi, S.: On realizability of a set of integers as degrees of the vertices of a linear graph. I. J SIAM. **10**(3), 496–506 (1962)
25. Gould, R.V., Fernandez, R.M.: Structures of mediation: a formal approach to brokerage in transaction networks. Sociol. Methodol. **1**, 89–126 (1989)
26. Fernandez, R.M., Gould, R.V.: A dilemma of state power: brokerage and influence in the national health policy domain. Am. J. Sociol. **99**(6), 1455–1491 (1994)
27. Burt, R.S.: Structural holes and good ideas. Am. J. Sociol. **110**, 349–399 (1995)
28. Granovetter, M.: The strength of weak ties: a network theory revisited. Sociol Theory **1**, 201–233 (1983)

Predicting Absorbance for Different Concentration of AgNPs Using Artificial Neural Network

Leo Benolić[1,2], Safi Ur Rehman Qamar[1,2(✉)], Lemana Spahić[1,2], and Nenad Filipović[1,2(✉)]

[1] Bioengineering Research and Development Centre (BioIRC), Prvoslava Stojanovica 6, 34000 Kragujevac, Serbia
surqamar@gmail.com, fica@kg.ac.rs

[2] Faculty of Engineering, The University of Kragujevac, Sestre Janjica 6, 34000 Kragujevac, Serbia

Abstract. The integration of artificial intelligence (AI) and nanotechnology offers significant potential to enhance the understanding and development of nanoparticles, particularly silver nanoparticles (AgNPs), which are pivotal in fields like biomedicine, electronics, and environmental remediation. This study demonstrates the use of neural networks to predict the absorbance of AgNPs across different concentrations, utilizing a dataset enriched by UV-Visible spectroscopy measurements. A custom-designed neural network, featuring a dual-layer architecture and ReLU activation, was trained on interpolated spectral data spanning various concentrations. The network achieved a near-perfect R2 score of 0.9898247, indicating exceptional accuracy in predicting absorbance values that align closely with experimental data. This predictive model not only reduces the need for extensive empirical testing, but also significantly enhances the efficiency of nanoparticle research and development. By enabling rapid and accurate predictions of AgNPs' optical properties, this approach could revolutionize nanoparticle synthesis and application, setting a new standard in the integration of AI within nanotechnology and material sciences. This study not only advances the application of neural networks in predictive modeling, but also underscores the transformative potential of AI in accelerating scientific discovery and innovation in nanoparticle technology.

Keywords: silver nanoparticles · neural networks · absorbance prediction · machine learning applications · artificial intelligence in nanotechnology

1 Introduction

The application of artificial intelligence (AI) in nanotechnology presents a transformative potential for enhancing the predictive capabilities and optimizing the synthesis processes of nanoparticles (NPs). Among various nanoparticles, silver nanoparticles (AgNPs) have garnered significant attention due to their extensive applications in biomedicine, electronics, and environmental remediation. Characterizing the optical properties of AgNPs,

© The Author(s), under exclusive license to Springer Nature Switzerland AG 2025
N. Filipović (Ed.): AAI 2024, LNNS 1446, pp. 272–278, 2025.
https://doi.org/10.1007/978-3-031-99201-8_28

such as absorbance, is crucial for understanding their interactions with light, which directly influences their efficiency and suitability in various applications. Typically, the absorbance of AgNPs can be determined experimentally using UV-Visible spectroscopy; however, this method can be time-consuming and resource-intensive when exploring a wide range of concentrations and conditions.

To address these challenges, this paper proposes the use of neural networks as a robust tool for predicting the absorbance of AgNPs across different concentrations measured in milligrams per liter (mg/L). Neural networks, particularly those architectures adept at regression tasks, offer a promising approach due to their ability to model complex nonlinear relationships and their adaptability to diverse data inputs. By training neural networks on datasets derived from experimental measurements, we aim to develop a predictive model that can efficiently estimate AgNP absorbance without the need for extensive empirical testing [1]. In the context of predicting spectral characteristics, the application of Artificial Neural Networks (ANNs) has proven useful in various research efforts [2]. For instance, some studies used ANNs to predict the emission spectra of fluorescent materials based on their absorbance spectra [3, 4]. While principal component analysis (PCA) can be utilized for more efficient implementation in that case, our approach applies more general uses of ANNs without specific pre-processing, focusing on modeling complex nonlinear relationships among data [5]. These approaches demonstrate how neural networks can be adapted and applied to different types of spectral data, including the absorbance spectra of AgNPs, without the need for specific preprocessing methods. Similarly, another study highlighted how tailoring the input parameters for ANNs, reducing them from 160 to just 10, could significantly speed up the training process and enhance prediction accuracy, as in the case of predicting chemical oxygen demand (COD) in wastewater [6]. Although our work does not involve drastic reduction of input parameters, the principle of adapting input features based on the specific needs of applications is directly applicable to our model for predicting AgNPs' absorbance spectra, demonstrating the flexibility and adaptability of ANNs across different applications. Through these examples from the literature, it is evident that neural networks can be broadly applied in predicting spectral properties, from fluorescent materials to wastewater, each time adapting to specific requirements and environments. This work extends that application to silver nanoparticles, exploring how general ANN architectures can effectively model the absorbance properties of AgNPs, thus contributing to the development of rapid and precise spectroscopic tools in nanotechnology.

This research not only explores the intersection of AI and nanotechnology but also seeks to push the boundaries of how machine learning can accelerate and enhance the study of nanoparticle properties. The integration of neural network models could significantly reduce experimental overhead, provide instant insights into the behavior of nanoparticles under various conditions, and facilitate rapid adjustments in synthesis processes to achieve desired outcomes. Ultimately, this study aims to establish a predictive framework that could serve as a valuable tool for researchers and industries engaged in the development and application of AgNPs, enhancing the speed and precision of nanomaterial engineering.

2 Material and Method

In this study, the data collection process involved measuring the absorbance of silver nanoparticles (AgNPs) at three distinct concentrations: 1.1 mg/ml, 1.5 mg/ml, and 2.0 mg/ml. These measurements were taken using UV-Visible spectroscopy across a wavelength range from 290 nm to 400 nm [1]. To enhance the dataset for neural network training, an interpolation technique was utilized, expanding the wavelength data to encompass 2200 data points per concentration. This method provided a comprehensive set of data points, capturing a detailed spectrum of absorbance values which are essential for training the model effectively.

The artificial neural network (ANN) designed for predicting the absorbance of AgNPs was structured with two input features: the solution concentration in mg/ml and the wavelength in nm. The network comprised three layers, including two hidden layers with ten neurons each, and one output layer with a single neuron, as illustrated in Fig. 1.

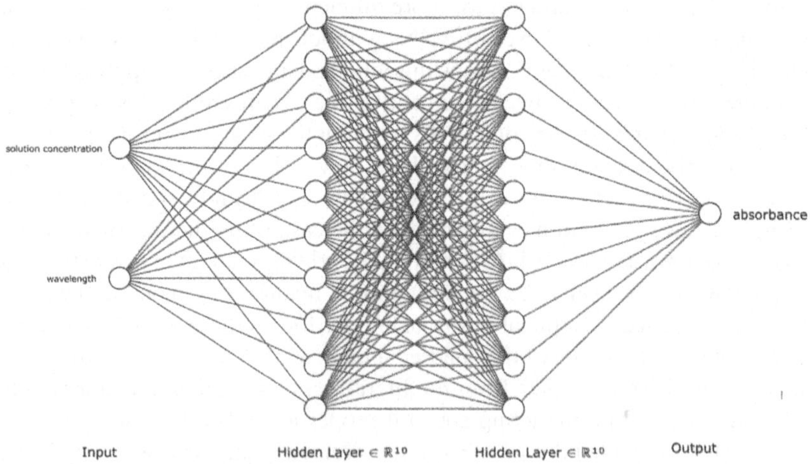

Fig. 1. Graphical representation of the ANN Architecture.

The ReLU activation function was employed in the hidden layers to introduce non-linearity, enabling the network to capture complex patterns in the data. The 'he_uniform' method was chosen for weight initialization, a strategy that maintains the variance of activations throughout the network and is particularly effective for layers using the ReLU function. The Adam optimizer was utilized for training the network due to its capability in handling sparse gradients and its efficiency on noisy datasets. The mean squared error (MSE) was used as the loss function to minimize the difference between predicted and actual absorbance values, with the network undergoing training over 1,000 epochs to achieve optimal performance without overfitting [7].

The dataset was initially split into two main parts: 90% of the data was used for the development phase and the remaining 10% was reserved for final testing. Within the development portion, the data were further divided into a training set and a validation set. The training set comprised 80% of the development data, which is equivalent to 72% of the total data, while the validation set accounted for the remaining 20% of the development data, equivalent to 18% of the total data. This split facilitated the effective monitoring of the model's performance and the adjustment of hyperparameters during the training phase, ensuring the model did not overfit and performed well on unseen data (see Fig. 2).

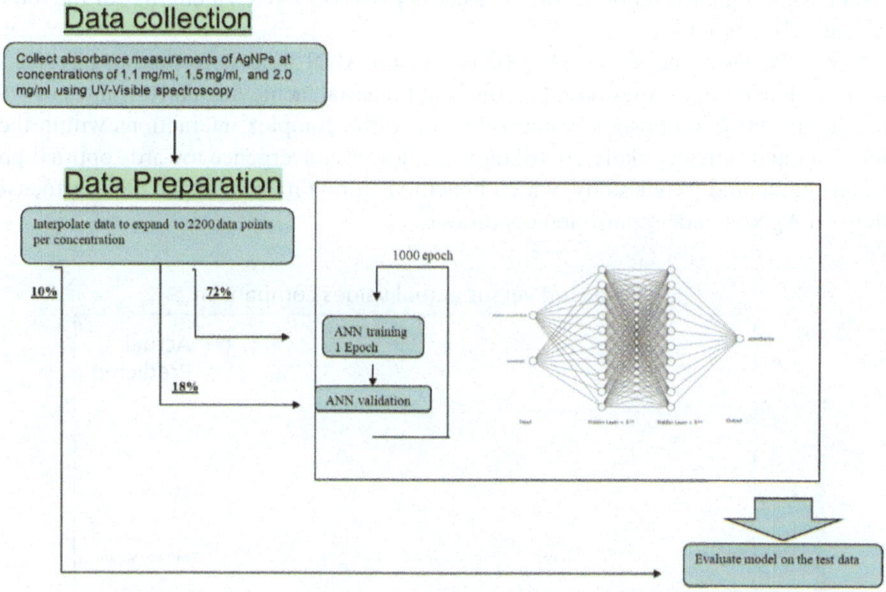

Fig. 2. Neural Network Workflow for predicting absorbance of AgNPs.

The overarching objective of the ANN was to facilitate the prediction of absorbance at various concentrations and wavelengths not directly measured during the initial experiments. This capability is critical for predicting the optical properties of AgNPs efficiently and accurately, thereby assisting researchers in determining optimal concentrations for specific applications without the need for extensive experimental testing. By enabling rapid predictions, the model significantly improves the efficiency of research and development processes in the field of nanotechnology, aiding in the exploration and application of AgNPs in various domains.

3 Results and Discussion

The neural network used in this study to predict the absorbance of silver nanoparticles (AgNPs) across various concentrations achieved an R2 score of 0.9998247, indicating an exceptionally high level of accuracy in predicting the experimental data. This high score is corroborated by graphical evidence, as shown in Fig. 3, where the scatter plot of the predicted against the actual values reveals a tight clustering of data points along the identity line. This close alignment demonstrates that the predictions made by the network closely match the actual measurements.

This exceptional accuracy can be contrasted with findings from similar studies, such as El-Naggar et al. (2023), who reported an R^2 score of 0.9885 in predicting the size distribution of gold nanoparticles using a similar ANN architecture [8]. The difference in accuracy can largely be attributed to the variability and complexity of the data sets used. In their study, the data encompassed a wide range of particle sizes and shapes, introducing a higher level of intrinsic variability that is more challenging for the model to accurately predict.

Similarly, Di Francesco et al. (2023) utilized ANN models to predict the size of curcumin-loaded liposomes based on the zeta potential, achieving an R^2 values of 0.98 [9]. The diversity in liposome compositions and the complex interactions within their chemical environments likely contributed to a lower convergence towards optimal predictions compared to our study, which benefited from a more uniform dataset focused solely on AgNPs under controlled conditions.

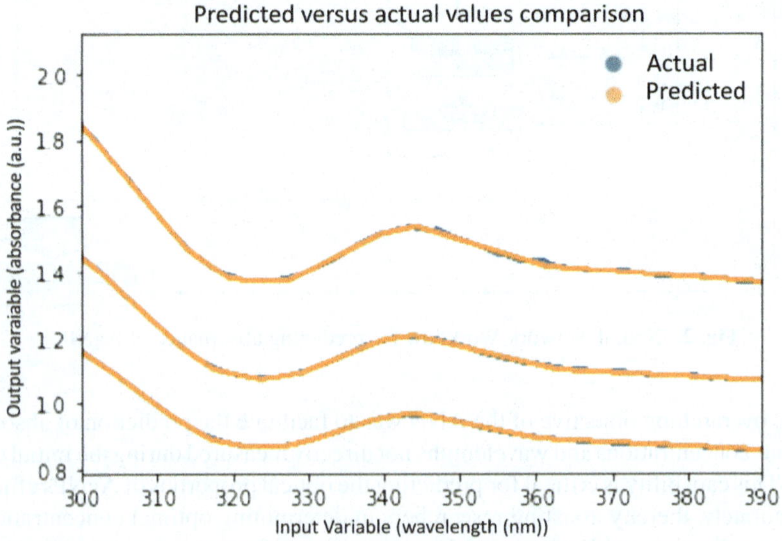

Fig. 3. Predicted values for 1.1 mg/ml, 1.5 mg/ml, and 2 mg/ml concentrations compared to the actual values.

These examples illustrate how the characteristics of the dataset—such as diversity in particle types, sizes, and experimental conditions—play a crucial role in the performance of predictive models. The architecture of our network, which includes two layers each with 10 fully connected neurons, was particularly well-suited to our dataset, enabling it to effectively capture and model the intricate relationships within a more homogeneously defined set of data.

Figure 4, which predicted absorbance for 9 different AgNPs concentrations, provides a detailed visual representation of the model's predictions across nine distinct concentrations of silver nanoparticles. This figure showcases the neural network's ability to predict absorbance, including concentrations that were not part of the original dataset, highlighting the model's capacity to generalize beyond the training data. Notably, an anomaly is observed at the concentration of 1.6 mg/ml, where a significant deviation in the curve suggests a potential structural change in the nanoparticles, such as coagulation, or indicates a limitation of the model that could be improved with more data. This insight emphasizes the importance of further expanding the dataset to capture such phenomena and enhance the model's predictive power across a wider range of concentrations.

To further improve the robustness of our neural network, it would be beneficial to expand the dataset to include a broader range of AgNP concentrations. Incorporating additional concentrations would not only help in validating the model's effectiveness across a wider spectrum but also enhances its predictive accuracy and generalizability. This expansion could potentially reveal additional complex patterns and interactions that are not observable within the current dataset limits, thereby providing deeper insights and more reliable predictions across various application scenarios.

The success of this neural network model, evidenced by the high R2 score and detailed graphical analysis, suggests that the model could be applied to other predictive tasks in nanotechnology and beyond. This modeling approach could potentially be adapted for predicting other important nanoparticle properties, which are crucial for applications in fields like medicine, environmental science, and materials engineering. Moreover, accurate predictions in these areas could greatly enhance experimental efficiency, reduce costs, and accelerate the pace of scientific discovery.

By discussing the comparative performance of similar models in the context of data variability, we highlight the adaptability and potential limitations of ANNs, underscoring the need for careful consideration of data selection and preprocessing to optimize model

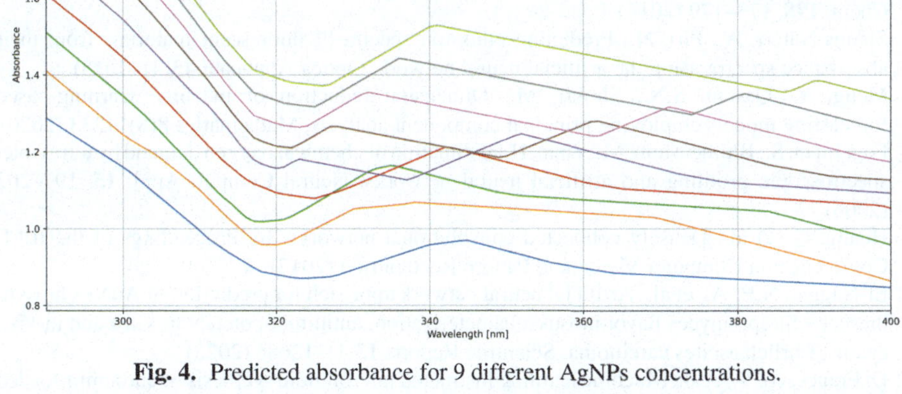

Fig. 4. Predicted absorbance for 9 different AgNPs concentrations.

performance. This analysis not only aligns with established research, but also paves the way for future innovations in the field of nanotechnology predictive modeling.

4 Conclusion

This study effectively demonstrated the application of neural networks to predict the absorbance of silver nanoparticles (AgNPs) across various concentrations, achieving a remarkably high R2 score of 0.9898247. The dual-layer neural network architecture, with ten neurons in each layer, enabled precise modeling of the complex interactions between light and nanoparticles. This approach not only enhances the efficiency of nanoparticle design in fields like biomedicine and environmental remediation, but also opens possibilities for adapting these methods to predict other crucial nanoparticle properties. Furthermore, this predictive model paves the way for broader applications of machine learning in materials science, reducing the reliance on extensive empirical testing and accelerating the pace of scientific discovery. The integration of AI with nanotechnology, as showcased in this study, sets new standards for nanoparticle engineering and highlights the transformative potential of interdisciplinary approaches in modern scientific research.

Acknowledgment. This research is supported by the DECODE project that has received funding from the European Union's Horizon 2020 research and innovation programme under the Marie Skłodowska-Curie grant agreement No 956470. This article reflects only the author's view. The Commission is not responsible for any use that may be made of the information it contains.

References

1. Qamar, S.U.R., et al.: Silver nanoparticles from Ocimum basilicum L. tea: A green route with potent anticancer efficacy. Colloid and Interface Science Communications **59**, 100771 (2024)
2. Maltarollo, V.G., Honório, K.M., da Silva, A.B.F.: Applications of artificial neural networks in chemical problems. Artificial Neural Networks-Architectures and Applications, pp. 203–223 (2013)
3. Cancilla, J.C., et al.: Artificial neural networks applied to fluorescence studies for accurate determination of N-butylpyridinium chloride concentration in aqueous solution. Sens. Actuators, B Chem. **198**, 173–179 (2014)
4. Shams-Nateri, A., Piri, N.: Predicting emission spectra of fluorescent materials from their absorbance spectra using the artificial neural network. Optica Applicata **45**(4) (2015)
5. Meng, Y., Qasem, S.N., Shokri, M.: Dimension reduction of machine learning-based forecasting models employing principal component analysis. Mathematics **8**(8), 1233 (2020)
6. Fogelman, S., Blumenstein, M., Zhao, H.: Estimation of chemical oxygen demand by ultraviolet spectroscopic profiling and artificial neural networks. Neural Comput. Appl. **15**, 197–203 (2006)
7. Huang, G., et al.: Densely connected convolutional networks. In: Proceedings of the IEEE Conference on Computer Vision and Pattern Recognition (2017)
8. El-Naggar, N.E.-A., et al.: Artificial neural network approach for prediction of AuNPs biosynthesis by Streptomyces flavolimosus, characterization, antitumor potency in-vitro and in-vivo against Ehrlich ascites carcinoma. Scientific Reports **13**(1), 12686 (2023)
9. Di Francesco, V., et al.: Machine learning instructed microfluidic synthesis of curcumin-loaded liposomes. Biomed. Microdevice **25**(3), 29 (2023)

Influence of Dimensionality Reduction Approaches on Various Machine Learning Models for a Biomedical High-Dimension Dataset

Sandi Baressi Šegota(✉), Nikola Anđelić, Jelena Štifanić, Daniel Štifanić, and Zlatan Car

Faculty of Engineering, University of Rijeka, Vukovarska 58, 51000 Rijeka, Croatia
sandi.segota@riteh.uniri.hr

Abstract. Dimensionality reduction is a preprocessing approach that serves to assist in reducing the number of variables (features) in the dataset, based on statistical approaches. This is extremely beneficial in machine learning (ML) applications, as it can serve to simplify the modeling process and improve the performance. In this paper, the authors apply 14 different techniques to reduce the dimensionality of a dataset with over 1200 features and test the influence across five techniques. Results indicate that some techniques have a significantly higher sensitivity to the dimensionality reduction technique applied. For example, a neural network-based technique shows a nearly 20% improvement, compared to the insignificant improvement shown by the support vector-based technique. A sensitivity difference is shown in the type of dimensionality reduction applied as well, with NB demonstrating a significant increase in performance with feature selection techniques, as opposed to decomposition techniques. Execution times are also tested, demonstrating low computational cost of most applied dimensionality reduction techniques. This indicates the need for more detailed research, to test which techniques are affected the most across different datasets.

Keywords: dataset decomposition · dimensionality reduction · feature selection · machine learning

1 Introduction

High-dimensionality datasets are those datasets that contain a large number of variables (features) in each data point. These datasets can be challenging to model using machine learning and other modeling approaches, especially when a relatively low amount of data points are present in the dataset. A large number of features may require larger models for full analysis, and this can significantly increase the computational complexity. In addition to this, using unnecessary features may have a negative influence on the model performance, especially when aiming for well-generalizing models [1]. To address this, two approaches are common – the first is data decomposition, and the second is

© The Author(s), under exclusive license to Springer Nature Switzerland AG 2025
N. Filipović (Ed.): AAI 2024, LNNS 1446, pp. 279–288, 2025.
https://doi.org/10.1007/978-3-031-99201-8_29

feature selection. Feature selection, sometimes also known as feature pruning, is based on determining feature importance metrics, usually the influence on the output, and then eliminating those features that have a low influence on the output [2]. On the other hand, dimensionality reduction works by recalculating new features, based on the combinations of existing features [3]. There are many examples of dimensionality reduction being used to improve the results of the dataset value adjustment in biomedicine [4, 5], but also in different fields such as engineering [6] and robotics [7]. While a lot of work exists, there is not a definite step toward defining which of the many techniques used by researchers for dimensionality reduction should be applied. There is also a lack of discussion on whether dimensionality reduction should be always applied, or just when appropriate to the method used [8].

In the presented paper, the authors test multiple approaches to the feature selection and dimension reduction on the publicly available dataset. Then, the authors train five different classification methods on each dataset to compare the performance against differently preprocessed datasets on different methods. The goal of this paper is to determine if there is a bias toward the score based on the applied dimensionality reduction based on the method applied for classification.

2 Dataset

The used dataset is the Toxicity dataset [9]. The dataset consists of 1203 features, with the targeted output being whether the molecule is toxic or not. There is a total of 171 data points in the dataset, making it a dataset with an extremely high number of dimensions compared to the amount of data points. The 1203 features are molecular descriptors, and the authors suggest the application of feature selection, because some features may be redundant. There are no missing values in the dataset – but there is a large amount of negative values which can cause issues with some dimensionality reduction functions, and outliers. Outliers are determined per the following equation for the variable x, in the set X with $\overline{x}$ being the mean value of the dataset and σ_X:

$$x - |\overline{x}| > 3\sigma_X, \tag{1}$$

The outliers are present in 1028 variables, making this a hard dataset to model, especially with the dimensionality reduction preprocessing approach [9]. Only preprocessing applied to the dataset, except for the data dimensionality reduction preprocessing methods which will be described in the continuation of the paper, is the encoding of the output labels. Originally given as "Toxic" and "NonToxic", they are encoded as binary labels – "1" and "0" respectively.

3 Methods

This section presents the methods used – briefly listing the dataset pruning methods (both data decomposition and feature selection preprocessing techniques), along with the applied machine learning techniques and the model evaluation process.

3.1 Preprocessing Methods

Two types of preprocessing methods are utilized in this paper – the first is decomposition methods, while the second is feature selection methods. Decomposition methods break down complex structures into newly created simpler components. The ones used in this paper are: independent component analysis (ICA) which separates a multivariate signal into additive independent components; factor analysis (FA) which explains observed variables by latent factors, assuming linear relationships; dictionary learning (DL) which decomposes data into a sparse linear combination of atoms from a learned dictionary, principal component analysis (PCA) which identifies the orthogonal exes that capture the maximum variance within the data; and truncated singular value decomposition (TSVD) which approximates a matrix by retaining only the largest singular values and corresponding vectors [10].

In addition to the above, six feature selection methods are used. These feature selection methods do not create new variables, but instead eliminate the ones that have a low influence on the output or share information with other variables. The used methods are F-score for classification (FC) - a statistical method that evaluates ANOVA F-value between each feature and the target variable, variance threshold (VT) which removes features with low variance, mutual info classification (MIC) which removes features based on mutual information between each feature and the target variable); recursive feature elimination (RFE) - a technique which removes features based on their importance based on the feature importance score which is based on the tree-based mode, feature permutation importance (FPI) - a technique which randomly shuffles values and eliminates them if they are shown not to influence the score of the tree-based model when shuffled, mean decrease in impurity (MDI) - which removes factors based on how much their introduction to the tree-based model lowers the overall impurity of the model.

Finally, a simple correlation is used, with three classification metrics, eliminating those variables that show a low correlation to the output. Three correlation metrics used are Pearson's correlation – a measure of linear relationship between two variables, calculated as:

$$r = \frac{\sum_{i=1}^{n}(x_i - \overline{x})(y_i - \overline{y})}{\sqrt{\sum_{i=1}^{n}(x_i - \overline{x})^2 \sum_{i=1}^{n}(y_i - \overline{y})^2}}, \tag{2}$$

Spearman's rank correlation – a measure of the monotonic relationship between two variables, given per:

$$\rho = 1 - \frac{6\sum d_i^2}{n(n^2 - 1)}, \tag{3}$$

and Kendall's correlation, a rank-based coefficient that measures the ordinal association between the two variables, expressed with:

$$\tau = \frac{2}{n(n-1)}\sum\nolimits_{i<j} \text{sign}(x_i - x_j)\text{sign}(y_i - y_j). \tag{4}$$

All of the methods are adjusted to return the same shape of the dataset – a total of 50 variables are returned. For the methods that have parameters that can be adjusted, the default parameters are used, as provided by the scikit-learn library [11]. While the different methods may benefit from generating a different number of points in the preprocessed dataset, the decision was made to keep this value equal across all methods, since a different amount of data vectors could affect the classification methods as well [12]. The methods were generally selected as those methods that are commonly used, with the pitfall that they can be used on the type of data that is presented – namely data with negative values. It should be noted that the negative values in data can be preprocessed and eliminated through transformations, but in the presented research authors have decided to utilize only the data as originally provided.

3.2 Classifiers

Five ML-based classification methods are used – a multilayer perceptron (MLP), which is a neural network model consisting of multiple layers of interconnected artificial neurons; random forest classifier (RFC) - an ensemble learning method that constructs multiple decision trees during training and outputs the mode of the classes of individual trees; naïve Bayes (NB) – a probabilistic classifier based on Bayes' theorem assuming the independence between features; support vector classifier (SVC) – a model that finds a hyperplane that best separates classes in a high-dimensional class, and gradient boosting classifier (GBC) – an ensemble learning methods building a sequence of decision trees, with each tree correcting the error of its predecessor [13]. MLP was selected due to the dimensionality of data having a large influence on the MLP models, as the shapes of MLP models are commonly suggested to be adjusted based on the shape of data. RFC was selected due to it being the basis of some of the methods in the feature selection category – RFE, MDI, and FPI, with GBC being added as an improved, but similar, method to RFE. SVC was selected because dimensionality reduction should have a low impact on the results, as the method prefers high-dimensional data. Finally, NB was selected due to data decomposition methods possibly lowering the amount of independent information available to it. All of the methods are trained using default hyperparameters, as defined per scikit-learn [11]. Five-fold cross-validation is applied to determine the performance on the entirety of the dataset [13], with the presented results being given as the average result across folds. This method works by splitting the dataset into 5 equal parts (in the case of the presented dataset four subsets of 34 points and one subset of 35 points). The training is then repeated five times, in such a way that a different subset (fold) is used for testing in each iteration and the remaining folds are combined into the training set [13]. The score used for evaluation was accuracy – the percentage of correct predictions versus all predictions. Although not a perfect evaluation metric, as it does not provide information on the type of error exhibited by the model, it is a commonly used metric for model performance and a good general metric when comparing models [14].

4 Results and Discussion

Figure 1 shows the results of the methods on different datasets. The best average improvement is achieved with the RFE technique. Per method, the biggest improvement for MLP is achieved with FC – 0.18. For RFC, PCA showed the best improvement of 0.08 with DL. NB showed the best improvement of 0.020 when the dataset was processed with any feature selection method. SVC showed no significant improvement, with some results being slightly worse than the original dataset, without any features removed. GBC demonstrates the largest improvement when MDI is applied at 0.09. Overall, it can be noted that, except for SVC, all methods show an improvement in results when dimensionality reduction techniques are applied to the dataset as a preprocessing step.

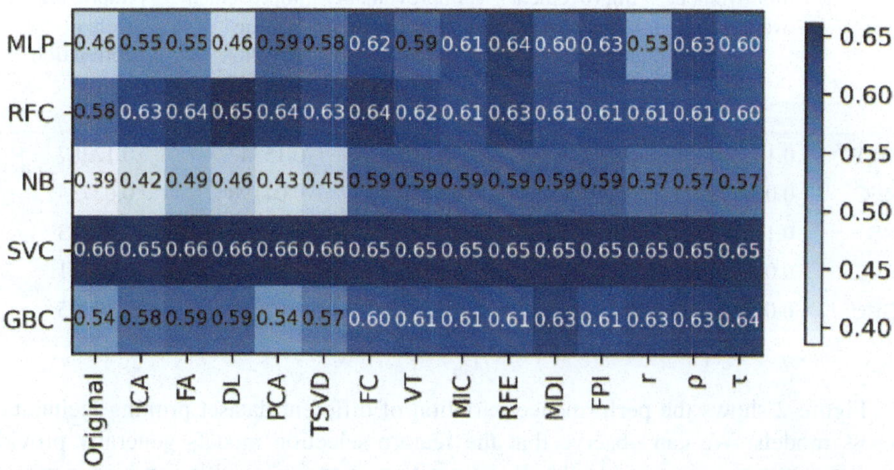

Fig. 1. The accuracy across different models when dimensionality reduction is applied.

It is important to note that overall scores are very low. This is the result of the lack of hyperparameter tuning in the used classification methods. As hyperparameters have an extremely high influence on the models' performance, not tuning them is expected to result in poor scores [13]. This was acceptable within the study design since the point of the study was to test how much different dimensionality reduction techniques improve performance for a given classification technique – and not to achieve a well-performing model.

As shown in Table 1, the average improvement per method is the greatest for MLP, especially when the results of feature selection without correlation are considered. FS without correlation overall gave the best improvement is the case for MLP, NB, and SVC. GBC achieved the best improvement using only correlation methods. SVC overall shows the lowest improvement in general. Observing the average improvement between feature selection methods and correlation-based feature selection, it can be seen that they are not always equal. Except for GBC and SVC, correlation-based feature selection shows lower average improvement than with other feature selection-based methods. Still, the correlation does offer performance improvements. Observing the difference between decomposition and feature selection, again – some methods prefer one and some the other. MLP, NB, and GBC show a much larger average improvement when applying feature selection. SVC and RFC prefer decomposition methods. This is especially interesting in the case of RFC, since the method was used to perform feature selection, as the base regressor in the case of FPI, MDI, and RFE methods which require a model to analyze. Still, overall, all the classification models showed overall improvement in scores, no matter which type of dimensionality reduction was applied.

Table 1. Average improvement for various techniques. FS – Feature selection.

Method	Average improvement overall	Average improvement - decomposition methods	Average improvement – Feature selection and correlation	Average improvement - Feature selection without correlation	Average improvement – only correlation
MLP	0.12264	0.08491	0.14170	0.15347	0.12387
RFC	0.04412	0.05684	0.03366	0.04190	0.02734
NB	0.15073	0.06336	0.19855	0.20504	0.18773
SVC	0.00450	0.00232	0.00571	0.00571	0.00571
GBC	0.06502	0.03496	0.08382	0.07560	0.09395

Figure 2 shows the performance variation of different dataset pruning techniques across models. We can observe that the feature selection models generally provide smaller variation compared to the decomposition methods, resulting in higher means. The best results on the dataset are overall the same. The largest variation is shown for the DL and PCA, including ICA and TSVD if the outliers are included. This indicates that the feature selection methods result in a lower score variance across methods, on the given dataset. The same can be observed with median scores, as they show a tendency to be higher for feature selection methods. Still, all the dataset's dimensionality reduction methods are shown to lower the variance in scores across models and improve the median score.

A similar can be observed in Fig. 3, which demonstrates the change in performance in models across different techniques used. It clearly shows how using different dimensionality reduction methods has the lowest influence on SVC and RFC methods, with GBC variance across methods being middling. MLP and NB both show very large variance, indicating that the selection of the dimensionality reduction approach for this should be performed with care.

4.1 Training and Execution Times

One of the key issues raised concerning the application of data preprocessing methods is the addition of processing time and computational complexity. To address this, the

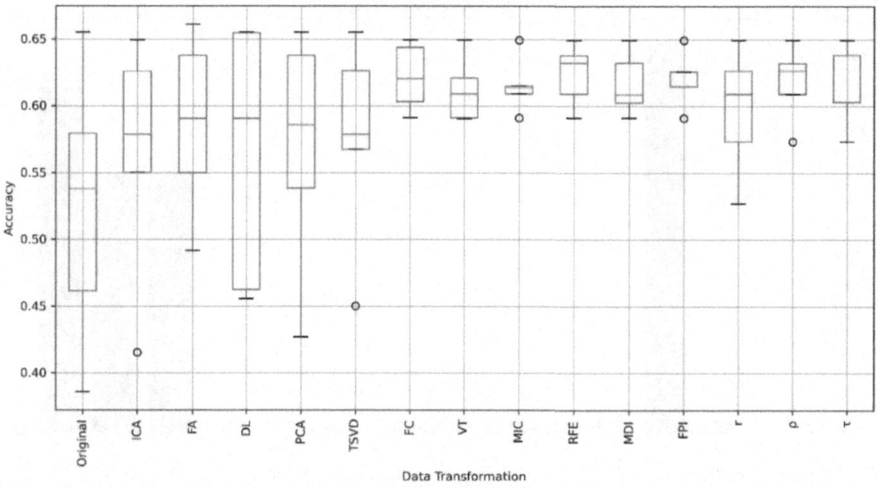

Fig. 2. A variation in scores across models, per each dimensionality reduction technique

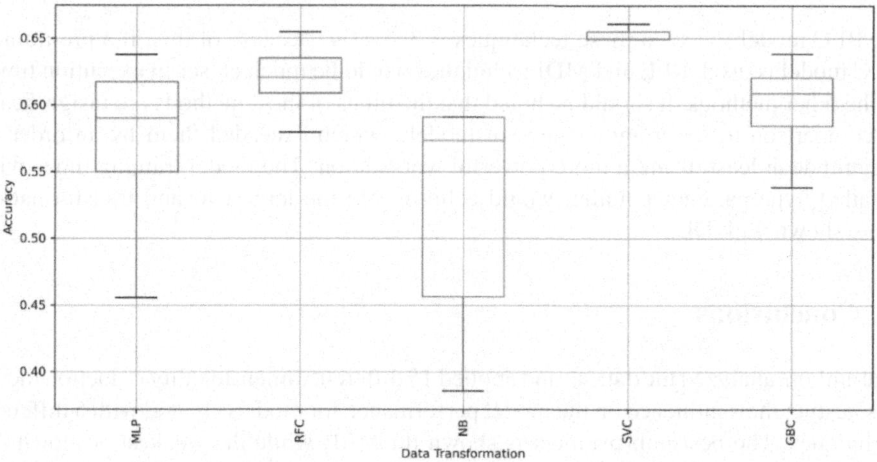

Fig. 3. A variation in scores across dimensionality reduction techniques, per each model

authors present the average times, across ten iterations, for each of the methods used. The methods are run on a system consisting of Intel® Core™ i5-9400 COU (six cores, six threads, clocked at 3.84 GHz) and 32 GB of RAM. All times include the times necessary to perform the model fitting and data transformation using the created model. These times are shown in Fig. 4.

As shown in Fig. 4, most methods take a negligible amount of time to process the dataset, fit it, and perform the dataset transformation. The only significant outliers are the DL method, and the RFC-based methods – RFE, MDI, and FPI. The latter three methods require training the RF model, and it should be noted that the average time of training

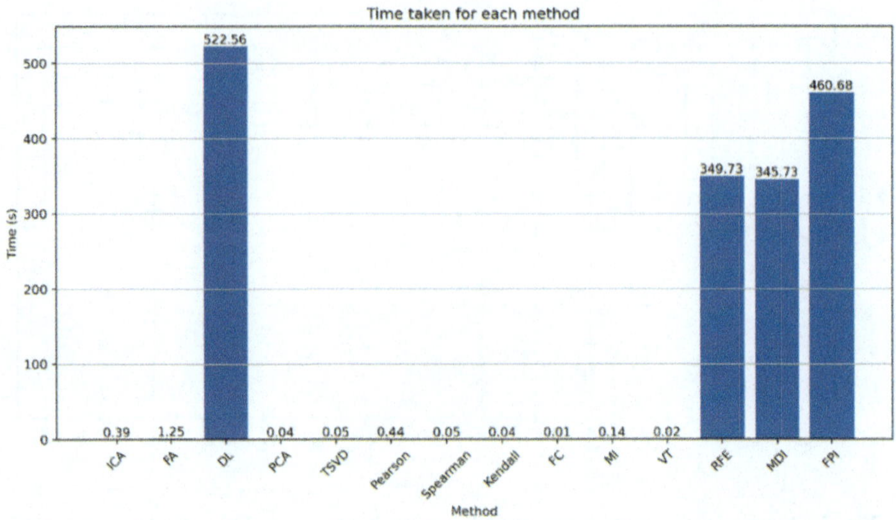

Fig. 4. Fitting and transformation times of dimensionality reduction methods on the used dataset

the RFC model to use in these techniques is 345.41 s. Because of this, if a pre-trained RFC model is used, RFE and MDI techniques would be much closer in execution times to the other methods. It should be noted that the times of most methods are insignificant in comparison to the training times of models, which exceeded them by an order of magnitude at least, using a more powerful workstation. The model training times with detailed hyperparameter tuning would eclipse even the largest fit and transformation time, shown with DL.

5 Conclusions

The authors analyzed the dataset and applied 15 different dimensionality reduction methods, testing their influence on the model performance for models created with 5 different techniques. The best improvement is shown for MLP, while the weakest is shown for SVC. It is also notable that some methods benefit from decomposition more than feature selection and vice-versa. In general, the data show that dimensionality reduction can and should be applied to the datasets, especially large dimensionality ones. No matter what method was used for classification modeling and the dimensionality reduction technique applied to the dataset, the results either did not change significantly or showed an improvement. Considering the methods are not computationally complex, there is no clear indication that dimensionality reduction should not be applied as a preprocessing step. Feature selection methods show a lower variance in scores across methods, compared to the decomposition methods, with higher means. Regarding the methods, all are shown not to be computationally expensive, except for DL and tree-based dimensionality reduction methods, indicating low cost to their use.

Limitations of the presented study are apparent – the testing is limited to a single dataset, meaning that the results cannot be generalized across a larger amount of data sets. The number of features after pruning is also fixed in this study, although it could have a significant influence, especially with feature selection techniques. In addition to that, the hyperparameter tuning of methods could have an influence, and cross-testing different hyperparameter combinations with different methods of dataset pruning could yield positive results.

From the results it can be concluded that different models are affected differently by dimensionality reduction methods, indicating that further research is necessary – mainly focusing on other datasets, and better tuning of hyperparameters of individual methods.

Acknowledgments. This research has been (partly) supported by the CEEPUS network CIII-HR-0108, the Erasmus+ projects WICT, under the grant 2021-1-HR01-KA220-HED-000031177, and AISE, under the grant KA220-SCH; and the University of Rijeka, under scientific grant uniri-tehnic-18-275-1447.

References

1. Anđelić, N., et al.: The development of symbolic expressions for the detection of hepatitis C patients and the disease progression from blood parameters using genetic programming-symbolic classification algorithm. Appl. Sci. **13**(1) (2022)
2. Schmid, P.: Dynamic mode decomposition and its variants. Annu. Rev. Fluid Mech. **54**, 225–254 (2022)
3. Baressi Šegota, S., et al.: Cancer rates per country-determining the importance of country level factors using random forest regressor. In: 1st Serbian Conference on Applied Artificial Intelligence (2022)
4. Anđelić, N., et al.: Development of symbolic expressions ensemble for breast cancer type classification using genetic programming symbolic classifier and decision tree classifier. Cancers **15**(13) (2023)
5. Remeseiro, B., et al.: A review of feature selection methods in medical applications. Comput. Biol. Med. (2019)
6. Baressi Šegota, S., et al.: Improvement of marine steam turbine conventional exergy analysis by neural network application. J. Mar. Sci. Eng. **8**(11) (2020)
7. Baressi Šegota, S., et al.: Regression model for the prediction of total motor power used by an industrial robot manipulator during operation. Machines **12**(4) (2024)
8. Jović, A., et al.: A review of feature selection methods with applications. In: 2015 38th International Convention on Information and Communication Technology, Electronics and Microelectronics (MIPRO). IEEE (2015)
9. Gul, S., et al.: Structure-based design and classifications of small molecules regulating the circadian rhythm period. Sci. Rep. **11**(1) (2021)
10. Saheed, Y.: Effective dimensionality reduction model with machine learning classification for microarray gene expression data, 153–164. Academic Press (2023)
11. Pedregosa, F., et al.: Scikit-learn: machine learning in Python. J. Mach. Learn. Res. **12** (2011)
12. Sugumaran, V., et al.: Effect of a number of features on classification of roller bearing faults using SVM and PSVM. Expert Syst. Appl. **38**(4) (2011)
13. Sevšek, L., et al.: Determining the influence and correlation for parameters of flexible forming using the random forest method. Appl. Soft Comput. **144** (2023)

14. Sidey-Gibbons, J., et al.: Machine learning in medicine: a practical introduction. BMC Med. Res. Methodol. **19**, 1–18 (2019)
15. Ahamad, G., et al.: Influence of optimal hyperparameters on the performance of machine learning algorithms for predicting heart disease. Processes **11**(3), 734 (2023)

Evaluating GDPR and HIPAA in the Integration of ML/AI for Future-Proofing Healthcare

Saso Nikolovski(✉) and Anita Petreska

Faculty of Information and Communication Technologies, Bitola University, "St. Kliment Ohridski", Bitola, Republic of Macedonia
sasnik@gmail.com, petreska.anita@uklo.edu.mk

Abstract. The implementation of machine learning and artificial intelligence as contemporary technologies and their integration in healthcare systems raises questions concerning data privacy and protection, which are regulated by the General Data Protection Regulation in the European Union and the Health Insurance Portability and Accountability Act in the United States. This paper is focused on examination of these two regulatory frameworks and their impact on the legal and technical incorporation in healthcare settings. It discusses the necessary measures for ensuring data privacy, security, and ethical standards that healthcare institutions must consider in the implementation of advanced methods and technologies for improving patient care, diagnostics, and operational efficiency. The comparative analysis highlights significant similarities and differences between these two regulations, particularly in the areas such as encryption, access management, and international data transfers. Additionally, the study underscores the need for new technologies to adhere to ethical principles, promoting transparency, fairness, and accountability in their decision-making processes.

Finally, the paper addresses potential future paths for aligning technological advancements with evolving regulatory requirements, aiming to support compliance while fostering the safe and efficient use of modern technologies in healthcare.

Keywords: Machine Learning · Artificial Intelligence · GDPR · HIPAA · Data Protection · Privacy Protection · Diagnostics · Ethics · Compliance · Algorithms

1 Introduction

As machine learning (ML) and artificial intelligence (AI) increasingly revolutionize healthcare, they bring enormous potential for advancing patient care, diagnostics, and operational efficiency. However, as these technologies become central to healthcare systems, there is a critical need to assess how they align with existing data protection and privacy regulations.

This research focuses on the intersection of two key regulatory frameworks - the General Data Protection Regulation (GDPR) in the European Union and the Health Insurance Portability and Accountability Act (HIPAA) in the United States. These regulations face the challenge of ensuring that ML and AI are safely and legally integrated into

© The Author(s), under exclusive license to Springer Nature Switzerland AG 2025
N. Filipović (Ed.): AAI 2024, LNNS 1446, pp. 289–313, 2025.
https://doi.org/10.1007/978-3-031-99201-8_30

healthcare while maintaining the highest standards of privacy and data security. While AI and ML offer transformative opportunities, improving care, enhancing diagnostics, and streamlining operations, the complexities of data protection and privacy present a significant hurdle. Both GDPR and HIPAA impose strict requirements for safeguarding personal health information, aiming to protect individuals' rights to privacy while ensuring secure data handling practices. The objective of this research is to thoroughly analyze the regulatory frameworks governing data protection in healthcare. By evaluating the integration of AI and ML within these regulations, we provide insights into the key security measures and standards organizations must adopt to ensure compliance. This includes a detailed examination of both technical and organizational safeguards required under GDPR and HIPAA. The paper begins with an overview of the fundamental data protection measures covered by both regulations, followed by a comprehensive analysis of the security standards they mandate. Additionally, we present a comparison of GDPR and HIPAA, highlighting the areas where they overlap and diverge in terms of data security and privacy. In a dedicated section, we explore AI and ML as modern approaches to decision-making in healthcare, illustrating their potential while discussing the regulatory considerations necessary to ensure ethical and legal implementation.

2 Data Protection Regulations and Challenges of Integrating AI and ML into Healthcare

The global need to protect sensitive personal data from security threats and privacy violations has led to the creation of various regulatory frameworks designed to ensure effective data protection. Integrating ML and AI into healthcare while maintaining compliance with the GDPR and HIPAA presents several key challenges for the future. Based on existing research, these challenges can be examined from multiple perspectives. Data privacy and security become one of the primary challenges in integrating AI and ML into healthcare because of the need for data privacy and security. This concern is highlighted by studies on the evolution of medical information management systems [1]. Compliance with GDPR and HIPAA mandates strict control over how patient data is accessed, stored, and utilized. From the perspective of the ethical Considerations, ethical issues are at the forefront of AI integration in healthcare, focusing on aspects such as privacy, transparency, trust, accountability, bias, and the quality of data [2]. Addressing these ethical concerns is crucial for the successful deployment of AI technologies. Regulatory compliance ensuring that AI and ML systems meet the regulatory requirements of GDPR and HIPAA is a substantial challenge. These regulations set high standards for protecting patient data and maintaining privacy [3]. The complexity of developing ethical frameworks and regulatory structures for AI in healthcare points to a significant future challenge in compliance.

Another critical issue is data sharing and reliability, which is an aspect that involves data sharing and ensuring data reliability, security, and stability. Emerging technologies like Blockchain have been suggested as a means to enhance secure data management addressing these concerns [4]. When we talk about the integration of these two technologies, we mean the integration into existing healthcare systems. The adaptation of ML and AI to existing healthcare systems, especially in specialized fields such as ophthalmology, presents unique challenges [5]. These challenges relate to how effectively these technologies can be incorporated into current healthcare infrastructure and practices. Across Diverse Healthcare Contexts, the implementation of AI and ML technologies varies across different healthcare systems globally, each posing distinct challenges [6]. These differences must be accounted for when integrating AI into various healthcare environments. Addressing these challenges will require a comprehensive approach that combines technological innovation, ethical guidelines, and regulatory adaptation to fit the diverse needs of healthcare systems worldwide.

2.1 GDPR and Its Global Implications for Data Processing

GDPR is a comprehensive legal framework within the EU that impacts companies worldwide when processing the personal data of EU citizens [7]. Applicable in all EU member states, the GDPR's reach extends beyond Europe, affecting global operations and setting a high standard for data protection [8]. This regulation is designed to ensure safe and transparent handling of personal data, which it defines as any information through which individuals can be identified. During processing, personal data undergoes several stages, each raising critical concerns about its security and privacy. This becomes particularly significant when dealing with sensitive data such as health records.

The complexity of managing personal data stems from factors such as its origin, intended use, storage duration, and accuracy requirements (Fig. 1).

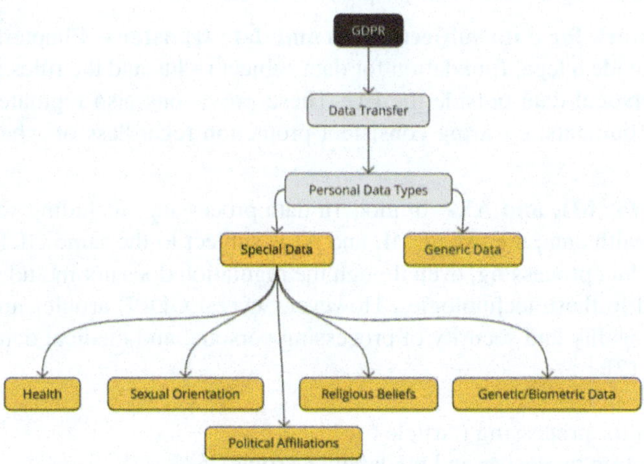

Fig. 1. Manipulation/transfer different types of personal data

Whether collected from individuals or corporate clients, data must have a defined purpose, strict storage guidelines, and be regularly updated to ensure its accuracy, which is especially vital for health-related data.

Key legal foundations under GDPR - Article 6 of the GDPR outlines the legal bases for processing personal data, emphasizing the high level of responsibility placed on all participants in the data processing chain [9]. These legal foundations underscore the regulation's objective of maintaining a uniform level of data protection across the EU.

To safeguard privacy, GDPR highlights two critical principles: "Privacy by Design" and "Privacy by Default." These principles require organizations to incorporate privacy measures from the very beginning of data collection and processing, ensuring that every stage complies with privacy and security standards. In the event of a data security breach, GDPR mandates immediate and transparent reporting, aiming to minimize any harm to the data subjects [10]. Organizations are obligated to implement both technical and organizational measures to mitigate the risk of such incidents.

In cases of technical safeguards and cross-border data transfers, pseudonymization and de-identification are technical methods promoted by GDPR to add layers of protection. Despite the removal of identifying fields, this information is still treated as personal data under the regulation, further emphasizing the need for comprehensive data protection strategies [13]. For organizations that transfer health data to third countries, binding corporate rules must be established to legitimize the transfer and ensure compliance with GDPR's global data protection standards. Although encryption is not explicitly required, it is recognized in specific articles (Article 5(1)(f) and Article 32) as an effective tool for ensuring data security [15].

Data storage and cloud services - In today's digital landscape, data storage frequently involves cloud infrastructures. While GDPR does not take a specific stance on cloud storage, it outlines obligations for organizations to establish stringent data protection measures for cloud-based services [17]. Articles 5, 6, and 32 detail technical procedures for secure data storage, which apply equally to cloud solutions.

Legal framework for data subject rights and data transfer - Chapters III and V of the GDPR provide a legal foundation for data subject rights and the rules governing the transfer of personal data outside the EU. These provisions also regulate data storage in cloud environments, ensuring consistent protection regardless of where the data is stored.

Implications for ML and AI - In modern data processing, including sensitive information like health data, the use of ML and AI is subject to the same GDPR provisions as traditional data processing, even though the regulation does not include specific articles dedicated to these technologies. However, several GDPR articles are relevant for ensuring the legality and security of processing personal and medical data with AI and ML systems [12]:

- Legal basis for processing (Article 6).
- Data protection by design and by default (Article 25).
- Data minimization (Article 5(1)(c)).
- Transparency (Articles 12–14).
- Automated decision-making and profiling (Article 22).
- Security of processing (Article 32).
- Data protection impact assessments (Article 35).

Implications for ML and AI in compliance with GDPR - In the case of AI-driven medical diagnostic systems, compliance with GDPR would require that organizations first establish a clear legal basis for processing patient data (Article 6). Moreover, by implementing Privacy by Design principles (Article 25), AI models should be built to minimize data use (Article 5(1)(c)) and incorporate mechanisms to ensure transparency in how data is processed and decisions are made (Articles 12–14). Importantly, where AI systems make automated decisions that affect patients, such as diagnosis or treatment recommendations, organizations must ensure that data subjects are informed and have the right to request human intervention (Article 22). Additionally, organizations must conduct Data Protection Impact Assessments (Article 35) to evaluate the risks to data privacy before implementing AI systems.

AI and ML systems compliance with GDPR - Organizations using AI systems to process personal data must comply with the general data processing requirements outlined in GDPR [25]. By consistently applying these standards (such as privacy by design, data minimization, and transparency) organizations can ensure that their AI systems protect the privacy rights of individuals whose data is being processed, while also maintaining regulatory compliance.

2.2 Health Insurance Portability and Accountability Act (HIPAA)

HIPAA is a vital U.S. law that sets standards for the protection of patient health information (PHI) and regulates its use, storage, transmission, and processing across the healthcare sector. HIPAA applies to both covered entities and their business associates, who must comply with its strict guidelines to maintain PHI security and privacy.

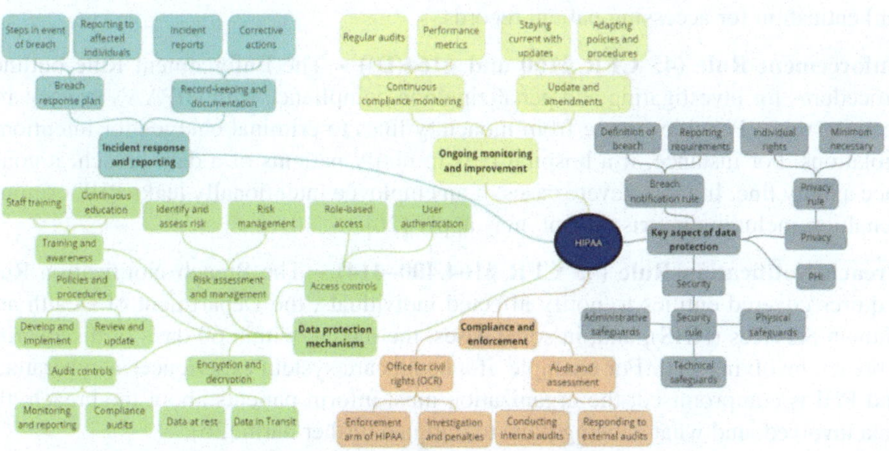

Fig. 2. Critical areas of data management in HIPAA

As illustrated in Fig. 2, HIPAA's framework addresses several critical areas of data management, each of which impacts the handling of health and personal information. HIPAA consists of four main regulatory components.

Privacy Rule (45 CFR §160 and §164.500–164.534) - The privacy rule is central to HIPAA, establishing comprehensive standards for protecting patient records and other identifiable health data. It governs when and how PHI can be used or disclosed by covered entities, such as healthcare providers and insurance companies.

For example, under the privacy rule, a healthcare provider must obtain patient consent before sharing medical information with a third-party research company. However, in cases related to treatment, payment, or healthcare operations, data may be shared without patient consent to ensure the necessary flow of health information for patient care [26]. Patients also have the right to access, review, and request corrections to their health data, ensuring transparency and control over their personal information.

Security Rule (45 CFR §160 and §164.300–164.318) - The Security Rule complements the privacy rule by mandating safeguards to protect electronic protected health information (ePHI).

These safeguards are divided into three categories:

- Administrative safeguards that include policies and procedures for managing the implementation of security measures;
- Physical safeguards that are measures to protect electronic systems and data from unauthorized access and physical threats like natural disasters; and
- Technical safeguards that present technology-based protections such as encryption and secure access controls to protect ePHI during transmission and storage [26].

As a practical example of using these safeguards can be the usage of encryption in hospitals to secure patient data stored on its servers and requiring multifactor authentication for accessing patient records.

Enforcement Rule (45 CFR §160 and §164(D)) - The Enforcement Rule outlines procedures for investigating and penalizing non-compliance with HIPAA's privacy and security rules. Penalties range from monetary fines to criminal charges for intentional violations. For instance, if a hospital fails to notify patients of a data breach, it could face a hefty fine. In more severe cases, if an employee intentionally leaks PHI, criminal penalties, including imprisonment, may apply [26].

Breach Notification Rule (45 CFR §164.400–414) - The Breach Notification Rule requires covered entities to notify affected individuals, the Department of Health and Human Services (HHS), and, in some cases, the media, within 60 days of discovering a breach involving PHI. For example, if a healthcare system experiences a cyberattack and PHI is compromised, the organization must inform patients about the breach, the data involved, and what is being done to mitigate further harm [26].

HIPAA Omnibus Rule and Updates - In response to evolving technologies and security risks, the HIPAA Omnibus Rule, introduced in 2013, updated and strengthened existing HIPAA regulations. It expanded HIPAA's reach to include business associates (third-party vendors handling PHI) and imposed stricter breach notification and enforcement measures [27]. Key updates include:

- Expanded breach definition - Organizations must now perform risk assessments to determine whether unauthorized use or disclosure of PHI qualifies as a breach.
- Business associate accountability - Business associates, like cloud service providers that store PHI, are now directly accountable for HIPAA compliance.
- Right to request restrictions - Patients can request restrictions on PHI disclosure, particularly for out-of-pocket paid services [27].

Compliance in the Context of AI and ML in Healthcare - The growing use of AI and ML in healthcare presents unique challenges for HIPAA compliance, as these technologies handle vast amounts of ePHI for tasks like patient diagnostics and predictive analytics. In healthcare, AI systems rely on large datasets to function, making it essential to comply with HIPAA's data minimization principle, which mandates that only the minimum necessary amount of data be used. For instance, a healthcare provider using an AI-driven diagnostic tool must ensure that the system only accesses the specific data needed for diagnosis and not unnecessary patient details.

Additionally, under the Privacy Rule, healthcare organizations must inform patients about how their data will be used by AI systems, including any automated decision-making processes. For example, an AI system used to predict patient outcomes must disclose how it uses PHI to generate those predictions [26].

Security Challenges and Technological Safeguards - AI and ML systems must also adhere to HIPAA's security rule, which requires technical safeguards such as encryption and access controls. For example, when AI systems process ePHI for clinical decision support, all data transfers should be encrypted to protect it from cyber threats. Furthermore, organizations must conduct regular risk assessments to identify vulnerabilities in AI systems. For example, a healthcare provider deploying AI for predictive analytics should routinely assess the system's security measures to prevent unauthorized access to sensitive patient data [26].

3 Safeguards in the Regulations

Privacy, traditionally defined as the right to be left alone, is recognized as a fundamental human right by the United Nations in the Universal Declaration of Human Rights. However, in today's digital age, where daily life and work are increasingly connected online, the concept of privacy has become more subjective. It is now interpreted and applied differently by each state or country [19].

The privacy of clinical data, both historically and in modern contexts, remains a topic of extensive research. One persistent challenge is determining the ownership of this data (specifically, how much of it belongs to the patient). This ambiguity raises important questions, such as whether patient consent is necessary when clinical data is used for research purposes [20–22].

When analyzing regulations such as the GDPR and the HIPAA, it is crucial to differentiate between privacy rules and security rules. Privacy rules and policies focus on protecting health data and personal information, ensuring confidentiality, and regulating its use. In contrast, security rules and policies are concerned with safeguarding data during its processing and analysis for health or scientific purposes, ensuring that the data remains protected from unauthorized access and breaches throughout its lifecycle.

3.1 Security Measures Under GDPR

The GDPR outlines a comprehensive set of security measures aimed at ensuring a high level of data protection. A critical step prior to implementing these measures is the recommendation for conducting a risk assessment to identify potential risks and determine appropriate safeguards based on the specific data processing activities of controllers and data processors [18].

The security measures outlined in Article 32 of GDPR are categorized into two main groups as technical measures and organizational measures.

- Technical measures are focused on both physical and digital aspects of data security. Key measures include ensuring the physical security of systems where data is stored and controlling user access to the data. This involves not only securing the infrastructure but also implementing network security protocols within the organization [14]. One important technical measure is access control, which is enhanced by the use of user profiles that are centrally authenticated and authorized at the organizational level [23]. Through this system, access to data is tightly restricted based on assigned roles and memberships, ensuring only authorized personnel can handle sensitive information. Data encryption is another fundamental technical measure. It is mandatory during the transmission of data to protect it from unauthorized access. Whether data is being processed or stored, encryption ensures that the information remains secure and inaccessible to unauthorized users. To prevent data loss or damage (whether through user error, hardware failures, or environmental disasters), organizations are required to implement regular backup procedures. These backups are a critical component of the Business Continuity Plan (BCP) and are stored both on-site and off-site to ensure rapid recovery of data in case of any incidents [24].
- Organizational Measures - from an organizational standpoint, safeguards are designed to ensure that personal data is processed lawfully and securely. These measures not only protect individuals' privacy rights but also ensure compliance with GDPR principles, as outlined in Article 5. One crucial organizational security measure is the Incident Response Plan (IRP), which is covered in Articles 33 and 34. The IRP outlines the steps organizations must take in the event of a security breach, including identifying and notifying affected individuals and reporting the incident to supervisory authorities. This plan ensures prompt action to mitigate any damage resulting from a data breach. Additionally, GDPR emphasizes the ongoing education of employees in relation to data privacy (as mentioned in Articles 37 and 39). This education covers critical topics such as password management, phishing scams, and social engineering, all of which aim to reduce the risks associated with mishandling personal data.

In conclusion, for the safeguards in the GDPR, the combination of technical and organizational measures under GDPR provides a robust framework for securing personal data, safeguarding privacy, and ensuring compliance with the regulation's high standards of data protection.

3.2 Security Measures Under HIPAA

Unlike the GDPR, which covers a broad range of personal data, the HIPAA focuses specifically on personal health data and the security measures required to protect it during processing, storage, and transmission. In today's healthcare landscape, providers use a variety of clinical platforms for managing EHR and systems for radiology, pharmacy, and laboratory results.

Under HIPAA, e-PHI includes any individually identifiable health information that is created, received, stored, or transmitted electronically. This consists of 18 identifiers defined in the regulation, which include data points such as a patient's name, address, phone number, social security number, medical record number, biometric data, and personal photos or recordings that could identify the patient [28, 29].

HIPAA's security standards are divided into three main categories, each addressing a different aspect of safeguarding e-PHI (as specified by rule 45 CFR §164.304) [30]:

- **Administrative Measures** - Outlined in rule 45 CFR §164.308, administrative measures focus on the organizational processes necessary to meet HIPAA's security requirements. These include assigning security responsibilities to specific individuals or teams within the organization and ensuring all personnel receive proper security training to protect e-PHI. For example, healthcare organizations are required to designate a security officer responsible for developing and implementing security policies and conducting regular risk assessments.
- **Physical Measures** - Described in rule 45 CFR §164.310, physical measures protect electronic systems, equipment, and the data they store from physical threats, environmental hazards, and unauthorized access. These measures govern the physical security of systems and facilities, including controlling access to locations where e-PHI is stored and managing the creation and storage of backup data. For example, restricting access to server rooms where EHR data is stored and ensuring that backup copies of e-PHI are kept at secure off-site locations to protect against natural disasters.
- **Technical Measures** - Technical measures, covered in rule 45 CFR §164.312, involve automated processes designed to secure e-PHI and control access to it. This includes the use of authentication protocols to verify the identity of individuals accessing e-PHI, as well as encryption and decryption methods to protect data while it is stored or transmitted. For instance, healthcare providers must implement encryption for e-PHI when transmitting data to external locations or other systems to ensure its security during transit.

3.3 Comparative Review of Regulations

As previously mentioned, GDPR and HIPAA are two distinct regulatory frameworks that, when applied together, enhance security and privacy, particularly in the realm of information and data protection. While GDPR governs the use and protection of all personal data of individuals within its scope, HIPAA has a more specific focus, applying only to PHI. To highlight the key differences, Table 1 shows a comparison between GDPR and HIPAA.

Despite the differences between these two regulations regarding the protection of sensitive personal data, a detailed comparison reveals several areas of overlap in their

Table 1. Key differences between GDPR and HIPAA.

	GDPR	HIPAA
Protected data	Any data that relates to, or can lead to the identification of a living person	Any information about health status, care, or payment that is created or collected by a HIPAA Covered Entity (or a Business Associate of a Covered Entity), that can be linked to a specific individual
Scope	Sets compliance standards for all entities that fall within its scope	Sets standards for covered entities and their business associates
Consent	Explicit consent is mandatory for the processing of personal health data. However, data may be processed without consent if it meets one of the conditions of processing in Article 9 of the GDPR and a legal basis applies	Allows disclosure of some PHI for "treatment purposes" without the consent of the individual
Right to be forgotten	Individuals have the right to be forgotten (or to have their data deleted upon request)	This right is not granted in HIPAA
Data breaches	Notification of supervisory authority in 72 hours together with affected persons	Organizations must protect PHI and limit disclosure under the HIPAA Privacy Rule. Covered entities must also notify affected individuals of security breaches. If more than 500 people are affected, both affected individuals and the Department of Health must be informed within 60 days

implementation of effective and efficient data protection. Both regulations require the following:

- Controlled access to sensitive data;
- The use of methods to detect unauthorized changes to PHI;
- The encryption of PHI during storage and transfer;
- The appointment of a Data Protection Officer (DPO);
- Ensuring an appropriate level of privacy for both customers and employees.

While GDPR and HIPAA have different rules and scopes, the overlapping measures and standards in data protection ensure comprehensive protection of personal data. This includes data of a special nature, such as that outlined in Articles 7 and 9 of the GDPR.

4 Integration of ML and AI in Healthcare

ML and AI is revolutionizing healthcare by transforming diagnostics, treatment planning, and operational efficiency. AI technologies mimic human cognitive functions (such as learning, reasoning, and self-correction) to perform tasks that typically require human intelligence. These tasks include analyzing genetic data, interpreting patient records, and identifying patterns in vast datasets to support informed decision-making. AI's capacity to process and manage large volumes of data makes it a crucial tool in healthcare, offering insights that were previously beyond human reach. One of the most significant impacts of AI and ML in healthcare is the advancement of personalized medicine, enabling more precise and customized treatments while reducing the trial-and-error approach of traditional medical practices [31].

Medical imaging is one field that has greatly benefited from AI and ML integration. Technologies such as MRI with AI, CT scans, and X-rays now offer faster and more accurate interpretations, enhancing early disease detection and speeding up the diagnostic process. Additionally, AI and ML play a critical role in optimizing the operational aspects of healthcare institutions. From managing patient flow to predicting hospital admissions and optimizing inventory, AI systems help streamline processes, reduce costs, and improve the overall patient experience.

However, the integration of AI and ML in healthcare also presents challenges, including ethical concerns, data privacy issues, and the need for significant infrastructure investments. Furthermore, the success of AI in healthcare hinges on the quality and availability of data, emphasizing the need for strong data management practices.

4.1 Protocols and Standards for Health Data Management

Effective health data management is essential in today's healthcare systems, ensuring the secure handling, storage, and exchange of sensitive health information. The adoption of standardized protocols and frameworks is crucial for maintaining privacy, security, and interoperability in health data management [44]. These standards enable seamless communication and data exchange between various health information systems, thereby enhancing the quality of patient care and operational efficiency. Below are some of the most widely recognized and implemented standards in the healthcare industry:

- HL7 (Health Level Seven International) - HL7 is a set of international standards providing guidelines for the exchange, management, and integration of electronic health information. It promotes interoperability and ensures seamless data sharing between different healthcare systems, facilitating efficient patient care and administrative processes.
- FHIR (Fast Healthcare Interoperability Resources) - FHIR represents a significant advancement in healthcare informatics, designed for the electronic exchange of healthcare information. It uses a modern, web-based approach with modular resources that can be combined to address clinical and administrative challenges quickly and efficiently.
- DICOM (Digital Imaging and Communications in Medicine) - DICOM is a specialized standard for medical imaging that standardizes the communication and management of image data. It ensures that medical images and associated information can be reliably and securely exchanged between different imaging devices and systems.
- CDA (Clinical Document Architecture) - CDA is an XML-based standard that supports the exchange of clinical documents, such as discharge summaries and progress notes, in a structured and standardized format. This ensures that these documents can be shared across healthcare settings while preserving their integrity and confidentiality.
- SNOMED CT (Systematized Nomenclature of Medical Clinical Terms) - is a comprehensive and multilingual clinical healthcare terminology that enables consistent representation of clinical content in health records. It improves data analysis and enhances the accuracy of recording and sharing patient health information.
- CCDA (Consolidated Clinical Document Architecture) - CCDA builds on the CDA standard and specifies the structure of clinical documents for exchange. It supports the meaningful use of electronic health records (EHRs) by improving communication between healthcare providers.

- IHE (Integrating the Healthcare Enterprise) - IHE is an initiative designed to enhance the interoperability of healthcare information systems. It promotes the coordinated use of established standards, such as HL7 and DICOM, to improve workflows and data access across different systems and healthcare settings.
- ISO 27001 (Information Security Management System) - ISO 27001 is an international standard that outlines best practices for an information security management system (ISMS). It provides a systematic approach to managing sensitive health information, ensuring data privacy and security.
- CEN/ISO EN 13606 (European EHR Standard) - This European standard defines a framework for the communication and interoperability of electronic health records across borders within Europe. It ensures the secure and seamless exchange of health information, supporting a unified approach to EHR management across different countries.

Together, these protocols and standards are essential for managing health data in a secure, efficient, and interoperable manner. They support the delivery of high-quality health services while ensuring patient privacy and data protection across the global healthcare ecosystem.

4.2 Harnessing Big Data and Data Mining in Healthcare

The healthcare industry continuously generates vast amounts of data, characterized by its immense volume and variety. Despite this, much of the potential value in this data remains untapped. The 10 Vs framework (Fig. 3) offers a comprehensive approach to unlocking the power of medical big data, transforming raw information into actionable insights [32].

With data originating from diverse sources such as eEHRs, medical imaging, wearable devices, and patient-generated data, descriptive analytics plays a vital role in summarizing and interpreting historical data to identify relevant patterns and trends.

Predictive analytics, on the other hand, uses statistical algorithms and ML models to forecast future trends and outcomes, enabling a proactive approach to healthcare management. Machine learning algorithms are becoming particularly valuable in diagnostics, employing NLP to extract meaningful information from unstructured texts like clinical notes and medical literature.

The application of data science in healthcare holds great promise for improving patient care, outcomes, and system efficiency. However, challenges related to data quality, model interpretability, and ethical considerations must be addressed to fully harness the potential of big data and data mining in transforming healthcare delivery and patient outcomes.

4.3 Opportunities and Data Privacy Concerns in the Process of ML Model Development in Healthcare

The development of ML models in healthcare represents a groundbreaking advance in predicting and managing health outcomes. However, it also introduces significant risks related to patient data privacy and security.

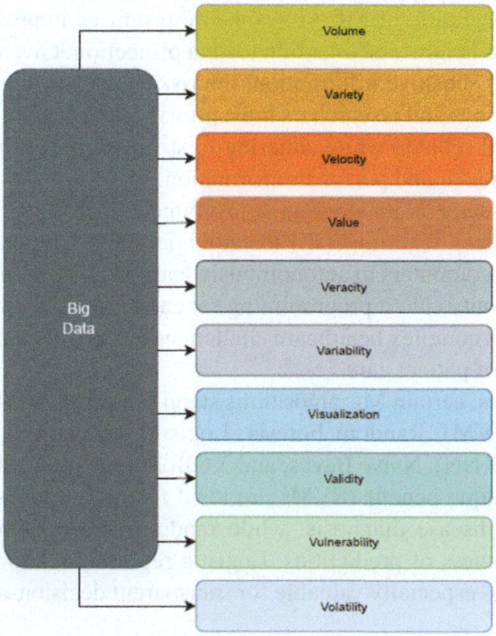

Fig. 3. 10 V's in Big Data manipulation process

The first step in building ML models is Exploratory Data Analysis (EDA), a critical phase where data is examined to uncover patterns, relationships, and characteristics [33]. Visualization techniques such as histograms, box plots, and scatter plots are commonly used to transform numerical data into graphical representations, making it easier to identify outliers, understand data distributions, and explore relationships between variables [34]. While this process is essential for model building, it raises concerns regarding the exposure of sensitive health information, as outliers (often unique patient cases) can inadvertently reveal private details, leading to potential privacy violations. Developing accurate, reliable, and ethically sound ML models in healthcare follows a structured process, starting from data collection and processing through to model training, validation, and deployment. Each stage requires handling sensitive health data, raising the risk of compromising patient privacy. For instance, dealing with outliers and data normalization, though essential for model accuracy, can lead to the re-identification of individuals. Partitioning data into training, validation, and testing sets is fundamental for creating robust models, but in the context of health data, this step heightens the importance of data protection to prevent unintentional disclosures. Overfitting not only reduces model performance on new datasets but also poses privacy risks when sensitive data patterns are closely embedded in the model, potentially exposing private information.

The need for transparency in interpreting and explaining model predictions is crucial in healthcare. However, such transparency must not compromise patient privacy. Techniques for model interpretation should be designed to ensure that no sensitive health data is inadvertently revealed during this process.

Although the integration of ML in healthcare promises improved patient care and outcomes, it demands heightened attention to data protection. Given that healthcare data often contains highly sensitive information, the risk of privacy breaches is substantial. Healthcare organizations and researchers must prioritize the development of models that are both accurate and reliable while adhering to stringent data protection standards to safeguard patients' health and personal information.

The healthcare sector is undergoing a profound transformation through the integration of ML and deep learning (DL) technologies [35]. These AI systems use ML algorithms to enable computers to autonomously learn from data, make predictions, and make decisions without explicit programming for each task. The ability of different ML algorithms to address complex healthcare challenges has greatly enhanced the accuracy and personalization of patient care.

In health analytics, certain ML algorithms stand out for their effectiveness: Support Vector Machines (SVM), Random Forests, Logistic Regression, Decision Trees, K-Nearest Neighbors (KNN), Naive Bayes, and XGBoost are among the most widely used [36]. Each offers unique benefits SVMs are ideal for high-dimensional data analysis, which is crucial in disease diagnosis, while random forests help reduce overfitting, enhancing the robustness of predictions. Logistic regression, known for its simplicity and interpretability, is especially valuable for transparent decision-making in healthcare settings.

DL, a subset of ML, further expands AI's capabilities in healthcare by allowing systems to learn from examples in a manner similar to human cognition. Neural networks, which form the backbone of DL, come in various architectures such as deep neural networks (DNNs), convolutional neural networks (CNNs), and recurrent neural networks (RNNs), each providing specialized benefits. CNNs, for instance, are crucial for improving the diagnostic accuracy of medical images, while RNNs excel in handling sequential data like patient records.

More advanced neural network models, such as long-short-term memory (LSTM) networks, generative adversarial networks (GANs), and CapsNets, offer even greater sophistication. LSTMs are excellent at modeling sequences, making them invaluable for monitoring chronic conditions or predicting disease trajectories. GANs are revolutionizing the generation of synthetic medical images, enhancing dataset availability, and aiding in healthcare professional training. CapsNets improve the recognition of hierarchical structures in data, which is particularly useful in diagnosing complex diseases.

The application of transfer learning and autoencoders in healthcare showcases the adaptability and versatility of DL models. Transfer learning enables the use of pre-trained networks on large datasets for specific healthcare tasks, optimizing performance even with limited labeled data. Autoencoders excel at feature learning and dimensionality reduction, making it easier to detect patterns and anomalies in healthcare data. Additionally, deep reinforcement learning is showing promise in optimizing treatment plans and resource allocation by learning optimal strategies through trial and error.

Selecting the Right ML Algorithm for Healthcare Analytics

Choosing the appropriate ML algorithm for healthcare analytics involves a process of experimentation, comparison, and optimization, often using techniques like cross-validation to evaluate model performance [37]. This iterative approach, guided by the combined expertise of data scientists and healthcare domain experts, ensures that the selected model fits the data well and produces actionable insights. Key factors influencing this choice include a deep understanding of the problem at hand, the nature and characteristics of the data, and the specific goals of the analysis.

The first step in selecting an ML algorithm is to clearly define the problem. Whether the objective is prediction, classification, or pattern discovery, the algorithm must align with the task to ensure both efficiency and relevance. This alignment provides a strong foundation for model development and evaluation.

Linear Models (for simplicity and interpretability) are models, such as linear regression, assume that the relationship between input features and output is linear. These models are valued for their simplicity, computational efficiency, and ease of interpretation. Linear models are particularly useful when relationships between variables are well understood and linear in nature, making them an ideal benchmark in many healthcare analytics tasks. They offer a straightforward method for understanding relationships and making predictions, such as predicting a patient's risk for a particular condition based on measurable factors like age or blood pressure.

Decision Trees and ensemble methods (for handling Complexity) - for tasks involving more complex and nonlinear relationships, algorithms like decision trees and ensemble methods (including random forests and gradient boosting) are highly effective. These methods excel at modeling intricate interactions between variables, which is often the case in healthcare analytics when the underlying patterns are not easily identifiable. Ensemble methods, which combine predictions from multiple models, enhance accuracy and robustness, making them suitable for tasks like disease risk prediction or outcome forecasting where precision is crucial.

Support Vector Machines (for high-dimensional data) - when dealing with high-dimensional data or complex classification and regression tasks, support vector machines (SVMs) are a solid choice. SVMs define the optimal boundary between different classes, maximizing the margin between data points of varying categories. This capability is invaluable in healthcare, where subtle distinctions between different health states or outcomes (such as identifying stages of a disease) can be critical.

Unsupervised learning algorithms, such as k-means clustering and hierarchical clustering, are essential for exploring data without predefined labels. These techniques are powerful tools for discovering natural groupings or hidden structures within the data, which can lead to new hypotheses in healthcare research. For instance, clustering algorithms can help uncover patient subgroups with similar characteristics, enabling more targeted treatments or interventions.

Neural networks, particularly their deep learning variants, have revolutionized ML by offering unmatched capabilities in handling large datasets and solving complex problems. Their architecture, inspired by the human brain, allows them to model nonlinear relationships through multiple layers of processing. In healthcare analytics, neural networks have demonstrated remarkable performance in tasks such as image recognition, NLP, and predictive modeling. For example, CNNs (convolutional neural networks) are widely used for analyzing medical images like X-rays or MRIs, while NLP models extract insights from clinical notes. However, neural networks typically require substantial computational resources and expertise, making them more demanding to implement.

To make balance between performance and interpretability, in healthcare, the choice of an ML algorithm is not solely based on performance metrics, interpretability is equally important. The ability to explain and understand how a model arrives at its decisions is critical, especially when those decisions could have significant implications for patient care. While more complex models like neural networks may offer higher accuracy, simpler models such as linear regression or decision trees are often preferred when interpretability is key.

Striking a balance between model complexity, accuracy, and interpretability is crucial in healthcare. This balance ensures that ML models are not only effective but also transparent and trustworthy, supporting better decision-making in critical healthcare scenarios.

Protecting PII in Machine Learning Models

The protection of personally identifiable information (PII) in ML models is a critical challenge in the realm of data privacy and security [38]. When ML models are trained on datasets containing sensitive information (whether from primary or external sources) there is a risk of inadvertently extracting or learning PII. To mitigate this risk, data scientists and analysts must employ rigorous data anonymization techniques and adhere to strict data preprocessing protocols to prevent the unintentional retention of PII in the model.

A significant risk in training ML models is overfitting, where the model memorizes specific details from the training data, including "noise" or unique data points, potentially compromising privacy. Regularization techniques play a vital role in preventing overfitting, ensuring that the model generalizes well to new data without memorizing specifics that might include PII.

Another key concern is the correlation between non-PII and PII elements within the dataset. These correlations can cause the model to inadvertently use PII in its decision-making process. To minimize this risk, thorough data analysis and preprocessing are essential. By effectively separating non-identifiable features from sensitive information, the likelihood of PII influencing the model's predictions is significantly reduced.

The responsibility of protecting PII extends beyond the initial data handling process. It also includes careful examination of the model's outputs and internal parameters. Without proper safeguards, these outputs or behaviors might unintentionally reveal PII, either directly or through inference based on the model's decisions.

Ensuring the confidentiality and integrity of PII in ML models requires a multi-layered approach and includes:

o Careful data preparation to anonymize and preprocess data;
o Application of regularization techniques to prevent overfitting and memorization of sensitive information;
o Ongoing monitoring of model outputs and internal parameters to ensure no PII is revealed.

These steps are crucial for developing ML models that achieve their intended purposes while safeguarding the privacy and security of individuals' sensitive information.

4.4 Comparing Traditional Cybersecurity Approaches to ML Methods

The evolution of cybersecurity strategies has marked a significant shift from traditional approaches to incorporating ML techniques [39, 40]. Historically, cybersecurity relied on rules-based and signature-based detection systems as the first line of defense against cyberattacks. These traditional methods work by using predefined rules to identify and mitigate known threats. While effective against familiar attack types, their major limitation lies in their static nature; they struggle to adapt to new, evolving threats that fall outside their predefined criteria. This limitation increases the risk of missing zero-day attacks and contributes to high rates of false positives and false negatives.

In contrast, ML offers a more dynamic and adaptive approach to cybersecurity. Unlike traditional systems that rely on fixed rules, ML algorithms learn from data, allowing them to detect patterns and anomalies that signal potential threats (even without prior knowledge of specific attack signatures). This adaptability improves the detection of both known and unknown threats, providing a robust framework for identifying new and evolving cyberattacks. Furthermore, the adaptive learning process inherent in ML models allows continuous updates and improvements as new data becomes available, reducing the need for manual intervention and regular system updates.

ML models excel at analyzing behavioral patterns and network traffic, detecting deviations from normal activities that could indicate a security threat. This ability is particularly valuable in identifying sophisticated attacks that are designed to blend in with legitimate activity or exploit unknown vulnerabilities. By processing and analyzing large volumes of data at high speed, ML enables faster and more effective responses to potential incidents, minimizing the opportunity for attackers to exploit vulnerabilities.

Deep learning, a subset of ML, is especially effective in detecting malware and other advanced cyber threats due to its ability to process complex data structures. The self-learning and adaptive capabilities of deep learning models mean that as cyber threats evolve, so too do the models' abilities to detect and neutralize them.

The transition from traditional cybersecurity methods to ML-based approaches is more than just a technological advancement (it's a necessary adaptation to the increasingly complex and rapidly changing cyber threat landscape).

Vulnerability Landscape in Machine Learning Algorithms

Vulnerabilities in ML algorithms are generally associated with three main categories of attacks: those aimed at extracting training data, those targeting the ML models themselves, and those designed to trigger misclassification errors [41, 42].

One of the most prominent threats is membership inference attacks, where attackers create "shadow" models to extract PII from the training data. These attacks exploit differences in the model's responses to trained versus untrained data, posing significant risks in both open (white-box) and closed (black-box) model scenarios. This can lead to the exposure of sensitive information, especially when the model behaves differently depending on whether the input data was part of the training set.

Data poisoning is another critical threat. In this scenario, malicious data is intentionally introduced into the training set, distorting the ML model's learning process. This manipulation can result in incorrect predictions or classifications, undermining the integrity of the model. These attacks test the model's resistance to external manipulation, highlighting the importance of robust data validation and monitoring processes.

The growing use of ML as a Service (MLaaS) introduces additional vulnerabilities, particularly when outsourcing ML training. Sensitive training data might be exposed or altered by adversaries, making secure data handling and processing protocols essential to protect against unauthorized access or manipulation in these environments.

Further complicating the security landscape are pattern inversion and extraction attacks, which seek to reconstruct or reveal sensitive information from training or input data. Similarly, model inversion attacks focus on embedding secret information into the model's parameters, potentially leading to unintended disclosure of private data. These attack vectors blur the line between data privacy and model security, challenging traditional cybersecurity approaches.

To defend against these multifaceted threats, a range of defense strategies is necessary. These include:

- Anonymization and encryption techniques to protect sensitive data;
- The use of secure multi-party computing environments to safeguard collaborative learning efforts;
- Applying differential privacy measures to maintain data anonymity while ensuring model utility;
- Implementing knowledge distillation, which transfers knowledge to more compact models to reduce attack surfaces;
- Using anomaly detection mechanisms to identify abnormal behavior in the model, signaling potential security breaches.

Confronting the vulnerabilities inherent in ML algorithms requires a multi-layered defense strategy that addresses both data privacy and model security. This approach not only mitigates risks from direct attacks but also enhances the overall robustness of ML systems against an evolving threat landscape.

Ensuring the implementation of comprehensive security measures is crucial, especially in applications handling sensitive or private data. Doing so preserves the integrity and reliability of ML-driven solutions, protecting both the models and the data they rely on.

Legal and Risk Considerations in the Integration of AI in Healthcare Systems
As AI and data-driven technologies become increasingly embedded in healthcare practices, the need to protect both commercial interests and patient data from cyber threats has intensified [43]. The potential for cyberattacks to compromise sensitive patient information, disrupt healthcare operations, and endanger patient safety underscores the importance of implementing strong cybersecurity measures within healthcare systems.

From an ethical perspective, the integration of AI in healthcare requires a careful examination of its implications for all stakeholders. Ethical considerations must emphasize awareness, transparency, and accountability to navigate the complex impact of digital technologies on healthcare delivery. These principles are essential for maintaining trust in AI applications, ensuring they align with the overarching goals of patient care and privacy protection. Many users remain unaware of the extent to which their apps and devices collect, analyze, and share data, often inadvertently exposing themselves to privacy risks. This issue is exacerbated by the growing demand for remote monitoring systems and health monitoring applications using wearable devices. While these technologies enhance patient care, they also heighten the risk of exploiting personal data, including sensitive health information.

To mitigate these risks, healthcare systems must invest in diverse and robust AI algorithms, reducing reliance on single-solution frameworks that may present security vulnerabilities. Diversifying AI frameworks enhances system resilience against cyber threats, contributing to the overall security of health data and infrastructure.

The integration of AI and big data into healthcare brings significant legal, ethical, and security considerations to the forefront. Addressing these challenges requires a multidisciplinary approach, combining robust cybersecurity practices, ethical oversight, and legal compliance. As healthcare continues to embrace technological advancements, balancing the benefits of AI and big data with the need to protect patient privacy and ensure system security will be crucial for the sustainable and responsible development of healthcare services.

Ethical Imperatives in the Advancement of ML and AI in Healthcare

Maximizing the potential of ML and AI in healthcare is not without its challenges. Key obstacles include non-standardized medical records, limited access to well-structured datasets, and stringent legal and ethical mandates focused on patient privacy protection. Navigating these challenges, especially the ethical considerations inherent in legal compliance, is crucial for the successful integration of ML and AI technologies into healthcare systems. One of the central issues is developing methods for safe and ethical data sharing that balance the innovative applications of AI in healthcare with the critical importance of maintaining patient confidentiality. The adoption of AI technologies must be accompanied by security measures and ethical guidelines to mitigate risks (particularly those associated with model training and testing phases, where sensitive data is frequently involved).

From a technical standpoint, ensuring ethical access to health data requires strict control mechanisms. One key approach is the use of Role-Based Access Control (RBAC) within healthcare IT infrastructures, which limits data access to authorized personnel only, ensuring ethical standards are upheld. Moreover, encryption of data during transmission and storage, whether on-site or in the cloud, has become an essential technical measure to ethically safeguard sensitive patient information.

In terms of physical security, healthcare organizations are responsible for enhancing ethical safeguards for data centers, servers, and other critical data repositories. This includes ethical practices in monitoring and controlling access to these areas, ensuring that the integrity and confidentiality of health data are preserved. Regularly updating systems and software solutions, including EHRs, with the latest security measures is an ethical obligation to prevent breaches and unauthorized access.

To reinforce these ethical imperatives, it is essential to foster ethical awareness and responsibility among healthcare professionals. This can be achieved through continuous training on data protection policies and security practices. Developing and implementing ethically driven data handling policies, regularly reviewed and updated to reflect evolving regulations and technologies, is critical to minimizing the risks associated with health data security and ensuring ongoing ethical compliance.

5 Future Research Directions for Reliable AI/ML Integration

As AI and ML continue to transform the healthcare landscape, future research must address several key areas to ensure these technologies are fully optimized while safeguarding patient privacy, maintaining ethical standards, and enhancing system security with:

- Enhancing data privacy and security protocols - One of the foremost areas for future research is the development of more advanced privacy-preserving techniques in ML. Current methods such as differential privacy, anonymization, and encryption have proven to be effective, but with the increasing volume and complexity of healthcare data, new techniques are needed to ensure robust data protection. Research into privacy-preserving machine learning (PPML) and homomorphic encryption, which allows computations on encrypted data without needing to decrypt it, holds promise for further enhancing patient data security. Additionally, future research can explore how federated learning can be leveraged to enable decentralized AI model training while keeping patient data local and secure.
- Ethical AI development and decision-making in healthcare - As AI becomes more integrated into clinical decision-making, future research should focus on improving the transparency and interpretability of AI models, particularly in high-stakes healthcare environments. Researchers need to develop frameworks that balance algorithmic transparency with the performance and complexity of deep learning models, ensuring that healthcare providers can understand how AI systems arrive at specific decisions. In parallel, explainable AI (XAI) should be further explored, with an emphasis on making complex models, like neural networks, more interpretable for clinicians and patients alike. Moreover, ethical considerations around AI bias and fairness must continue to be a focal point of research. AI models are often trained on datasets that may contain inherent biases, which could lead to discriminatory outcomes in healthcare. Future research should aim to mitigate these risks by developing bias detection and correction methods that can be integrated into AI development pipelines to ensure equitable care for all patients, regardless of race, gender, or socioeconomic background.
- Improving interoperability and standardization of health data - The lack of standardization in health data formats continues to be a significant barrier to the effective use of AI in healthcare. Future research should focus on creating more seamless methods for data standardization and interoperability across different health information systems. Protocols like FHIR, HL7, and DICOM have already laid the groundwork, but further advancements are needed to ensure AI systems can efficiently interact with diverse healthcare datasets from multiple sources. Exploring the role of blockchain technology in enhancing data transparency, auditability, and security across interoperable health systems could provide new avenues for ensuring seamless and secure data exchange.
- Resilience against adversarial attacks on ML models - Future research must also focus on addressing the vulnerabilities of ML models to adversarial attacks, such as data poisoning, model inversion, and membership inference. Although existing defense mechanisms like knowledge distillation and anomaly detection provide a

layer of protection, they need to be continuously improved to keep up with increasingly sophisticated attacks. Research into adversarial machine learning should explore new techniques that can harden models against these threats while maintaining their predictive accuracy. Additionally, understanding the long-term impact of these attacks on healthcare systems and patient safety should be a priority, with new frameworks developed for continuous monitoring and real-time threat detection.

- AI Integration in clinical workflows and patient care - While AI has demonstrated potential in automating diagnostics, enhancing imaging analysis, and predicting patient outcomes, more research is needed on how to seamlessly integrate AI into clinical workflows. Future research should focus on optimizing the human-AI collaboration, ensuring that AI tools are integrated in ways that enhance rather than disrupt existing clinical practices. Understanding how clinicians interact with AI systems and identifying best practices for training healthcare professionals on AI applications will be key areas of study. Additionally, research should investigate the long-term impact of AI tools on patient care outcomes, with particular attention to patient satisfaction and the overall quality of care.
- AI governance and regulatory compliance - As the adoption of AI in healthcare continues to expand, future research must focus on developing robust governance frameworks that ensure compliance with GDPR, HIPAA, and other data protection regulations. This will require interdisciplinary research involving legal scholars, ethicists, data scientists, and healthcare providers to create comprehensive AI governance structures that address issues of accountability, consent, and liability. Exploring how AI systems can be certified for regulatory compliance without stifling innovation will be crucial, as will identifying metrics and benchmarks for AI performance, fairness, and safety in healthcare settings.
- Expanding AI applications in personalized and preventive medicine - With the advancement of personalized medicine (the tailoring of medical treatment to the individual characteristics of each patient), future research should focus on expanding AI's capabilities to analyze genomic data, lifestyle factors, and environmental influences to offer more precise and personalized treatment plans. AI's role in predictive analytics and preventive healthcare should also be explored further, with the aim of developing models that can predict the onset of diseases before symptoms appear, enabling earlier intervention and more proactive patient care.
- Ethical frameworks for remote healthcare and wearable devices - As wearable health technologies and remote monitoring systems become more widespread, future research should investigate the ethical implications of these technologies, particularly in terms of data ownership and consent. The continuous collection of personal health data via wearables introduces new risks related to data exploitation and privacy violations. Research must focus on creating ethical frameworks that ensure patients retain control over their data while still benefiting from the predictive capabilities of AI-driven health monitoring systems. Additionally, the development of secure communication protocols for wearable devices will be essential in preventing unauthorized access to sensitive health data.

With these directions for possible development, the integration of AI and ML into healthcare presents tremendous opportunities but also requires substantial research into ethical, legal, and technical challenges. By addressing these issues, future research can help ensure that AI technologies are deployed responsibly, ethically, and securely, contributing to more effective and equitable healthcare systems worldwide.

6 Conclusion

The integration of machine learning and artificial intelligence into healthcare systems offers unprecedented opportunities to enhance patient care, improve diagnostic accuracy, and streamline operational efficiency. However, as these technologies become more integral to healthcare, they introduce significant challenges related to legal, ethical, and security considerations. These challenges must be addressed carefully to ensure that the benefits of AI and ML are realized without compromising the privacy, security, and trust that underpin the healthcare sector.

One of the most critical concerns in the deployment of AI and ML in healthcare is the protection of patient privacy. Healthcare data often includes sensitive information, such as medical histories, genetic data, and personally identifiable information (PII). The risk of cyberattacks, data breaches, and unauthorized access to such data can have severe consequences, not only for individual patients but also for the healthcare system as a whole. Legal frameworks, such as the General Data Protection Regulation and the Health Insurance Portability and Accountability Act, provide a foundation for protecting patient data, but ensuring compliance with these regulations requires ongoing vigilance and adaptation as technology evolves.

The legal and risk considerations associated with AI in healthcare are multifaceted. As AI-driven technologies are increasingly integrated into healthcare systems, there is a growing imperative to protect commercial interests and safeguard patient data from cyber threats. The potential for cyberattacks to compromise patient data, disrupt healthcare operations, and jeopardize patient safety is a constant threat that demands strong cybersecurity measures. The complexity of cybersecurity in healthcare is further heightened by the fact that many healthcare providers and institutions are rapidly adopting remote monitoring systems and wearable health devices, which generate and share vast amounts of data. While these technologies improve patient outcomes, they also introduce new risks related to data security and privacy.

In addition to the legal and risk considerations, the ethical implications of AI in healthcare are profound. Ethical assessments must focus on maintaining transparency, accountability, and trust across all levels of the healthcare system. Many patients are unaware of the extent to which their data is collected, analyzed, and shared by the applications and devices they use. This lack of awareness can lead to privacy risks that are often overlooked in the push for technological advancement. Ethical frameworks are essential to ensure that AI systems not only improve healthcare delivery but also respect patients' rights to privacy and confidentiality. Healthcare providers, policymakers, and technologists must work together to establish and enforce ethical guidelines that prioritize patient welfare and uphold the core values of healthcare.

A key challenge in the ethical integration of AI and ML into healthcare is the need to ensure safe and ethical data sharing. The interoperability of health data is vital for improving patient care and outcomes, but it also raises concerns about the security and privacy of sensitive information. Data anonymization, encryption, and access control mechanisms such as Role-Based Access Control are critical technical measures to ensure that only authorized personnel can access sensitive health data. Furthermore, the importance of physical security in safeguarding data centers, servers, and other repositories of health data cannot be overstated. Ensuring that these infrastructures are secure from both cyber and physical threats is essential for maintaining the integrity of healthcare systems.

Another important aspect of AI integration in healthcare is the vulnerability landscape associated with ML algorithms. ML models are susceptible to various types of attacks, including membership inference, data poisoning, and pattern extraction. These attacks can compromise the privacy and security of training data, potentially leading to the disclosure of sensitive patient information. Defending against these threats requires a multi-layered security approach, which includes the use of differential privacy, anonymization, and encryption techniques. Additionally, knowledge distillation and anomaly detection can help identify and mitigate model vulnerabilities, ensuring that ML systems remain robust in the face of evolving cyber threats.

The legal frameworks governing health data management are crucial to ensuring that healthcare providers can safely and securely integrate AI technologies into their operations. Standards such as HL7, FHIR, DICOM, and SNOMED CT play a vital role in facilitating the secure exchange of health data between different healthcare systems. These standards promote interoperability, enabling seamless communication and data sharing while maintaining the confidentiality and integrity of patient information. In Europe, CEN/ISO EN 13606 defines a framework for the secure exchange of electronic health records, ensuring that health data can be safely shared across borders. Similarly, ISO 27001 provides guidelines for managing sensitive health data within an information security management system (ISMS), offering a systematic approach to safeguarding data privacy and security.

The advancement of AI and ML in healthcare also depends on the ability to choose the right algorithms for specific healthcare tasks. Different algorithms, from linear models to deep learning neural networks, offer unique advantages and trade-offs in terms of accuracy, complexity, and interpretability. In healthcare, interpretability is particularly important, as decisions made by AI systems can have significant implications for patient outcomes. Therefore, healthcare providers must balance the performance of ML models with the need for transparency and explainability, ensuring that AI systems can be trusted to make ethical and effective decisions.

Finaly we can conclude that the successful integration of AI and ML into healthcare systems requires a comprehensive approach that addresses legal, ethical, and security concerns. Protecting patient data, ensuring the security of healthcare infrastructures, and maintaining ethical standards are all critical to the sustainable advancement of AI in healthcare. As healthcare continues to evolve, driven by technological innovation, it is essential that all stakeholders (healthcare providers, policymakers, technologists, and patients) work together to build AI-driven solutions that are not only innovative but also secure, ethical, and trustworthy. By doing so, healthcare systems can harness the full potential of AI and ML to improve patient outcomes, optimize healthcare delivery, and ensure the long-term integrity of the healthcare ecosystem.

References

1. Epizitone, A., Moyane, S.P., Agbehadji, I.E.: A systematic literature review of health information systems for healthcare. Healthcare **11**(7). MDPI (2023)
2. Jeyaraman, M., et al.: Unraveling the ethical enigma: artificial intelligence in healthcare. Cureus **15**(8) (2023)
3. Husnain, A., et al.: AI'S healing touch: examining machine learning's transformative effects on healthcare. J. World Sci. **2**(10), 1681–1695 (2023)
4. Hassan, M., et al.: Adoption of blockchain-based artificial intelligence in healthcare. In: 2022 5th International Conference on Artificial Intelligence and Big Data (ICAIBD). IEEE (2022)
5. Ong, J., Selvam, A., Chhablani, J.: Artificial intelligence in ophthalmology: optimization of machine learning for ophthalmic care and research. Clin. Experiment. Ophthalmol. **49**(5), 413–415 (2021)
6. Umer, L., Khan, M.H., Ayaz, Y.: Transforming healthcare with artificial intelligence in Pakistan: a comprehensive overview. Pakistan Armed Forces Med. J. **73**(4), 955–963 (2023)
7. Ke, T.T., Sudhir, K.: Privacy rights and data security: GDPR and personal data markets. Manag. Sci. **69**(8), 4389–4412 (2023)
8. Kuner, C.: Protecting EU data outside EU borders under the GDPR. Common Mark. Law Rev. **60**, 77 (2023)
9. Kurapati, S., Gilli, L.: Synthetic data: a convergence between Innovation and GDPR. J. Open Access L. **11**, 1 (2023)
10. Soysal, T.: Can the right to explanation in GDPR be a remedy for algorithmic discrimination?. In: Kılıç, M., Bozkuş Kahyaoğlu, S. (eds.) Algorithmic Discrimination and Ethical Perspective of Artificial Intelligence. Accounting, Finance, Sustainability, Governance & Fraud: Theory and Application. Springer, Singapore (2024). https://doi.org/10.1007/978-981-99-6327-0_5
11. Blind, K., Niebel, C., Rammer, C.: The impact of the EU General data protection regulation on product innovation. Indust. Innov. 1–41 (2023)
12. Świtała, K.: Medical data in the digital era-legal challenges related to providing information security, applying GDPR and respecting the professional secrecy. In: 2023 46th MIPRO ICT and Electronics Convention (MIPRO). IEEE (2023)
13. Rill, J.C., et al.: Data privacy and anonymization for GDPR compliance in location-based mobility applications: a case study of G-LocON. SEATUC J. Sci. Eng. **4**(1), 46–51 (2023)
14. Di Martino, M., et al.: Revisiting identification issues in GDPR 'Right Of Access' policies: a technical and longitudinal analysis (2022)
15. Politou, E., et al.: The "right to be forgotten" in the GDPR: implementation challenges and potential solutions. Privacy and Data Protection Challenges in the Distributed Era 41–68 (2022)
16. Volpe, S., ed. Health Informatics: Multidisciplinary Approaches for Current and Future Professionals. CRC Press (2022)
17. Ahmadi, S.: Security and privacy challenges in cloud-based data warehousing: a comprehensive review. Int. J. Comput. Sci. Trends Technology (IJCST) **11** (2023)
18. Diamantopoulou, V., et al.: Eu gdpr: toward a regulatory initiative for deploying a private digital era. Modern Socio-Tech. Perspect. Privacy **427** (2022)
19. Kayaalp, M.: Patient privacy in the era of big data. Balkan Med. J. **35**(1), 8–17 (2018)
20. National Research Council: Networking health: prescriptions for the internet (2000)
21. Richter, G., et al.: Patient views on research use of clinical data without consent: legal, but also acceptable?. Europ. J. Hum. Genetics **27**(6), 841–847 (2019)
22. Shalowitz, D.I., Miller, F.G.: Disclosing individual results of clinical research: implications of respect for participants. JAMA **294**(6), 737–740 (2005)

23. Nikolovski, S.: Authentication of users in heterogeneous information environment. Horizons Series **B4** (2017)
24. Nikolovski, S., Mitrevski P.: On the requirements for successful business continuity in the context of disaster recovery. In: 2022 57th International Scientific Conference on Information, Communication and Energy Systems and Technologies (ICEST). IEEE (2022)
25. Jeon, S.J., Seok Go, M., Namgung, J.H.: Use of personal information for artificial intelligence learning data under the Personal Information Protection Act: the case of Lee-Luda, an artificial-intelligence chatbot in South Korea. Asia Pacific Law Rev. **31**(1), 55–72 (2023)
26. https://www.ecfr.gov/current/title-45. Accessed 04 Feb 2024
27. Choi, Y.B., Williams, C.E.: A HIPAA security and privacy compliance audit and risk assessment mitigation approach. In: Research Anthology on Securing Medical Systems and Records, pp. 706–725. IGI Global (2022)
28. Shumin, B.: A Digital Healthcare Application for Patient Monitoring and Assessment (2022)
29. Pillai, C.S., et al.: Patient data privacy, protected health information, and ethics of real-world evidence. In: Real-World Evidence in a Patient-Centric Digital Era, pp. 51–72. Chapman and Hall/CRC (2022)
30. https://www.ecfr.gov/current/title-45/subtitle-A/subchapter-C/part-164. Accessed 04 Feb 2024
31. Alowais, S.A., et al.: Revolutionizing healthcare: the role of artificial intelligence in clinical practice. BMC Med. Educ. **23**(1), 689 (2023)
32. Guo, C., Chen, J.: Big data analytics in healthcare. Knowledge Technology and Systems: Toward Establishing Knowledge Systems Science, pp. 27–70. Springer, Singapore (2023)
33. Dhany, H.W., Sutarman, S., Izhari, F.: Exploratory Data Analysis (EDA) methods for healthcare classification. J. Intell. Decis. Supp. Syst. (IDSS) **6**(4), 209–215 (2023)
34. Tariq, Dr T.H.S., Aithal, P.S.: Visualization and explorative data analysis. Int. J. Enhanced Res. Sci. Technol. Eng. **12**(3), 11–21 (2023)
35. Rubinger, L., et al.: Machine learning and artificial intelligence in research and healthcare. Injury **54**, S69–S73 (2023)
36. Jiwani, N., Gupta, K., Whig, P.: Machine learning approaches for analysis in smart healthcare informatics. In: Machine Learning and Artificial Intelligence in Healthcare Systems, pp. 129–154. CRC Press (2023)
37. Loftus, T.J., et al.: Ideal algorithms in healthcare: explainable, dynamic, precise, autonomous, fair, and reproducible. PLOS Digital Health **1**(1), e0000006 (2022)
38. Shah, A., Delima, N., Patra, S.: Anonymizing personally identifiable information in stored data. U.S. Patent Application No. 17/406, 622 (2023)
39. Apruzzese, G., et al.: The role of machine learning in cybersecurity. Digital Threats Res. Pract. **4**(1), 1–38 (2023)
40. Bharadiya, J.: Machine learning in cybersecurity: techniques and challenges. Euro. J. Technol. **7**(2), 1–14 (2023)
41. Chivukula, A.S., et al.: Adversarial Machine Learning: Attack Surfaces, Defence Mechanisms, Learning Theories in Artificial Intelligence. Springer Nature (2023)
42. Chen, H., Ali Babar, M.: Security for machine learning-based software systems: a survey of threats, practices and challenges. ACM Comput. Surv. (2022)
43. Naik, N., et al.: Legal and ethical consideration in artificial intelligence in healthcare: who takes responsibility?. Front. Surg. **9**, 266 (2022)
44. Sadhu, P.K., et al.: Prospect of internet of medical things: a review on security requirements and solutions. Sensors **22**(15), 5517 (2022)

Artificial Intelligence-Based Anomaly Detection Traffic Patterns Associated with the Ddos Attack in Software-Defined Networking

Stefan Ž. Biševac[1](✉), Aleksandar Atanasijević[2], Aleksandar Jokić[1], and Marko Šarac[1]

[1] Singidunum University, Danijelova 32, 11000 Belgrade, Serbia
stefan.bisevac@kg.ac.rs, aleksandar.jokic.22@singimail.rs, msarac@singidunum.rs

[2] Faculty of Engineering, University of Kragujevac, 6 Sestre Janjica Street, 34000 Kragujevac, Serbia
aatanasijevic@kg.ac.rs

Abstract. Software-Defined Networking (SDN) has emerged as a revolutionary paradigm for network management, offering dynamic control and programmability. However, the open and programmable nature of SDN introduces new challenges, particularly in terms of security. Abnormal traffic does not necessarily mean an attack, but the detection of anomalous patterns and their mitigation in abnormal traffic is a critical aspect in terms of effective protection of SDN entities. In this context, artificial intelligence (AI) techniques have shown great promise in enhancing network security by enabling the identification and mitigation of abnormal traffic patterns. This paper explores the use of AI-based anomaly detection algorithms in SDN environments to detect and mitigate network attacks patterns in abnormal traffic effectively. Initially, specific features were extracted from SDN environment under normal conditions and during DDoS attack scenarios to create a dataset that contains over 1.000.000 records which is a substantial amount for conducting analysis. Dataset features are extracted from the IP header from each packet that passes through the SDN datapath, taking into account that different parameters are important for different protocols. Since different protocols differ in their structure, the entire data set were divided into four sub-datasets according to their specific characteristics (ARP, ICMP, TCP, UDP). As the basis of AI, we employed a variety of machine learning (ML) techniques, including Support Vector Machine (SVM), Decision Tree (DT), Logistic Regression (LR), Random Forest (RF), and Naive Bayes (NB), to train and classify different types of anomalies. The results we have obtained underscore the ability of artificial intelligence to effectively recognize and detect anomalies in abnormal traffic. Our results highlight the effectiveness of artificial intelligence in recognizing and detecting anomalies within abnormal traffic, paving the way for improved SDN security measures. Moving forward, further research will focus on refining AI models, exploring advanced anomaly detection algorithms, and implementing real-time response mechanisms to swiftly mitigate identified threats, ensuring robust and adaptive SDN security.

© The Author(s), under exclusive license to Springer Nature Switzerland AG 2025
N. Filipović (Ed.): AAI 2024, LNNS 1446, pp. 314–324, 2025.
https://doi.org/10.1007/978-3-031-99201-8_31

Keywords: Software Defined Networking · Machine Learning · Artificial Intelligence · Abnormal Traffic · Anomaly Detection

1 Introduction

In the current digital arena, computer networks serve as the backbone of modern communication, facilitating the exchange of data between different devices and systems. As the demands placed on networks continue to grow, traditional networking approaches struggle to keep up with the dynamic demands of modern applications and services. This is where software-defined networking (SDN) emerges as a transformative paradigm, offering unparalleled flexibility, agility and efficiency in network management and operation. At its core, SDN represents a fundamental change in the way networks are designed, deployed and managed. Unlike traditional network architectures, which rely on the tightly integrated connection of control and data forwarding functions within network devices, SDN separates these functions, thereby centralizing control and enabling programming through software-defined interfaces. This abstraction of network control from the underlying infrastructure enables dynamic and centralized management of network resources, paving the way for greater automation, scalability, and customization.

Along with the abstraction and separation of the control from the forwarding plane come challenges that SDN does not bypass. One of the most critical challenges facing SDN environments is the detection and mitigation of abnormal traffic patterns between those two planes. Abnormal traffic, often indicative of malicious activity or network anomalies, can pose serious threats to network performance, reliability, and security. Detecting and mitigating such abnormal traffic is paramount to ensuring the integrity and resilience of SDN infrastructures. This paper should show a solution in which way we can apply certain ML algorithms in order to protect the network from unwanted load.

2 Related Work, Platform and Machine Learning Algorithms

This chapter reviews the literature related to SDNs and machine learning. In addition, there is a brief overview of the SDN platform and ML algorithms used in this research.

2.1 Literature Review

Analyzing the available literature, one comes across a large number of topics dealing with the problem of identifying potential attacks in computer networks. More specifically, the topic of software-defined networks and the integration of different AI algorithms is the subject of research by many scientists who managed to unite and integrate computer networks and machine learning, as a new direction in the development of artificial intelligence.

The inspiration for starting the research was a topic covered by colleagues Mohammad Reza Parsaei et al. [1]. In addition to showing an approach to traffic categorization using neural networks and the success of these networks in anomaly detection, they also defined one of the ways to integrate the ML module, which is used to detect suspicious traffic. By training ML models with data from a previously recorded data set, Saurav Nanda et al. have developed several algorithms to try to predict which host will be attacked next [2]. Some authors have proposed anomaly-based methods employing machine learning algorithms on uninterrupted data streams for identifying intrusions within an SDN setting [3]. This strategy seeks to tackle the primary hurdles faced in constructing an anomaly-based system employing machine learning algorithms. In order to delineate these anomalies, they've scrutinized a specific form of DDoS assault categorized as an infrastructure attack, which takes into account the repercussions of both bandwidth and resource depletions. Several research papers have delved into the security challenges posed by deep learning-based Software-Defined Networking (SDN) in a native SDN environment. These studies focus on adapting deep learning intrusion detection modules specifically for SDN environments [4–6]. The study examines the issue at hand and suggests employing various machine learning algorithms, such as Support Vector Machine, Multi-Layer Perceptron, Decision Tree, and Random Forest. The goal is to categorize Distributed Denial of Service (DDoS) attacks in a simulated Software-Defined Networking (SDN) setup. The research endeavors to demonstrate the effectiveness of Support Vector Machine (SVM), Multilayer Perceptron (MLP), Decision Tree, and Random Forest in analyzing the problem and recommends their utilization for this purpose. The goal is to classify Distributed Denial of Service (DDoS) attacks within a simulated Software Defined Network (SDN) environment [7].

The focus of the aforementioned works is the processing of collected datasets by applying different machine learning algorithms as the basis of AI. This paper aims to explain the method of collecting important data and the way in which the dataset is formed that will be processed in order to understand the possible patterns in DDoS traffic. Furthermore, the paper aims to show the way of training the model and its application in the production environment.

2.2 SDN Platform

The platform on which the research is carried out unites Open vSwitch as one of the datapath representatives, Ryu python-based software controller located in the control plane domain and OpenFlow protocol that connects control and forwarding plane.

OpenFlow is a foundational protocol in Software-Defined Networking (SDN) that allows direct manipulation of the forwarding plane of network devices like switches and routers. By decoupling the control plane from the data plane, OpenFlow enables centralized control of network traffic through an SDN controller, which communicates with OpenFlow-compatible devices using standardized commands. This separation facilitates more flexible and dynamic network management, promoting innovations in network configurations, policy implementations, and overall efficiency [8].

Ryu is an open-source SDN controller platform designed to facilitate the development and deployment of SDN applications [9]. It provides a flexible and programmable framework for managing network resources by offering comprehensive APIs and libraries that support various SDN protocols, including OpenFlow. Written in Python, Ryu is highly extensible and user-friendly, making it accessible for developers to create custom network applications and services. Its modular architecture allows for easy integration with other SDN components, enabling seamless control over network devices and traffic flows.

Open vSwitch (OVS) [10] is an open-source virtual switch designed to provide high-performance networking capabilities for virtualized environments. It facilitates seamless integration between virtual machines and the physical network, supporting a wide range of network protocols and standards. OVS is highly programmable, enabling network administrators to automate and customize network configurations through standard interfaces such as OpenFlow. Its flexibility, scalability, and robust performance make Open vSwitch a critical component for modern cloud infrastructures and SDN deployments, empowering users to build dynamic, efficient, and secure virtual networks.

2.3 Machine Learning Algorithms Used in Research

Support Vector Machine (SVM) stands as a prevalent supervised machine learning tool employed for classification and regression, with a primary focus on classification tasks. Its fundamental concept revolves around identifying the optimal hyperplane that effectively segregates data points belonging to distinct classes within the feature space. This optimization is attained by maximizing the margin, denoting the separation distance between the hyperplane and the nearest data points from any class, known as support vectors. SVM exhibits versatility in handling both linear and non-linear classification through the utilization of kernel functions like polynomial and radial basis function, enabling data transformation into higher-dimensional spaces conducive to linear separation. The algorithm is effective in high-dimensional spaces and is robust to overfitting, making it suitable for applications like image and text classification, bioinformatics, and handwriting recognition. The performance of SVM depends on the careful selection of the kernel function and regularization parameter [11].

Another supervised algorithm, the Decision Tree, serves purposes in both classification and regression tasks. Its methodology involves iteratively dividing the data into subsets based on input feature values, constructing a tree-like representation of decision pathways. Each node within the tree corresponds to a feature, while each branch signifies a decision rule, and every leaf node indicates an outcome or class label. The objective is to devise a model capable of predicting the target variable's value by discerning straightforward decision rules from the data's features. Decision Trees offer simplicity in comprehension and interpretation, resembling human decision-making processes. They accommodate both numerical and categorical data and necessitate minimal data preprocessing. Nevertheless, they are susceptible to overfitting, particularly when excessively deep and intricate. Techniques such as pruning, which entails eliminating sections of the tree lacking added predictive power for target variables, can mitigate overfitting. Furthermore, Decision Trees may exhibit instability, wherein minor data fluctuations could yield entirely disparate tree structures [11].

The next supervised machine learning technique to discuss is Logistic Regression, primarily applied in binary classification tasks. Its function is to estimate the probability that a given input belongs to a specific class. Unlike linear regression, which forecasts continuous outcomes, logistic regression predicts the probability of categorical outcomes utilizing the logistic function. This function maps predicted values onto probabilities, ensuring they lie between 0 and 1. The model is trained by determining the most suitable parameters that minimize the variance between the forecasted and actual outcomes through a process known as maximum likelihood estimation. Logistic Regression boasts simplicity in implementation, efficiency in training, and interpretability, rendering it a favored option for initial classification endeavors. It exhibits proficiency when classes are linearly separable but may encounter challenges in handling more intricate, non-linear relationships without employing transformations or kernel methods [12].

Random Forest represents an ensemble learning technique utilized for both classification and regression tasks. Its operation involves the creation of numerous decision trees during training and then determining the mode of classes (classification) or the mean prediction (regression) derived from these individual trees. Each tree within the forest is trained on a distinct subset of the data, generated through random sampling with replacement (bootstrap sampling), and considers a random subset of features for node splitting, thereby instilling diversity among the trees. This injection of randomness aids in curbing overfitting while enhancing the model's ability to generalize to new, unseen data. Renowned for its high accuracy, resilience to noise, and efficacy in managing extensive datasets with heightened dimensionality, Random Forest also furnishes an estimate of feature importance, facilitating the comprehension of underlying data patterns. Nonetheless, this method can be computationally demanding and less interpretable compared to a singular decision tree. Despite these complexities, Random Forest finds widespread application across various domains such as image and speech recognition, financial modeling, and bioinformatics, attributed to its adaptability and robust performance [11].

Naive Bayes represents a group of probabilistic algorithms rooted in the application of Bayes' theorem under the assumption of feature independence. Primarily tailored for classification tasks, it garners praise for its simplicity and efficiency. Despite its straightforwardness, Naive Bayes frequently surpasses more intricate models, particularly in scenarios involving extensive datasets and text classification dilemmas like spam detection and sentiment analysis. The model computes the posterior probability for each class based on the input features and then selects the class with the highest probability. Various types of Naive Bayes classifiers exist, including Gaussian, Multinomial, and Bernoulli, each tailored for distinct data types. Gaussian Naive Bayes presumes that continuous feature values adhere to a Gaussian distribution. Multinomial Naive Bayes finds utility in document classification, where features denote word frequencies. Bernoulli Naive Bayes, on the other hand, suits binary feature vectors. Despite the independence assumption often diverging from reality, Naive Bayes remains effective and enjoys widespread usage owing to its rapidity, scalability, and adeptness across numerous applications [11].

3 Our Proposal

The following section describes the existing packet flow, the methods used on the controller to manipulate incoming packets, as well as the proposed solution regarding the integration of machine learning models into the SDN platform.

3.1 Existing Packet Flow in SDN Controller

The Ryu, python-based software controller, was used for this research. In order to improve the existing application that realizes the basic functionality of each Layer 2 switch with new machine learning methods, it is necessary to show how the Ryu controller manages incoming and regulates outgoing packets. The working principle is shown in Fig. 1.

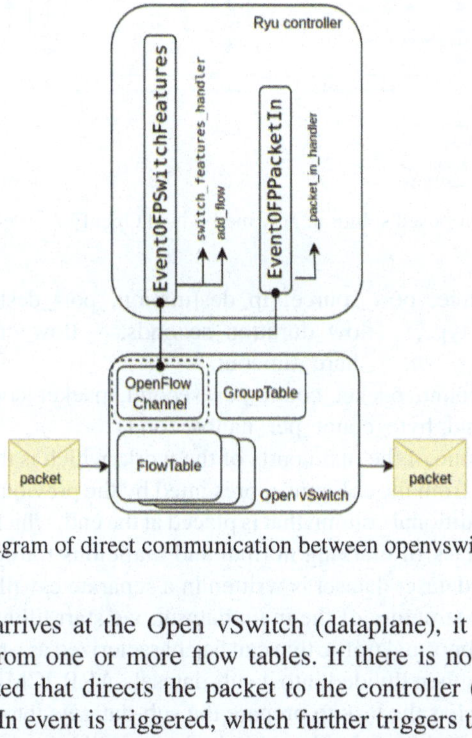

Fig. 1. Block diagram of direct communication between openvswitch and controller

When a packet arrives at the Open vSwitch (dataplane), it is checked to match one of the records from one or more flow tables. If there is no match, then a table-miss record is initiated that directs the packet to the controller (control plane). Then the EventOFPPacketIn event is triggered, which further triggers the packet_in_handler method. Its task is to record the original MAC address and FLOOD the given packet on all ports except the one from which the packet arrived. If there is a response to a given packet, the controller records it on the switch using the global Add_flow function to prevent the controller from being called again for the same activity. If there is no response, the controller discards the given packet.

3.2 Packet Flow Modification in SDN Controller

Based on the existing packet processing algorithm in the Ryu controller, we introduce the changes given in Fig. 2. This gives us the ability to record and at the same time classify packets, after which we can do further analysis and pattern learning in order to detect abnormal traffic that may be part of DDOS attacks.

One of the predefined OpenFlow event classes named EventOFPFlowStatsReplay has the method _flow_stats_reply_handler which we improved to perform extraction of the following parameters from the packet header: timepoint, data_packet_id,

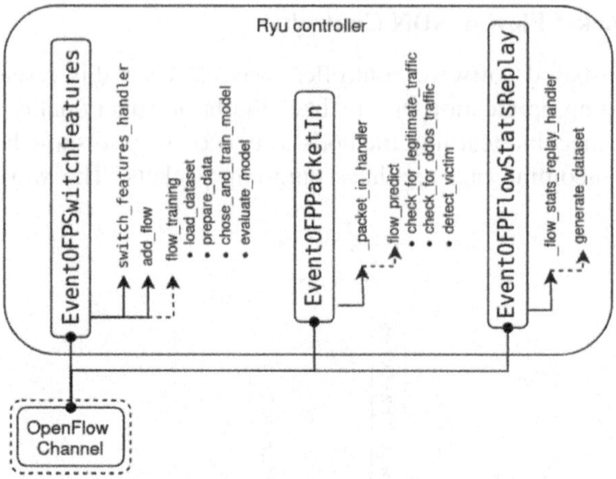

Fig. 2. Proposed solution: new methods in OpenFlow event classes

data_flow_id, ip_source, port_source, ip_destination, port_destination, ip_protocol, icmp_code, icmp_type, flow_duration_seconds, flow_duration_nanoseconds, idle_timeout, hard_timeout, communication_flags, packet_count, byte_count, packet_count_per_second, packet_count_per_nanosecond, byte_count_per_second, byte_count_per_nanosecond.

Here we come to one of the main parts of the work, which is the creation of dataset. The dataset is formed from the columns represented by the previous extraction of header parameters with an additional column that is placed at the end, which is the status column. This column has the task of marking normal and malicious traffic patterns. By calling the method generate_dataset dataset is written in a separate csv file.

After forming the structure of the main dataset, we start filling it with normal type traffic and then with abnormal traffic that can be characterized as a potential DDoS attack pattern. The dataset is then divided into 4 sub-datasets ARP, ICMP, TCP and UDP. The flow_training method has the task to prepare the sub-datasets into test and training data and then train the desired models. We include the trained models in the controller and monitor their success in detecting unusual patterns in abnormal traffic conditions by flow_predict method.

4 Case Study

The SDN environment in which state recording and measurements are realized consists of two virtual machines on the University cluster which has cluster of 5 servers with 24 x Intel(R) Xeon(R)E5–2650 v4 @ 2.20GHz processors. One of the machines is intended for hosting the controller, while on the other a mininet with 6 switches and 18 hosts is executed, so that 3 hosts are connected to each switch. The switches are interconnected in one row. After establishing the topology, a python web server is started on one of the hosts, which will be used to test the detection of potential malicious patterns. Case study protocols are precisely those that were used to create the dataset, with the fact that in this case a different tool is used for testing a malicious attack DoSinator [13].

The study begins by training the models SVM, DT, LR, RF and NB on sub-datasets previously recorded and related to ARP, ICMP, TCP and UDP protocols. The time for which the training is performed in relation to the number of records in the dataset is given in Table 1.

Table 1. Learning and Predicting time

Protocol/records	SVM	DT	LR	RF	NB
ARP (470412)	0:02:02.11s	0:00:16.33s	0:00:07.09s	0:00:10.60s	0:00:03.56s
ICMP (598744)	0:02:20.40s	0:00:18.70s	0:00:08.76s	0:00:12.33s	0:00:05.32s
TCP (1049468)	0:03:59.70s	0:00:24.91s	0:00:11.30s	0:00:12.31s	0:00:07.42s
UDP (609911)	0:02:34.13s	0:00:11.58s	0:00:05.90s	0:00:06.04s	0:00:04.40s

Although the results show that Naive Bayes is a fast-learning model, this method still does not give the best results for smaller datasets as we can see in the following diagram. For the models defined above, the evaluation of the basic parameters given in Fig. 3 and Table 2 was performed. It can be seen from the attached that for the defined dataset, the best models are based on the decision trees, i.e. DT and RF.

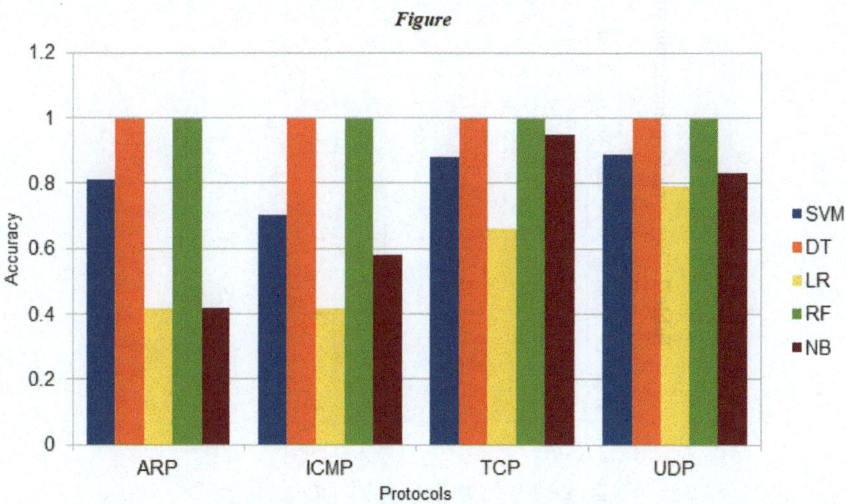

Fig. 3. Performance evalutaion: Accuracy per ML method in relation to the protocol dataset

From the previous table, we see that some values are 0. This means that the data model we have chosen is not suitable for training the given model. Here it is important to point out that in such a situation one can approach the correction of the dataset or the correction of the model. Models based on the decision tree (DT, RF) performed best for the given dataset. Given that this paper should demonstrate the power of AI

Table 2. Basic performance evaluation – precision, recall, f1_score

	LR-precision	LR-recall	LR-F1_Score	RF-precision	RF-recall	RF-F1_Score	NB-precision	NB-recall	NB-F1_Score	DT-precision	DT-recall	DT-F1_Score	SVM-precision	SVM-recall	SVM-F1_Score
Normal ARP	0.41	1	0.6	**1**	**1**	**1**	0.47	0.06	0.12	**1**	**1**	**1**	0.61	1	0.8
DDoS ARP	0	0	0	**1**	**1**	**1**	0.55	0.91	0.69	**1**	**1**	**1**	0	0	0
Normal ICMP	0.42	1	0.59	**1**	**1**	**1**	0.49	0.08	0.14	**1**	**1**	**1**	0.51	1	0.81
DDoS ICMP	0	0	0	**1**	**1**	**1**	0.59	0.94	0.72	**1**	**1**	**1**	0	0	0
Normal TCP	0	0	0	**1**	**1**	**1**	0.61	1	0.76	**1**	**1**	**1**	0.88	1	0.93
DDoS TCP	0.79	1	0.88	**1**	**1**	**1**	1	0.93	0.96	**1**	**1**	**1**	0	0	0
Normal UDP	0	0	0	**1**	**1**	**1**	0.54	1	0.7	**1**	**1**	**1**	0.88	1	0.91
DDoS UDP	0.66	1	0.8	**1**	**1**	**1**	1	0.57	0.72	**1**	**1**	**1**	0	1	0

methods in anomaly detection, for further testing one algorithm that has extremely good performance (DT) and one that does not (NB) will be deliberately used. At the end, the integration of these methods into the controller is performed as explained earlier and the topology of hosts and switches is started.

As with any other ML-related project, it is extremely important to scan each input flow and make a prediction based on it. For the purpose of final testing, we focus on launching a DDoS attack and analyze the state on the controller. Every packet that arrives at the controller goes through a prediction method, based on which it is decided whether the given packet is legitimate or malicious. The numbers show the statistics given in Tables 3 and 4.

Table 3. Decision Tree model report in detecting DDoS attacks

	Protocol	Attack	pkt_sent	pkt_detected
DT	ARP	ARP-spoof	522	522
	ICMP	Ping-flood	3550	3550
	TCP	Syn-flood	10155	10155
	UDP	UDP-flood	600	600

Table 4. Naive Bayes model report in detecting DDoS attacks

	Protocol	Attack	pkt_sent	pkt_detected
NB	ARP	ARP-spoof	522	287
	ICMP	Ping-flood	3550	2013
	TCP	Syn-flood	10155	10151
	UDP	UDP-flood	600	554

The tables shown above shows the relationship between the protocol, the type of attack, the packets sent during the attack and the number of detected packets. It is concluded that Tables 3 and 4 is in direct relation with Table 2, that is, ML methods that have good performance show a better ability to detect anomalies.

5 Conclusions

This work aimed to provide a comprehensive method for creating a dataset and treating malicious patterns by applying various machine learning algorithms. The applied techniques demonstrate that the proper creation of the dataset and the careful selection of the ML algorithm are crucial for the successful detection of potential DDoS patterns. The process involved meticulous feature extraction, data preprocessing, and algorithm evaluation to ensure that the resulting models were both accurate and efficient in identifying anomalous traffic. The result we obtained from this work shows that decision-tree based ML algorithms are the best for detecting and mitigating anomalies, which we also support with tests in which we showed the maximum efficiency of the decision tree or random forest algorithm, compared to Navie Bayes, which has a detection efficiency of about 60- 80% depending on the protocol being processed. Future research should focus on enhancing this solution by integrating advanced time-series analysis methods, developing continuously updating datasets, and implementing novel AI-based solutions that operate in real-time. These improvements will further refine the detection capabilities and enable more robust protection against evolving DDoS threats, ensuring the security and resilience of SDN environments.

References

1. Parsaei, M.R., et al.: Network traffic classification using machine learning techniques over software defined networks. Int. J. Adv. Comput. Sci. Appl. **8**(7) (2017)
2. Nanda, S., et al.: Predicting network attack patterns in SDN using machine learning approach. In: IEEE Conference on Network Function Virtualization and Software Defined Networks (NFV-SDN) (2016)
3. Ribeiro, A., et al.: Anomaly detection technique for intrusion detection in SDN environment using continuous data stream machine learning algorithms. In: IEEE International Systems Conference (SysCon), pp. 1–7 (2021)
4. Isa, M.M., Mhamdi, L.: Native SDN intrusion detection using machine learning. In: IEEE Eighth International Conference on Communications and Networking (ComNet) (2020)
5. Tang, T.A., et al.: Deep recurrent neural network for intrusion detection in SDN-based networks. In: 4th IEEE Conference on Network Softwarization and Workshops NetSoft, pp. 462–469 (2018)
6. Shone, N., et al.: A deep learning approach to network intrusion detection. IEEE Trans. Emerg. Topics Comput. Intell. **2**(1), 41–50 (2018)
7. Santos, R.S., et al.: Machine learning algorithms to detect DDoS attacks in SDN. Concurr. Comput. Pract Exper (2020)
8. Lara, A., Kolasani, A., Ramamurthy, B.: Network innovation using OpenFlow: a survey. IEEE Commun. Surv. Tutor. **16**(1), 493–512 (2014)
9. https://osrg.github.io/ryu-book/en/html/, April 2024
10. https://docs.openvswitch.org/en/latest/topics/, April 2024
11. Kotsiantis, S.B., et al.: Supervised machine learning: a review of classification techniques. Emer. Artif. Intell. Appl. Comput. Eng. **160**, 3–24 (2007)
12. Ullah, M., et al.: Detecting distributed Denial of service attack using Logistic regression and SVM methods. In: The 2019 International Conference on Applied Information Technology and Innovation (ICAITI 2019) (2019)
13. https://github.com/HalilDeniz/Dosinator, April 2024

The Role of Artificial Intelligence in Transforming Hotels in Developing Countries (DCs) with a Special Focus on the Republic of Serbia

Tamara Gajić[1](✉) and Andrea Ivanišević[2]

[1] Geographical Institute Jovan Cvijić, SASA, Belgrade, Republic of Serbia
tamara.gajic.1977@gmail.com

[2] Faculty of Technical Sciences, University of Novi Sad, Novi Sad, Republic of Serbia
andrea.ivanisevic@gmail.com

Abstract. This study explores the role of artificial intelligence (AI) in transforming hotels in developing countries (DCs), with a special focus on the Republic of Serbia. The research was conducted among hotel management in the Republic of Serbia, encompassing 134 respondents. The results indicate a high level of support for the implementation of AI, highlighting its potential to enhance efficiency, guest experience, and competitiveness. The study emphasizes the importance of technological innovation in the hospitality industry and provides insights into the industry's readiness to adopt and implement AI. This research contributes to understanding current trends and future perspectives of AI application in the hospitality sector, with particular attention to the challenges and opportunities in developing countries such as Serbia. The findings offer useful guidelines for hotel managers and decision-makers in the hospitality sector, as well as for researchers interested in the application of new technologies in tourism and hospitality.

Keywords: AI · hospitality · developing countries (DCs) · Republic of Serbia

1 Introduction

The hospitality industry is increasingly integrating artificial intelligence (AI) to enhance operational efficiency, improve guest experiences, and promote sustainability (Tihanyi and DeCelles, 2021; Gajić et al. 2024a, 2024b). This transformation is particularly significant in developing countries, where technology can play a crucial role in overcoming various challenges. The Republic of Serbia serves as a good example of a country facing numerous opportunities and challenges in applying AI in the hotel industry (Petrović et al. 2023a, 2023b). The application of artificial intelligence in hotels includes automating reservations, personalizing guest services, optimizing pricing, and enhancing security (Tripura and Avi, 2021). Such innovations not only increase business efficiency, but also improve the overall guest experience, which is crucial for competitiveness in the modern tourism market (Pavlović et al. 2021; Wellman et al. 2023).

© The Author(s), under exclusive license to Springer Nature Switzerland AG 2025
N. Filipović (Ed.): AAI 2024, LNNS 1446, pp. 325–343, 2025.
https://doi.org/10.1007/978-3-031-99201-8_32

The aim of this research is to examine the role of artificial intelligence in transforming hotels in the Republic of Serbia. The research will analyze hotel management's perceptions and readiness to implement AI, as well as identify key factors influencing the successful implementation of this technology. Although there are many studies exploring the application of AI in the hospitality industry, the lack of research focused on developing countries, especially Serbia, represents a significant gap in the literature. Most of the existing research focuses on developed countries, leaving the challenges and opportunities specific to developing countries underexplored. Therefore, the goal is to fill these gaps in the literature with the results of this research. This research is innovative because it provides a comprehensive analysis of the specific challenges and opportunities for applying AI in the hotel industry in Serbia, a developing country with a unique economic, cultural, and infrastructural context. Additionally, the research will offer practical guidelines for hotel operators and decision-makers in the hospitality sector. The results of this research will contribute to a better understanding of current trends and future perspectives of AI application in the hospitality sector, with a particular focus on the challenges and opportunities in developing countries such as Serbia. Furthermore, the research will provide valuable insights for hotel managers, policymakers, and researchers interested in the application of new technologies in tourism and hospitality.

Considering all the above, the research will include an analysis of perceptions, readiness, and key factors for AI implementation among hotel management and staff across Serbia. It is expected that this research will provide useful guidelines for improving the application of AI in the hotel industry of Serbia and other developing countries.

2 Theoretical Background

Artificial intelligence (AI) has a wide range of applications in the hotel industry, including automating reservations, personalizing guest services, optimizing pricing, and enhancing security (Katić et al. 2019; Xu et al. 2021; Citak et al. 2021; Sharma et al. 2022; Limna, 2023; Indora 2024). These technologies enable hotels to operate more efficiently and provide better experiences for their guests. According to Štilić et al. (2023), AI can significantly improve operational processes in hotels, enhance guest experiences, and contribute to business sustainability. This is especially important for hotels in developing countries, where resources are often limited. Aderibigbe et al. (2023) emphasize that successful AI implementation requires capacity building, public-private partnerships, and tailored policies. Their study highlights the importance of adapting AI technology to local needs and conditions. Talukder et al. (2023) explain how sentiment analysis and AI-driven marketing techniques can help hotels respond to individual guest preferences and optimize revenues. Sharma et al. (2022) point out that AI technologies ensure a safe and hygienic stay for guests, which has become crucial during the COVID-19 pandemic.

Technologies such as chatbots, virtual assistants, and predictive analytics are particularly significant in enhancing guest experiences and streamlining hotel operations (Wreford et al. 2019; Davenport et al. 2020). Infante-Moro (2021) and Marković (2020) emphasize the importance of management support, technology reliability, and security in decision-making regarding AI implementation in hotels. Said (2023) highlights the significance of AI in personalizing guest services, while Kumar (2021) and Limna (2023) provide a broader overview of AI and robotics' impact on tourism and hospitality, including challenges like job displacement and the need for privacy protection. Singh (2022) notes the growing interest in researching AI and robotics in the hotel industry, underscoring the importance of these technologies. This study explores how AI can transform the hotel industry, particularly in developing countries such as Serbia. It emphasizes the need for technological innovations and local adaptations to achieve the best results in efficiency, guest experience, and sustainability.

The application of artificial intelligence (AI) in the hotel industry can be analyzed through several key hypotheses covering different aspects of this technology (Pillai and Sivathanu, 2020; Tuomi and Ascenção, 2023). The first hypothesis relates to the impact of capacity building and education through AI on the transformation of hotel operations. Capacity building and education are crucial factors enabling staff to effectively use AI technologies (Leber et al. 2018; Demirović et al. 2023a, 2023b; Gajić et al. 2024c). Successful education allows employees to better understand and apply AI technologies, increasing the efficiency and competitiveness of hotels.

H1: Capacity building and education through artificial intelligence (CBEAI) positively affect the transformation of hotel operations (THB).

The second hypothesis examines how sustainable tourism through AI can positively impact the transformation of hotel operations. AI can play a key role in promoting sustainability by optimizing resource use, promoting local products, and reducing waste (Gupta, 2017; Gajić et al. 2023c; Kovačić et al. 2023). This capability of AI to contribute to sustainable tourism is particularly important for hotels aiming to attract environmentally conscious guests and reduce their ecological footprint (Khatri, 2021; Hidayati et al. 2022).

H2: Sustainable tourism through artificial intelligence (STAI) positively affects the transformation of hotel operations (THB).

The third hypothesis focuses on the impact of technological integration on the transformation of hotel operations. Technological integration includes the use of advanced AI technologies such as chatbots, virtual assistants and predictive analytics. These technologies not only improve operational efficiency, but also significantly enhance guest experiences. Said (2023) highlights that AI can personalize guest services in unprecedented ways, contributing to higher guest satisfaction and loyalty. Infante-Moro (2021) and Marković (2020) add that technology reliability and management support play a crucial role in the successful implementation of AI.

H3: Technological integration (TI) positively affects the transformation of hotel operations (THB).

However, the use of artificial intelligence in the industry also raises concerns regarding job displacement, data privacy, and security (Li and Bartley, 2014; Martínez-Martínez et al. 2019; Mingotto et al. 2021). To address these issues, the industry must consider the ethical implications of AI and ensure fairness, accountability, sustainability, and transparency in its use (Pizzi et al. 2022). While AI can enhance customer service, improve operational efficiency, and personalize guest experiences (Katić et al. 2018; Oermann and Knafl, 2021), it also raises concerns about job insecurity and potential negative impacts on employees (Koo et al. 2021; Naqvi et al. 2023).

The implementation of artificial intelligence may lead to job shifts in low-tech sectors and create risks for privacy and security (Drexler and Lapré, 2019; Gajić et al. 2022, 2023a, b; Indora 2024). The impact of artificial intelligence on customer satisfaction and loyalty is complex, with some studies suggesting that employee service quality remains crucial (Talukder and Iva, 2024). Anthropomorphism in AI devices can enhance social presence, but may also trigger identity threats, affecting customer attitudes (Leković et al., 2013; Sterne, 2017). As artificial intelligence becomes more prevalent in hotels, finding the right balance between automation and personalized human services remains a key challenge (Zhou et al. 2014; Katić et al. 2020; Tong-On et al. 2021).

3 Methodology

3.1 Participants and Procedure

The research was conducted from December 2023 to February 2024. A total of 134 hotel managers from across Serbia participated in the study. Random sampling was used to reduce bias and ensure the representativeness of the sample. Managers were randomly selected from a list of hotels that agreed to participate in the research. The G*Power test indicated that a sample of 129 managers was sufficient to achieve the desired statistical power of 0.95 with an effect size of 0.3. These results confirm that a sample of 134 is adequate for conducting SEM analysis and that the results can be considered valid and reliable.

First, a list of hotels that agreed to participate in the research was created. Hotel managers were contacted via email and phone to arrange their participation. Questionnaires were distributed to the managers through online platforms and included questions related to demographic data, as well as specific statements concerning the application of artificial intelligence in their operations. All collected data were processed anonymously and analyzed using statistical methods. The demographic characteristics of the respondents include age ranging from 30 to 50 years, with 60% male and 40% female participants. Regarding education, 90% of the respondents have a higher education degree, while 10% have a secondary education degree. In terms of management level, 50% of the respondents belong to middle management, 30% to senior management, and 20% to lower management (Table 1).

Table 1. Demographic characteristics of respondents

Characteristics	Description
Age	30–50 years
Gender	60% male, 40% female
Education	90% with a higher education degree, 10% with a secondary education degree
Management level	50% middle management, 30% senior management, 20% lower management

3.2 Study Area

The research encompassed hotels of various categories and types, located in the main tourist destinations of Serbia: Kopaonik, Divčibare, Zlatibor, Belgrade and Novi Sad. The hotels included ski resort hotels with 4–5 stars in Kopaonik, mountain hotels with 3–4 stars in Divčibare, wellness and spa hotels with 4–5 stars in Zlatibor, city luxury hotels with 4–5 stars in Belgrade, and business hotels with 3–4 stars in Novi Sad (Table 2).

Table 2. Characteristics of hotel properties

Location	Hotel type	Star rating
Kopaonik	Ski resort hotels	4–5 stars
Divčibare	Mountain hotels	3–4 stars
Zlatibor	Wellness & spa hotels	4–5 stars
Belgrade	City luxury hotels	4–5 stars
Novi Sad	Business hotels	3–4 stars

3.3 Questionnaire Design

The statements were adapted and modified according to our research from the authors Štilić, Ničić and Puška (2023), Infante-Moro, Infante-Moro and Gallardo-Pérez (2021), Marković, Raspor Janković and Zubović (2020), Kumar, Kumar and Attri (2021), and Limna (2023). The factor Technological integration contains 5 statements related to improving service personalization, more efficient reservation management, enhancing operational efficiency, rapid response to changes in guest preferences, and improving marketing strategies.

The factor Sustainable tourism through AI has 5 statements focusing on optimizing resource usage, promoting local products and resources, educating employees about sustainable tourism, monitoring and reducing waste, and improving partnerships with local authorities. Capacity building and education through AI includes 5 statements related to educating employees for successful implementation, collaborating with the academic community, raising awareness of ethical implications, enhancing technical infrastructure, and educating the local population about sustainable practices. The factor Transformation of hotel operations contains 5 statements related to improving service quality, reducing operational costs, increasing guest satisfaction, better tracking of hotel performance, and assisting in strategic decision-making.

3.4 Data Analysis

The data in this study were analyzed using the software packages SPSS 23.00 and SmartPLS. Before detailed analysis, the data were cleaned to ensure accuracy and completeness by removing incomplete and incorrect responses, thereby reducing the risk of bias (Bollen, 1989). Descriptive statistics were conducted to summarize the basic characteristics of the data, including means, standard deviations, and frequency distributions for the main variables (McDonald, 2002).

The reliability of the scales was assessed using Cronbach's alpha coefficient (α). All items had values above the recommended threshold of 0.7, indicating high internal consistency and reliability of the questionnaire (Morris, 2011). Exploratory factor analysis (EFA) identified the underlying structure of the data. The Kaiser-Meyer-Olkin (KMO) measure was 0.791, and Bartlett's test of sphericity was significant ($\chi 2 = 1.024$, $p < 0.001$, $df = 15$), confirming the suitability of the data for factor analysis (Hancock and Mueller, 2006). Promax rotation was used for the interpretation of factors (Sobel, 1982).

Structural equation modeling (SEM) was conducted using SmartPLS software to test the hypotheses and research model (Baron and Kelyy, 1986). Sensitivity analysis assessed the impact of changes in key assumptions on the model results (Hu and Bentler, 1999). Cross-validation techniques further confirmed the predictive power of the model (Raykov and Marcoulides, 2006). The dataset was split into training and testing sets using k-fold cross-validation ($k = 10$), where the model was trained on nine subsets and tested on the remaining one, repeating the process ten times. This allowed for the evaluation of the model's performance on different segments of the data, confirming its generalization (Civelek, 2018).

The standardized root mean square residual (SRMR) value of 0.002 indicates an excellent fit, as values closer to zero are generally better, and values below 0.08 typically indicate a good fit (Gerbing and Anderson, 1988). The d_ULS value of 0.187 and d_G value of 0.632 are specific measures used in PLS-SEM to assess model fit, with lower values generally indicating better fit (Hair et al., 2017). The chi-square ($\chi 2$) value of 1.642 suggests minimal difference between the observed and expected covariance matrices. The comparative fit index (CFI = 0.97), root mean square error of approximation (RMSEA = 0.03), and Tucker-Lewis index (TLI = 0.96) indicate a good fit of the model.

The R^2 values were 0.464, indicating an adequate explanation of the variance of the dependent variables. The normalized fit index (NFI) value of 0.972 indicates excellent model fit, as values closer to 1.0 indicate better fit (Hu and Bentler, 1999). The Fornell-Larcker Criterion and HTMT were conducted to ensure the validity and reliability of the constructs, as well as to achieve compliance with relevant statistical standards and industry best practices (Fornell and Larcker, 1981).

4 Results

The results of the factor analysis indicate several key aspects of the application of artificial intelligence (AI) in the hospitality industry, which are grouped into four main factors: Technological integration (TI), Sustainable tourism through AI (STAI), Capacity building and education through AI (CBEAI), and Transformation of hotel operations (THB) (Table 3).

Factor Technological integration (TI) with high reliability ($\alpha = 0.823$) and a percentage of variance of 27.090%, emphasizes how AI can improve service personalization for guests ($m = 3.12$), enable more efficient reservation management ($m = 3.82$), enhance the operational efficiency of the hotel (m = 3.51), allow quick responses to changes in guest preferences (m = 3.36), and improve marketing strategies (m = 3.53). These statements are supported by high factor loadings, indicating a strong correlation between these statements and the TI factor.

Factor Sustainable tourism through AI (STAI) with a reliability of α = 0.825 and a percentage of variance of 12.974%, focuses on optimizing resource usage (m = 2.29), promoting local products and resources (m = 2.82), educating employees about sustainable tourism (m = 2.46), monitoring and reducing waste (m = 2.65), and improving

Table 3. Descriptive values of the statements and factors

Factors α = 0.821	Statements	m	sd	α	FL
Technological Integration TI α = 0.823 sd = 1.376 % variance = 27.090 CR = 0.860 AVE = 0.607	AI improves the personalization of guest services	3.12	1.330	0.824	0.724
	AI enables more efficient booking management	3.82	1.339	0.811	0.715
	AI improves hotel operational efficiency	3.51	1.425	0.816	0.702
	AI enables a quick reaction to changes in guest preferences	3.36	1.390	0.813	0.799
	AI improves marketing strategies and attracts more guests	3.53	1.397	0.813	0.772
Sustainable tourism through –STAI α = 0.825 sd = 0.702 % variance = 12.974 CR = 0.849 AVE = 0.614	AI optimizes the use of resources	2.29	1.298	0.815	0.870
	AI promotes local products and resources	2.82	1.276	0.813	0.809
	AI educates employees about sustainable tourism	2.46	1.685	0.818	0.772
	AI monitors and reduces waste	2.65	0.045	0.836	0.738
	AI improves partnerships with local authorities	2.24	0.207	0.842	0.713
Capacity building and education through AI – CBEAI α = 0.818 sd = 0.854 % variance = 7.754 CR = 0.844 AVE = 0.692	AI educates employees for successful implementation	3.54	1.978	0.804	0.728

(continued)

Table 3. (*continued*)

Factors α = 0.821	Statements	m	sd	α	FL
	AI helps collaborate with the academic community	3.21	0.001	0.800	0.788
	AI raises awareness of ethical implications	2.92	1.989	0.811	0.754
	AI advances technical infrastructure	3.37	0.138	0.804	0.724
	AI educates local people about sustainable practices	3.55	0.164	0.796	0.711
Transformation of hotel business – THB α = 0.808 sd = 0.834 % variance = 6.817 CR = 0.848 AVE = 0.652	AI improves the quality of services in the hotel	2.21	0.125	0.810	0.899
	AI reduces operating costs	2.30	0.088	0.807	0.872
	AI increases guest satisfaction	2.76	1.997	0.813	0.800
	AI enables better monitoring of hotel performance	2.91	1.914	0.801	0.834
	AI helps in making strategic decisions	3.05	1.047	0.808	0.826

Note: m – arithmetic mean, sd – standard deviation, α – Cronbach alpha, FA – factor loadings, CR – Composite reliability, AVE - Average variance extracted

partnerships with local authorities (m = 2.24). These statements highlight key aspects of sustainable development that AI can support in the hospitality industry.

Capacity building and education through AI (CBEAI) with a reliability of α = 0.818 and a percentage of variance of 7.754%, emphasizes the importance of educating employees for successful AI implementation (m = 3.54), collaborating with the academic community (m = 3.21), increasing awareness of ethical implications (m = 2.92), enhancing technical infrastructure (m = 3.37), and educating the local population about sustainable practices (m = 3.55). These aspects are crucial for the long-term adoption and integration of AI in hotel operations.

Factor Transformation of hotel operations (THB) with a reliability of α = 0.808 and a percentage of variance of 6.817%, emphasizes how AI can improve service quality in the hotel (m = 2.21), reduce operational costs (m = 2.30), increase guest satisfaction (m = 2.76), enable better tracking of hotel performance (m = 2.91), and assist in strategic decision-making (m = 3.05). These results indicate the significant impact of AI on the efficiency and strategy of hotel operations.

The construct reliability and validity analysis shown in Table 4 demonstrates that all factors in the study are reliable and valid, confirming the consistency and adequacy of the measurement scales used.

Capacity building and Education through artificial intelligence (AI) shows solid results with a Cronbach's alpha of 0.775, indicating good internal consistency. The values of rho_A (0.796) and composite reliability (0.846) further confirm the reliability

Table 4. Construct reliability and validity

	Cronbach's Alpha	rho_A	Composite Reliability	Average Variance Extracted (AVE)
Capacity building and education through AI	0.775	0.796	0.846	0.526
Sustainable tourism through AI	0.730	0.788	0.807	0.574
Technological integration	0.890	0.725	0.863	0.608
Transformation of hotel business	0.822	0.836	0.876	0.587

of this factor. The average variance extracted (AVE) is 0.526, which is above the threshold of 0.50, indicating adequate convergent validity. These results highlight the importance of education and capacity building for the successful implementation of AI.

Sustainable tourism through AI has a Cronbach's alpha of 0.730, which also shows good internal consistency. The values of rho_A (0.788) and composite reliability (0.807) confirm the construct's reliability, while the AVE is 0.574, indicating good convergent validity. This factor emphasizes how AI can contribute to sustainability through resource optimization, promotion of local products, and employee education.

Technological integration shows the highest value of Cronbach's alpha (0.890), indicating extremely high internal consistency. Although the value of rho_A is slightly lower (0.725), the composite reliability (0.863) and AVE (0.608) confirm high reliability and adequate convergent validity. These results highlight the importance of technological integration for improving service personalization, more efficient reservation management, and enhancing operational efficiency.

Transformation of hotel operations with a Cronbach's alpha of 0.822, shows high internal consistency. The values of rho_A (0.836) and composite reliability (0.876) further confirm the reliability of the factor, while the AVE is 0.587, indicating good convergent validity. This factor emphasizes how AI can improve service quality, reduce operational costs, increase guest satisfaction, and enable better tracking of hotel performance.

To assess discriminant validity, the Heterotrait-Monotrait (HTMT) ratio was used. The HTMT values were below the recommended threshold of 0.85, indicating adequate discriminant validity of the constructs. Convergent validity was assessed by analyzing the loadings and average variance extracted (AVE). All loading values were above 0.70, while the AVE values exceeded the threshold of 0.50, confirming convergent validity (Table 5). Values below the diagonal represent the Fornell-Larcker Criterion, and values above the diagonal represent the HTMT ratios.

Multicollinearity was tested using the variance inflation factor (VIF). All VIF values were below 3, indicating the absence of multicollinearity issues among the predictor variables (Fig. 1).

Table 5. Combined Fornell-Larcker criterion and HTMT

Construct	Capacity building and education through AI	Sustainable tourism through AI	Technological integration	Transformation of hotel business
Capacity building and education through AI	0.725	0.292	0.431	0.729
Sustainable tourism through AI	0.244	0.689	0.833	0.474
Technological integration	0.258	0.517	0.555	0.625
Transformation of hotel business	0.613	0.408	0.392	0.766

Table 6 presents the results of the path analysis, which examines the impact of various factors on the transformation of hotel operations (THB).

Capacity building and education through artificial intelligence (CBEAI) has a strong and positive effect on the transformation of hotel operations, with an effect size of 0.525. The mean value is 0.532, the standard deviation is 0.067, and the t-value is 7.774, which is statistically significant with a p-value less than 0.001. These results confirm that capacity building and education through artificial intelligence positively impact the transformation of hotel operations, supporting hypothesis H1.

Sustainable tourism through artificial intelligence (STAI) also has a positive impact on the transformation of hotel operations, but to a lesser extent, with an effect size of 0.301. The mean value is 0.298, the standard deviation is 0.073, and the t-value is 2.743, which is statistically significant with a p-value of 0.006. These results confirm hypothesis H2 that sustainable tourism through artificial intelligence positively impacts the transformation of hotel operations.

Technological integration (TI) also shows a positive impact on the transformation of hotel operations, with an effect size of 0.452. The mean value is 0.447, the standard deviation is 0.081, and the t-value is 1.882, which is statistically significant with a p-value of 0.040. These results support hypothesis H3 that technological integration positively impacts the transformation of hotel operations.

The structural model depicted in Fig. 2 illustrates the interrelationships between various factors in the study of artificial intelligence (AI) application in the hotel industry. The model showcases the latent constructs, their indicators, and the strength of the relationships among them.

From the figure, it is evident that technological integration (TI), sustainable tourism through AI (STAI), and capacity building and education through AI (CBEAI) are significant contributors to the transformation of hotel business (THB). The path coefficients demonstrate the relative influence of each construct on the transformation process.

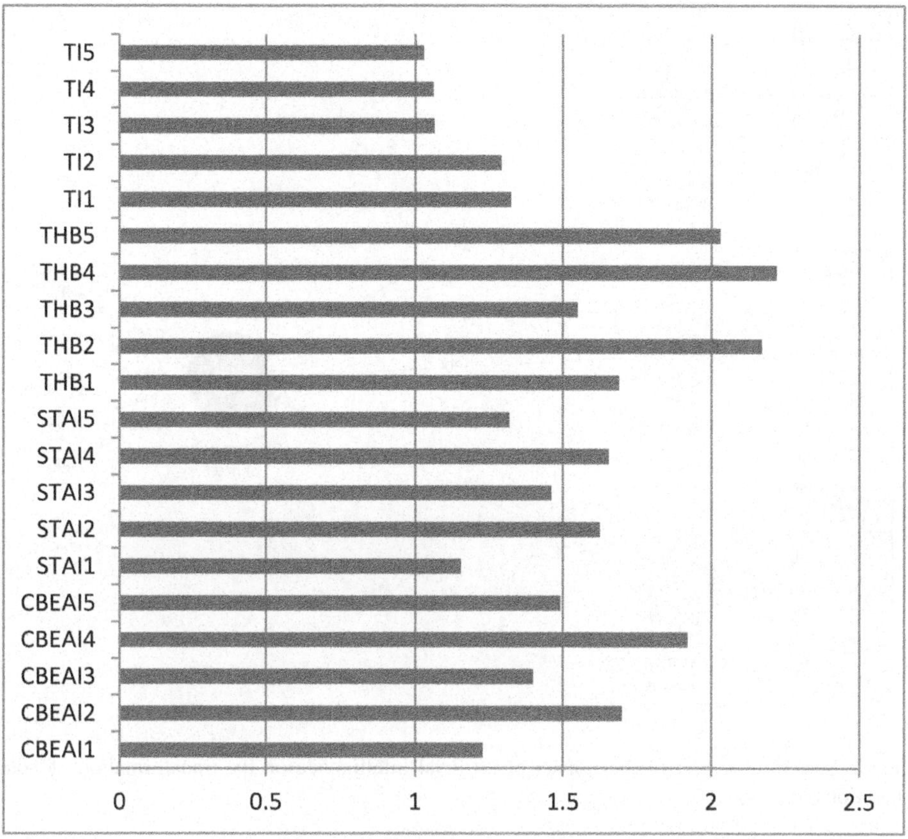

Fig. 1. VIF values for statements

Table 6. Path analysis results

Path	Effects	m	sd	t values	p values	Confirmation	
CBEAI → THB	0.525	0.532	0.067	7.774	0.000	H1	confirmed
STAI → THB	0.301	0.298	0.073	2.743	0.006	H2	confirmed
TI → THB	0.452	0.447	0.081	2.882	0.040	H3	confirmed

Note: CBEAI - Capacity building and education through AI, STAI - Sustainable tourism through AI, TI - Technological integration, THB - Transformation of hotel business, m – arithmetic mean, sd – standard deviation, t – t statistic, p – significance

Technological integration has a substantial impact on hotel transformation with a path coefficient of 0.452, highlighting its critical role in enhancing operational efficiency and guest experiences. Sustainable tourism through AI, although influential, has a lower path coefficient of 0.301, indicating its moderate contribution to hotel transformation. Capacity building and education through AI exhibit the strongest influence on hotel

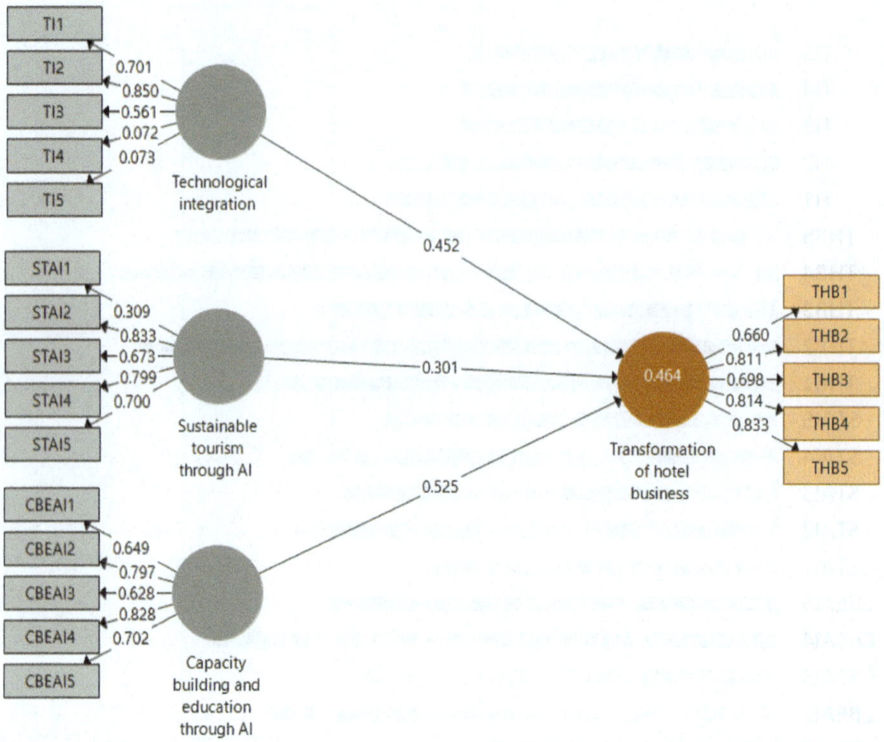

Fig. 2. Structural model of the impact of artificial intelligence on the transformation of hotel operations.

transformation with a path coefficient of 0.525, underscoring the importance of training and skill development for successful AI implementation.

The model also presents the factor loadings for each indicator within the constructs, demonstrating the reliability and validity of the measures used. For instance, indicators like TI2 (0.850) and CBEAI4 (0.828) show high factor loadings, suggesting their strong representation of the respective constructs.

5 Discussion

This study provides significant insights into the role of artificial intelligence (AI) in transforming hotel operations, particularly in the context of developing countries such as the Republic of Serbia. The research results clearly indicate a high level of support for the implementation of AI technologies among hotel management in Serbia, highlighting its potential to enhance operational efficiency, guest experience, and competitiveness.

The analysis of the results shows that capacity building and education through AI (CBEAI) are crucial for the successful transformation of hotel operations. This aligns with the findings of Ivanov et al. (2017), who emphasize the importance of employee education and skill development for the effective application of AI. Our study further confirms that education and adaptation to local conditions are key factors for AI implementation, especially in countries with limited resources.

Additionally, sustainable tourism through AI (STAI) shows a positive impact on the transformation of hotel operations, albeit to a lesser extent compared to capacity building and education. AI technologies such as resource monitoring and optimization, promotion of local products, and waste reduction contribute to sustainability and environmental awareness in the hospitality industry. Chiang & Trimi (2020) explain that AI can help hotels track and analyze resource consumption, reducing costs and negative environmental impact, which is particularly important for hotels aiming to attract environmentally conscious guests.

Technological integration (TI) also has a significant positive impact on the transformation of hotel operations. The use of advanced AI technologies such as chatbots, virtual assistants, and predictive analytics not only improves operational efficiency but also significantly enhances guest experiences. Prentice et al. (2020) highlight that AI can personalize guest services in unprecedented ways, contributing to higher guest satisfaction and loyalty. Kuo et al. (2017) further emphasize the importance of technology reliability and management support for the successful implementation of AI.

However, the use of artificial intelligence also brings certain challenges, including concerns about job displacement, data privacy, and security (Nikoličić et al. 2015; Ivanišević et al. 2024; Gajić et al. 2024a). To address these issues, the industry must consider the ethical implications of AI and ensure fairness, accountability, sustainability, and transparency in its use (Katić et al. 2017; Gretzel and Murphy, 2019). While AI can enhance customer service, operational efficiency, and personalize guest experiences, it also raises concerns about job insecurity and potential negative impacts on employees. The implementation of artificial intelligence may lead to job shifts in low-tech sectors and create risks for privacy and security (Murphy et al. 2017). The impact of artificial intelligence on guest satisfaction and loyalty is complex, with some studies suggesting that the quality of employee service remains crucial. As artificial intelligence becomes more prevalent in hotels, finding the right balance between automation and personalized human services remains a key challenge (Gajić et al. 2024b).

6 Conclusion

This study confirms the significant potential of AI to enhance operational efficiency, guest experience, and sustainability in the hospitality industry, especially in developing countries. The findings provide a solid foundation for further application and research of AI, with a particular focus on adapting technology to local conditions and needs. The application of AI technologies not only improves work processes and reduces costs, but also enables hotels to better understand and respond to their guests' needs. This research contributes to a broader understanding of how AI can play a key role in modernizing the hotel sector, particularly in the context of the specific challenges faced by developing countries.

6.1 Theoretical Implications

Theoretically, this study contributes to the literature on the application of AI in hospitality by providing evidence of its effectiveness and utility in the context of developing countries. It adds value to existing research by emphasizing the importance of capacity building and employee education, as well as promoting sustainable tourism through AI technologies. These findings expand existing theoretical frameworks by integrating specific factors significant for developing countries, such as local economic conditions, cultural specifics, and infrastructural limitations.

6.2 Practical Implications

Practically, the results of this study provide useful guidelines for hotel managers and decision-makers. It is crucial for hotels in developing countries, such as Serbia, to invest in training and skill development of employees to effectively utilize AI technologies. Additionally, hotels should focus on implementing AI solutions that promote sustainability, such as resource optimization and waste reduction, to attract environmentally conscious guests. Technological integration of AI, including the use of chatbots and predictive analytics, can significantly enhance operational efficiency and service personalization, increasing guest satisfaction and loyalty. The results also suggest the need for establishing public-private partnerships and creating policies that support the implementation of AI in the hospitality sector.

6.3 Limitations

While this study provides important insights, it has certain limitations. First, the research was conducted only in the Republic of Serbia, which may limit the generalizability of the results to other developing countries. Second, the sample included only hotel managers, whereas including a broader spectrum of employees could provide a more comprehensive view of AI implementation. Third, the research was conducted over a specific time period, so longitudinal studies could provide deeper understanding of the long-term effects of AI in hospitality. Fourth, the study did not consider different types of hotels (e.g. luxury, mid-range, budget) which may have different needs and challenges in AI implementation.

6.4 Recommendations for Future Research

Future research should focus on comparative studies between different developing countries to identify specific challenges and opportunities in AI implementation. Additionally, longitudinal studies tracking the impact of AI over a longer period can provide deeper insights into the long-term benefits and challenges. Finally, research involving different levels of employees, as well as guest perspectives, can provide a more comprehensive understanding of the impact of AI on the hospitality industry. It is also recommended to explore the interaction between different AI technologies and their combined effect on operational efficiency and guest satisfaction. Furthermore, future research can focus on developing specific AI implementation models tailored to the cultural and economic contexts of individual countries.

Acknowledgments. This research was supported by the Ministry of Science, Technological Development and Innovation of the Republic of Serbia (Contract No. 451-03-136/2025-03/200172).

References

Aderibigbe, A.O., Ohenhen, P.E., Nwaobia, N.K., Gidiagba, J.O., Ani, E.C.: Artificial intelligence in developing countries: bridging the gap between potential and implementation. Comput. Sci. IT Res. J. (2023). https://doi.org/10.51594/csitrj.v4i3.629

Baron, R.M., Kenny, D.A.: The moderator-mediator variable distinction in social psychological research: conceptual, strategic, and statistical considerations. J. Pers. Soc. Psychol. **51**, 1173–1182 (1986). https://doi.org/10.1037/0022-3514.51.6.1173

Bollen, K.: Structural Equations with Latent Variables. Wiley Structural Equations with Latent Variables | Wiley Online Books (1989)

Chiang, A.H., Trimi, S.: Impacts of service robots on service quality. Serv. Bus. **14**(3), 439–459 (2020). https://doi.org/10.1007/s11628-020-00421-y

Citak, J., Owoc, M., Weichbroth, P.: A note on the applications of artificial intelligence in the hospitality industry: preliminary results of a survey. Int. Conf. Knowl.-Based Intell. Inform. Eng. Syst. (2021). https://doi.org/10.1016/j.procs.2021.09.233

Civelek, M.: The mediator roles of attitude toward the web site and user satisfaction on the effect of system quality on net benefit: a structural equation model on web site success. Euro. Sci. J. 61–73 (2018). https://doi.org/10.19044/esj.2016.v12n10p61

Davenport, T., Guha, A., Grewal, D., Bressgott, T.: How artificial intelligence will change the future of marketing. J. Acad. Mark. Sci. **48**(1), 24–42 (2020). https://doi.org/10.1007/s11747-019-00696-0

Demirović Bajrami, D., Cimbaljević, M., Petrović, M.D., Radovanović, M.M., Gajić, T.: Internal marketing and employees' personality traits toward green innovative hospitality. Tourism Rev. (2023). https://doi.org/10.1108/TR-05-2023-0307b

Demirović Bajrami, D., Cimbaljević, M., Petrović, M., Syromiatnikova, Y.A., Gajić, T.: Feeling ready to volunteer after COVID-19? the role of psychological capital and mental health in predicting intention to continue doing volunteer tourism activities. J. Hosp. Tour. Manag. **54**, 500–5112 (2023). https://doi.org/10.1016/j.jhtm.2023.02.009

Drexler, N., Lapré, V.B.: For better or for worse: shaping the hospitality industry through robotics and artificial intelligence. Res. Hospital. Manag. **9**(2), 117–120 (2019). https://doi.org/10.1080/22243534.2019.1689701

Fornell, C., Larcker, D.: Evaluating structural equation models with unobservable variables and measurement error. J. Mark. Res. **18**(1), 39–50 (1981). https://doi.org/10.2307/3151312

Gajić, T., Blešić, I., Petrović, M.D.: Digitization in the hotel industry: sustainability of contactless communication. In: Almeida, F.L., Morais, J.C., Santos, J.D. (Eds.) Digital sustainability: Inclusion and transformation. ISPGAYA 2022. Springer Proceedings in Earth and Environmental Sciences. Springer, Cham (2024a). https://doi.org/10.1007/978-3-031-57650-8_21

Gajić, T., et al.: Stereotypes and prejudices as (non) attractors for willingness to revisit tourist-spatial hotspots in Serbia. Sustainability **15**, 5130 (2023). https://doi.org/10.3390/su15065130

Gajić, T., Minasyan, L.A., Petrović, M.D., Bakhtin, V.A., Kaneeva, A.V., Wiegel, N.L.: Travelers' (in) Resilience to environmental risks emphasized in the media and their redirecting to medical destinations: enhancing sustainability. Sustainability **15**(21), 15297 (2023). https://doi.org/10.3390/su152115297

Gajić, T., et al.: Tourists' willingness to adopt AI in hospitality—Assumption of sustainability in developing countries. Sustainability **16**(9), 3663 (2024). https://doi.org/10.3390/su16093663

Gajić, T., et al.: The adoption of artificial intelligence in Serbian hospitality: a potential path to sustainable practice. Sustainability **16**, 3172 (2024). https://doi.org/10.3390/su16083172

Gajić, T., et al.: Risks in the role of co-creating the future of tourism in "stigmatized" destinations. Sustainability **14**, 15530 (2022). https://doi.org/10.3390/su142315530

Gerbing, D., Anderson, J.: An updated paradigm for scale development incorporating unidimensionality and its. J. Mark. Res. **25**(2), 186–192 (1988). https://doi.org/10.1177/002224378802500207

Williams, R.B., Slak-Valek, N.: Pokémon GO is serious leisure that increases the touristic engagement, physical activity and sense of happiness of players. Inform. Technol. Tourism **21**(4), 515–533 (2019). https://doi.org/10.1007/s40558-019-00153-2

Gupta, A.: Editorial - the year in review and path forward. Inf. Syst. Res. (2017). https://doi.org/10.1287/isre.2017.0761

Hair, J., Hult, G., Ringle, C., Sarstedt, M.: A primer on partial least squares structural equation modeling (PLS-SEM). Los Angeles: SAGE. Available online: (1) (PDF) A Primer on Partial Least Squares Structural Equation Modeling (PLS-SEM) (researchgate.net) (2017)

Hancock, G.R., Mueller, R.O.: Structural Equation Modeling: A Second Course. Information Age Publishing, Inc., Greenwich, CT (2006)s. https://doi.org/10.1111/j.1744-6570.2007.00081_13.x

Hidayati, H., Sutikno, A.N., Erawanto, S.: The pivotal issues of human rights: a literature review by eight world researchers. Jurnal Studi Sosial dan Politik (2022). https://doi.org/10.19109/jssp.v6i1.10198

Hu, L.T., Bentler, P.M.: Cutoff criteria for fit indexes in covariance structure analysis: conventional criteria versus new alternatives. Struct. Equa. Model. **6**, 1–55 (1999). https://doi.org/10.1080/10705519909540118

Indora, A.: Artificial Intelligence (AI) in the hospitality industry: a review article. Int. J. Multidis. Res. https://doi.org/10.36948/ijfmr.2024.v06i03.19393

Infante-Moro, A., Infante-Moro, J., Gallardo-Pérez, J.: Key factors in the process of acceptance and implementation of artificial intelligence in the hotel sector. Adv. Bus. Inform. Syst. Anal. (2021). https://doi.org/10.4018/978-1-7998-6985-6.ch014

Ivanišević, A., Katić, I., Marjan, L., Buchmeister, B.: Business plan feedback for cost effective business processes. Adv. Prod. Eng. Manag. ISSN 1854-6250 (Print), ISSN 1855-6531 (Online)

Ivanov, S., Webster, C., Berezina, K.: Robotics in tourism and hospitality. Ann. Tour. Res. **69**, 205–222 (2017). https://doi.org/10.1016/j.annals.2017.03.003

Katić, I., Knežević, T., Berber, N., Ivanišević, A., Leber, M.: The impact of stress on life, working, and management styles: how to make an organization healthier? Sustainability **11**(15), 4026 (2019). https://doi.org/10.3390/su11154026

Katić, I., Ivanišević, A., Grubić-Nešić, L., Penezić, N.: Effects of sociodemographic characteristics and personality traits on career development. Int. J. Aging Hum. Dev. (2017). https://doi.org/10.1177/0091415017743008

Katić, I., Berber, N., Slavić, A., Ivanišević, A.: The relations between investment in employees' development and organizational productivity and service quality. Tehnički vjesnik-technical gazette **27**(4), 1077–1083 (2020). https://doi.org/10.17559/TV-20181121101314

Katić, I., Ivanišević, A., Grubić-Nešić, L., Penezić, N.: Effects of sociodemographic characteristics and personality traits on career development. Int. J. Aging Hum. Dev. **87**(2), 201–216 (2018). https://doi.org/10.1177/0091415017743008

Khatri, M.: How digital marketing along with artificial intelligence is transforming consumer behaviour? Int. J. Res. Appl. Sci. Eng. Technol. **9**(7), 523–527 (2021). https://doi.org/10.22214/ijraset.2021.36287

Koo, C., Xiang, Z., Gretzel, U., Sigala, M.: Artificial Intelligence (AI) and robotics in travel, hospitality, and leisure. Electro. Market. **31**(3), 473–476 (2021). s12525-021-00494-z.pdf (springer.com)

Kovačić, S., et al.: How has COVID-19 changed the way we travel? exploring tourist personality, reactions to the perceived risk and change in travel behavior. Sustainability **15**, 1951 (2023). https://doi.org/10.3390/su15031951

Kumar, S., Kumar, V., Attri, K.: Impact of artificial intelligence and service robots in tourism and hospitality sector: current use & future trends. Admin. Dev. J. HIPA, Shimla (2021). https://doi.org/10.53338/adhipa2021.v08.si01.04

Kuo, C.M., Chen, L.C., Tseng, C.Y.: Investigating an innovative service with hospitality robots. Int. J. Contemp. Hosp. Manag. **29**(5), 1305–1321 (2017). https://doi.org/10.1108/IJCHM-08-2015-0414

Leber, M., Ivanišević, A., Borocki, J., Radišić, M., Ślusarczyk, B.: Fostering alliances with customers for the sustainable product creation. Sustainability **10**(9), 3204 (2018). https://doi.org/10.3390/su10093204

Leković, B., Ivanišević, A., Marić, B., Demko-Rihter, J.: Assessment of the most significant impacts of environment on the changes in company cost structure. Ekonomska istraživanja-Econ. Res. **26**(1), 225–242 (2013). https://doi.org/10.1080/1331677X.2013.11517599

Li, T., Bartley, G.: Publishing systematic reviews in ophthalmology: new guidance for authors. Ophthalmology (Rochester, Minn.) (2014). https://doi.org/10.1016/j.ophtha.2013.10.003

Limna, P.: Artificial Intelligence (AI) in the hospitality industry: a review article. Int. J. Comput. Sci. Res. (2023). https://doi.org/10.25147/ijcsr.2017.001.1.103

Marković, S., Raspor Janković, S., Zubović, V.: The impacts of robots and artificial intelligence on service quality in the hotel industry (2020). https://doi.org/10.31410/balkans.jetss.2020.3.2.163-170

Martínez-Martínez, A., Cegarra-Navarro, J.G., Garcia-Perez, A., Wensley, A.: Knowledge agents as drivers of environmental sustainability and business performance in the hospitality sector. Tour. Manage. **70**, 381–389 (2019). https://doi.org/10.1016/j.tourman.2018.08.030

McDonald, R.P., Ho, M.H.: Principles and practice in reporting structural equation analyses. Psychol. Methods **7**, 64–82 (2002). https://doi.org/10.1037/1082-989X.7.1.64

Mingotto, E., Montaguti, F., Tamma, M.: Challenges in re-designing operations and jobs to embody AI and robotics in services. Findings from a case in the hospitality industry. Elec. Markets **31**(3), 493–510 (2021). https://doi.org/10.1007/s12525-020-00439-y

Morris, N.J., Elston, R.C., Stein, C.M.: A framework for structural equation models in general pedigrees. Hum. Hered. **70**, 278–286 (2011). https://doi.org/10.1159/000322885

Murphy, H.C., Hofacker, C., Gretzel, U.: Dawning of the age of robots in hospitality and tourism: challenges for teaching and research. Euro. J. Tourism Res. **15**, 104–111 (2017). https://doi.org/10.54055/ejtr.v15i.265

Naqvi, G., Hamza, M., Saddique, M., Sarim, S., Riyan, U.: Application of AI in hotel management and its benefits. Stat. Comput. Interdisciplin. Res. (2023). https://doi.org/10.52700/scir.v3i2.141

Nikoličić, S., Kilibarda, M., Atanasković, P., Duđak, L., Ivanišević, A.: Impact of RFID technology on logistic process efficiency in retail supply chains. Traf. Transp. **27**(2), 137–146 (2015). ISSN 0353-5320 (Print), ISSN 1848-4069 (Online)

Oermann, M., Knafl, K.: Strategies for completing a successful integrative review. Nurse Author Ed. (2021). https://doi.org/10.1111/nae2.30

Pavlović, A., et al.: The impact of foreign direct investments and economic growth on environmental degradation: the case of the Balkans. Energies **14**(3), 566 (2021). https://doi.org/10.3390/en14030566

Petrović, M., et al.: How can tufa deposits contribute to the geotourism offer? the outcomes from the first UNESCO Global Geopark in Serbia. Land **12**(2), 285 (2023). https://doi.org/10.3390/land12020285

Petrović, M., et al.: The degree of environmental risk and attractiveness as a criterion for visiting a tourist destination. Sustainability **15**, 14215 (2023). https://doi.org/10.3390/su151914215

Pillai, R., Sivathanu, B.: Adoption of AI-based chatbots for hospitality and tourism. Int. J. Contemp. Hosp. Manag. **32**(10), 3199–3226 (2020). https://doi.org/10.1108/IJCHM-04-2020-0259

Pizzi, G., Vannucci, V., Shukla, Y., Aiello, G.: Privacy concerns and justice perceptions with the disclosure of biometric versus behavioural data for personalized pricing tell me who you are; I'll tell you how much you pay. Consumers' fairness and privacy perceptions with personalised pricing. J. Bus. Res. **148**, 420–432 (2022). https://doi.org/10.1016/j.jbusres.2022.04.072

Prentice, C., Dominique Lopes, S., Wang, X.: The impact of artificial intelligence and employee service quality on customer satisfaction and loyalty. J. Hosp. Market. Manag. **29**(7), 739–756 (2020). https://doi.org/10.1080/19368623.2020.1685057

Raykov, T., Marcoulides, G.: A First Course in Structural Equation Modeling. Lawrence Erlbaum Associates, Mahwah (2006). https://doi.org/10.4324/9780203930687

Said, S.: The role of Artificial Intelligence (AI) and data analytics in enhancing guest personalization in hospitality. J. Mod. Hospital. (2023). https://doi.org/10.47941/jmh.1556

Sharma, S., Rawal, Y., Pal, S., Dani, R.: Fairness, Accountability, Sustainability, Transparency (FAST) of artificial intelligence in terms of hospitality industry. ICT Anal. Appl. (2022). https://doi.org/10.1007/978-981-16-5655-2_48

Singh, A., Tyagi, P., Singh, A.K., Tyagi, P., Kapure, S., Singh, E.R.: Robotics and artificial intelligence in the hotel industry: a systematic literature review. In: 2022 8th International Conference on Advanced Computing and Communication Systems (ICACCS) (2022). https://doi.org/10.1109/ICACCS54159.2022.9785257

Sobel, M.: Asymptotic confidence intervals for indirect effects in structural equation models. Sociol. Methodol. **13**, 290–312 (1982). https://doi.org/10.2307/270723

Sterne, J.: Artificial intelligence for marketing: practical applications. John Wiley & Sons (2017). https://doi.org/10.1002/9781119406341

Talukder, B., Iva, R.D.: The technology impacts and AI solutions in hospitality. i-manager's J. Artif. Intell. Mach. Learn. (2024). https://doi.org/10.26634/jaim.2.1.20291

Talukder, M., Kabir, F., Muhsina, K., Das, I.R.: Emerging concepts of artificial intelligence in the hotel industry: a conceptual paper. Int. J. Res. Publ. Rev. (2023). https://doi.org/10.55248/gengpi.4.923.92451

Tihanyi, L., DeCelles, K.A.: Publishing original research in AMJ: advice to prospective authors. Acad. Manag. J. (2021). https://doi.org/10.5465/amj.2021.4003

Tong-On, P., Siripipatthanakul, S., Phayaphrom, B.: The implementation of business intelligence using data analytics and its effects towards on performance in the hotel industry in Thailand. Int. J. Behav. Anal. **1**(2), 91 (2021). https://ssrn.com/abstract=3944077

Tripura, K., Avi, M.A.R.: Application of innovative technologies in the tourism and hospitality industry of Bangladesh: role analysis of the public and private institutions. Technol. Appl. Tourism Hospital. Indust. Bangladesh 215–226 (2021). https://doi.org/10.1007/978-981-16-2434-6_13

Tuomi, A., Ascenção, M.P.: Intelligent automation in hospitality: exploring the relative automatability of frontline food service tasks. J. Hospital. Tourism Insights **6**(1), 151–173 (2023). https://doi.org/10.1108/JHTI-07-2021-0175

Wellman, N., Tröster, C., Grimes, M.G., Roberson, Q., Rink, F., Gruber, M.: Publishing multimethod research in AMJ: a review and best-practice recommendations. Acad. Manag. J. (2023). https://doi.org/10.5465/amj.2023.4004

Wreford, O., Williams, N.L., Ferdinand, N.: Together alone: an exploration of the virtual event experience. Event Manag. **23**(4–5), 721–732 (2019). https://doi.org/10.3727/152599519X15506259855625

Xu, F.Z., Zhang, Y., Zhang, T., Wang, J.: Facial recognition check-in services at hotels. J. Hosp. Market. Manag. **30**(3), 373–393 (2021). https://doi.org/10.1080/19368623.2020.1813670

Zhou, S., Wu, Z., Li, J., Zhang, X.P.: Real-time energy control approach for smart home energy management system. Elec. Power Compo. Syst. **42**(3–4), 315–326 (2014). https://doi.org/10.1002/adfm.202003619

Štilić, A., Ničić, M., Puška, A.: Check-in to the future: exploring the impact of contemporary information technologies and artificial intelligence on the hotel industry. Turističko Poslovanje (2023). https://doi.org/10.5937/turpos0-43739

Identification of Potential Biomarkers and Pathways in Dilated Cardiomyopathy Using Bioinformatics Analysis

Tamara Mladenović[1,2](✉), Sanja Matić[1], Jelena Pavić[1,2], Katarina Virijević[1,2], and Nenad Filipović[2,3]

[1] Institute for Information Technologies, Department of Science, University of Kragujevac, Jovana Cvijića bb, 34000 Kragujevac, Serbia
{tamara.mladenovic,jelena.pavic, katarina.virijevic}@uni.kg.ac.rs, sanjamatic@kg.ac.rs

[2] Faculty of Engineering, University of Kragujevac, Sestre Janjić 6, 34000 Kragujevac, Serbia
fica@kg.ac.rs

[3] Bioengineering Research and Development Center (BioIRC), Prvoslava Stojanovića 6, 34000 Kragujevac, Serbia

Abstract. Dilated cardiomyopathy (DCM) is one of the most common causes of sudden cardiac arrest and ranks as the second most frequent cause of heart failure. The objective of this study was to identify potential biomarkers and explore the underlying mechanisms involved in DCM, using various bioinformatics approaches. The GSE3585 microarray datachip from GEO omnibus database was analyzed using the Limma function in RStudio and identified 101 differentially expressed genes (DEGs). The results showed 68 up-regulated and 33 down-regulated DEGs. All DEGs were functionally enriched and pathway analyzed using DAVID and Metascape. The PPI network was created in STRING and visualized in Cytoscape. Hub genes were retrieved using CytoHUBBA plugin, and GeneMANIA determined their roles and co-expressed genes. From 101 DEGs, 10 were hub genes: STAT3, HSP90AB1, SMAD7, CCN2, H2AZ1, H1–0, SMAD6, EEF1A1, HMGN2, XPO1. STAT3, SMAD6, SMAD7, HSP90AB1, and CTGF play complex molecular roles in TGF-β signalling, cardiac fibrosis advancement, heart response to high oxidative stress, and cardiac muscle contraction weakness. Histone genes H1–0, H2AZ1 and non-histone gene HMGN2 play a role in epigenetic modelling of gene expression. EEF1A1 is expressed in the heart, although its effect on DCM pathophysiology is unknown. XPO1 mediates nucleocytoplasmic transport and affects DCM cardiac remodeling. Despite valuable insights, this study has limitations. The microarray data, sourced from a public database, has a small sample size, and experimental validation is needed to confirm the obtained results. These findings may lead to therapeutic biomarkers and a deeper understanding of DCM pathophysiology.

Keywords: dilated cardiomyopathy · bioinformatics · hub genes · differentially expressed genes · functional enrichment analysis · Rstudio

© The Author(s), under exclusive license to Springer Nature Switzerland AG 2025
N. Filipović (Ed.): AAI 2024, LNNS 1446, pp. 344–364, 2025.
https://doi.org/10.1007/978-3-031-99201-8_33

1 Introduction

Heart failure (HF) is a complex medical condition and is considered to be the final stage of chronic heart disease [1]. The most frequent causes include myocardial infarction followed by muscle necrosis and chronic myocardial hypoperfusion [2]. In advanced countries, HF is now the most common reason for adult hospitalization [1]. All types of cardiomyopathies (CMs), including hypertrophic, dilated, and restricted, can result in HF, whereas DCM is the most prevalent cause of chronic congestive HF and is regarded as the main global indication for heart transplantation [1, 2]. CMs are categorized as cardiac disorders that affect the heart muscle and are characterized by a variety of anatomical and functional abnormalities [3]. While secondary CMs are brought on by systemic conditions accompanied by cardiac failure, primary CMs are limited to the myocardium [4]. DCM affects 1 in 2,500 people, but these statistics are based on clinical criteria discovered during the genetic era. Males are affected more commonly than females, the frequency is 7 per 100,000. It is more likely that recent estimates, which are based on the information from sources including systematic family screening, early phenotypes, and genetic testing, would reveal that DMC is more prevalent [5, 6].

DCM is a myocardial condition defined by systolic dysfunction and left ventricular or biventricular dilatation in the absence of hypertension or coronary artery disease sufficient to explain the dysfunction [7]–[9]. Nuclear DNA (nDNA) mutations (Mendelian inheritance) or mitochondrial DNA (mtDNA) mutations (matrilineal inheritance) can also result in DCM [5]. Numerous instances of Mendelian genetic abnormalities lead to DCM; however, there is a chance that DCM is brought on by a person's unique genetic susceptibility combination with other factors including alcoholism or drug addiction, inflammation, and tachyarrhythmias, autoimmune diseases, endocrine or neuromuscular causes [5, 10]. According to estimates, 21–36% of cases of DCM in high-income nations are caused by alcohol addiction, and 40% of DCM cases may be inherited, primarily autosomal dominant [4, 8].

Genetic variations in more than 40 genes encoding diverse range of proteins are essential for DCM [11]. For the etiologic diagnosis of DCM, genetic and immunologic markers are used, although sometimes high levels of biomarkers might be brought on by other cardiovascular disorders [12]. Due to genetic variability, not all DCM molecular pathways are understood [13]. As a result, the search for potential genes involved in DCM and the study of the underlying mechanisms of DCM are of great importance for human health.

The objective of this study was to identify potential biomarkers and explore the underlying mechanisms involved in DCM, using various bioinformatics approaches. Gene expression microarray analysis, an excellent bioinformatics-based method, is frequently used to identify potentially important genes connected to DCM as well as possible molecular functions and biological pathways, which are then employed as biological markers for diagnosis and prognosis [14]. Therefore, we first accessed the Gene Expression Omnibus (GEO) website from where we downloaded the microarray gene expression profile labeled GSE3585 according to Barth et al. [14]. Then, DEGs were identified between the control samples and the samples suffering from DCM. Functional enrichment analysis and protein-protein interactions were then determined for these DEGs. In addition, key genes that could potentially be very important for the progression of DCM were determined. Afterwards, an analysis of those genes was performed to identify the function of co-expressed genes.

2 Materials and Methods

2.1 Microarray Data

The gene expression profile for DataSeries GSE3585, generated by Barth et al. [14], was retrieved from the GEO (http://www.ncbi.nlm.nih.gov/geo), a repository at the National Center for Biotechnology Information archive [15]. GEO offers the scientific community free access to next-generation functional genomic datasets and gene expression data produced using DNA microarray technology. The GSE3585 dataset, which uses single channel hybridizations of 12 various cardiac biopsy samples, is based on the Affymetrix Human Genome U133A Array GPL96 platform. The 12 different subendocardial left ventricular tissue samples and the five non-failing (NF) donor hearts that were not transplanted because of evident coronary calcifications, came from seven DCM patients at the time of transplantation. Technical or biological replicas were not provided. Using oligonucleotide array analysis, the samples were examined to discover a typical gene expression profile in DCM [14].

2.2 Data Preprocessing and Obtaining the Differentially Expressed Genes

The initial step involved downloading and normalizing the raw microarray data from the chip (Fig. 1). This was done using the Geoquery and Limma package [17] in the RStudio programming language [15]. The reason for normalization was to account for variations in the intrinsic backgrounds of different chips, which could affect the calculation of expression values. The specific version used was 2.68.0, which can be accessed at (http://www.bioconductor.org/packages/release/bioc/httml/GEOquery.html). Subsequently, the expression data underwent a log2 transformation, and a linear regression model was built to compare the gene expression between DCM and control samples. The false-positive results from the microarray were corrected using the adjusted p-value (adj. p). The adj. p-value was utilized to correct for the presence of false positive outcomes by applying the Benjamini and Hochberg false discovery rate method as the default, resulting in the calculation of the false discovery rate (FDR) [16]. We used the RStudio language and the Limma package [17] to find the DEGs in samples with DCM and samples that were not affected by DCM. The threshold for identifying DEGs was set at FDR of less than 0.05 and log2 fold- change $\neq 0$ (logFC). Genes were initially filtered using the adj. p-value (FDR) and then further refined based on the logFC. In our circumstances, it was more logical to initially select the genes based on their adj. p-value rather than relying on the fold change alone, which may not meet the usual fold change cut-off threshold. The typical threshold for fold change is a logarithmic value of -2 or -1 for down-regulated genes and a value greater than 2 or 1 for genes that are up-regulated. The fold change [18] criteria quantifies the magnitude of the differential expression effect, providing deeper biological insights than statistical information. Nevertheless, if the gene has markedly distinct levels of expression, it is already justifiable to explore further, as even small differences in gene expression might yield a substantial influence.

2.3 Functional and Pathway Enrichment Analysis

The functional enrichment analysis of the acquired DEGs was conducted using two online bioinformatics tools, namely Database for Annotation, Visualization and Integrated Discovery (DAVID) and Metascape. This was done to ensure mutual verification of the results, as both tools employ distinct algorithms for enrichment analysis. Firstly, the list of 108 identified DEGs was uploaded to the online software DAVID (https://david.ncifcrf.gov/) and Gene Ontology (GO) enrichment analysis and Kyoto Encyclopedia of Genes and Genomes (KEGG) pathway analysis were performed, to identify the overrepresented GO categories and KEGG pathways. DAVID, a free bioinformatics software, incorporates all main public resources and can be used for functional annotation and gene enrichment analysis [19]. Similar to DAVID, Metascape (https://metascape.org/) is an online bioinformatics tool for gene list annotation, gene ontology, and pathway enrichment analysis. It incorporates over 40 knowledge bases for gene annotation, functional enrichment, membership search, and interactome analysis [20]. GO enrichment analysis of significantly enriched DEGs between DCM and healthy control biopsies was categorized into biological processes, cellular components and molecular functions. Biological processes (BP) are the largest of the three ontology aspects in GO enrichment because they represent the larger processes carried out by numerous molecular activities. Molecular function (MF) refers to processes carried out by a gene product, respectively, while cellular component (CC) refers to a location in cellular structures where a gene product performs a molecular function [21]. $P < 0.05$ was considered statistically significant.

2.4 PPI Network Construction and Identification of the Hub Genes

We used Search Tool for the Retrieval of Interacting Genes/Proteins (STRING) (https://string-db.org/) online database to predict protein-protein interactions (PPI) and better comprehend the probable connections between the proteins encoded by the identified DEGs. STRING is a multi-source database that is used to collect and combine PPIs, functional associations, and physical interactions systematically [22].

The acquired outcomes were subsequently integrated into a software application called Cytoscape (version 3.7.0) to generate a network model. To discover hub genes, we used the Cytohubba plug-in within the Cytoscape software. Among the available methods, we opted for the Degree method to identify the top 10 hub genes and construct a PPI network. This choice was made due to the method's ability to yield more precise predictions of important proteins. The co-expression network of the identified hub genes was analyzed using the GeneMANIA web tool, utilizing the GeneMANIA database. GeneMANIA [23] is a renowned web program that serves as a prediction server, enabling users to input a gene list. Consequently, it generates a compilation of genes that are functionally associated by utilizing pre-existing genomes and proteomics data.

3 Results

3.1 Identification of Differentially Expressed Genes (DEGs)

Based on the corrected p-value ($p < 0.05$), we were able to identify 101 DEGs, of which 68 are up-regulated genes and 33 are down-regulated genes, from our data acquired from the GPL96 oligonucleotide microarray platform, which is made up of 10,664 probe sets. The logFC values for up-regulated genes vary from 4.5 to 0.3, whereas for down-regulated genes, the values range from -2.3 to -0.3. Details about some of the DEGs are displayed in Table 1. PHLDA1, NPPB, NPPA and CTFG which are already recognized as significant genes implicated in the pathogenesis of dilated cardiomyopathy, are found to be significantly upregulated. Subsequently, the ggplot2 package was used in RStudio to create a volcano plot to visualize the DEGs (Fig. 2).

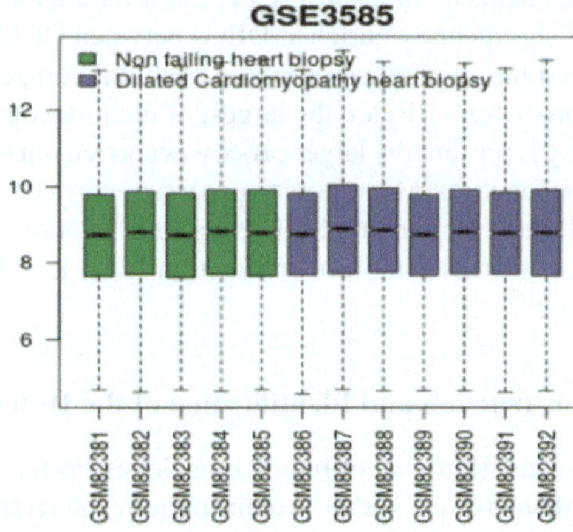

Fig. 1. Boxplot shows that median-centered values are indicative that the data are normalized and cross-comparable.

3.2 GO Term Enrichment Analysis and Pathways

A total of 101 DEGs that were enriched between DCM and non-failing cardiac tissues were uploaded to the DAVID database and subsequently to Metascape. The results of the enrichment analysis obtained from the DAVID database revealed that there was no enrichment in terms of BP associated with down-regulated genes. For CC, the enriched terms were extracellular exosome and nucleosome. In terms of MF, the enrichment analysis identified prenylated protein tyrosine phosphatase activity, structural constituent of chromatin, chromatin DNA binding, and selenium binding. The most significantly enriched terms for up-regulated genes in the BP category were negative regulation of transforming growth factor beta receptor signalling pathway, cell differentiation, vasodilation, negative regulation of BMP signalling pathway, response to laminar fluid shear stress, neural crest cell differentiation, positive regulation of cardiac muscle contraction,

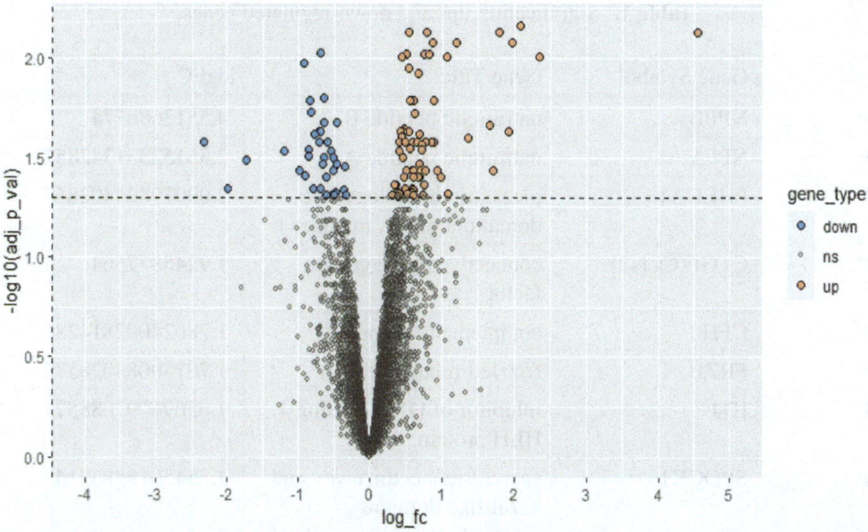

Fig. 2. Volcano plot displays statistical significance (-log10 P value) versus magnitude of change (log2 fold change).

neuroepithelial cell differentiation, negative regulation of protein phosphorylation, and negative regulation of SMAD protein import into the nucleus. In the CC category, the enriched terms were cytoplasm, macromolecular complex, perinuclear region of cytoplasm, nucleus, SMAD2-SMAD3 protein complex, stress fiber, and extracellular space. In the MF category, the enriched terms were protein binding, hormone receptor binding, cytoskeletal protein binding, type I transforming growth factor beta receptor binding, and actin binding. The results are shown in Tables 2 and 3.

Finally, the results derived from Metascape are shown in Fig. 3 showcasing the top 20 enrichment clusters that hold substantial importance. BPs such as the enzyme-linked receptor protein pathway, negative regulation of microtubule polymerization or depolymerization, astrocyte differentiation, regulation of growth, chromatin organization, regulation of systemic arterial blood pressure, response to leukemia inhibitory factor, kidney development, and regulation of proteolysis were found to be the most enriched. The perinuclear region of cytoplasm is the most enriched term linked to CC. Regarding MFs, DEGs exhibited enrichment in insulin-like growth factor binding, chromatin DNA binding, molecular function inhibitor activity, protein tyrosine/serine/threonine phosphatase activity, protein homodimerization activity, and molecular adaptor activity.

DAVID KEGG pathway analysis results showed enrichment in TGF-β signalling pathway (Table 2) for up-regulated genes and no pathway enrichment for down-regulated genes (Table 3). Pathway enrichment results obtained from Metascape analysis showed signalling by Rho GTPases, glutathione metabolism and TGF-β signalling pathway (Figs. 3 and 4).

Table 1. Significantly up- and down-regulated genes.

	Gene Symbol	Gene Title	logFC
UP	NPPB	natriuretic peptide B	4.5513686774
UP	NPPA	natriuretic peptide A	2.35152331342857
UP	PHLDA1	pleckstrin homology like domain family A member 1	2.09079891402857
UP	CTGF(CCN2)	connective tissue growth factor	1.9245877564
UP	CFH	complement factor H	1.78624002814286
UP	FRZB	frizzled-related protein	1.70780083028572
UP	ID4	inhibitor of DNA binding 4, HLH protein	1.65696792088571
UP	SPOCK1	sparc/osteonectin, cwcv and kazal-like domains proteoglycan (testican) 1	1.35450185837143
UP	ODC1	ornithine decarboxylase 1	1.19691683451429
UP	DUSP6	dual specificity phosphatase 6	1.08458717765714
UP	KLHL3	kelch like family member 3	1.07022309168571
UP	SNCA	synuclein alpha	0.984564600142859
UP	SSPN	Sarcospan	0.927598992085713
DOWN	RARRES1	retinoic acid receptor responder 1	-2.34009080802857
DOWN	S100A8	S100 calcium binding protein A8	-1.99514280471428
DOWN	C1orf105	chromosome 1 open reading frame 105	1.19851295994286
DOWN	PTP4A3	protein tyrosine hosphatase type IVA, member 3	-0.982246958285717
DOWN	IDH2	isocitrate dehydrogenase (NADP(+)) 2, mitochondrial	-0.923849411285717
DOWN	CCDC69	coiled-coil domain containing 69	-0.901952554342861
DOWN	SELENBP1	selenium binding protein 1	-0.858077363971429
DOWN	MID1IP1	MID1 interacting protein 1	-0.854360626371429
DOWN	STAT3	signal transducer and activator of transcription 3	-0.846692372285712

Table 2. GO biological functional enrichment and KEGG pathway for up-regulated DEGs obtained from DAVID database.

UP-REGULATED	GO ID	GO Term	P-Value	Genes
BP Category	GO:0030512	negative regulation of transforming growth factor beta receptor signalling pathway	0.007	BAMBI, SMAD6, SMAD7, SPRED2
	GO:0030154	cell differentiation	0.017	CTGF, SMAD6, SMAD7...
	GO:0042311	Vasodilation	0.019	EXT1, NPPA, NPPB
	GO:0030514	negative regulation of BMP signalling pathway	0.021	BAMBI, SMAD6, SMAD7
	GO:0034616	response to laminar fluid shear stress	0.037	SMAD6, SMAD7
	GO:0014033	neural crest cell differentiation	0.042	EXT1, FRZB
	GO:0060452	positive regulation of cardiac muscle contraction	0.047	CTGF, NPPA
	GO:0060563	neuroepithelial cell differentiation	0.047	DLG5, TMEM231
CC Category	GO:0005737	Cytoplasm	0.0025	SMAD6, SMAD7, EFFA1, XPO1, HMGN2, NPPA, NPPB...
	GO:0032991	macromolecular complex	0.0029	XPO1, HSP90AB1, PIK3CA, PHLDA1, ROR1...
	GO:0048471	perinuclear region of cytoplasm	0.012	CTGF, SNCA...
	GO:0005634	Nucleus	0.014	ETV5, MYH10, SMAD6, SMAD7
	GO:0071144	SMAD2-SMAD3 protein complex	0.04	SMAD6, SMAD7
	GO:0001725	stress fiber	0.042	MYH10, ROR1, TPM1

(continued)

Table 2. (*continued*)

UP-REGULATED	GO ID	GO Term	P-Value	Genes
MF Category	GO:0005515	protein binding	0.0011	CCN2, EEF1A1, XPO1, HSP90AB1, HMGN2, MYH10, PHLDA1, ETV5, SPOCK1…
	GO:0051427	hormone receptor binding	0.02	NPPA, NPPB
	GO:0008092	cytoskeletal protein binding	0.034	NCKIPSD, DLG5, TPM1
	GO:0034713	type I transforming growth factor beta receptor binding	0.045	SMAD6, SMAD7
KEGG pathway	hsa04350	TGF-beta signaling pathway	0.0015	BAMBI, SMAD6, SMAD7, ID4, LTBP1

Table 3. GO biological functional enrichment for down-regulated DEGs obtained from DAVID database.

DOWN-REGULATED	GO ID	GO Term	P-Value	Genes
CC Category	GO:0070062	extracellular exosome	0.016	S100A8, RARRES1, SELENBP1…
	GO:0000786	nucleosome	0.022	H1–0, H2AZ1, H2BC5
MF Category	GO:0004727	prenylated protein tyrosine phosphatase activity	0.0048	PTP4A2, PTP4A3
	GO:0030527	structural constituent of chromatin	0.011	H1–0, H2AZ1, H2BC5
	GO:0031490	chromatin DNA binding	0.014	H1–0, H2AZ1, STAT3
	GO:0008430	selenium binding	0.021	GPX3, SELENBP1

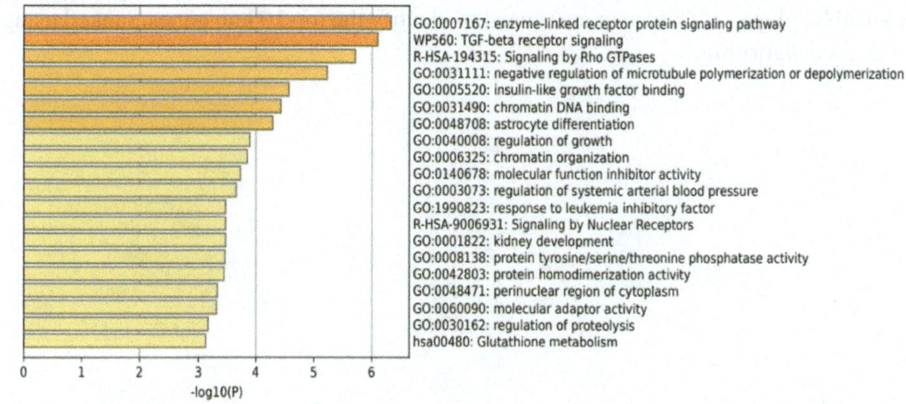

Fig. 3. Metascape chart bar of 20 biological enriched terms and pathways based on the P-value and the percentage of genes.

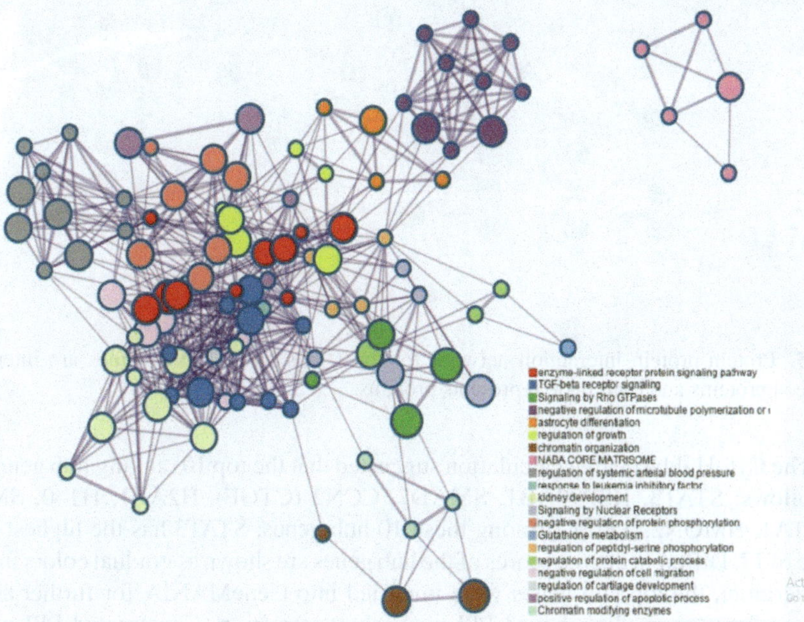

Fig. 4. Network of enriched Metascape terms colored by cluster ID, nodes that share the same cluser ID are close to each other.

3.3 Building the PPI Network and Obtaining the Hub Genes

To obtain the interaction data, all DEGs were entered into the STRING database, and the obtained PPI network with 57 nodes and 91 edges (combined scores of > 0.4) was then imported into Cytoscape for further analysis and visualization (Figs. 5 and 6). Then,

CytoHubba plugin of Cytoscape was used to identify the top 10 hub genes, ranked using the Degree algorithm.

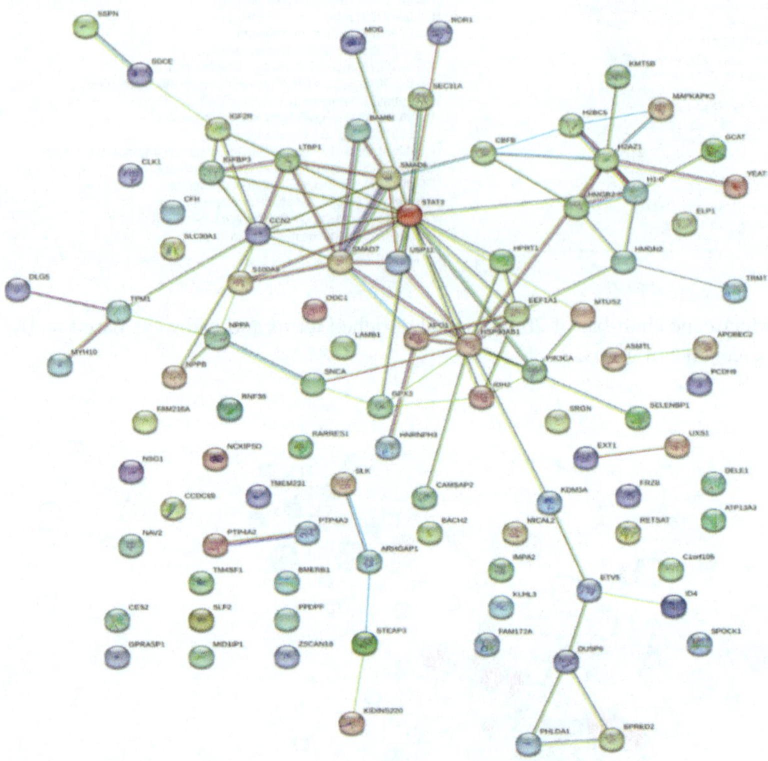

Fig. 5. Protein-protein interaction network for the DEGs by STRING. Lines are interactions between proteins and each node represents proteins.

The CytoHubba degree calculation suggested that the top 10 ranking hub genes were as follows: STAT3, HSP90AB1, SMAD7, CCN2 (CTGF), H2AZ1, H1–0, SMAD6, EEF1A1, HMGN2, XPO1. Among these 10 hub genes, STAT3 has the highest degree score of 17. Different degree scores of the hub genes are shown as gradual colors in Fig. 7. Furthermore, the top hub genes were imported into GeneMANIA for further analysis of gene function prediction and PPI network construction. Constructed PPI network in GeneMANIA identified the function of the co-expressed genes, which are primarily related to nucleocytoplasmic carrier activity, molecular carrier activity, nuclear transport and nucleocytoplasmic transport with FDR < 0.05 (Fig. 8).

4 Discussion

DCM is one of the most common causes of sudden cardiac arrest and ranks as the second most frequent cause of HF [24, 25]. The primary challenge lies in the complexity of early detection of DCM, given that the underlying causes and molecular mechanisms

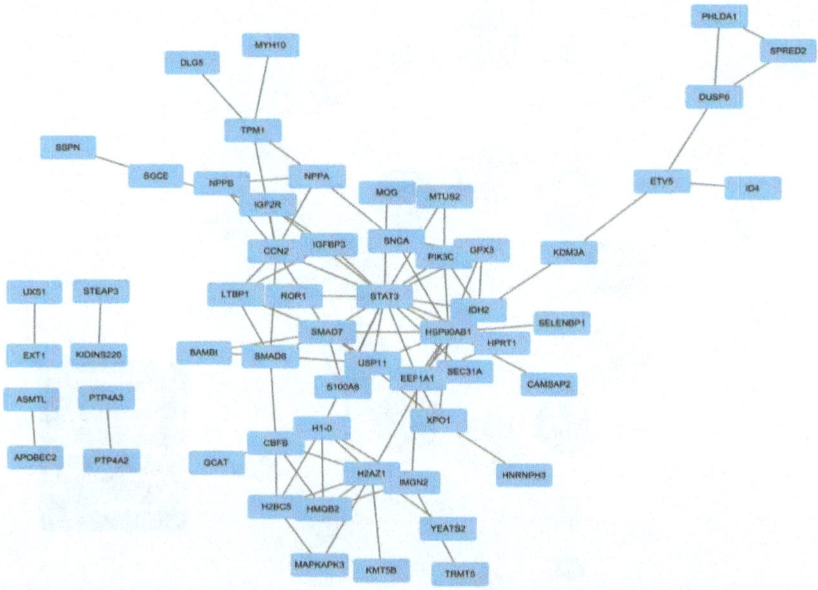

Fig. 6. Network of differentially expressed genes created by STRING and Cytoscape. Disconnected nodes were excluded.

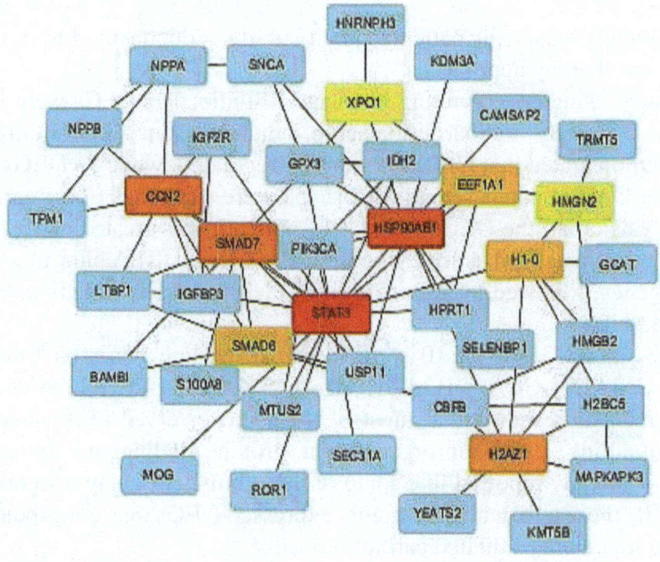

Fig. 7. The top 10 hub genes ranked by the Degree algorithm and their neighbors among all the DEGs. The darker the color, the stronger the association with other genes

of this condition remain incompletely comprehended [26, 27]. Hence, it is imperative to identify novel biomarkers and precise molecular pathways to aid in early detection of

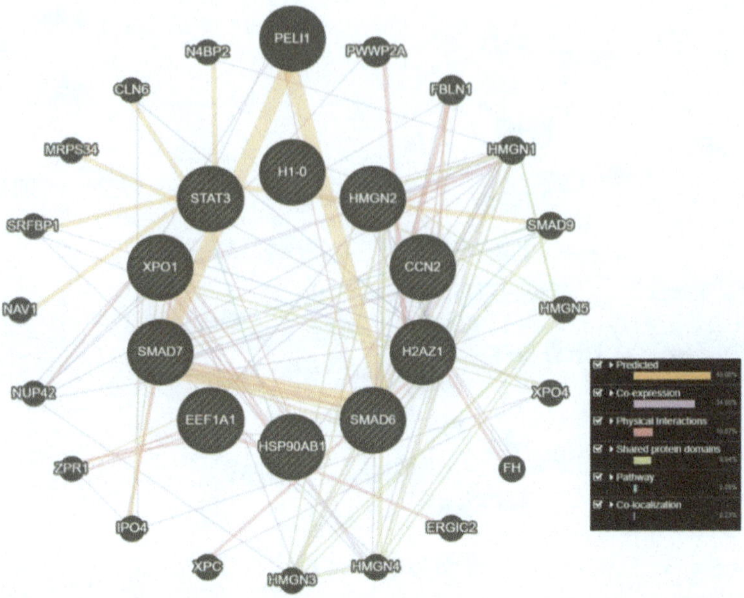

Fig. 8. PPI network of the top 10 hub genes. Inner circles are hub genes and outer circles correspond to predicted genes by GeneMania

DCM, consequently enhancing patients' quality of life, reducing morbidity, and enabling more efficacious therapy approaches.

In this study, using programming language RStudio, 101 DEGs were identified by analyzing the GSE3585 microarray datachip, obtained from GEO omnibus database. The results demonstrated that 68 DEGs were up-regulated, while 33 DEGs were down-regulated. The DAVID database and Metascape were used to perform enrichment and pathway analysis on all the DEGs found. PPI network was created using the STRING database and then visualized using Cytoscape. The CytoHUBBA plug-in was utilized to acquire hub genes, followed by the retrieval of their functions and co-expressed genes using GeneMANIA.

From 108 obtained DEGs, 10 were identified as hub genes as follows STAT3, HSP90AB1, SMAD7, CCN2, H2AZ1, H1–0, SMAD6, EEF1A1, HMGN2, XPO1. Enrichment analysis revealed that most of them are involved in regulation of TGF-β pathway signalling, chromatin organization, protein binding and signalling by Rho GTP-ases. It is already reported that some of these hub genes play a significant role in DCM. Studying these hub genes and highly expressed DEGs may contribute to the early diagnosis and treatment of dilated cardiomyopathy.

Signal Transducer and Activator of Transcription 3 (STAT3) is one of the most important signalling molecules involved in the maintenance of normal cardiac function under different physiological and pathological conditions through various molecular mechanisms [28–30]. As a highly down-regulated gene, it is associated with enrichment terms such as "chromatin-DNA binding", "negative regulation of cell migration" and "response to growth factors". As a transcription factor, it has a protective effect by promoting the expression of several genes that prevent cell death (anti-apoptotic genes), reduce oxidative stress (anti-oxidative genes) and influence the response of mitochondria to low oxygen levels (hypoxia) in cardiac pathological remodeling. STAT3 can bind to DNA structures that are involved in the structure and assembly of nucleosomes and could thus influence chromatin organization [31]. In agreement with our findings, Liu et al. [24] and Liu et al. [32] also identified STAT3 as a hub gene and an important biomarker through bioinformatic analyses. Studies have shown that doxorubicin-induced CM in an animal model leads to the increased expression of STAT3 as a protective mechanism for the heart. Conversely, reduced expression of STAT3 in the diseased heart leads to HF, as the protective function of this molecule is absent [33, 34]. Podewski et al. [35] showed that the last stage of DCM is characterized by a reduced expression of STAT3. These findings are consistent with our results. When developing a therapeutic approach to enhance STAT3 to protect the myocardium, it is essential to ensure that the activation of STAT3 is specific to the heart and only transient. This is because continuous activation of STAT3 in other tissues has been shown to lead to various types of malignancies [36]. Thus, the literature review is consistent with the results obtained, suggesting that STAT3 is an important biomarker for monitoring DCM. Furthermore, some evidence suggests that the link between STAT3 and the spectrin-based cytoskeleton may potentially be critical for maintaining the balance in STAT3 activation during the treatment of chronic cardiac stress. In addition, it is important to include some canonical signalling pathways such as TGF-β/SMAD and others [37] that play a role in the regulation of cardiac fibrosis and arrhythmia.

Regarding this information, both SMAD6 and SMAD7 are also hub genes. Functional enrichment analysis revealed that these hub genes are associated with negative regulation of TGF-β receptor signalling, negative regulation of BMP signalling, laminar fluid shear stress response, SMAD2-SMAD3 protein complex, and transforming growth factor beta receptor type I binding. SMAD proteins, which belong to the TGF-β superfamily, play an important role in the control of several processes associated with heart disease. These mechanisms include the treatment of myocardial infarction, aortopathies, deposition of extracellular matrix, regulation of cardiac fibrotic responses, and facilitation of cardiac healing [38–41]. SMADs transduce signals from TGF-β ligands located outside the cell to the nucleus, where they initiate gene transcription [42]. Inhibitory SMAD proteins, a crucial group of proteins, work together to inhibit TGF-β signalling [41]. It was found that SMAD6 specifically inhibits the transmission of bone morphogenetic protein (BMP) signals. On the other hand, SMAD7 inhibits both TGF-β/activin- and BMP-mediated signalling pathways [40, 41]. The canonical response to TGF-β is mediated by SMAD proteins, while SMAD6 and SMAD7 are responsible for inhibiting signal transduction and play a crucial role in the development of cardiac fibrosis [40]. Cardiac fibrosis is associated with non-ischemic DCM and is mainly caused by TGF-β and other factors such as CD4 + and CD8 + T cells, endothelial cells, macrophages and mast cells. These factors directly activate myofibroblasts [43, 44]. Interstitial fibrosis is a form of fibrosis that develops in DCM due to an increase in extracellular matrix and collagen deposition [44]. The extent of the contribution of SMAD6 to cardiac fibrosis is not yet fully investigated and understood, while the importance of SMAD7 in the regulation of cardiac fibrosis is known [40]. Humeres et al. [41] conducted a study in which they made numerous observations on the function of the SMAD7 protein in cardiac fibrosis. This protein inhibits the expression of fibrosis-related genes in cardiac fibroblasts by blocking TGF-β/SMAD signalling pathways and acts as an antifibrotic factor through other pathways. Both SMAD6 and SMAD7 function to inhibit the BMP-SMAD signalling pathway, which has been shown to have a significant impact on cardiovascular disease [45, 46]. The upregulation of SMAD6 in atherosclerotic blood vessels may contribute as a secondary factor to the development of DCM [47]. Nevertheless, Niu et al. [48] conducted a study on the expression of SMAD6 and TGF-β in cardiovascular disease produced by obesity. Elevated expression is observed at the early stage of HF, and it has been demonstrated to contribute to the restoration of the damaged heart. Malgija et al. [49] conducted a study to examine the significance of cardiac fibrosis in hypertrophic and DCM. The findings corroborate our own, indicating an elevated presence of SMAD6 and SMAD7 proteins in DCM. Consequently, the TGF-β 1-Smad pathway is inhibited. The literature provides evidence highlighting the significance of SMAD6 and SMAD7 proteins in DCM, which aligns with our results.

HSP90AB1 is a chaperone protein that binds to client proteins, ensuring their stability and correct folding, particularly in response to different forms of stress [50]. Slightly up-regulated, it is associated with perinuclear region of cytoplasm, regulation of TGF-β receptor signalling pathway, protein binding and macromolecular complex. HSP90AB1 has been demonstrated to contribute to pathological cardiac remodeling associated with ventricular hypertrophy and HF. TGF-β I receptor works in conjunction with HSP90, so by suppressing it, it is possible to decrease collagen formation and activation of the TGF-β cascade [51]. Datta et al. [52] discovered that HSP90 derived from myocytes influences the activation of STAT3 in cardiac fibroblasts, resulting in the overproduction of collagen and subsequent cardiac dysfunction. This suggests that HSP90 plays a significant role in the pathophysiology of the heart during periods of high-pressure stress.

Moreover, connective tissue growth factor (CTGF, CCN2) exhibits a significant level of expression and is also one of the hub genes. It is well-established that its elevated expression is associated with cardiac tissue remodelling in cases of HF [53]. CTGF is enriched in terms such as positive regulation of cardiac muscle contraction, perinuclear region of cytoplasm, response to growth factor and regulation of cytoskeleton organization. Literature survey shows that overexpressed CTGF is associated with HF, cardiac fibrosis and cardiac injury, which is the main characteristic of DCM [53]–[56]. Research has shown that inhibition of CTGF improves the function of the damaged heart [53]. Furthermore, another study confirmed that CTGF has an influence on the progression of cardiac fibrosis and in the pathological remodeling of the heart [55]–[57]. However, some studies contradict these studies where it has been shown that the increased expression of CTGF has little or no effect on cardiac remodeling and in promoting cardiac fibrosis, nor is TGF-β an effector [58, 59]. So, the latest research examines the role of CTGF as a pro-fibrotic factor, and there is a possibility that in addition to CTGF, some other factors influence cardiac fibrosis [59]. Dorn et al. [54] explained in their studies these somewhat discrepancies and concluded that CTGF is an autocrine factor and affects cardiac fibrosis only if it is secreted by the fibroblasts themselves, which suggests that there may be post-translational modifications of CTGF that can affect its change functions. Thus, CTGF can be considered a very important biomarker in DCM [53]–[56].

Further results showed that hub genes also include the down-regulated histone genes, H1–0, and H2AZ1. Enrichment results showed that these histones are most involved in processes such as DNA packaging, chromatin DNA binding, regulation of gene silencing, epigenetics, nucleosome, structural constituent of chromatin, chromatin organization, etc. Histones are essential proteins that play a crucial role in the transition between active and inactive chromatin, whereby histone variants have evolved to significantly influence gene regulation and epigenetic imprinting [60]. It has been proven that chromatin remodeling in the form of histone modifications can affect the reprogramming of gene expression, which leads to epigenetic changes in cardiac cells and can lead to cardiac fibrosis and hypertrophy in end-stage HF [3, 61–64]. In vivo transient depletion of histone H1–0, in response to extracellular stress, leads to chromatin decompaction and prevents fibrosis in the cardiac muscle [65]. Also, it is assumed that the reduced expression of these histone genes can lead to the imprinting of specific genes and thereby cause increased expression of other genes that affect the progression of DCM via nucleosomes [66].

On the other hand, high mobility group N (HMGN) is a family of non-histone proteins that bind to nucleosomes, the building blocks of chromatin and play a role in modeling chromosome structure and influences gene expression [67, 68]. The HMGN2 isoform, which was overexpressed in our study, otherwise has an impact on epigenetic changes through histone modifications [67]. It is enriched in terms of chromatin binding, nucleosomal DNA binding, chromatin organization, cytoplasm, and protein binding. Li et al. [69] identified HMGN2 as one of the biomarkers in the progression of HF, while Camargo et al. [70] also obtained HMGN2 as a biomarker in the progression of HF and more specifically DCM, but she emphasized that Barth et al. [14] also obtained this gene as differentially expressed. Further research and analysis are needed to determine the precise role of this gene in DCM.

Jiang et al. [71] also obtained EEF1A1 as a possible biomarker for DCM by bioinformatic analysis. Eukaryotic translation elongation factor-A1 (EEF1A1) is one of the two EEF1A1 isoforms and is expressed almost everywhere, including in heart cells. It is thought to be expressed during prenatal development, after which its level decreases in cardiomyocytes. So far, not a single disorder of this gene variant has been found in humans, and if it occurs, it would probably be fatal [72].

Studies have shown that some nuclear-cytoplasmic transport molecules are related to HF or ventricular dysfunction [73]. This includes XPO1, which is an important exportin that regulates nucleocytoplasmic transport. In accordance, up-regulated XPO1 was enriched in protein binding and signalling by Rho GTPases, macromolecular complex, and cytoplasm. One study found that XPO1 is overexpressed in cardiomyocytes in DCM, which may be related to the ability of this gene to repair the heart [74]. Studies have shown that XPO1 is important for nuclear-cytoplasmic transport and plays an important role in DCM, which is up-regulated and plays a role in cardiac development [73–76]. XPO1 is the main mediator in the transport of several proteins, among them RhoU, which belongs to the family of Rho GTPases, which, through complex mechanisms, affect the functioning and development of the heart [66, 75, 76]. It could be noted that the obtained hub genes and their enrichments and pathways play a crucial role in the pathology of DCM, so they could become future therapeutic targets for clinical treatments.

5 Conclusion

In this study, we applied different bioinformatics approaches to find and analyze possible biomarkers for DCM. The results of our study emphasize a number of hub genes and genes that are expressed at high levels, which are strongly linked to the development and advancement of DCM. This analytical methodology has not only identified crucial genes that may contribute to the advancement of cardiomyopathy, but it has also shown possible targets for treatment and biomarkers for early identification. These hub genes, which play important roles as central points within complex biological networks, present potential targets for therapeutic intervention and biomarkers for early detection. Moreover, this analysis deepens our comprehension of the regulatory processes that contribute to heart dysfunction and the advancement of disease. In addition, through an extensive review of the existing literature, we have verified the significance and impact of the identified hub genes in the pathological restructuring of the heart and the advancement of DCM. This, to some extent, confirms our predictive model. However, the study also recognizes the limitations inherent in bioinformatics research, including the small sample size, the microarray data obtained from public database and the need for validation of these biomarkers in clinical settings. To build on the foundational work presented here, future research should focus on longitudinal studies to verify the clinical relevance of these markers and to explore their utility in real-world diagnostic processes. Ultimately, the continued refinement of these bioinformatic strategies, in combination with experimental validation, will pave the way for more personalized and effective treatment strategies for DCM, significantly impacting patient outcomes and quality of life.

Acknowledgements. This work is supported by the Serbian Ministry of Science, Technological Development, and Innovation Agreement No. 451-03-66/2024-03/200378.

Data Availability. The datasets generated and analyzed during the current study are available in the public Gene Expression Omnibus database (https://www.ncbi.nlm.nih.gov/geo/query/acc.cgi?acc=GSE3585).

References

1. Chen, X., Zhang, Q., Zhang, Q.: Predicting potential biomarkers and immune infiltration characteristics in heart failure. Math. Biosci. Eng. **19**, 8671–8688 (2022)
2. Tayal, U., Prasad, S., Cook, S.A.: Genetics and genomics of DCM and systolic heart failure. Genome Med. **9**, 1–14 (2017)
3. Yu, J., Zeng, C., Wang, Y.: Epigenetics in DCM. Curr. Opin. Cardiol. **34**, 260–269 (2019)
4. Weintraub, R.G., Semsarian, C., Macdonald, P.: DCM. Lancet **390**, 400–414 (2017)
5. Favalli, V., Serio, A., Grasso, M., Arbustini, E.: Genetic causes of DCM. Heart **102**, 2004–2014 (2016)
6. Pérez-Serra, A., et al.: Genetic basis of DCM. Int. J. Cardiol. **224**, 461–472 (2016)
7. Schultheiss, H.-P., et al.: DCM. Nat. Rev. Dis. Primer. **5**, 32 (2019)
8. Lakdawala, N.K., Winterfield, J.R., Funke, B.H.: DCM. Circ. Arrhythm. Electrophysiol. **6**, 228–237 (2013)
9. Merlo, M., Cannatà, A., Gobbo, M., Stolfo, D., Elliott, P.M., Sinagra, G.: Evolving concepts in DCM. Eur. J. Heart Fail. **20**, 228–239 (2018)
10. Sinagra, G., Elliott, P.M., Merlo, M.: DCM: so many cardiomyopathies! Eur. Heart J. **41**, 3784–3786 (2020)
11. Escobar-Lopez, L., et al.: Association of genetic variants with outcomes in patients with nonischemic DCM. J. Am. Coll. Cardiol. **78**, 1682–1699 (2021)
12. Zhang, M., et al.: Integrated bioinformatics analysis for identifying key genes and pathways in female and male patients with DCM. Sci. Rep. **13**, 8977 (2023)
13. Wang, L., et al.: Analyzing gene expression profiles in DCM via bioinformatics methods. Braz. J. Med. Biol. Res. **49**, 4897 (2016)
14. Barth, A.S., et al.: Identification of a common gene expression signature in DCM across independent microarray studies. J. Am. Coll. Cardiol. **2006**, 1610–1617 (2006)
15. Barrett, T., Edgar, R.: [19] Gene expression omnibus: microarray data storage, submission, retrieval, and analysis. Methods Enzymol. **411**, 352–369 (2006)
16. Davis, S., Meltzer, P.S.: GEOquery: a bridge between the Gene Expression Omnibus (GEO) and BioConductor. Bioinformatics **23**, 1846–1847 (2007)
17. Smyth, G.K.: Linear models and empirical Bayes methods for assessing differential expression in microarray experiments. Stat. Appl. Genet. Mol. Biol. **3**, 1–25 (2004)
18. Barber, R.F., Ramdas, A.: The *p* -filter: multilayer false discovery rate control for grouped hypotheses. J. R. Stat. Soc. Ser. B Stat. Methodol. **79**, 1247–1268 (2017)
19. Sherman, B.T., et al.: DAVID: a web server for functional enrichment analysis and functional annotation of gene lists (2021 update). Nucleic Acids Res. **50**(2022), W216–W221 (2021)
20. Zhou, Y., et al.: Metascape provides a biologist-oriented resource for the analysis of systems-level datasets. Nat. Commun. **10**, 1523 (2019)
21. Thomas, P.D.: The Gene Ontology Handbook, Methods in Molecular Biology, vol. 1446 (2017)
22. Szklarczyk, D., et al.: Protein–protein association networks and functional enrichment analyses for any sequenced genome of interest. Nucleic Acids Res. **51**(2023), D638–D646 (2023)
23. Montojo, J., Zuberi, K., Rodriguez, H., Bader, G.D., Morris, Q.: GeneMANIA: fast gene network construction and function prediction for Cytoscape, F1000Research, vol. 3, p. 153 (2014)
24. Liu, Z., Song, Y.-N., Chen, K.-Y., Gao, W.-L., Chen, H.-J., Liang, G.-Y.: Bioinformatics prediction of potential mechanisms and biomarkers underlying DCM, World. J. Cardiol. **14**, 282–296 (2022)

25. Rosenbaum, A.N., Agre, K.E., Pereira, N.L.: Genetics of DCM: practical implications for heart failure management. Nat. Rev. Cardiol. **17**, 286–297 (2020)
26. Zhou, L., Liu, C., Zou, Y., Chen, Z.: Development and verification of the nomogram for DCM gene diagnosis. Sci. Rep. **12**, 8908 (2022)
27. Zheng, Y., et al.: Exploring key genes to construct a diagnosis model of DCM. Front. Cardiovasc. Med. **9**, 865096 (2022)
28. Harhous, Z., Booz, G.W., Ovize, M., Bidaux, G., Kurdi, M.: An update on the multifaceted roles of STAT3 in the heart. Front. Cardiovasc. Med. **6**, 150 (2019)
29. Pipicz, M., Demján, V., Sárközy, M., Csont, T.: Effects of cardiovascular risk factors on cardiac STAT3. Int. J. Mol. Sci. **19**, 3572 (2018)
30. Ricke-Hoch, M., Stapel, B., Gorst, I., Haghikia, A., Hilfiker-Kleiner, D.: The STAT3 pathway and downstream mechanisms in cardiac remodeling: Friend or Foe. In: Jugdutt, B., Dhalla, N. (eds.) Cardiac Remodeling. Advances in Biochemistry in Health and Disease, vol. 5. Springer, New York, NY (2013). https://doi.org/10.1007/978-1-4614-5930-9_20
31. Kurdi, M., Zgheib, C., Booz, G.W.: Recent developments on the crosstalk between STAT3 and inflammation in heart function and disease. Front. Immunol. **9**, 3029 (2018)
32. Liu, X., Zhang, W., Han, Z.: Decreased circulating follicular regulatory T cells in patients with DCM. Braz. J. Med. Biol. Res. **54**, e11232 (2021)
33. Haghikia, A., Ricke-Hoch, M., Stapel, B., Gorst, I., Hilfiker-Kleiner, D.: STAT3, a key regulator of cell-to-cell communication in the heart. Cardiovasc. Res. **102**, 281–289 (2014)
34. Kääb, S., et al.: Global gene expression in human myocardium-oligonucleotide microarray analysis of regional diversity and transcriptional regulation in heart failure. J. Mol. Med. **82**, 308–316 (2004)
35. Podewski, E.K., et al.: Alterations in Janus Kinase (JAK)-Signal Transducers and Activators of Transcription (STAT) signalling in patients with end-stage DCM. Circulation **107**, 798–802 (2003)
36. O'Sullivan, K.E., Breen, E.P., Gallagher, H.C., Buggy, D.J., Hurley, J.P.: Understanding STAT3 signalling in cardiac ischemia. Basic Res. Cardiol. **111**, 27 (2016)
37. Patel, N.J., Nassal, D.M., Gratz, D., Hund, T.J.: Emerging therapeutic targets for cardiac arrhythmias: role of STAT3 in regulating cardiac fibroblast function. Expert Opin. Ther. Targets **21**, 63–73 (2021)
38. Hanna, A., Humeres, C., Frangogiannis, N.G.: The role of Smad signalling cascades in cardiac fibrosis. Cell. Signal. **77**, 109826 (2021)
39. Yousefi, F., et al.: TGF-β and WNT signalling pathways in cardiac fibrosis: non-coding RNAs come into focus. Cell Commun. Signal **18**, 87 (2020)
40. Działo, E., Tkacz, K., Błyszczuk, P.: Crosstalk between TGF-β and WNT signalling pathways during cardiac fibrogenesis. Acta Biochim. Pol. **65**, 341–349 (2018)
41. Chen, Q., et al.: Smad7 is required for the development and function of the heart. J. Biol. Chem. **284**, 292–300 (2009)
42. Duan, Y., Zhu, W., Liu, M., Arshad, M., Su, M.: The expression of Smad signalling pathway in myocardium and potential therapeutic effects. Histol. Histopathol. **32**, 651–659 (2017)
43. Cojan-Minzat, B.O., Zlibut, A., Agoston-Coldea, L.: Non-ischemic DCM and cardiac fibrosis. Heart Fail. Rev. **26**, 1081–1101 (2021)
44. Eijgenraam, T.R., Sillјé, H.H.W., De Boer, R.A.: Current understanding of fibrosis in genetic cardiomyopathies. Trends Cardiovasc. Med. **30**, 353–361 (2020)
45. Herrera, B., Inman, G.J.: A rapid and sensitive bioassay for the simultaneous measurement of multiple bone morphogenetic proteins. Identification and quantification of BMP4, BMP6 and BMP9 in bovine and human serum. BMC Cell Biol. **10**, 20 (2009)
46. Itoh, F.: Promoting bone morphogenetic protein signalling through negative regulation of inhibitory Smads. EMBO J. **20**, 4132–4142 (2001)

47. Miyazawa, K., Miyazono, K.: Regulation of TGF-β Family Signalling by Inhibitory Smads. Cold Spring Harb. Perspect. Biol. **9**, a022095 (2017)
48. Niu, H.M., Liu, C.L.: The aberrant expression of Smad6 and TGF-beta in obesity linked cardiac disease. Eur. Rev. Med. Pharmacol. Sci. **21**, 138–142 (2017)
49. Malgija, B., Kumar, N.S., Piramanayagam, S.: Collective transcriptomic deregulation of hypertrophic and DCM – importance of fibrotic mechanism in heart failure. Comput. Biol. Chem. **73**, 85–94 (2018)
50. Haase, M., Fitze, G.: HSP90AB1: helping the good and the bad. Gene **575**, 171–186 (2016)
51. García, R., et al.: Extracellular heat shock protein 90 binding to TGFβ receptor I participates in TGFβ-mediated collagen production in myocardial fibroblasts. Cell. Signal. **28**, 1563–1579 (2016)
52. Datta, R., et al.: Myocyte-derived Hsp90 modulates collagen upregulation via biphasic activation of STAT-3 in fibroblasts during cardiac hypertrophy. Mol. Cell. Biol. **37**, e00611–16 (2017)
53. Koshman, Y.E., et al.: Connective tissue growth factor regulates cardiac function and tissue remodeling in a mouse model of DCM. J. Mol. Cell. Cardiol. **89**, 214–222 (2015)
54. Dorn, L.E., Petrosino, J.M., Wright, P., Accornero, F.: CTGF/CCN2 is an autocrine regulator of cardiac fibrosis. J. Mol. Cell. Cardiol. **121**, 205–211 (2018)
55. Lipson, K.E., Wong, C., Teng, Y., Spong, S.: CTGF is a central mediator of tissue remodeling and fibrosis and its inhibition can reverse the process of fibrosis. Fibrogenesis Tissue Repair. **5**, S24 (2012)
56. Zhang, Y., Yan, H., Guang, G., Deng, Z.: Overexpressed connective tissue growth factor in cardiomyocytes attenuates left ventricular remodeling induced by angiotensin II perfusion. Clin. Exp. Hypertens. **39**, 168–174 (2017)
57. Zhang, C., et al.: Human umbilical cord mesenchymal stem cells alleviate interstitial fibrosis and cardiac dysfunction in a DCM rat model by inhibiting TNF-α and TGF-β1/ERK1/2 signalling pathways. Mol. Med. Rep. **17**, 71–78 (2017)
58. Accornero, F., et al.: Genetic analysis of connective tissue growth factor as an effector of transforming growth factor β signalling and cardiac remodeling. Mol. Cell Biol. **35**, 2154–2164 (2015)
59. Fontes, M.S.C., et al.: CTGF knockout does not affect cardiac hypertrophy and fibrosis formation upon chronic pressure overload. J. Mol. Cell. Cardiol. **88**, 82–90 (2015)
60. Mariño-Ramírez, L., Kann, M.G., Shoemaker, B.A., Landsman, D.: Histone structure and nucleosome stability. Expert Rev. Proteomics **2**, 719–729 (2005)
61. Han, P., Yang, J., Shang, C., Chang, C.-P.: Chromatin remodeling in heart failure, epigenetics in cardiac disease. In: Backs, J., McKinsey, T.A. Eds. in Cardiac and Vascular Biology, vol. 1, pp. 103–124 (2016)
62. Movassagh, M., et al.: Distinct epigenomic features in end-stage failing human hearts. Circulation **124**, 2411–2422 (2011)
63. Papait, R., Serio, S., Condorelli, G.: Role of the epigenome in heart failure. Physiol. Rev. **100**, 1753–1777 (2020)
64. Mahmoud, S.A., Poizat, C.: Epigenetics and chromatin remodeling in adult cardiomyopathy. J. Pathol. **231**, 147–157 (2013)
65. Parvatiyar, M.S., Lopez, T.D., Franklin, S., Vondriska, T.M.: Abstract 199: distinctive roles of linker histone H1 variants in the hypertrophic response of cardiomyocytes. Circ. Res. **113**, 2411–2422 (2013)
66. Zhang, H., Yu, Z., He, J., Hua, B., Zhang, G.: Identification of the molecular mechanisms underlying DCM via bioinformatic analysis of gene expression profiles. Exp. Ther. Med. **13**, 273–279 (2017)
67. Garza-Manero, S., et al.: Maintenance of active chromatin states by HMGN2 is required for stem cell identity in a pluripotent stem cell model. Epigenetics Chromatin. **12**, 73 (2019)

68. Nanduri, R., Furusawa, T., Bustin, M.: Biological functions of HMGN chromosomal proteins. Int. J. Mol. Sci. **21**, 449 (2020)
69. Li, D., Lin, H., Li, L.: Multiple feature selection strategies identified novel cardiac gene expression signature for heart failure. Front. Physiol. **11**, 604241 (2020)
70. Camargo, A., Azuaje, F.: Identification of DCM signature genes through gene expression and network data integration. Genomics **92**, 404–413 (2008)
71. Jiang, C., Zhong, G.: [Identify the characteristic genes of DCM and immune cell infiltration based on bioinformatics and machine learning], Xi Bao Yu Fen Zi Mian Yi Xue Za Zhi Chin. J. Cell. Mol. Immunol. **39**, 26–33 (2023)
72. Kaneko, M., Rosser, T., Raca, G.: DCM in a patient with autosomal dominant EEF1A2-related neurodevelopmental disorder. Eur. J. Med. Genet. **64**, 104121 (2021)
73. García-Manzanares, M., et al.: XPO1 gene therapy restores cardiac function in rats with chronic induced myocardial infarction, Molecular Biology. bioRxiv 575340 (2019)
74. Rivera, J., et al.: Nucleocytoplasmic transport and changes in gene expression in patients with heart failure. J. Am. Coll. Cardiol. **59**, E964 (2012)
75. Molina-Navarro, M.M., et al.: Heart failure entails significant changes in human nucleocytoplasmic transport gene expression. Int. J. Cardiol. **168**, 2837–2843 (2013)
76. Molina-Navarro, M.M., et al.: Functional networks of nucleocytoplasmic transport-related genes differentiate ischemic and dilated cardiomyopathies. New Therapeut. Opport. **9**, e104709 (2014)

Automatization of 3D Reconstruction of Coronary Arteries from Angiography Projections Using AI-Enhanced Segmentation Techniques

Tijana Djukic[1,2](✉), Ognjen Pavic[1,2], Lazar Dasic[1,2], Tijana Geroski[2,3], and Nenad Filipovic[2,3]

[1] Institute for Information Technologies, University of Kragujevac, Jovana Cvijica bb, 34000 Kragujevac, Serbia
tijana@kg.ac.rs

[2] Bioengineering Research and Development Center, BioIRC, Prvoslava Stojanovica 6, 34000 Kragujevac, Serbia

[3] Faculty of Engineering, University of Kragujevac, Sestre Janjica 6, 34000 Kragujevac, Serbia

Abstract. Fractional flow reserve (FFR) is a clinical diagnostics measurement performed to assess the physiological significance of stenosis potentially present in coronary arteries. Virtual fractional flow reserve (vFFR) is an alternative non-invasive approach that involves performing a 3D reconstruction of the patient-specific coronary artery and afterwards applying techniques of computational fluid dynamics (CFD) to obtain the vFFR value. Within this paper, a new approach is presented that combines the AI-enhanced segmentation of DICOM images obtained during X-ray angiography (XRA) examination and 3D reconstruction, and finite element (FE) mesh generation. This enables an automated reconstruction and numerical simulations of blood flow through patient-specific coronary arteries. The developed software is accurate, executes in a timely manner and is intuitive to use, which makes it a useful tool for clinicians to perform hemodynamic analyses of the state of coronary arteries. It can thus aid in the treatment planning that is adapted to the specific patient.

Keywords: X-ray angiography · image segmentation · 3D reconstruction · CFD · FFR calculation

1 Introduction

Fractional flow reserve (FFR) is a clinical diagnostics measurement performed to assess the physiological significance of stenosis potentially present in coronary arteries. Virtual fractional flow reserve (vFFR) is an alternative non-invasive approach that consists of performing a 3D reconstruction of the patient-specific coronary artery and afterwards applying techniques of computational fluid dynamics (CFD) to obtain the vFFR value. Several studies in the literature have proposed approaches to perform this task [1–3].

© The Author(s), under exclusive license to Springer Nature Switzerland AG 2025
N. Filipović (Ed.): AAI 2024, LNNS 1446, pp. 365–370, 2025.
https://doi.org/10.1007/978-3-031-99201-8_34

However, most of the previously proposed techniques require manual intervention from the clinician. In this study, an approach is presented that combines the AI-enhanced segmentation of DICOM images obtained during an X-ray angiography (XRA) examination and 3D reconstruction, and finite element (FE) mesh generation. In order to calculate the FFR equivalent, the approach proposed in the literature [4] is used. The so-called virtual functional assessment index (vFAI) is calculated. In the literature, it has been shown that this approach can accurately predict the clinically measured values of FFR [4, 5]. The entire combined approach is implemented in software that is fully automated and enables the user to analyze the results of the simulation of blood flow.

The paper is organized as follows. Section 2 presents the combined approach, the developed software is presented in Sect. 3, and Sect. 4 concludes the paper.

2 Materials and Methods

The segmentation of patient-specific DICOM images obtained during the XRA examination is initially performed using image filtering techniques to extract the region of interest. Image filtering was conducted in seven consecutive phases using different filters and image transformations. Namely, loaded XRA images were passed through a contrast limited adaptive histogram equalization (CLAHE) filter, followed by color inversion and top-hat transformation for creating a new base image in which coronary arteries were more distinguished from the surrounding heart tissue. A ridge detection hessian matrix filter was applied to determine the standout vessel bodies, followed by a Gaussian blur filter to reduce discontinuations along the main vessels and eliminate small disturbances in the image background. After image blurring, a system of 32 Gabor filters with different combinations of orientations and wavelength amplitudes was applied for edge detection. The final step in image filtering was determining the largest extracted continuous bodies of foreground pixels and merger of all other foreground pixels, that do not belong to this body with the background of the image since they do not belong to the main coronary artery tree or are too small to have an impact on the overall FFR calculation. This image filtering pipeline coupled with binary image skeletonization resulted in coronary artery centerline information necessary for the reconstruction in a 3D space. Additionally, branch length equalization based on multipoint matching and duplicate removal was conducted in order to avoid manual point annotation in multiple X-Ray views. Epipolar geometry based on XRA C-ARM position and matching point pair positions in different views was used to create the centerline and surface point cloud reconstructions in a 3D space.

This extracted data was then used in the 3D reconstruction process. The centerline points obtained during segmentation were used to define the parameterized centerline, as a non-uniform B-spline [6]. The surface point cloud was used to create the cross-sections (patches) along the parameterized centerline, which were further used to define the parametric NURBS surface (Non-Uniform Rational Basis Splines) [6]. The 3D FE mesh of the coronary artery was then created using the NURBS surface [7, 8].

The FFR value is clinically measured during the coronary angiography examination, under the induction of hyperemia, by measuring two pressure values. When the measured pressure distal to the arterial stenosis (P_d) is divided by the aortic pressure (P_a), the value of FFR is obtained. The vFAI index used in this study is calculated by performing two numerical simulations. These simulations were performed using the FE method [9], more specifically using the in-house developed software PakF [10]. The validation of this software was performed in many papers published in the literature [9–11] and during numerous EU-financed projects. In order to simulate the 3D blood flow, the Navier-Stokes equations and continuity equation were solved within the mentioned software. The blood is considered a viscous, incompressible fluid. The zero velocity is prescribed on all nodes on the wall of the artery. The two mentioned simulations that had to be performed in order to calculate the vFAI value differ only in the prescribed boundary conditions. In both cases the pressure of 100 mmHg is prescribed at the inlet, while the flow rate is prescribed at the outlet. For the first simulation, the flow rate is equal to 1 ml/s, and for the second the value of prescribed flow rate is 3 ml/s. These precise values are prescribed because they correspond to the average blood flow during rest and under stress, respectively. The pressure gradient is obtained for both considered cases, and then the relationship between pressure gradient ΔP and flow rate Q is defined as follows:

$$\Delta P = 0 + f_v Q + f_s Q^2 \tag{1}$$

where f_v and f_s are the coefficients of pressure loss due to viscous friction and flow separation, respectively [12]. The values of pressure gradient calculated for the two considered cases are used to calculate these coefficients, and then the following relation can be defined:

$$\frac{P_d}{P_a} = 1 - f_v \frac{Q}{P_a} - f_s \frac{Q^2}{P_a} \tag{2}$$

Using this relation, a graph of dependence of the value P_d / P_a from blood flow can be drawn and the vFAI can be calculated as the area under this curve, compared to the reference area. More details about this procedure can be found in the cited literature [4, 5], where this approach to calculate the FFR alternative was also extensively validated, demonstrating that it agrees well with clinically measured values of FFR for large clinical datasets.

3 Results

The proposed approach is implemented within the in-house developed software. The AI-enhanced segmentation is implemented in Python, primarily using the image processing libraries SkImage and OpenCV, as well as the numerical operation simplification library NumPy, while the 3D reconstruction is implemented in C++, with the interface and visualization implemented with the use of OpenGL and GLUT libraries.

Figure 1 shows the results of the segmentation, i.e. extracted centerlines and patches placed along the centerlines. On the left, the original patient-specific DICOM XRA images are shown.

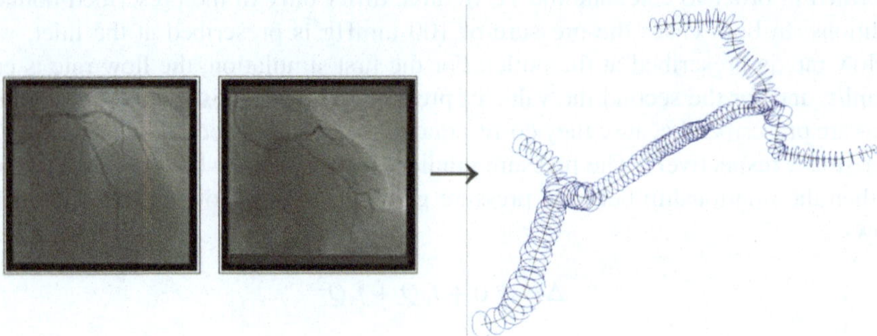

Fig. 1. Results of the automated AI-enhanced segmentation. The original patient-specific DICOM images (on the left) and the extracted centerline data together with obtained patches.

Figure 2 shows a screenshot of the developed software, after the performed numerical simulation. Within the numerical simulation, not only the value of FFR equivalent is calculated, but the full blood flow simulation is performed which enables the visualization of the hemodynamic quantities. In Fig. 2 the distribution of wall shear stress (WSS) is shown.

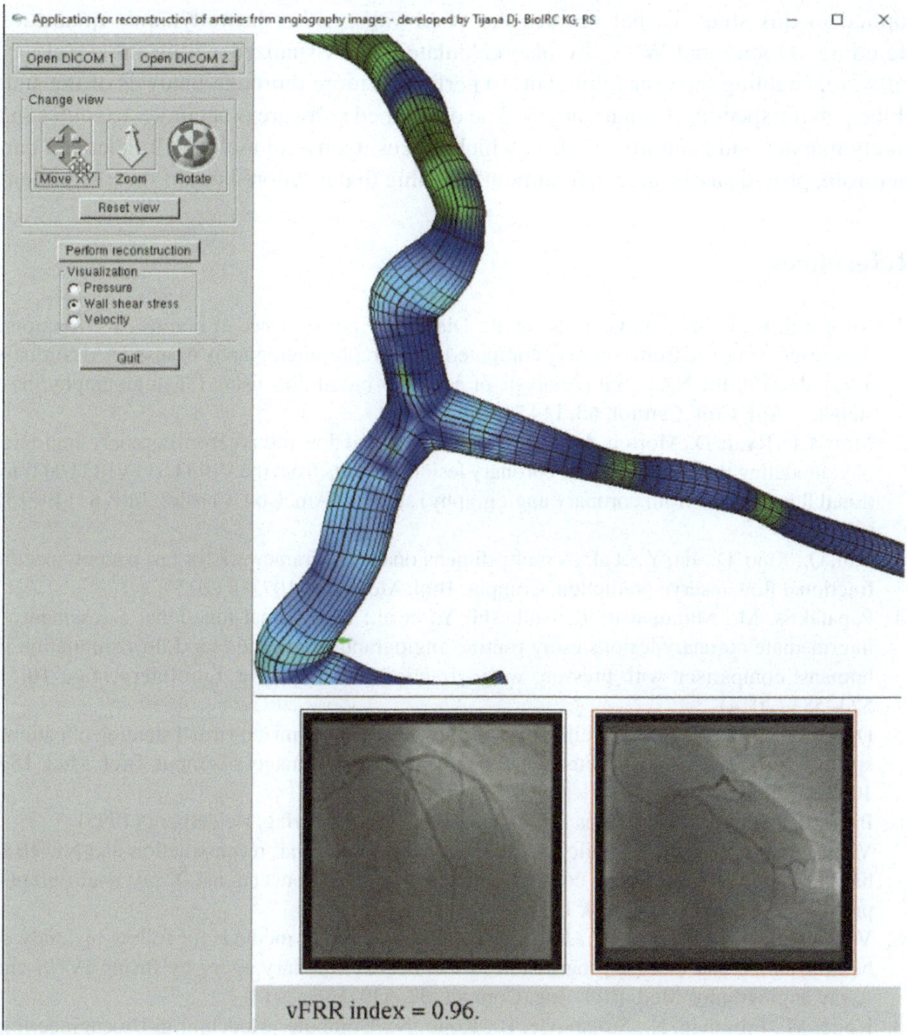

Fig. 2. Screenshot of the developed software, with calculated FFR equivalent value and shown distribution of WSS.

4 Conclusion

Within this study, an approach for automated segmentation and 3D reconstruction of patient-specific coronary angiography images is presented, with the aim to calculate the equivalent of invasively measured FFR, called vFAI. An important benefit of the vFAI calculations applied in the presented approach is that the results can be obtained within a few minutes, making it very applicable for automated software that executes almost in real time, as presented in this study. Another benefit of the numerical simulations applied in this study is that besides the vFAI value, other hemodynamic quantities, including velocity and WSS, are also calculated and visualized within the developed software, enabling the user (clinician) to perform a more thorough analysis of the state of the patient-specific coronary artery. The developed software is accurate, executes in a timely manner and is intuitive to use, which makes it a useful tool for clinicians. It can, therefore, provide assistance in treatment planning that is tailored to the specific patient.

References

1. Norgaard, B., Leipsic, J., Gaur, S., et al.: Diagnostic performance of noninvasive fractional flow reserve derived from coronary computed tomography angiography in suspected coronary artery disease: the NXT trial (Analysis of coronary blood flow using CT angiography: next steps). J. Am. Coll. Cardiol. **63**, 1145–1155 (2014)
2. Morris, P., Ryan, D., Morton, A., et al.: Virtual fractional flow reserve from coronary angiography: modeling the significance of coronary lesions: results from the VIRTU-1 (VIRTUal fractional flow reserve from coronary angiography) study. J. Am. Coll. Cardiol. Intv. **6**, 149–157 (2013)
3. Yan, Q., Xiao, D., Jia, Y., et al.: A multi-dimensional CFD framework for fast patient-specific fractional flow reserve prediction. Comput. Biol. Med. **168**, 107718 (2024)
4. Papafaklis, M., Muramatsu, T., Ishibashi, Y., et al.: Fast virtual functional assessment of intermediate coronary lesions using routine angiographic data and blood flow simulation in humans: comparison with pressure wire - fractional flow reserve. EuroIntervention **10**(5), 574–583 (2014)
5. Djukic, T., Tomasevic, S., Saveljic, I., et al.: Software for optimized virtual stenting of patient-specific coronary arteries reconstructed from angiography images. Comput. Biol. Med. **183**, 109311 (2024)
6. Piegl, L., Tiller, W.: The Nurbs Book, 2nd edn. Springer-Verlag, Heidelberg (1995)
7. Vukicevic, A., Çimen, S., Jagic, N., et al.: Three-dimensional reconstruction and NURBS-based structured meshing of coronary arteries from the conventional X-ray angi-ography projection images. Sci. Rep. **8**, 1711 (2018)
8. Vukicevic, A., Stepanovic, N., Jovicic, G., et al.: Computer methods for follow-up study of hemodynamic and disease progression in the stented coronary artery by fusing IVUS and X-ray angiography. Med. Biol. Eng. Comput. **52**, 539–556 (2014)
9. Kojic, M., Filipovic, N., Stojanovic, B., Kojic, N.: Computer Modeling in Bioengineering: Theoretical Background, Examples and Software. John Wiley and Sons, Chichester (2008)
10. Filipovic, N., Mijailovic, S., Tsuda, A., Kojic, M.: An implicit algorithm within the Arbitrary Lagrangian-Eulerian formulation for solving incompressible fluid flow with large boundary motions. Comp. Meth. Appl. Mech. Engrg. **195**, 6347–6361 (2006)
11. Parodi, O., Exarchos, T., Marraccini, P., et al.: Patient-specific prediction of coronary plaque growth from CTA angiography: a multiscale model for plaque formation and progression. IEEE Trans. Inf. Technol. Biomed. **16**(5), 952–956 (2012)
12. Gould, K.L.: Pressure-flow characteristics of coronary stenoses in unsedated dogs at rest and during coronary vasodilation. Circ. Res. **43**, 242–253 (1978)

Cardiac Segmentation Using UNETR: A Transformer Based Deep Learning Approach on the ACDC Dataset

Tijana Geroski[1,2,4(✉)], Ognjen Pavić[2,3], Lazar Dašić[2,3], Amir Amini[4], and Nenad Filipović[1,2]

[1] Faculty of Engineering, University of Kragujevac, Sestre Janjic 6, 34000 Kragujevac, Serbia
{tijanas,fica}@kg.ac.rs

[2] Bioengineering Research and Development Center (BioIRC), Prvoslava Stojanovica 6, 34000 Kragujevac, Serbia
{opavic,lazar.dasic}@kg.ac.rs

[3] Institute for Information Technologies Kragujevac, University of Kragujevac, Jovana Cvijica bb, 34000 Kragujevac, Serbia

[4] Department of Electrical and Computer Engineering, University of Louisville, Louisville, KY 40292, USA
amir.amini@louisville.edu

Abstract. Cardiac segmentation is essential for accurate diagnosis and treatment planning in cardiovascular diseases. Deep learning methods, particularly convolutional neural networks (CNNs), have improved the efficiency of segmentation by capturing local spatial features. However, transformers offer better modeling of long-range dependencies and global context, making them suitable for image processing tasks. This study utilizes the UNet Transformer (UNETR) architecture for cardiac segmentation on the Automatic Cardiac Diagnosis Challenge (ACDC) dataset, which includes cine-MRI images of the heart. The dataset consisted of 150 patients, with manual annotations for three semantic classes: left ventricle (LV), myocardium (MYO), and right ventricle (RV). The model was fine-tuned on the ACDC dataset with pre-trained ImageNet weights. UNETR's integration of transformers into the traditional UNet framework shows potential; however, data-hungry transformer-based structure does not capture complex dependencies with a smaller size dataset. Through extensive experiments and comparison with foundational models (Segment Anything Model (SAM) and MedSAM), our approach demonstrates competitive performance and generalization capabilities.

Keywords: deep learning · heart structure · transformers · medical image processing · multiclass cardiac segmentation

1 Introduction

Cardiovascular diseases have become the leading cause of death globally, highlighting the need for accurate diagnostic methods [1]. Accurate creation of personalized heart models is crucial for their clinical application, but this process is mainly performed through manual segmentation by experienced experts. Manual segmentation

© The Author(s), under exclusive license to Springer Nature Switzerland AG 2025
N. Filipović (Ed.): AAI 2024, LNNS 1446, pp. 371–385, 2025.
https://doi.org/10.1007/978-3-031-99201-8_35

can be subjective, inconsistent, and time-consuming, while timely diagnosis is critical in clinical settings. Therefore, minimizing the time required for heart modeling makes the development of automated segmentation methods essential for effectively implementing personalized heart models in clinical practice.

Accurate segmentation of cardiac structures in medical imaging, such as Magnetic Resonance Imaging (MRI) and Computed Tomography (CT), is crucial for evaluating heart function, guiding treatment plans, and tracking disease progression. This process typically involves detecting key structures such as the left ventricle (LV), right ventricle (RV), and myocardium (MYO), which are the basis for assessing cardiac health by calculating metrics such as ejection fraction (EF), myocardial mass, and ventricular volumes [1]. These metrics are used for diagnosing various heart conditions, such as heart failure. Advanced segmentation methods can improve the precision of these evaluations and support more personalized treatment strategies [2]. However, obtaining high-quality segmentation often depends on large, annotated datasets, which are difficult to gather due to the significant cost and specialized expertise required in medical imaging.

In addition, the challenges in achieving high accuracy segmentation include difficulties in capturing complex cardiac structures, insufficient contextual information, unclear edges, and the challenge of balancing global and local details. Deep learning, particularly convolutional neural networks (CNNs), have become the state-of-the-art of cardiac MRI segmentation. This success is primarily driven by the appearance of encoder-decoder networks such as the Fully Convolutional Network (FCN) [3] and UNet [4]. Many variations of U-shaped networks have been developed since the baseline UNet, including UNet++ [5], UNet3+ [6], Attention UNet [7], Swin UNet [8] etc. Models like UNet and its variations excel in capturing fine details and preserving contextual information, making them the standard for medical image segmentation.

After the introduction of UNet [4], researchers began focusing on optimizing the U-shaped encoder-decoder architecture. In UNet++ [5], the architecture was improved by incorporating dense skip connections between modules. However, Huang et al. [6] found that these dense skip connections in UNet++ only focused on high-level features, limiting the network's ability to capture full-scale information. This limited its capability to accurately learn organ locations and boundaries. To address the limitation, they introduced full-scale skip connections in UNet3+, connecting every encoder layer to all corresponding layers in both the encoder and decoder. Meanwhile, Oktay et al. introduced Attention UNet [7], the first UNet variant to integrate attention mechanisms. They incorporated Attention Gates (AG) before the feature fusion stage in the skip-connected decoders, generating a gating signal that suppresses irrelevant regions in the image, allowing the model to highlight essential features for specific tasks while reducing noise. Additionally, several UNet variants integrate unique residual blocks designed to enhance feature extraction [9, 10].

Recently, transformer-based models have gained attention by improving the ability to capture long-range dependencies in images, offering potential for improved segmentation accuracy [8]. Despite these advancements, some challenges remain, such as anatomical variability, imaging artifacts, and the need for models that generalize well across diverse populations and imaging conditions.

Building upon the success of transformers in Natural Language Processing (NLP), transformers have quickly advanced in visual understanding tasks. Dosovitskiy et al. [11] introduced the Visual Transformer (ViT), which utilizes a self-attention mechanism to capture global information, achieving state-of-the-art results in image classification. To improve efficiency and reduce dependence on large datasets for generalization, several adaptations of the Visual Transformer have been proposed [12, 13]. Swin UNet [8] is the first architecture specifically designed for medical image processing using a fully Transformer-based approach. In this case, "fully" means that the Transformer layers handle feature extraction and processing directly, without depending on any pretrained backbone architecture. However, ViT is limited by the drawback that its performance of using spatial context is not as good as CNNs. Therefore, this limitation could be overcome combining ViTs and CNN architecture [14] which achieved state-of-the-art performance on the multi-organ segmentation dataset.

Recently, the Segment Anything Model (SAM) was introduced as a groundbreaking foundational model for image segmentation [15]. Built upon the Vision Transformer architecture [11], SAM was trained on an extensive dataset of 11 million images containing 1 billion masks. The key advantage of SAM lies in its strong zero-shot segmentation performance, enabling it to handle effectively previously unseen datasets and tasks. This is achieved through various prompts, such as points and boxes, which guide pixel-level semantics and region-level positions for target objects. SAM has demonstrated remarkable performance in a wide range of segmentation challenges [15].

Using foundational SAM's pre-trained models, several studies have further explored its performance in diverse zero-shot segmentation scenarios. Most studies though investigate only one mode (automatic, point prompts or box prompts), as well as investigate its application in segmenting everyday objects. Two studies examined SAM's performance in Everything mode (automatic mode) for segmenting camouflaged objects [16, 17]. The findings revealed that SAM struggled in these scenarios, such as detecting camouflaged animals that blend into their natural surroundings. The authors noted that SAM also failed to identify occluding defects in industrial contexts. Additionally, [18] investigated three testing methods for SAM (points, boxes, and everything) across various applications. Their tasks included natural image segmentation (focusing on noticeable, but also, camouflaged, and transparent objects, as well as shadow detection), agriculture (crop segmentation, pest monitoring, and leaf disease detection), manufacturing (anomaly and surface defect identification), and remote sensing (building and road extraction). They concluded that, while SAM performs well in certain scenarios, such as segmentation of objects that are mostly noticeable, it falls short in other applications. Furthermore, they demonstrated that human prompts can significantly enhance segmentation results compared to the automatic Everything (automatic mode) approach.

To evaluate SAM's performance on medical data, [19] compiled and organized 53 open-source datasets, resulting in a large medical segmentation dataset that includes 18 modalities, 84 objects, 125 object-modality paired targets, 1,050,000 2D images, and 6,033,000 masks. In addition, [20] concluded that performance of SAM varies significantly across the 19 evaluated medical imaging datasets, showing optimal results when box annotations are provided for each object component, outperforming other methods in interactive prompt base mode, and indicating its potential as a valuable tool for medical image segmentation when applied correctly.

The simplest method for adapting SAM to medical image segmentation is to fine-tune the model directly on the specific task [21]. In that sense, MedSAM [22] has been developed for universal medical image segmentation, adapting SAM by collecting a diverse and comprehensive dataset that includes over one million medical image-mask pairs across 11 modalities. MedSAM achieved median Dice Similarity Coefficient (DSC) scores of 94.0%, 94.4%, 81.5%, and 98.4% for segmentation tasks involving intracranial hemorrhage in CT scans, glioma in MR T1 scans, pneumothorax in chest X-rays, and polyps in endoscopy images, respectively, outperforming specialized UNet models. However, they concluded that MedSAM encounters difficulties when segmenting vessel-like branching structures, as the ambiguity in bounding box prompts can complicate these scenarios.

Therefore, transformer-based networks have a limitation in their ability to extract low-level features, similar to convolutional operations [14], potentially leading to the omission of some details. To address this issue, we investigate UNETR, a medical image segmentation framework that integrates Transformers and CNNs with UNet. To utilize the advantages of transformers, we incorporate transformer blocks to capture high-resolution features and spatial location information during the encoder feature learning phase, while CNN blocks are employed in the decoder phase to establish long-range dependencies. We also investigate the depth and number of transformers, as well as other hyperparameters, to perform segmentation on ACDC cardiac dataset. Additionally, we investigate the capability of the network to recognize the slices where cardiac structures of interest are not present. Our results demonstrate that this architecture benefits from using both convolutional and transformer advantages, improving medical image segmentation. Experimental results are compared with state-of-the-art foundational models such as Segment Anything Model (SAM) and MedSAM.

2 Materials and Methods

The materials and methods section gives details regarding the architecture and implementation of the UNETR, including the specific components utilized in the encoder and decoder, as well as the dataset description and evaluation metrics used for segmentation of cardiac structures. The hardware setup used was NVIDIA Quadro RTX 6000 GPU, 64 GB of RAM, and an Intel(R) Xeon(R) Gold 6240R CPU operating at 2.40 GHz. The network was developed using the Python programming language, utilizing TensorFlow and Keras libraries for implementation.

2.1 Dataset

This study was conducted retrospectively using publicly available human subject data - Automated Cardiac Diagnosis Challenge (ACDC) dataset [23]. The dataset was originally collected for the MICCAI 2017 (Medical Image Computing and Computer-Assisted Intervention) challenge.

The ACDC dataset is a well-established benchmark for cardiac structure segmentation in MRI scans, consisting of cine-MRIs from 150 patients. It includes short-axis MRI cine slices covering both the left and right ventricles (LV and RV), from the base to the apex of the heart, with slice thicknesses between 5 to 8 mm and inter-slice gaps of 5 to 10 mm. The spatial resolution ranged from 1.37 to 1.68 mm per pixel. Each cardiac cycle was recorded across 28 to 40 frames, capturing both end-diastole (ED) and end-systole (ES) phases, and the data was stored in NIfTI format.

The patients in the dataset represent a range of conditions, including normal cardiac function (NOR), dilated cardiomyopathy (DCM), hypertrophic cardiomyopathy (HCM), myocardial infarction-induced heart failure (MINF), and abnormal right ventricle (ARV). Each MRI scan is annotated with three key cardiac structures: left ventricle (LV), right ventricle (RV), and myocardium (MYO) (Fig. 1).

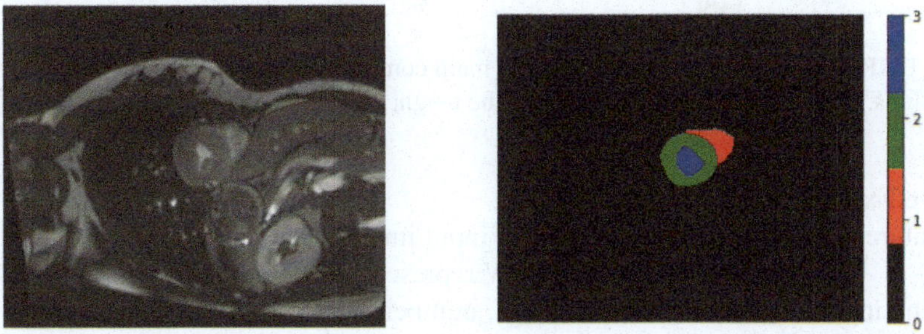

Fig. 1. An example of a cardiac MRI image from the ACDC dataset, along with manually labelled target regions (ground truth) - background, left ventricle (LV), myocardium (MYO), and right ventricle (RV).

For neural network processing, the 3D DICOM images were converted into 2D slices. The dataset was already split into 100 patients for training and 50 for testing by the challenge rules. The training dataset contained 2 446 images, which were further split into training and validation subset in the ratio of 80–20%. The testing dataset was kept separate from training and validation and contained 532 images. The slices also included images where no cardiac structure is present, enabling the investigation of whether the model would be able to recognize these cases as well.

2.2 Proposed Methodology

The architecture of UNETR is inspired and has the basis in U-shaped design, consisting of two key parts: the encoder and the decoder. The encoder is responsible for extracting features and consists of feature extraction blocks, transformer blocks, and downsampling layers. On the decoder side, it uses classical convolutional neural network (CNN), meaning no transformer blocks, but instead convolutional and deconvolutional blocks, working together to generate the final output. To improve the accuracy and recovery of fine image details, cross connections are integrated symmetrically between corresponding feature maps in the encoder and decoder, ensuring more precise feature propagation across the network (Fig. 2).

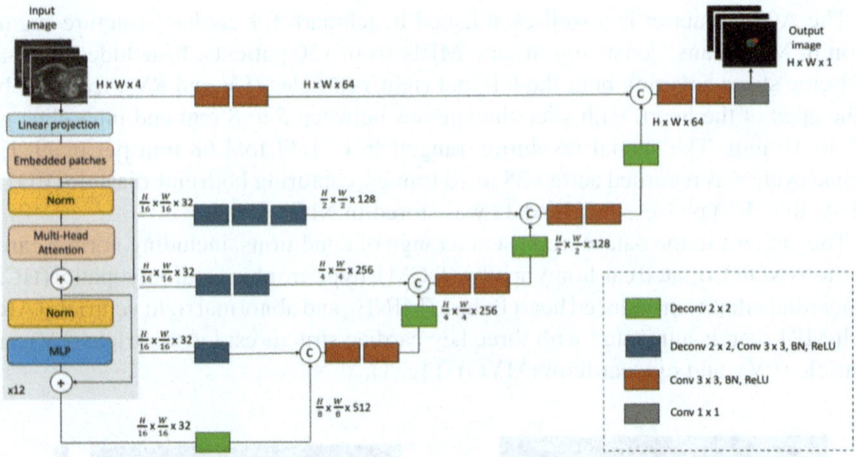

Fig. 2. UNRTR architecture consists of three main components: the encoder, decoder, and cross connections. Dimensions H and W represent the height and width, respectively.

Feature Extraction Block

The feature extraction block converts the input image I into a high-dimensional tensor $I \in R\frac{H}{4} \times \frac{W}{4}$, where dimensions H and W represent the height and width, respectively. Since the input is a 2-dimensional image, number of channels and sequences is 1. We follow the proposed idea by Chen et al. [14], to create linear projection of input image and create image embeddings.

Transformer Block

After the feature extraction block, embedded patches are passed into the Transformer block. This structure, colored in light grey, maximizes the transformer's ability to capture long-range dependencies by linking the high-resolution features from the previous layer with the multi-scale features obtained through convolutional downsampling in the next layer. As shown in Fig. 2, each Transformer block includes a LayerNorm (Norm) layer, a multi-headed self-attention module, a skip connection, LayerNorm (Norm) layer again, followed by multilayer perceptron (MLP) with skip connection and GELU activation. Transformer blocks are then stacked in consecutive order for a number of times noted in the left down corner of Transformer block. For the self-attention calculation, we adopt the method proposed by [24, 25], incorporating a relative position bias to enhance the model's spatial awareness.

Convolutional Up-Sampling

The up-sampling process is designed in such a way to follow the convolutional and deconvolutional structure. Stepwise deconvolution is employed to convert the feature map into the output, effectively reconstructing the adjacent feature map dimensions. This reduces the feature dimension to half its original size in each layer. Furthermore, skip connections are utilized to merge the features extracted by the transformer encoder with the up-sampled features from the decoder. A final convolution operation 1×1 is applied to produce the final output.

Skip Connection
Similar to UNet [4], skip connections are employed to integrate multiscale features from the encoder with the up-sampled features in the decoder. This approach merges both shallow and deep features, effectively minimizing the loss of spatial information that typically occurs during the down-sampling process [14].

2.3 Segment Anything Model (SAM) and Medical Segment Anything Model (MedSAM)

The Segment Anything Model (SAM) is designed to segment a specific object in an image, based on prompts provided by the user. These prompts can include a single point, multiple points, a bounding box, or even a text. The model is expected to produce a valid segmentation mask. The underlying principle of this approach is that the model has learned the concept of an object, enabling it to segment any indicated object. This capability gives SAM significant potential to segment types of objects it has not previously encountered, demonstrating strong performance in zero-shot learning scenarios. While the model can technically operate without a prompt to identify any visible object, this approach is not anticipated to be effective for medical images, where multiple objects often exist alongside the one of interest. In the case of medical images, we are, however, interested in the automatic segmentation of the tissue of interest. We have introduced several tunable parameters in the automatic mask generation that control how densely points are sampled and what the thresholds are for removing low quality or duplicate masks. Additionally, generation was automatically run with post-processing to remove stray pixels and holes. Since this approach segments multiple objects, the choice of best object to correspond to LV, MYO and RV was performed in such a way to determine the object that has the least distance to the metric defined as a sum of Dice coefficient and object centroids between the generated objects by SAM and ground truth mask. It should be emphasized that we have used SAM as is, and these approaches did not involve fine tuning of the SAM/MedSAM.

2.4 Evaluation Metrics

Evaluation metrics for assessing the performance of the proposed methods were Dice Similarity Coefficient (DSC) and the Intersection over Union (IoU), also known as the Jaccard Index. These metrics quantify the degree of overlap between the predicted segmentation mask and the ground truth, helping to evaluate the accuracy and reliability of the segmentation model. They are particularly effective in measuring how well a model captures complex boundaries and structures, making them standard metrics used in segmentation tasks.

The Dice Similarity Coefficient (DSC) measures the overlap between the predicted and ground truth regions and is defined as:

$$DSC = \frac{2 \times |A \cap B|}{|A| + |B|} \tag{1}$$

where A and B represent the set of pixels in the predicted mask and the ground truth mask, respectively. The numerator $2 \times |A \cap B|$ captures the number of overlapping pixels, while the denominator $|A| + |B|$ accounts for the total number of pixels in both masks. The DSC ranges from 0 to 1, where 1 indicates perfect overlap and 0 indicates no overlap. A higher DSC value implies better segmentation performance, making it a useful metric for assessing binary or multi-class segmentation accuracy, especially when the goal is to balance sensitivity and specificity.

The Intersection over Union (IoU), also known as the Jaccard Index, is another metric that measures the ratio of the intersection to the union of the predicted and ground truth masks and is calculated as:

$$\mathrm{IOU} = \frac{|A \cap B|}{|A \cup B|} \tag{2}$$

In this formula, $|A \cap B|$ denotes the intersection between the predicted and ground truth masks, while $|A \cup B|$ indicates the union (combined area) of both. IoU values range from 0 to 1, where 1 means complete agreement between the predicted and actual regions. The IoU is generally considered stricter than the DSC because it penalizes false positives and false negatives more heavily. While DSC focuses on overlapping regions, IoU emphasizes the overall extent of overlap relative to the union, making it useful for assessing segmentation performance, particularly in datasets with high-class imbalance.

3 Results and Discussion

The default optimizer used for UNETR is Adam with a momentum of 0.99. For the loss function, we tested several different loss functions (categorical cross-entropy, Dice, IoU, Tversky). The model is trained for 25 epochs, with incorporated early stopping. The number of batches per epoch was set to 50 and the number of samples per batch was set to 32. The learning rate that proved to achieve the best performance was 10^{-1}.

In terms of optimal architecture, it was found that the best combinations were the number of filters equals to [64, 128, 256, 512], number of stacked ViTs was 12, number of attention heads was 12, number of MLP nodes was 12, activation function of MLP nodes was GELU, number of convolutional layers per downsampling level/block = 2, number of convolutional layers (after concatenation) per upsampling level/block = 2, activation was ReLU. The training and validation loss curve illustrates the model's convergence, with the training loss steadily decreasing over time and the validation loss following a similar trend, indicating effective learning and generalization (Fig. 3).

The visual outputs of the segmentation process for the left ventricle (LV), right ventricle (RV), and myocardium (MYO) are presented in Fig. 4, with each class clearly outlined by color-coded probability maps. In these visualizations, red indicates areas with a high probability of belonging to a particular class, while blue represents regions with no probability, which illustrates the model's confidence in segmentation for each cardiac structure.

The numerical results, including metrics such as Dice coefficient and IoU, demonstrate the model's high accuracy in segmenting the LV, RV, and MYO regions, with consistently strong performance across all classes (Table 1).

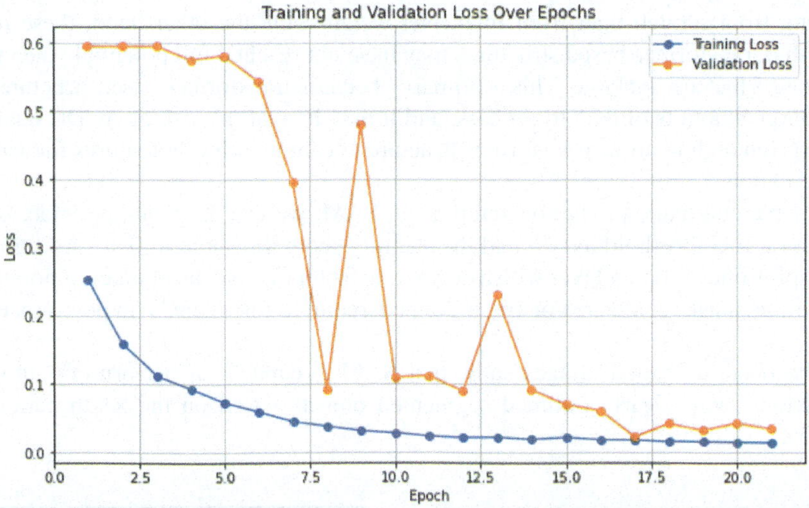

Fig. 3. Training and validation loss using UNETR architecture.

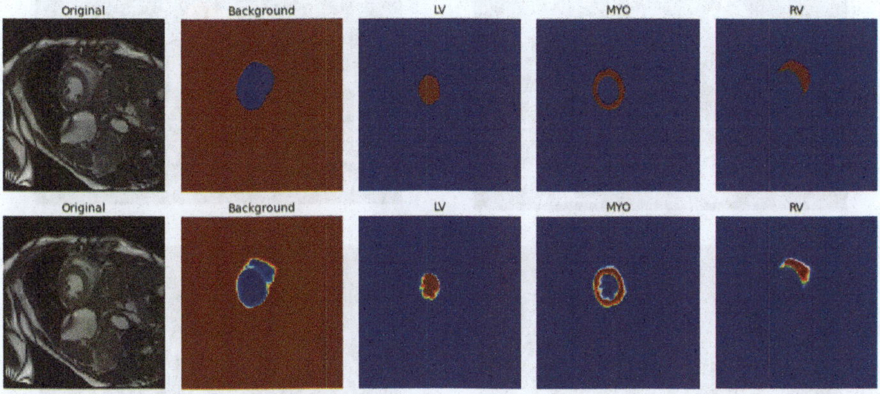

Fig. 4. Example of segmentation outputs using UNETR – original image, background, LV, MYO, RV (up – ground truth masks, down – predicted masks).

Table 1. Evaluation metrics for LV, RV, and MYO segmentation.

Baseline	IOU			Dice			HD (pixels)		
	LV	RV	MYO	LV	RV	MYO	LV	RV	MYO
Scratch	0.704	0.650	0.565	0.826	0.788	0.722	15.652	36.331	20.149
ImageNet	0.782	0.608	0.578	0.878	0.671	0.731	37.563	45.276	39.127

It can be seen that LV and MYO are better detected starting with ImageNet weights; however, RV is better detected when learning from scratch. Also, HD is smaller when learning from scratch than from ImageNet weights. On the other hand, these results show that transformer-based structures may have not reached full potential when tested on datasets that are not large. This is primarily because transformer-based structures rely upon large amounts of data to process, and it may be that the size of ACDC dataset is not large enough to allow transformer structure to capture relevant features for accurate segmentation.

For the automatic prediction setup using SAM, we use 32 points per side, set the prediction IoU threshold to 0.9, and the stability score threshold to 0.96. Additionally, we employ one cropping layer with a downscale factor of 2 for the number of points, and a minimum mask region area of 100 is defined, required for OpenCV in post-processing stage.

The resulting tested images show that SAM is capable of performing automatic predictions, with clearly outlined segmented objects based on the setup parameters (Fig. 5).

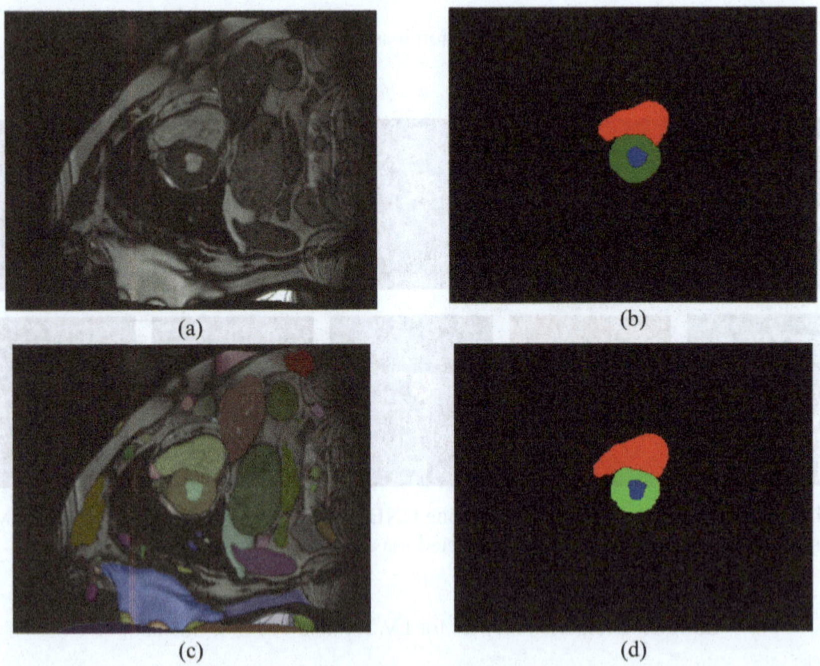

Fig. 5. Example of correct segmentation outputs using SAM – original image (a), target ground truth mask (b), object prediction using SAM (c), post processed image to extract only relevant objects LV, MYO, RV (d).

It was shown that foundational SAM model, used out of the box and tested on the ACDC dataset, was not always able to correctly extract relevant objects and often missed the class representing MYO (either combining it with RV or not detecting it at all). One such example of incorrect segmentation using SAM is presented in Fig. 6, where it did not capture the MYO, highlighting the limitations in accurately segmenting all relevant structures.

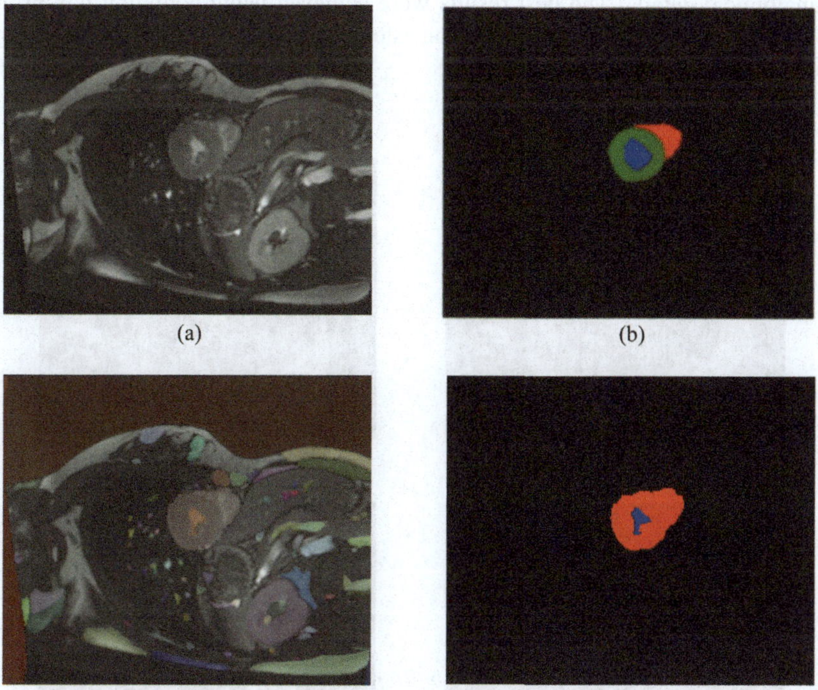

(a) (b) (c) (d)

Fig. 6. Example of incorrect segmentation outputs using SAM – original image (a), target ground truth mask (b), object prediction using SAM (c), post processed image to extract only relevant objects LV, MYO, RV (d).

The numerical results for the metrics Dice and IoU for testing SAM in automatic detection without manual prompting are presented in Table 2.

Table 2. Evaluation metrics for LV, RV, and MYO segmentation using SAM.

	DICE			IOU		
	RV	MYO	LV	RV	MYO	LV
Average	0.606	0.360	0.723	0.544	0.265	0.630
Standard deviation	0.391	0.305	0.280	0.393	0.248	0.300

In testing MedSAM without any prompts, the automatic segmentation results were poor, failing to accurately detect relevant structures in cardiac images. When using point-based prompting, the performance improved slightly but still struggled with complex anatomical details. The best results were achieved using bounding box prompts, which provided the most accurate and consistent segmentations across different image modalities (Fig. 7).

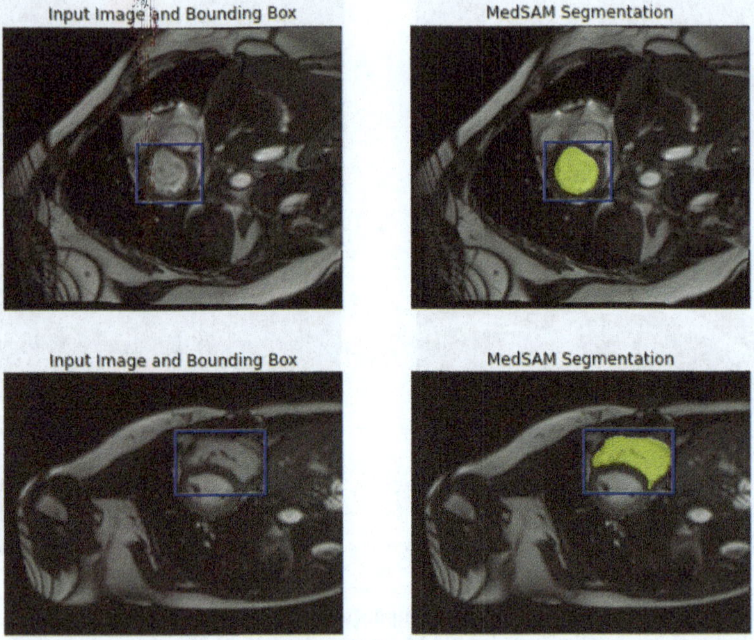

Fig. 7. Example of correct segmentation outputs using MedSAM – original image with bounding box as prompt, predicted segmentation (example for LV and RV).

Despite the improved performance with bounding box prompts, MedSAM still faced challenges in accurately segmenting certain structures such as MYO (Fig. 8).

Although the results from this paper seem promising, the proposed methodology also has some limitations. Even though transformers are excellent at capturing long-range dependencies, they may not capture fine, low-level features as effectively as CNNs. This can lead to challenges in segmenting small structures like thin boundaries in medical images (i.e. MYO). In addition, transformers are designed for larger datasets, so small size dataset such as ACDC may not contain enough information to extract relevant features to accurately segment complex and variable anatomical structures. The Segment Anything Model (SAM) and MedSAM, when tested without fine-tuning, did not perform well on certain cardiac structures such as MYO, in zero-shot automatic settings. This indicates that general-purpose models may require task-specific fine-tuning for medical applications.

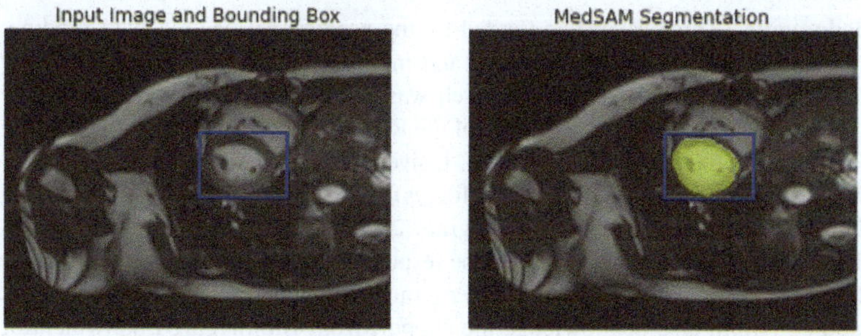

Fig. 8. Example of incorrect segmentation outputs using MedSAM – original image with bounding box as prompt and predicted segmentation (fails to differentiate between LV and MYO).

Future directions will be aimed at improving the integration between transformers and CNNs, particularly at the feature fusion stage. This could involve experimenting with advanced skip connections or hybrid architectures that optimize both local and global feature extraction. Fine-tuning SAM or MedSAM for specific medical imaging tasks, such as cardiac segmentation, could significantly improve their performance. Exploring more advanced prompt techniques, such as combining points, boxes, and masks in different ways, could also improve segmentation accuracy. These future directions provide a roadmap for enhancing the utility of the proposed models in clinical applications and overcoming current limitations in medical image segmentation.

4 Conclusions

This paper explores the use of a hybrid deep learning model combining convolutional neural networks (CNNs) and transformers for cardiac segmentation tasks. The focus of segmentation are key heart structures - left ventricle (LV), myocardium (MYO), and right ventricle (RV) using the UNETR architecture, which integrates transformers to capture global context and CNNs to capture fine details. The ACDC dataset, consisting of cine-MRI images, is used for model training and testing. The proposed UNETR model consists of an encoder that employs transformer blocks for global feature extraction and a decoder that utilizes CNNs for fine-grained segmentation. The skip connections between the encoder and decoder ensure precise feature propagation. The model's performance is compared with state-of-the-art approaches such as the Segment Anything Model (SAM) and MedSAM. The UNETR model demonstrated competitive performance, achieving high accuracy in segmenting LV, MYO, and RV structures. The DSC and IoU metrics for UNETR showed superior results compared to SAM, particularly in complex anatomical structures. While SAM and MedSAM struggled with the segmentation of MYO and performed inconsistently across the ACDC dataset, UNETR consistently achieved better segmentation accuracy due to its ability to balance global and local feature extraction. The integration of transformer blocks within the UNETR architecture enables the model to capture long-range dependencies and global context, which are critical for accurate medical image segmentation. Future work may explore further refinements and adaptations for broader clinical applications.

Acknowledgments. This research is funded by the project that has received funding from the European Union's Horizon Europe research and innovation programme under grant agreement No 101080905 (STRATIFYHF). This research was also supported by the Ministry of Science, Technological Development and Innovation of the Republic of Serbia, contract numbers [451-03-65/2024-03/200107 (Faculty of Engineering, University of Kragujevac)] and [451-03-66/2024-03/200378 (Institute for Information Technologies)]. T. Geroski acknowledges financial support by the Fulbright U.S. Visiting Scholar Program 2024–2025, which is sponsored by the U.S. Department of State. Its contents are solely the responsibility of the author and do not necessarily represent the official views of the Fulbright Program or the Government of the United States. T. Geroski and A. Amini also acknowledge the support from Bioimaging Endowment of Medical Imaging Lab of the University of Louisville.

Competing Interests. The authors declare they have no competing interests.

Ethics Approval. This study was conducted retrospectively using publicly available human subject data, specifically the Automated Cardiac Diagnosis Challenge (ACDC) dataset [23].

References

1. Chen, C., et al.: Deep learning for cardiac image segmentation: a review. Front. Cardiovasc. Med. **7**(25), 1–33 (2020)
2. Litjens, G., et al.: A survey on deep learning in medical image analysis. Med. Image Anal. **42**, 60–88 (2017)
3. Long, J., Shelhamer, E., Darrell, T.: Fully convolutional networks for semantic segmentation. In: Proceedings of the IEEE Conference on Computer Vision and Pattern Recognition (2015)
4. Ronneberger, O., Fischer, P., Brox, T.: U-net: Convolutional networks for biomedical image segmentation. In: Medical Image Computing and Computer-Assisted Intervention–MICCAI 2015: 18th International Conference, Munich, Germany (2015)
5. Zhou, Z., Rahman Siddiquee, M., Tajbakhsh, N., Liang, J.: UNet++: a nested u-net architecture for medical image segmentation. In: Deep Learning in Medical Image Analysis and Multimodal Learning for Clinical Decision Support: 4th International Workshop, DLMIA 2018, and 8th International Workshop, ML-CDS 2018, Held in Conjunction with MICCAI 2018, Granada, Spain (2018)
6. Huang, H., et al.: UNet 3+: A full-scale connected UNet for medical image segmentation. In: ICASSP 2020–2020 IEEE International Conference on Acoustics, Speech and Signal Processing (ICASSP) (2020)
7. Oktay, O., et al.: Attention U-Net: learning where to look for the pancreas. arXiv preprint arXiv:1804.03999 (2018)
8. Cao, H. et al.: Swin-Unet: Unet-like pure transformer for medical image segmentation. In: Karlinsky, L., Michaeli, T., Nishino, K. (eds.) Computer Vision – ECCV 2022 Workshops. ECCV 2022. Lecture Notes in Computer Science, vol. 13803, pp. 205–218. Springer, Cham (2023). https://doi.org/10.1007/978-3-031-25066-8_9
9. Xiao, X., Lian, S., Luo, Z., Li, S.: Weighted Res-UNet for high-quality retina vessel segmentation. In: Proceedings of the 9th International Conference on Information Technology in Medicine and Education (ITME). IEEE (2018)

10. Kaul, C.: Models of visual attention in deep residual CNNs. Doctoral dissertation, University of York (2019)
11. Dosovitskiy, A.: An image is worth 16 × 16 words: transformers for image recognition at scale. arXiv preprint arXiv:2010.11929 (2020)
12. Yuan, L., et al.: Tokens-to-Token ViT: training vision transformers from scratch on imagenet. In: Proceedings of the IEEE/CVF International Conference on Computer Vision (2021)
13. Xia, Z., Pan, X., Song, S., Li, L., Huang, G.: Vision transformer with deformable attention. In: Proceedings of the IEEE/CVF Conference on Computer Vision and Pattern Recognition (2022)
14. Chen, J., et al.: TransUNet: transformers make strong encoders for medical image segmentation. arXiv preprint arXiv:2102.04306 (2021)
15. Kirillov, A., et al.: Segment anything. In: Proceedings of the IEEE/CVF International Conference on Computer Vision (2023)
16. Tang, L., Xiao, H., Li, B.: Can SAM segment anything? When SAM meets camouflaged object detection. arXiv preprint, arXiv:2304.04709 (2023)
17. Ji, G., Fan, D., Xu, P., Cheng, M., Zhou, B., Van Gool, L.: SAM struggles in concealed scenes–empirical study on "Segment Anything". arXiv preprint, arXiv:2304.06022 (2023)
18. Ji, W., et al.: Segment anything is not always perfect: an investigation of SAM on different real-world applications. Mach. Intell. Res. **21**, 617–630 (2024)
19. Huang, Y., et al.: Segment anything model for medical images? Med. Image Anal. **92**, 103061 (2024)
20. Mazurowski, M., Dong, H., Gu, H., Yang, J., Konz, N., Zhang, Y.: Segment anything model for medical image analysis: an experimental study. Med. Image Anal. **89**, 102918 (2023)
21. Zhang, Y., Shen, Z., Jiao, R.: Segment anything model for medical image segmentation: current applications and future directions. Comput. Biol. Med., 108238 (2024)
22. Ma, J., He, Y., Li, F., Han, L., You, C., Wang, B.: Segment anything in medical images. Nat. Commun. **15**(1), 654 (2024)
23. Bernard, O., et al.: Deep learning techniques for automatic MRI cardiac multi-structures segmentation and diagnosis: is the problem solved? IEEE Trans. Med. Imaging **37**(11), 2514–2525 (2018)
24. Hu, H., Gu, J., Zhang, Z., Dai, J., Wei, Y.: Relation networks for object detection. In: Proceedings of the IEEE Conference on Computer Vision and Pattern Recognition (2018)
25. Hu, H., Zhang, Z., Xie, Z., Lin, S.: Local relation networks for image recognition. In: Proceedings of the IEEE/CVF International Conference on Computer Vision (2019)

Evaluation of Using Balanced and Unbalanced Data for Smart City Solution Based on IoT Using Classification and AdaboostM1

Valerij Grasic(✉)

Telekom Slovenia, Cigaletova 17, 1000 Ljubljana, Slovenia
valerij.grasic@telekom.si

Abstract. Smart cities are still at their beginnings. They add intelligence to the existing city systems, enabling more to be achieved with less. Artificial intelligence can be helpful in this domain. The result is applications with transparent real-time information that aid in making better choices. This can be useful for everyday cases, but also for many natural conditions like earthquakes, tsunamis, floods, and sleet, which cause major disasters. The question is how to set up the concept of a smart city so that the city would be more resilient to disasters and climate change. The problem that needs to be addressed is to forecast the daily number of incoming calls based on data from IoT sensor devices, which includes river flows and weather data. The proposal is made for a safe city system, which is a smart city subsystem. The data refer to Slovenia as a whole and consist of both balanced and imbalanced data. AdaBoostM1 classification algorithm is used. The given evaluation will help in a more detailed analysis and in preparation of a comprehensive concept for the blueprint of a smart city.

Keywords: IoT · AdaBoostM1 · AdaBoost · machine learning · smart city · classification · safe city

1 Introduction

Smart cities are still in their nascent stages. They add intelligence to the existing city systems, enabling more to be achieved with less. The outcome is a spectrum of applications providing transparent, real-time information to facilitate informed decision-making. Artificial intelligence (AI) stands out as one of the most important key enablers in this transformation. This technological advancement holds promise not only for addressing everyday urban challenges, but also for mitigating the impact of natural disasters such as earthquakes, tsunamis, floods, and inclement weather conditions, which pose significant threats to urban resilience. All these introduce novel perspectives and challenges. Given work is motivated by a desire to advance the discourse surrounding smart city development, aiming to provide fresh insights into the blueprint of a resilient urban ecosystem.

The question is how to set up the concept of a smart city so that it would be more resilient to disasters and climate change. The paper is addressing the daily number

© The Author(s), under exclusive license to Springer Nature Switzerland AG 2025
N. Filipović (Ed.): AAI 2024, LNNS 1446, pp. 386–401, 2025.
https://doi.org/10.1007/978-3-031-99201-8_36

of incoming calls to the smart city system, more precisely to a safe city system as a subsystem. The main evaluation is based on the use of both balanced and imbalanced data sets. The AdaBoostM1 is used to make models for different periods, both annually and for individual quarters. The idea is to do a detailed analysis based on different data options and models for a given case. Input data consist of 176 different data points on a daily base and include mainly river flows and weather data from Internet of things (IoT) sensor devices. The data points refer to Slovenia as a whole, which acts like a smart city or community.

The problem statement is delineated, followed by an overview of related work. Subsequently, the methodology is outlined, detailing the steps undertaken in data collection, data preprocessing, and model preparations. The results of the evaluation are then presented, offering insights into the efficacy of various approaches. Finally, the findings are discussed, highlighting conclusions and future research proposals.

2 Smart Cities

The topic of smart cities is becoming more and more popular. Especially due to the development of technology such as IoT, cloud, sensors, and digitalization as such. A very important topic within the area is also artificial intelligence.

2.1 The Concept of Smart Cities

According to the definition of the European Commission a smart city is a place where traditional networks and services are made more efficient using digital and telecommunication technologies, for the benefit of its residents and businesses [1]. According to the definition of the English government, the concept of smart cities is not static [2]. There is no absolute definition of a smart city, but rather a process or series of steps that make cities more alive and flexible and, therefore, able to respond more quickly to new challenges. The guidelines for smart cities are going in the direction that problem-solving should be more innovative and intelligent [3] and that systems and solutions within should be made more intelligent [4].

2.2 Smart Cities and Increasing Resilience

Smart city solutions can increase the resilience of cities and their citizens. The solutions for smart cities are different and cover different areas. Solutions in smart cities involve monitoring, observability, theoretical models, control systems, event management, alarming, and other related aspects. It can also be related to public safety, like 112 or 911 systems themselves.

In the case of 911 system, the question can be related to incoming calls for each of the special services separately, which can be ambulance, firefighters or the police. The subject of the research is the prediction of incoming 911 calls for firefighters in case of fires [5]. The authors analyzed fire-related call data for Victoria, Australia. Using machine learning, the authors were able to track the progress of a fire with a spatial component in real-time. Predicting the number of crimes using 10 classes for the next day by a deep neural network was analyzed in the cities of Chicago (Illinois, USA) and Portland (Oregon, USA), where weather data, general data, and data for public transport [6] were considered. Achieved accuracy was 75.6% for Chicago and 65.3% for Portland. Deep Convolutional Neural Networks (CNN) were used to predict crime risk for the next day in the case of New York City (New York, USA) [7]. They achieved a value of 88% for F1 and 92% for ROC.

Research work also refers to the total number of incoming calls for all special services. This is an issue of classifying incoming calls for a smart city [8], with a proposal on how to prepare a more general model for such a case [9]. The issue is also how to increase the resilience [10].

3 Methodology Description

3.1 Predictive Modeling

The evaluation is based on machine learning. The term machine learning refers to the automated detection of significant patterns in data [11]. Classification is part of machine learning [12], where predicting a qualitative response for an observation can be characterized as classifying that observation because it involves assigning the observation to a category or class. By definition [13] predicting class labels is a classification which is different from predicting values (e.g. using regression techniques) which is called a prediction.

The proposal behind the evaluation is to convert the predicted number of incoming calls into the appropriate class. In October 2014, floods hit Slovenia. Figure 1 shows the movement of the number of incoming calls to the 112 system on a daily basis. Within the figure, a proposal is given on how to classify the number of incoming calls into four classes according to the number of incoming calls. The division is demarcated by three horizontal lines.

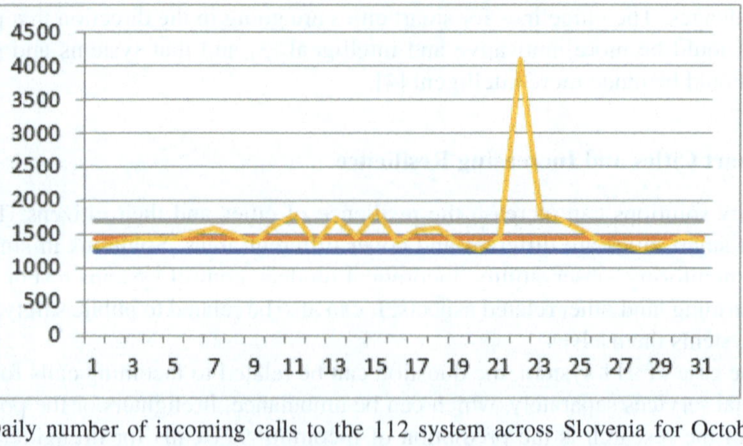

Fig. 1. Daily number of incoming calls to the 112 system across Slovenia for October 2014. Three horizontal lines give the proposal on how to classify the number of incoming calls into four classes.

3.2 Data Used

The motivation for preparing the data was to consider those factors that have caused the greatest damage in recent years, such as floods and ice. Another factor is that data should be from the whole Slovenia and that the data are distributed according to the whole Slovenia. The whole data set consists of 176 attributes. Weather data include 10 weather stations on each station with 16 attributes. For river flows, 16 attributes are used. Two attributes are for traffic accidents. The full dataset is available on GitHub [14].

Data for incoming calls are obtained from the StatKlic application [15]. Figure 2 shows the StatKlic application with basic data for the given day.

Statklic Poročila▾

Osnovni podatki za dan

Ura	0	1	2	3	4	5	6	7	8	9	10	11	12	13	14	15	16	17
Vse regije																		
Število odgovorjenih klicev	15	11	24	11	6	7	21	28	32	34	47	56	64	69	74	40	46	49
Št. operaterjev	2	1	2	1	1	1	2	2	2	2	2	2	3	3	3	2	2	2
Število neodgovojenih klicev	0	3	1	1	3	0	1	2	12	4	8	6	8	10	8	5	8	13
Število klicev	15	14	25	12	9	7	22	30	44	38	55	62	72	79	82	45	54	62

Fig. 2. The StatKlic application with basic data for the given day (StatKlic) [15].

The weather data and the river flow data are obtained from the open data system of the National Meteorological Service of Slovenia [16]. Data on traffic accidents are obtained from the Slovenian Police. Figure 3 shows an example of two automatic measuring stations for river flows [17].

Fig. 3. Automatic measuring stations for river flows [17] (photo by Marko Burger, 2011) (CC BY-ND 4.0 by ARSO).

For weather data, 16 attributes are prepared for each weather station: average, the highest and the lowest daily temperature (ºC), average wind speed (m/s), cloud cover

(%), humidity (%), precipitation amount (mm), snow depth (cm), height of new snow (cm), duration of solar radiation (hours), stormy wind flag (0 – no, 1 – yes), storm flag (0 – no, 1 – yes) and freezing rain flag, hail and frost (0 – no, 1 – yes). For river flows, 16 measuring points are used. At each measuring point as an attribute, the river flow is measured in m3/s. For traffic accidents, the number of accidents and the number of people involved in accidents are used.

3.3 AdaBoostM1

Boosting learning is one of the group methods that performs learning for basic classifiers on different samples. Reinforcement learning algorithms build multiple models from a data set, using some other possible models, such as decision trees or neural networks, which do not need to be very good at building models. The Adaptive Boost Learning Classifier (the AdaBoost variant) is an adaptive boost learning model of ensemble methods that works especially well in the case of decision trees. The most popular algorithm is AdaBoostM1, which was introduced in 1996 by Freund and Schapire [18].

In general, an ensemble method using weak learning systems can lead to a strong model. But, there are also exceptions because, in certain cases, adaptive reinforcement learning is not exactly the best. This applies to cases where there are not enough data or where weak models are too complex. The advantage of adaptive reinforcement learning is also that it is more robust to any data changes, and this is because it is a group method.

3.4 Data Preparation for the Classification

The basic characteristics of the data used for the evaluation are as follows:

- Data granulation is daily. Data classification is made into two and four classes.
- Two data sets are used, for the first and the second period. The data set for the first period contains data for the years 2013 to 2016, with 1,186 data instances. Both model and evaluation are made based on the data for the first period.
- The data set for the second period covers the year 2018, with 365 data instances. The evaluation is made on the models prepared for the first period.
- In the case of division into two classes, 10 to 15% of all calls are in the alarm class, and the rest are in the regular class. The limit between the two classes is set at 1,700 calls per day.
- In the case of four classes, the regular class is further divided into small, normal, and increased traffic classes, thereby obtaining additional information about the volume of the calls. The limit between classes is set to 1,250, 1,450 and 1,700 calls per day.

The tool used is Weka [19].

3.5 Data Imbalance Handling

An important issue is data imbalance. In the case of the two classes, the majority class contains between 80 and 90% of all data instances, which means the data are not completely balanced. Learning from imbalanced data is an important research question. The field has been actively studied since 2000. An overview survey contains over 220 references on the topic of imbalanced data [21]. Conventional algorithms are often biased towards the majority class, as their loss functions seek to optimize algorithm performance by minimizing error rates while ignoring the distribution of the data, i.e. the distribution of the minority class [22]. Models that have poor predictive performance for the minority class are also a problem because the minority class is usually more important and, therefore, more sensitive to imbalances than the majority class [23]. So far, many techniques have been tried, leading to different results and few clear answers. The question of what to do if the data are imbalanced does not have a precise answer, because everything depends on the data.

The practical guidelines for solving the problem in the case of imbalanced data propose four solutions [22]. These solutions are leaving it as they are, balancing the training data set, making changes at the algorithm level or developing a completely new algorithm. Following the proposal, the first and the second approach are being used. For the first approach, the data are kept as they are, which means they remain imbalanced. Additionally, stratification is used. For the second approach, balancing is performed, with the use of oversampling or under sampling.

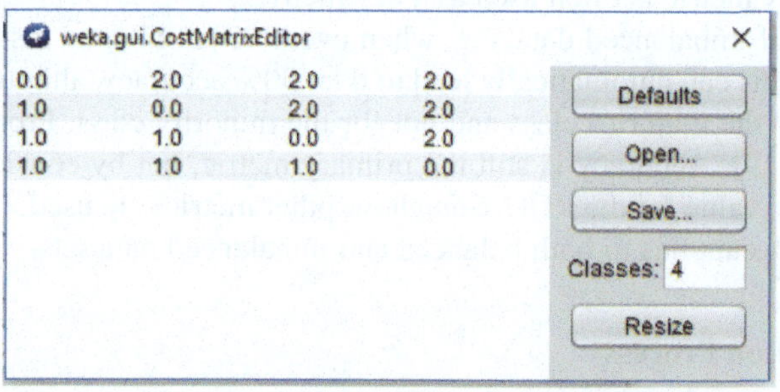

Fig. 4. Cost matrix for cost sensitivity in the case of four classes (by Weka).

Cost sensitivity must be taken into account in the case of balancing. The data to be set are parameters C1 and C2. The goal is to get and find such a combination of C1 and C2 where the accuracy is the highest. Figure 4 shows the cost matrix in the case of four classes as seen in Weka, by Knowledge Flow. The upper right parameters are C1, and the lower left parameters are C2. The given example shows the case when C1 = 2.0 and C2 = 1.0.

Table 1 shows a description of the whole selection process for C1 and C2. It is based on the first-period data. Such a selection process is valid for both two and four classes. Since in the case of two classes the ratio of the number of instances between the majority and minority class is 7.18 (1041/145), the optimal value is sought in twice this ratio. This means the optimal values for C1 and C2 are being sought in the range from 0.6 to 15.0. In the case of four classes, the ratio of the number of instances between the largest number of instances to the class with the smallest number of instances is 2.8 (411/145). Therefore, an optimal value for C1 and C2 in the range of 0.6 to 4.0 is sought.

Table 1. Selection process for *C1* and *C2* for the cost sensitivity for the first period data

Classes	Handling of *C1* and *C2*	Description of the selection process
2 classes	*C1* = 1.0 till 15.0, by step 1.0; *C2* = 1.0 *C1* = 0.6 till 3.5, by step 0,1; *C2* = 1.0 *C1* = 1.0; *C2* = 1.0 till 15.0, by step 1.0 *C1* = 1.0; *C2* = 0.6 till 3.5, by step 0.1	Choose *C1* and *C2*, where the accuracy is the best
4 classes	*C1* = 0.6 till 3.5 by step 0.1; *C2* = 1.0 *C1* = 1.0; *C2* = 0.6 till 3.5 by step 0.1	Choose *C1* and *C2*, where the accuracy is the best

3.6 Evaluation Metrics

For the evaluation, the typical verification metrics for the classification are used. These metrics are accuracy, precision, recall, F1 and ROC. ROC-curve is a two-dimensional graph that visually depicts the performance and performance gain for a classification model. A ROC area approaching 1 means the best classifier. If the value for ROC is 0.5, it means a random guess.

The accuracy metric is often used as a key metric.

In the case of imbalanced data, e.g. when over 90% of all data instances are in the majority class, this can automatically lead to over 90% accuracy, although such a result is valid only for the majority class and not for the minority class. Proposals are made for such cases [19]. Accuracy is still the primary metric, but by considering F1 and a sufficiently large value for the ROC. Somehow, other metrics are used: recall, precision, and ROC. All this applies to both balanced and imbalanced data sets.

3.7 Classification Process

The classification process covers both imbalanced and balanced data. In the case of evaluating models based on imbalanced data, the entire model evaluation process is simplified. First, data normalization is performed. Then classification is performed, and the associated model is built using cross-validation. The results of the obtained metrics are checked and compared with each other. The data used for each of the cross-validation is the same for each of the methods, which is provided in the Weka tool with the parameter seed = 1.

In the case of balanced data, data normalization is performed. This is followed by data balancing using the SMOTE technique [24]. Then a model is set up, using cost sensitivity. The optimal values of C1 and C2 are sought, where the accuracy is the best, in conjunction with the measure metric F1 and ROC.

When the preparation of data is over, the entire classification process goes on. For the first period (2013–2016), a classification for two and four classes is made. The classification is repeated also for the second period (2018). In the hope of obtaining even better classification results, the data of the first period are additionally divided by quarters. Models are made and model testing is performed, which applies to classes two and four, and for each quarter separately (Q1–Q4). All this is also done for the second period.

4 Evaluation

The evaluation for different cases for classification is given below. First, the evaluation of the setting of the P parameter for the purpose of data balancing is given. Then, different classification situations are presented, both for different classes as well as for different time periods. They include various data, both balanced and imbalanced. Finally, a discussion of the results is given.

4.1 Setting the P Value in the Case of Balancing (SMOTE)

Table 2 shows setting of P parameter in the case of balancing, which is done by using SMOTE. This is valid for the first period data. All different options are covered.

Table 2. Setting of the *P* parameter in the case of balancing (SMOTE) for the first period data

Class	Class index	Time period	Number of instances-original	Number of instances-new	P value for given class
2	0/1	year	1041/145	1041/1015	0/600
4	1/2/3/4	year	223/411/407/145	411	84.3/0/1.0/183.4
2	0/1	Q1	335/26	335	0/1188.5
	0/1	Q2	226/47	226	0/380.9
	0/1	Q3	228/48	228	0/375
	0/1	Q4	252/24	252	0/950
4	1/2/3/4	Q1	118/119/98/26	119	0.9/0/21.5/358.7
	1/2/3/4	Q2	19/73/134/47	134	605.3/83.6/0/185.2
	1/2/3/4	Q3	22/97/109/48	109	395.5/12.4/0/127.1
	1/2/3/4	Q4	64/122/66/24	122	90.7/0/84.9/408.4

4.2 Evaluation of Accuracy for Balanced Data for Two and Four Classes

Table 3 shows the evaluation for balanced data, for two classes. It refers to the key metric, which is accuracy, while also considering the F1 metric. The best results are in bold (where the value is more than 0.75). The data refer to the first (p1) or second period (p2), covering the entire period (y) or given quarter (Q1-Q4). Also, given are the values for C1 and C2, where the best results for accuracy were achieved.

In the case of the two classes, the conclusion is that it is possible to make models that provide good results. For the first period (2013–2016), good results are achieved for the entire period and by quarters. For the second period (2018), good results were also achieved in Q3 and Q4. The worse results for the second period are expected due to the lag in data and because, above all, weather influences are different for individual quarters.

Table 3. Evaluation of accuracy for balanced data for two classes based on the first (p1) or second (p2) period covering the whole period (y) or given quarter.

Data set	TP_1	TP_2	TP	$F1_1$	$F1_2$	F1	*C1/C2*
p1, y	**0.819**	0.739	**0.780**	**0.790**	**0.768**	**0.779**	1.1/1.0
p2, y	0.510	**0.792**	0.529	0.669	0.181	0.637	1.1/1.0
p1, Q1	**0.893**	**0.937**	**0.915**	**0.913**	**0.917**	**0.915**	1.0/1.2
p1, Q2	0.748	**0.912**	**0.830**	**0.814**	**0.843**	**0.828**	0.8/1.0
p1, Q3	**0.877**	**0.833**	**0.855**	**0.858**	**0.852**	**0.855**	1.0/1.1
p1, Q4	**0.873**	**0.917**	**0.895**	**0.892**	**0.897**	**0.895**	1.0/0.7
p2, Q1	0.667	0.333	0.656	**0.789**	0.061	0.765	1.0/1.2
p2, Q2	0.663	0.545	0.648	**0.768**	0.273	0.708	0.8/1.0
p2, Q3	**0.779**	0.667	**0.772**	**0.865**	0.276	**0.826**	1.0/1.1
p2, Q4	**0.955**	0.500	**0.935**	**0.966**	0.400	**0.941**	1.0/0.7

Table 4. Evaluation of accuracy of balanced data for four classes based on the first (p1) or second (p2) period covering the whole period (y) or given quarter.

Data set	TP_1	TP_2	TP_3	TP_4	TP	$F1_1$	$F1_2$	$F1_3$	$F1_4$	F1	C_1/C_2
p1, y	**0.971**	0.000	0.000	0.611	0.395	0.520	0.000	0.000	0.540	0.265	2.0/1.0
p2, y	0.735	0.000	0.000	**0.833**	0.282	0.615	0.000	0.000	0.172	0.202	2.0/1.0
p1, Q1	0.000	0.689	0.101	**0.849**	0.410	0.000	0.482	0.150	0.607	0.310	1.0/1.4
p1, Q2	**0.948**	0.000	0.000	**0.791**	0.435	0.564	0.000	0.000	0.599	0.291	1.0/1.3
p1, Q3	**0.954**	0.000	0.000	**0.908**	0.466	0.603	0.000	0.000	0.641	0.311	1.4/1.0
p1, Q4	0.533	0.000	0.000	**0.984**	0.379	0.568	0.000	0.000	0.477	0.261	1.0/2.7
p2, Q1	0.000	0.535	0.000	0.333	0.267	0.000	0.495	0.000	0.047	0.238	1.0/1.4
p2, Q2	0.333	0.000	0.000	**1.000**	0.165	0.364	0.000	0.000	0.239	0.077	1.0/1.3
p2, Q3	0.077	0.000	0.000	**1.000**	0.076	0.143	0.000	0.000	0.124	0.028	1.4/1.0
p2, Q4	0.059	0.000	0.000	**0.750**	0.065	0.087	0.000	0.000	0.077	0.052	1.0/2.7

Table 4 shows the evaluation for balanced data, for four classes. It refers to the key metric, which is accuracy, while also considering the F1 metric. There is no situation where the results are good.

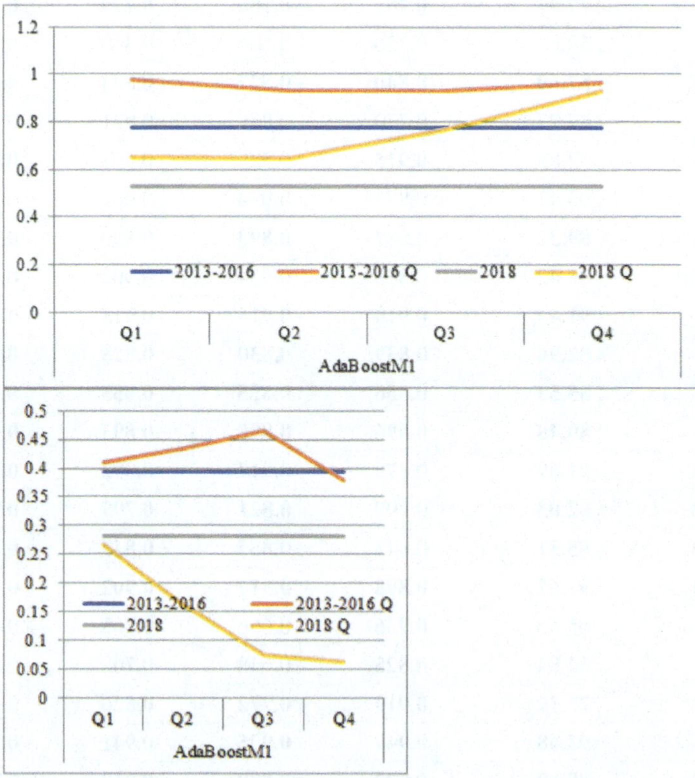

Fig. 5. Evaluation of accuracy on balanced data for two (left) and four (right) classes.

Figure 5 shows the evaluation of accuracy on balanced data for two (left) and four (right) classes. It shows that for the two classes, the accuracy results are good, both for the entire first period and by quartiles, as well as for Q3 and Q4 for the second period.

4.3 Detailed Evaluation in the Case of Two Classes

Table 5 shows a detailed evaluation of two classes based on the first (p1) or second (p2) period covering the whole period (y) or given quarter. Both balanced (B) and imbalanced (imB) data are covered. Good results where the value is above 0.7 are bolded. Additionally, for the whole period, ZeroR (Zero) is also included, which simply predicts the majority category (class) and has no predictability power. But, it is useful for determining baseline performance.

Main conclusions for the evaluation of two classes with data for Slovenia:

Table 5. Detailed evaluation of two classes based on the first (p1) or second (p2) period covering the whole period (y) or given quarter for both balanced (B) and imbalanced (imB) data.

Data set	Acc (%)	Prec	Rec	F1	ROC
p1, y, B	**77.97**	**0.781**	**0.780**	**0.779**	**0.861**
p1, Zero, B	50.63	0.256	0.506	0.340	0.498
p1, y, imB	**87.69**	**0.840**	**0.877**	**0.843**	**0.765**
p1, Zero, imB	87.77	0.770	0.878	0.821	0.491
p2, y, B	52.88	**0.915**	0.529	0.637	**0.723**
p2, Zero, B	93.42	0.873	0.934	0.902	0.500
p2, y, imB	**89.32**	0.887	**0.893**	**0.890**	**0.734**
p2, Zero, imB	93.42	0.873	0.934	0.902	0.500
p1, Q1, B	**91.49**	**0.916**	**0.915**	**0.915**	**0.963**
p1, Q2, B	**82.96**	**0.839**	**0.830**	**0.828**	**0.892**
p1, Q3, B	**85.53**	**0.856**	**0.855**	**0.855**	**0.923**
p1, Q4, B	**89.48**	**0.896**	**0.895**	**0.895**	**0.943**
p1, Q1, imB	**91.69**	0.875	**0.917**	**0.892**	**0.846**
p1, Q2, imB	**82.05**	0.790	**0.821**	**0.799**	**0.733**
p1, Q3, imB	**85.51**	0.843	**0.855**	**0.847**	**0.813**
p1, Q4, imB	**91.67**	**0.898**	**0.917**	**0.902**	**0.757**
p2, Q1, B	65.56	**0.936**	0.656	**0.765**	0.644
p2, Q2, B	64.84	**0.825**	0.648	**0.708**	0.669
p2, Q3, B	**77.17**	**0.919**	**0.772**	**0.826**	0.670
p2, Q4, B	**93.48**	**0.949**	**0.935**	**0.941**	**0.727**
p2, Q1, imB	**88.89**	**0.932**	**0.889**	**0.910**	0.628
p2, Q2, imB	**87.91**	**0.849**	**0.879**	**0.853**	**0.821**
p2, Q3, imB	**78.26**	**0.877**	**0.783**	**0.825**	0.685
p2, Q4, imB	**91.30**	**0.945**	**0.913**	**0.927**	**0.722**

- In the case of unbalanced data for the whole period, the results for the second period (p2) are even better than for balanced data, both accuracy and ROC.
- The results for balanced data are better for quarters than for the whole period. This is valid for the first period.
- For the first period, by individual quarters, the results for balanced data are better than for imbalanced data.
- In the case of the second period, by individual quarters, the results for unbalanced data in several cases are better than in the case for balanced data.

Figure 6 shows the evaluation of classification into two classes for the first period (p1) and unbalanced data for two classes in the case of data for Slovenia. Main metrics

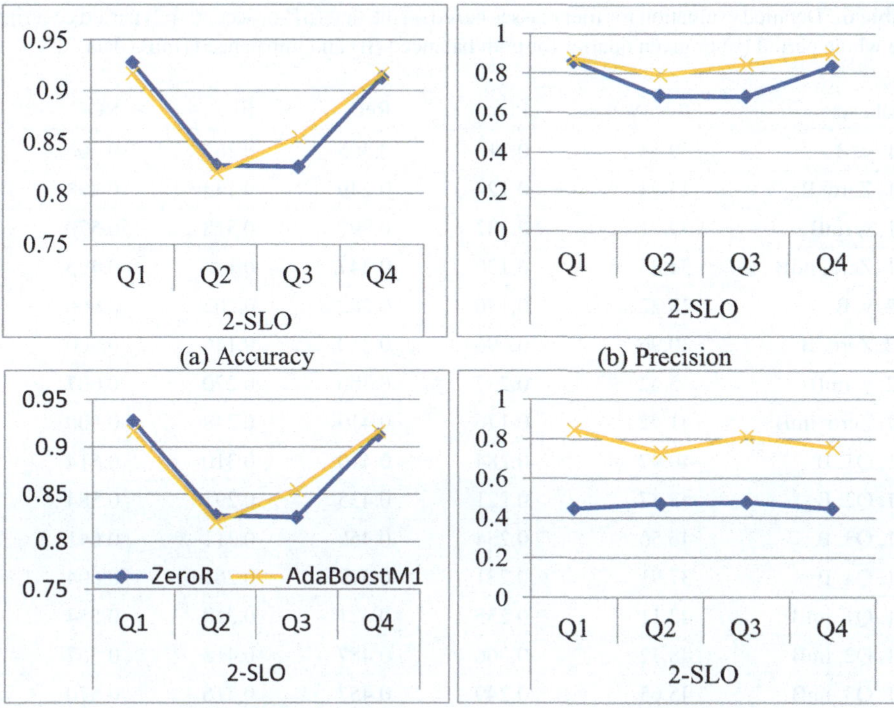

(a) Accuracy (b) Precision (c) Recall (d) ROC

Fig. 6. Evaluation of classification for the first period (p1) and unbalanced data for two classes, comparing AdaBoostM1 and baseliner ZeroR.

are used, and methods AdaBoostM1 and baseliner ZeroR are compared. The accuracy in the case of AdaBoostM1 is slightly better than the baseliner ZeroR, with the ROC for AdaBoostM1 being much larger.

4.4 Detailed Evaluation of Four Classes

Table 6 shows a detailed evaluation of four classes based on the first (p1) or second (p2) period covering the whole period (y) or given quarter. Both balanced (B) and imbalanced (imB) data are covered. For the whole period, ZeroR (Zero) is also included, which simply predicts the majority category (class) and has no predictability power. But, it is useful for determining baseline performance.

Main conclusions for the evaluation of four classes with data for Slovenia:

- In the case of four classes AdaBoostM1 is not a good choice. This is valid for both balanced and imbalanced data sets.
- However, it seems that, in certain cases, the results are still better than the baseline.
- The best results are around ROC = 0.6, and the accuracy is up to 40 to 50%.
- The results in the case of balanced or imbalanced data are similar.

Table 6. Detailed evaluation for four classes based on the first (p1) or second (p2) period covering the whole period (y) or given quarter for both balanced (B) and imbalanced (imB) data.

Data set	Acc (%)	Prec	Rec	F1	ROC
p1, y, B	39.54	0.210	0.395	0.265	0.606
p1, Zero, B	24.94	0.124	0.249	0.133	0.498
p1, y, imB	39.71	0.282	0.397	0.318	0.570
p1, Zero, imB	34.65	0.120	0.347	0.178	0.495
p2, y, B	28.22	0.170	0.282	0.202	0.606
p2, Zero, B	30,96	0,096	0,310	0,146	0,500
p2, y, imB	35.62	0.233	0.356	0.270	0.607
p2, Zero, imB	41.92	0.176	0.419	0.248	0.500
p1, Q1, B	40.97	0.284	0.410	0.310	0.614
p1, Q2, B	43.47	0.221	0.435	0.291	0.584
p1, Q3, B	46.56	0.234	0.466	0.311	0.641
p1, Q4, B	37.91	0.231	0.379	0.261	0.604
p1, Q1, imB	42.11	0.255	0.421	0.310	0.554
p1, Q2, imB	48.72	0.366	0.487	0.418	0.567
p1, Q3, imB	45.65	0.249	0.457	0.316	0.570
p1, Q4, imB	45.29	0.304	0.453	0.361	0.556
p2, Q1, B	26.67	0.221	0.267	0.238	0.497
p2, Q2, B	16.48	0.069	0.165	0.077	0.512
p2, Q3, B	7.61	0.146	0.076	0.028	0.503
p2, Q4, B	6.52	0.094	0.065	0.052	0.470
p2, Q1, imB	13.33	0.113	0.133	0.101	0.413
p2, Q2, imB	39.56	0.305	0.396	0.322	0.513
p2, Q3, imB	40.22	0.276	0.402	0.315	0.554
p2, Q4, imB	36.96	0.678	0.370	0.224	0.518

4.5 Evaluation for the Transition from Two to Four Classes by Quarters

Figure 7 shows the evaluation of the classification for the transition from two to four classes by quarters for data for Slovenia for the first period (p1) and imbalanced data (imB). This may be appropriate for the case of analyzing a change from a two-class system to a multi-class system. Graphs are presented for key metrics by quarter.

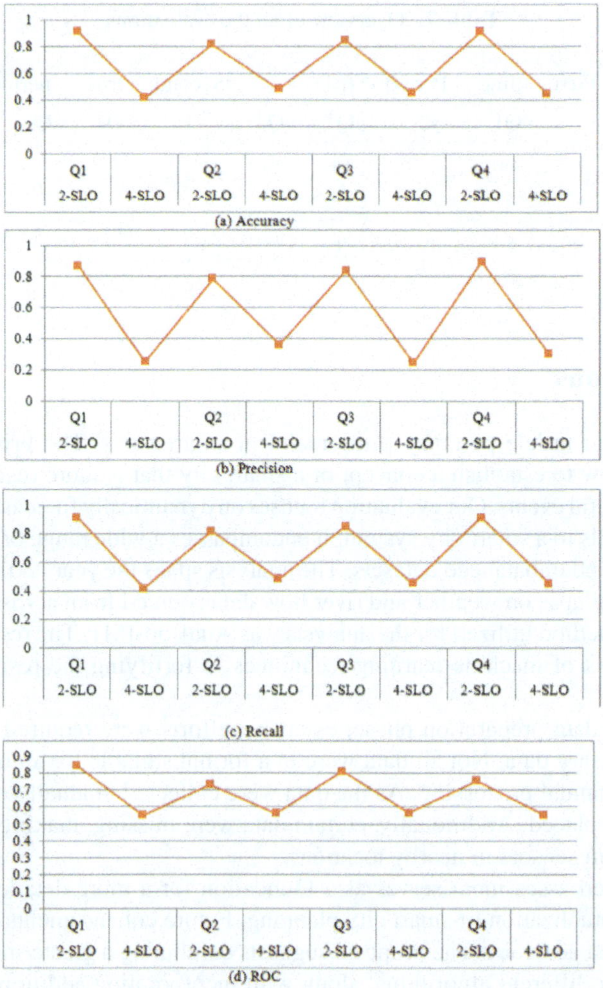

Fig. 7. Detailed evaluation of the classification for the transition from two to four classes by quarters for data for Slovenia for the first period (p1) and imbalanced data (imB) data.

4.6 Discussion

Table 7 shows an overview of all the evaluations according to quarters (Q1-Q4) or for the entire period (y), for balanced (B) and unbalanced data (imB), and for two (2) or four classes (4). Good results are marked by (+), and very good by (++).

The AdaBoostM1 method has proven to be a good choice for cases with two classes. The AdaBoostM1 method has good accuracy and ROC results on the second-period dataset. The results for the balanced data are much better. Surprisingly, in the case of two classes and a second period, the results for balanced data can be even better. On the other hand, the AdaBoostM1 method does not perform well in the case of four classes.

Table 7. Overview of all the evaluations.

Data	Class	First	time	Period	(p1)		Second	time	period	(p2)	
		y	Q1	Q2	Q3	Q4	y	Q1	Q2	Q3	Q4
B	2	+	++	+	++	++				+	++
	4										
imB	2	++	++	+	+	++	++	+	+	+	++
	4										

5 Conclusions

The topic of our article pertains to the realm of smart cities. The primary challenge addressed is how to establish a concept of a smart city that is more resilient to climate change and natural events. Our evaluation focuses on dynamically forecasting the number of incoming calls to a smart city system. It encompasses a wide range of data, including both balanced and unbalanced datasets. The analysis spans the years 2013 to 2018, with a particular emphasis on weather and river flow data sourced from across Slovenia. The classification method utilized for the analysis was AdaBoostM1. The results underscore the effectiveness of machine learning techniques in fortifying the resilience of smart cities.

During the data preparation phase, extensive efforts were required to prepare and clean the necessary data. Not all data were in a format suitable for automated transfer, necessitating manual processing. As the data were collected gradually over time, some sensors were replaced. Additionally, certain data were missing, making it necessary to source data from sensors in nearby locations.

The provided assessment serves as a foundation for a more detailed and comprehensive conceptualization of smart city planning. Future enhancements are expected to improve forecast accuracy. The proposal suggests combining a greater number of methods or utilizing different algorithms, along with incorporating additional data such as information on disease epidemics, holidays, vacations, and other relevant factors.

References

1. European Commission: Smart cities - smart living. Shaping Europe's digital future (2019). https://ec.europa.eu/digital-single-market/en/smart-cities
2. Department for Business innovation & Skills, Global Innovators: International Case Studies on Smart Cities Smart Cities Study (Bis research paper no. 135) (2013)
3. Komninos, N.: The age of intelligent cities : smart environments and innovation-for-all strategies, 1st edn. Routledge, Taylor & Francis Group, New York (2015)
4. Picon, A.: Smart Cities: A Spatialised Intelligence. Wiley, Chichester (2015)
5. Zhong, X., Duckham, M., Chong, D., Tolhurst, K.: Real-time estimation of wildfire perimeters from curated crowdsourcing. Sci. Rep. **6**, 1–10 (2016)
6. Stec, A., Klabjan, D.: Forecasting crime with deep learning. arXiv, arXiv:1806.07366v5 (2018)

7. Duan, G., Zhao, W., Hu, Z., Fang, D.: An improved model of regional flood disaster risk assessment based on remote sensing data. In: Proceedings of the 19th International Conference on Geoinformatics, pp. 1–6 (2011)
8. Grašič, V.: Napovedovanje števila dohodnih klicev na sistem za klic v sili 112 ob uporabi odprtih podatkov interneta stvari. Doctoral dissertation, Faculty of Information Studies (FIS) in Novo mesto (2021)
9. Grasic, V., Kos, A., Mileva-Boshkoska, B.: Classification of incoming calls for the capital city of Slovenia smart city 112 public safety system using open Internet of Things data. Int. J. Distrib. Sens. Netw. **14**(9) (2018)
10. Grašič, V., Robnik, A.: Increasing resilience and resistance of Smart Cities by forecasting the number of incoming emergency calls based on IoT open data. In: EENA Conference, Marseille, France, 27–29 April 2022
11. Shalev-Shwartz, S., Ben-David, S.: Understanding Machine Learning: From Theory to Algorithms. Cambridge University Press, New York (2014)
12. James, G., Witten, D., Hastie, T., Tibshirani, R.: An Introduction to Statistical Learning: with Applications in R. Springer, New York (2014)
13. Han, J., Kamber, M., Pei, J.: Data Mining: Concepts and Techniques, 3rd edn. Morgan Kaufmann (2011)
14. SafeCity112 (2022). https://github.com/SafeCity112/SafeCity112
15. Statklic: Administration of the Republic of Slovenia for Civil Protection and Disaster Relief. Ministry of defense (2022). http://statklic.sos112.si
16. Meteo.si: Slovenian Environmental Agency (2022). https://meteo.arso.gov.si
17. ARSO (photo by Burger, M.), Pregled hidroloških razmer površinskih voda v Sloveniji: Poročilo o monitoringu za leto 2012, ISSN 2335-3597, photo: Marko Burger 2011 (Figure 3, Page 5, 2011), by ARSO, Ljubljana, Slovenia, February 2014. https://arso.si/vode/publikacije%20in%20poro%C4%8Dila/Poro%C4%8Dilo%20o%20hidrolo%C5%A1kem%20monitoringu%20povr%C5%A1inskih%20voda%20za%20leto%202012.pdf
18. Freund, Y., Schapire, R.E.: Experiments with a new boosting algorithm. In: Proceedings of the Thirteenth International Conference on Machine Learning, ICML1996, Bari, Italy, 3–6 July, pp. 148–156 (1996)
19. Frank, E., Hall, M.A., Witten, I.H.: The WEKA Workbench. Online Appendix for Data Mining: Practical Machine Learning Tools and Techniques, 4th Edn. (2016)
20. Brownlee, J.: How to fix k-fold cross-validation for imbalanced classification. Machine Learning Mastery (2020). https://machinelearningmastery.com/cross-validation-for-imbalanced-classification/
21. Branco, P., Torgo, L., Ribeiro, R.P.: A survey of predictive modelling under imbalanced distributions. ACM Comput. Surv. **49**(2), Article id 31 (2016)
22. Fawcett, T.: Learning from imbalanced classes. Silicon Valley Data Science (2016). https://www.svds.com/learning-imbalanced-classes/
23. Brownlee, J.: A gentle introduction to imbalanced classification. Machine Learning Mastery (2019). https://machinelearningmastery.com/what-is-imbalanced-classification/
24. Chawla, N.V., Bowyer, K.W., Hall, L.O., Kegelmeyer, W.P.: SMOTE: synthetic minority over-sampling technique. J. Artif. Intell. Res.Intell. Res. **16**, 321–357 (2002)

Internet of Medical Things (IoMT): Smart Hearing Aids, Today and Tomorrow

Vesna Simikić[1,2](✉)

[1] Public Health Institution Hospital 'Sveti Vracevi' Bijeljina, Street Srpske vojske 53, 76300 Bijeljina, Republic of Srpska, Bosnia and Herzegovina
simikic.vesna.85@gmail.com

[2] Department of Information Technologies, Faculty of Technical Sciences, University of Kragujevac, 65 Svetog Save Street, 32000 Cacak, Serbia

Abstract. The history of wireless headphones represents the pinnacle of technological evolution in audiology, from the early days of radio technology to today's high-end smart headphones. As technology continues to advance, these devices will continue to empower people with hearing loss to fully engage and enjoy the world around them. Smart hearing aids offer a range of features and benefits that significantly improve the quality of life for people with hearing loss. Using the power of connectivity and intelligent technology, these devices enable users to stay connected, engaged and active in their daily lives. While the advent of smart hearing devices offers exciting opportunities to improve the lives of individuals with hearing loss, emphasis is placed on the major safety risks in their use. Even this modern technology is not immune to hacker attacks, so it is essential to take care of the cyber security of these devices in order to ensure their safety and privacy in an increasingly interconnected world. The integration of artificial intelligence into smart hearing aids holds great promise for improving the efficiency, usability, and accessibility of hearing aid technology. Using AI-driven innovation, these devices can provide users with personalized, adaptive and intelligent solutions that improve their hearing experience and quality of life.

Keywords: IoT · smart hearing aids · artificial intelligence · cyber security · healthcare

1 Instructions

People with hearing impairments often faced discrimination and limited opportunities in education and employment. However, modern times bring many advances in technology, education and social awareness, which help the hearing impaired to better integrate into society and live a full life. Wireless hearing aids are an excellent example of the application of modern technologies in order to improve the quality of life of people with hearing impairments. The idea of developing these devices came with the advent of wireless headphones as the pinnacle of technological innovation in the field of audiology [1]. The concept of wireless communication for headphones dates back to the late

© The Author(s), under exclusive license to Springer Nature Switzerland AG 2025
N. Filipović (Ed.): AAI 2024, LNNS 1446, pp. 402–411, 2025.
https://doi.org/10.1007/978-3-031-99201-8_37

19th and early 20th centuries, when pioneers like Alexander Graham Bell and Thomas Edison were experimenting with technologies to help the hearing impaired. The advent of radio technology at the beginning of the 20th century laid the foundation for wireless communication. In the 1920s, vacuum tubes were adapted to create the first wearable headphones [2]. These early devices, although bulky and awkward, represented a significant advance in amplifying sound without the need for cables. The development of Bluetooth technology in the 1990s revolutionized wireless communication [3].

Bluetooth-enabled headphones began appearing in the early 2000s, offering users the ability to wirelessly connect to a variety of audio sources such as smartphones, televisions, and computers. This connectivity has improved the user experience by ensuring seamless integration with devices used every day. Over the years, wireless headphones have become smaller and more discreet. Advances in miniaturization and digital signal processing have allowed manufacturers to create headphones that are nearly invisible when worn, improving comfort and aesthetics for users [4]. Recent innovations in wireless headphones include the integration of artificial intelligence (AI) and machine learning algorithms [5]. These smart headphones can adapt to different listening environments in real time, automatically adjusting settings to optimize sound quality and speech intelligibility [6]. Some models can even track health data such as heart rate and physical activity, giving users additional functionality beyond traditional hearing aid service. The future of wireless headsets lies in integration with wearable technology and the Internet of Things (IoT) enabling real-time speech translation and communication with other smart devices [7]. As technology continues to evolve, the opportunities to improve the lives of people with hearing loss are limitless.

The aim of this paper is to show the current state of application in the field of smart hearing devices and cochlear implants, the integration of artificial intelligence in assistive hearing devices, and their further development with reference to possible cyber security risks. The research was carried out by analyzing research papers and technical solutions in the field of application and development of smart hearing aids, cyber security of smart hearing aids, and application of artificial intelligence in the operation of smart hearing aids and cochlear implants. This research will provide an overview of the current state of smart hearing aids, their application, further development, and their cyber security status with recommendations for manufacturers and users, as well as integration with artificial intelligence algorithms. An overview of further research in the field of artificial intelligence application in smart hearing aids and cochlear implants will also be given.

2 Smart Hearing Aids

In the field of healthcare and assistive technology, smart IoT hearing devices have attracted significant attention for their potential to completely revolutionize the way hearing-impaired individuals communicate with each other and the world [8]. These devices use IoT technology to connect to other smart devices and platforms, enabling features such as remote control, real-time monitoring, and individual settings [9]. However, with the increasing number of interconnected devices and the growth of IoT networks, it is important to address the security vulnerabilities and cyber risks associated with smart IoT hearing devices [10]. With any type of digital transformation, security challenges must be considered in the early stages of design. Healthcare data is sensitive and any breach threatens patient privacy [11]. The use of artificial intelligence (AI) in cybersecurity can play a key role in protecting IoT networks used for healthcare. AI has the ability to analyze large amounts of data in real time, with the detection of potential cyber threats [9].

Cybersecurity systems with artificial intelligence can learn from patterns and behaviors, continuously improving their ability to identify and respond to new risks. AI can also improve the security of smart IoT hearing devices by applying advanced encryption techniques and authentication methods [12]. The application of the Internet of Things has had a positive impact on e-health, assisted living and the well-being of all users [13]. Integrating artificial intelligence into the security systems of smart IoT hearing devices can strengthen their protection against cyber threats. It is critical to ensure that smart IoT hearing devices are designed with robust security measures, including encryption and authentication protocols, to protect the sensitive data they collect and transmit. By addressing these safety concerns, smart IoT hearing devices can truly transform the lives of hearing-impaired individuals, providing them with greater accessibility and independence. The use of artificial intelligence in cyber security for smart IoT hearing devices can greatly improve their security status. With the help of artificial intelligence, smart IoT hearing devices can detect and respond to potential cyber threats in real time, ensuring the privacy and security of patient data while enabling seamless communication and interaction with the world.

Additionally, AI can also help identify patterns and anomalies in user behavior, enabling early detection of potential security breaches. Integrating AI into the security systems of smart IoT hearing devices improves their overall cyber security and ensures the protection of sensitive data, both collected and transmitted. Integrating AI into cybersecurity for smart IoT hearing devices is essential to address the security challenges associated with digital transformation in healthcare. The use of artificial intelligence can improve the security measures of smart IoT hearing devices, ensuring the privacy and security of patient data [14]. It is important to consider security challenges in the early stages of designing smart IoT hearing devices. This includes implementing encryption protocols, authentication, as well as using artificial intelligence-based cyber security systems that can continuously learn and adapt to emerging risks. Furthermore, the use of artificial intelligence in cybersecurity for smart IoT hearing devices can also address potential vulnerabilities and risks associated with wireless connectivity. IoT hearing devices in medicine refers to advanced hearing aids and other hearing support systems that use IoT technology to improve user experience and health care outcomes. These devices usually have smart features, such as the ability to connect to the Internet and other devices, allowing data exchange, remote adjustments by audiologists and personalized user settings.

IoT hearing devices can be tuned directly through a smartphone app, allowing users to tailor their hearing experience to different environments. They can also track hearing health over time, providing valuable data to healthcare professionals for better diagnosis and treatment. IoT connectivity can also facilitate features such as fall detection, caregiver alerts, and integration with other smart home devices, thereby offering a wider range of support for patients, especially the elderly or those with disabilities. IoT in healthcare, including hearing devices, contributes to improved patient engagement, cost savings, operational efficiency and delivery of telemedicine services, which can be particularly beneficial for patients in remote or underdeveloped areas [15].

A special category is represented by implantable hearing devices, more specifically cochlear implants. This technology makes it possible to overcome the problem of hearing loss in people whose problem is closely related to a problem with the inner ear. The implant assumes the role of hair cells and transmits all signals to the brain. Like a traditional hearing aid, a cochlear implant begins with sound reception. A small microphone in the external component of the implant picks up sounds from the environment. The sound is then processed by a speech processor, a small device worn behind the ear or on the body. The speech processor analyzes and digitizes sound into coded electrical signals. Coded electrical signals are sent wirelessly through the skin to the internal component of the implant through the transmitter coil. The internal component of the cochlear implant, which is surgically implanted under the skin behind the ear, consists of a receiver-stimulator and a series of electrodes. The receiver-stimulator receives coded electrical signals and sends them to a series of electrodes. An array of electrodes is inserted into the cochlea, the spiral-shaped cavity of the inner ear. Electrodes within the array stimulate auditory nerve fibers directly in response to coded signals received from the speech processor. Stimulated auditory nerve fibers transmit electrical signals to the brain, specifically the auditory cortex, where they are interpreted as sound. The brain perceives these electrical signals as sound, allowing the cochlear implant recipient to hear and understand speech, environmental sounds, and other auditory stimuli [16].

3 Cyber Security of Smart Hearing Aids

Wireless hearing aids offer a number of benefits, but as with all wireless devices, it is important to consider safety aspects, be well informed about any risks and take steps to mitigate them in order to protect personal information. Like other IoT devices, smart hearing aids are vulnerable to hacking attacks, which can lead to the theft of sensitive data, such as medical records. Along with benefits such as improved connectivity and functionality, these devices also introduce potential security vulnerabilities [17]. Hacking a smart hearing device can have serious consequences, from violating privacy to disrupting the functionality of the device. A hacker could exploit security flaws in the software or communication protocols of a smart hearing aid to gain unauthorized access. This could allow remote control of devices, access to sensitive user data, or eavesdropping on conversations [18].

Smart hearing devices often collect and store personal data, such as user preferences and usage patterns. A successful hacking attack could lead to data destruction, compromising this sensitive information and potentially exposing users to identity theft or other types of cybercrime. Hackers can alter the audio signals processed by a smart hearing aid, resulting in a distorted or misleading sound perception for the user. This can have serious implications, especially on individuals who rely on the device for communication or security purposes. By compromising a device's software or firmware, hackers could disrupt its normal functionality or even render it unusable. This could pose significant challenges for individuals who depend on hearing aids in their daily lives. Smart hearing aids may contain features such as Bluetooth connectivity and wireless data transmission, which hackers could exploit to intercept or manipulate audio streams. This can lead to risks regarding the privacy and confidentiality of patients' personal data. That is why it is recommended to use the authentication of the user of the medical device, if possible multi-factor and/or biometric, because in this way the protection of the privacy of the patients is ensured [19]. The use of various cryptosystems such as RSA, DSA or public key systems is always desirable, but in the case of IoMT it is difficult to apply due to their miniaturization (appearance and easy portability at the expense of security and strength) [20]. The simplicity of the design reduces the possibility of computing power, and, therefore, the power of the entire system.

To mitigate the risk of hacker attacks on smart hearing devices, manufacturers must prioritize cybersecurity in the product design and development processes themselves. This includes implementing robust encryption protocols, regularly updating device firmware to patch security vulnerabilities, and conducting thorough security assessments and audits [19, 20]. It is also necessary that the functionality of the device in terms of security and data privacy comply with the requirements of GDPR and HIPPAA, and in terms of cyber security it is necessary to meet the requirements of the ISO/IEC 27002 and ISO/IEC 27017 standards [21]. Special attention should also be paid to the security of the transmission itself between the device and the server, because communication mostly takes place via the Internet [19–21].

Users of smart hearing devices can take steps on their own to help protect both their devices and their data [22]. In addition to the following basic postulates of cyber hygiene such as using strong, unique passwords to access the device, two-factor authentication, avoiding unsecured Wi-Fi networks and being alert to any signs of suspicious activity or unauthorized access to their devices, it is necessary, before purchasing wireless hearing aids, to read and understand the manufacturer's privacy policy so that they know what data are collected and how they are used. It is also very important to choose models that offer data encryption, multi-factor and/or biometric authentication and regular software updates.

4 Smart Hearing Aids and Artificial Intelligence

Smart hearing aids represent a significant advance in assistive hearing technology, using connectivity and intelligent features to enhance the user experience and improve hearing outcomes [23]. The integration of Bluetooth technology allowed them to wirelessly connect to various audio sources such as smartphones, tablets, televisions and computers. This allows users to stream phone calls, music, podcasts and other audio content directly to their hearing aids, improving clarity and comfort. Many smart hearing aids come with accompanying mobile apps that allow users to remotely control and adjust settings [24]. Through the app, users can adjust volume, change programs, adjust sound preferences for different environments, and even locate lost hearing aids using GPS tracking [25].

Smart hearing aids and artificial intelligence (AI) are increasingly coming together to continue revolutionizing the field of audiology, offering innovative solutions for the hearing impaired [16–26]. AI algorithms analyze and interpret incoming sound signals in real time, allowing smart hearing aids to automatically adapt their settings to different listening environments. These adaptive audio processing capabilities improve speech intelligibility while reducing background noise. Smart hearing aids equipped with machine learning algorithms can learn and adapt to an individual's unique listening preferences and behavior over time. By continuously analyzing user feedback and usage patterns, these devices can personalize sound settings and optimize performance to better meet specific user needs. AI-based feedback cancellation algorithms can effectively identify and suppress feedback noise caused by acoustic feedback or improper fitting of the hearing aid. This ensures a clearer and more comfortable listening experience for users without interference such as whistling or buzzing. AI-based speech enhancement algorithms selectively amplify speech signals while suppressing background noise, making it easier for users to understand conversations in noisy environments [26]. These algorithms can distinguish between spoken and non-speech sounds, ensuring that important auditory information is prioritized.

Some smart hearing aids use artificial intelligence to monitor the wearer's health indicators, such as heart rate and physical activity [27]. By analyzing these data points, the hearing aid can provide insight into the wearer's overall health and even alert them to potential health problems or falls [28]. Artificial intelligence-based NLP technology enables smart hearing aids to understand and process spoken commands and instructions [29] and users to interact with their hearing aids using voice commands, control settings, make adjustments or access additional features hands-free [26–29]. AI-enabled smart hearing aids support teleaudiology services, allowing users to receive remote assistance and adjustments from audiologists via telecommunications platforms. AI algorithms can facilitate remote diagnosis, programming and troubleshooting, reducing the need for in-person appointments and improving accessibility to hearing healthcare services [30].

Unlike smart hearing aids, cochlear implants rely primarily on hardware components and electrode arrays to provide the sensation of sound to people with hearing loss. Advances in artificial intelligence have begun to play a significant role in improving their functionality and efficiency [16]. AI algorithms are being built into cochlear implants to improve the speech processing capabilities of cochlear implant speech processors. These algorithms also help improve speech understanding in noisy environments by selectively filtering background noise while preserving speech signals. AI-based adaptive signal processing algorithms continuously analyze incoming sound signals and adjust cochlear implant settings in real-time to optimize performance based on the listening environment and individual speech characteristics. Artificial intelligence algorithms can automatically classify different sound environments (e.g. quiet, noisy, speech, music) based on acoustic features extracted from incoming sound signals. This information is then used to adjust the settings of the cochlear implants according to optimal sound perception. Artificial intelligence techniques are used to optimize the pattern and intensity of electrical stimulation delivered by an array of electrodes to the auditory nerve. These algorithms take into account factors such as an individual's residual hearing, electrode-neuron interface properties, and speech processing strategies to maximize speech understanding and minimize power consumption. AI algorithms are used to suppress feedback and artifacts that can occur in cochlear implant systems, especially at high stimulation levels. By predicting and canceling feedback signals in real time, these algorithms help improve the clarity and stability of perceived sound. AI-powered adaptive learning algorithms track the user's response to electrical stimulation over time and automatically adjust stimulation parameters to accommodate changes in auditory perception and electrode-neuron interface dynamics. AI-powered remote monitoring systems allow clinicians to remotely monitor the performance of cochlear implant users, analyze data trends, and make adjustments to device settings without the need for in-person examinations. This increases the availability of care and enables personalized rehabilitation strategies. The integration of advanced AI algorithms into cochlear implants promises to greatly improve speech perception, sound quality, and user satisfaction for individuals with hearing loss, ultimately improving their communication abilities and quality of life [31].

5 Future Direction in Research of the Application of AI in Assistive Hearing Devices

Future research directions for the application of artificial intelligence in smart hearing devices will likely focus on improving the capabilities and effectiveness of these devices in addressing various challenges faced by individuals with hearing loss. The development of AI algorithms that can automatically adapt and personalize device settings based on individual hearing profiles, preferences and listening environments could involve the integration of machine learning techniques to continuously learn and refine personalized audio processing strategies. Real-time language translation within hearing aids by AI could enable people with hearing loss to better communicate with other speakers of different languages by providing simultaneous translation of spoken language into text or synthesized speech. The use of AI algorithms that can intelligently analyze contextual information, such as the user's location, activities and social interactions, can enable dynamic adjustment of device settings and provide customized hearing assistance in different situations (e.g. meetings, restaurants, outdoor environments).

Research into ways to improve the naturalness and quality of sound perception using advanced AI algorithms could include developing models that mimic the auditory processing mechanisms in the human brain to better simulate natural hearing experiences for hearing aid users. Research integrating AI-based emotion recognition and social cue detection capabilities into hearing devices, and by analyzing subtle acoustic cues and facial expressions, smart hearing devices can help users better understand the emotional context of conversations and social interactions. Exploring the use of AI algorithms for health monitoring and early detection of hearing-related conditions, such as hearing threshold changes, tinnitus, or auditory processing disorders, and by analyzing patterns in hearing data collected over time, can aid in early intervention and treatment of hearing problems.

The ability to integrate multimodal sensing technologies (e.g. audio, visual, physiological) with artificial intelligence algorithms would provide more comprehensive and accurate assessments of the user's auditory and communication needs and enable synergistic interactions between different sensor modalities to improve overall device performance. Research methodologies for incorporating user feedback and preferences into the design and development of AI hearing devices could include user-centered design approaches, participatory design workshops, and usability studies to ensure that the devices meet different user needs and preferences.

By focusing on these lines of research, smart hearing aids with AI have the potential to revolutionize the way people with hearing loss perceive and interact with the auditory world.

6 Conclusions

The application of artificial intelligence in smart hearing aids brings numerous advantages that improve the quality of life of people with hearing loss. Thanks to AI, these devices can provide a more personalized experience tailored to the individual needs of the user. Using signal processing algorithms, AI enables better recognition and filtering of sounds, improving speech understanding in noisy environments. By integrating with other smart devices, the hearing aid can provide healthcare professionals with insight into the user's (patient's) general health status. The use of AI can enable the automatic adjustment of hearing aid and implant settings in real time according to changes in the environment or the needs of the user. It can also help to better analyze contextual information such as the user's location, subtle acoustic cues and facial expressions, and provide more comprehensive and accurate assessments of the user's hearing and communication needs to improve the performance of smart hearing devices.

This paves the way for fuller participation in social activities and improves the quality of communication for people with hearing loss. However, it is important to emphasize the need and importance for the safe use of smart hearing devices, the existence of awareness of the need to protect the data that is collected and processed, as well as the need for an ethical and transparent approach in the development and application of this technology to ensure that it is used in a way which respects the user's privacy and integrity.

References

1. Hallenbeck, S.: Wireless technology: where have we been? Where are we now? Where are we going?. AudiologyOnline, Article 17925 (2016)
2. Burrows, K.: A not-so-short history of deaf technology. Lond. J. Press **22**(15), 1–14 (2022)
3. Zeadally, S., Siddiqui, F., Baig, Z.: 25 years of bluetooth technology. Future Internet **11**(9), 194 (2019)
4. Mills, M.: Hearing aids and the history of electronics miniaturization in IEEE. Ann. Hist. Comput. **33**(2), 24–45 (2011)
5. Xiaoran, F., et al.: HeadFi: bringing intelligence to all headphones. In: Proceedings of the 27th Annual International Conference on Mobile Computing and Networking (MobiCom 2021), pp. 147–159. Association for Computing Machinery, New York (2021)
6. Alexander, J.M.: Hearing aid technology to improve speech intelligibility in noise. Semin. Hear. **42**(3), 175–185 (2021)
7. Turchet, L., et al.: The internet of sounds: convergent trends, insights, and future directions. IEEE Internet Things J. **10**(13), 11264–11292 (2023)
8. Alsharif, B., Ilyas, M.: Internet of things technologies in healthcare for people with hearing impairments. In: Wang, S. (eds.) IoT and Big Data Technologies for Health Care. IoTCare 2022. Lecture Notes of the Institute for Computer Sciences, Social Informatics and Telecommunications Engineering, vol. 501, pp. 299–308. Springer, Cham (2023). https://doi.org/10.1007/978-3-031-33545-7_21

9. Chataut, R., Phoummalayvane, A., Akl, R.: Unleashing the power of IoT: a comprehensive review of IoT applications and future prospects in healthcare, agriculture, smart homes, smart cities, and industry 4.0. Sensors 23, 7194 (2023)
10. Lee, I.: Internet of Things (IoT) cybersecurity: literature review and IoT cyber risk management. Future Internet **12**, 157 (2020)
11. Keshta, I., Odeh, A.: Security and privacy of electronic health records: concerns and challenges. Egypt. Inf. J. **22**(2), 177–183 (2021)
12. Adeel, A., Ahmad, J., Larijani, H., Hussain, A.: A novel real-time, lightweight chaotic-encryption scheme for next-generation audio-visual hearing aids. Cogn. Comput. **12**(3), 589–601 (2020)
13. Marques, G., Saini, J., Pires, I.M., Miranda, N., Pitarma, R.: Internet of things for enhanced living environments, health and well-being: technologies, architectures and systems. In: Handbook of Wireless Sensor Networks: Issues and Challenges in Current Scenario's, pp. 616–631 (2020)
14. Sun, Y., Lo, F.P.-W., Lo, B.: Security and privacy for the internet of medical things enabled healthcare systems: a survey. IEEE Access **7**, 183339–183355 (2019)
15. Albahri, A.S., et al.: IoT-based telemedicine for disease prevention and health promotion: state-of-the-art. J. Netw. Comput. Appl. **173** (2021)
16. Carlyon, R.P., Goehring, T.: Cochlear implant research and development in the twenty-first century: a critical update. J. Assoc. Res. Otolaryngol. **22**(5), 481–508 (2021)
17. Algarni, M., Alkhelaiwi, M., Karrar, A.: Internet of things security: a review of enabled application challenges and solutions. (IJACSA) Int. J. Adv. Comput. Sci. Appl. **12**(3) (2021)
18. Grispos, G., Flynn, T., Glisson, W.B., Choo, K-K.R.: Investigating protected health information leakage from android medical applications. In: Proceedings of the 5th EAI International Conference on Future Access Enablers of Ubiquitous and Intelligent Infrastructures (2021)
19. Sun, W., Cai, Y., Li, Y., Liu, F., Fang, Sh., Wang, G.: Security and privacy in the medical internet of things: a review. Secur. Commun. Netw. **2018** (2018)
20. Suleski, T., Ahmed, M., Yang, W., Wang, E.: A review of multi-factor authentication in the Internet of Healthcare Things. Digit. Health (2023)
21. Mullet, V., Sondi, P., Ramat, E.: A review of cybersecurity guidelines for manufacturing factories in industry 4.0. IEEE Access **9**, 23235–23263 (2021)
22. Naveed, M., Usman, S.N., Satti, M.I., Aleshaiker, S., Anwar, A.: Intrusion detection in smart IoT devices for people with disabilities. In: IEEE International Smart Cities Conference (ISC2) (2022). 978-1-6654-8561-6/22.2022
23. Pradyumna, G.R., Hegde, R.B., Bommegowda, K.B., Tony, J., Naik, G.R.: Empowering healthcare with IoMT: evolution, machine learning integration, security, and interoperability challenges. IEEE Access **12**, 20603–20623 (2024)
24. Barapatre, P., Ingolikar, Y., Desai, P., Jajoo, P., Thakre, P.: A secured architecture for IoT-based healthcare system. 3C Empresa. Investigaci6n y pensamiento crítico, **11**(2), 222–230 (2022)
25. Obogo, J.U.: A Security and Privacy Architecture for Healthcare IoT Devices. Bournemouth University (2020)
26. Lesica, N.A., Mehta, N., Manjaly, J.G., et al.: Harnessing the power of artificial intelligence to transform hearing healthcare and research. Nat. Mach. Intell. **3**, 840–849 (2021)
27. AlAdwani, H., ALSiyabi, Z.: A systematic review of IoT integration on health monitoring system. Int. J. Eng. Manage. Res. **13**(1) (2023). e-ISSN: 2250-0758, p-ISSN: 2394-6962
28. Patra, M.K., Sahoo, B., Turuk, A.K.: Smart healthcare system using containerized internet of medical things. IGI Global (2022). https://doi.org/10.4018/978-1-6684-4580-8.ch014.2022
29. Juhn, Y., Liu, H.: Artificial intelligence approaches using natural language processing to advance EHR-based clinical research. J. Allergy Clin. Immunol. **145**(2), 463–469 (2020)

30. Waltzman, S.B., Kelsall, D.C.: The use of artificial intelligence to program cochlear implants. Otol. Neurotol. **41**(4), 452–457 (2020)
31. Essaid, B., Kheddar, H., Batel, N., Lakas, A., Chowdhury, M.E.: Advanced artificial intelligence algorithms in cochlear implants: review of healthcare strategies, challenges, and perspectives (2024)

Evaluation of Magnetized Micro- and Nano-Object Parameters Using Artificial Intelligence

Vitaly A. Goranov(✉) and Alexander Makhaniok

BioDevice Systems s.r.o., Bulharská, 10-Vršovice, 996/20, 10100 Praha, Czech Republic
info@biodevicesystems.com

Abstract. The development of magnetic micro- and nano-objects (MOs) holds significant potential in biotechnology and medicine, but there is a need for fast and efficient methods to measure a range of physical parameters. Our study proposes the use of artificial intelligence (AI) to assess the magnetization and hydrodynamic radius of MOs using a microfluidic chip. The methodology involves the use of a microfluidic chip and standard microscopy techniques to track the movement of MOs under the influence of an external magnetic field. The evaluation of cell magnetization using the AI application allows better prediction of average MOs values exceeding 10^5 MOs/cell than in the case of simulation using finite element analysis. Furthermore, the application of the neural network "U-Net" and the corresponding AI analysis offers the possibility to improve the evaluation of the hydrodynamic radius for MOs larger than 500 nm and for cells magnetized with MOs in the range of 10-100μg/ml culture medium.

Keywords: machine learning · neural network · magnetic nanoparticles · magnetic field · artificial intelligence · biomedical application

1 Introduction

Despite the significant potential of magnetic micro- and nano-objects (MOs) in biotechnology and medicine, there are few experimental methodologies for fast and efficient measuring some physical parameters, which are readily available. Such crucial parameters as magnetization and hydrodynamic radius play a central role in assessing the motility of MOs in a liquid environment under the influence of a magnetic field (MF). Magnetization dictates how nanoparticles respond to external magnetic fields and influences their behavior in magnetic environments. While the hydrodynamic radius determines the diffusion and convection properties of MOs in a liquid. In this study, we propose to use AI-based image analysis to evaluate these parameters using videos of experiments in which MOs were influenced by MF in a microfluidic chip. This is also particularly important because of the heterogeneity of recently prepared MOs [1] and the need to perform a rapid screening analysis of the predominant parameters characterising the suspension of MOs. In the context of biocompatibility, it is necessary to develop specific

© The Author(s), under exclusive license to Springer Nature Switzerland AG 2025
N. Filipović (Ed.): AAI 2024, LNNS 1446, pp. 412–418, 2025.
https://doi.org/10.1007/978-3-031-99201-8_38

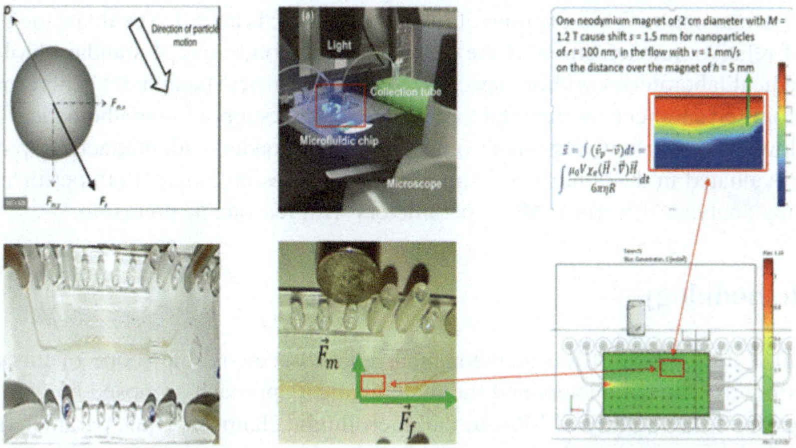

Fig. 1. Schematically outlined workflow of experimental and mathematical evaluation of MOs flow in the microfluidic chamber. a – scheme of the counterbalance of two forces: magnetic attraction and fluid drag force; b – experimental tool; c – FEM simulation example; d – microfluidic chip (microfluidic ChipShop); e – deviation (shift) of MO's flow under MF influence (the field for video-fixation for following AI-analysis outlined by red square); f - schematic designation of the area in the microfiber chip calculated using FEM and the corresponding field delimited for further analysis.

recommendations to reduce the amount of magnetised material and thus the potential cytotoxicity of the magnetic core, while the biophysical parameters of the functionalized MOs can be tailored in order to provide appropriate motility. This targeted approach is not only necessary to minimise the systemic distribution of cytotoxic substances in vivo, but also enables effective medical treatments with lower dosages.

Starting from video tracking of the diffusion dynamics of the tested magnetic particles, we developed a mathematical method based on machine learning early on to predict the magnetization of MOs in solution [2]. We analyzed the macroscopic behavior of a uniform solution of MOs or MOs/liposome mixtures contained in a quartz microfluidic capillary under the influence of an external magnetic field. The observed separation dynamics strongly depend on the magnetization and the hydrodynamic diameter of the particles. The results contribute to our understanding of the complex interplay between magnetic forces and fluid dynamics in such systems. However, this approach lack sufficient accuracy for objects larger than 100 nm. This is because the crucial decrease in Brownian motility and the phase transition of the MOs-solution, from a fluid of non-interacting Brownian particles to a ferrofluid, becomes challenging when an external magnetic field is applied. In such cases, both the interaction between the MOs themselves and their interaction with an external magnetic field come into play [3].

While the above-mentioned modelling is basically applicable for nano-objects with dimensions up to 150 nm, the evaluation of certain parameters becomes practically important for larger objects and cells loaded with MOs. In such cases, the diffusion of nano-objects becomes negligible for particles of several hundred nanometers and the previously proposed approach is hardly applicable. Moreover, considering an increasing interest in magnetic 3D bio-printing of living cells, there is a need to evaluate the parameters of relative magnetization of the cells under the conditions of standard biological and medical laboratories without specialized magnetometric equipment. Therefore, we have proposed another experimental tool (Fig. 1b.) and a supporting mathematical model that allows to assess how large nano-objects and cells loaded with magnetized particles can be evaluated in the context of its motility. It enables investigators to perform rapid screening analyses of various MO‘s parameters with reasonable precision.

2 Methodology

A detailed description of the mathematical model is beyond the scope of this paper. We mainly use the parameters and the mathematical approach proposed by [4] for the description of the motility of MOs in the microfluidic chamber. Thus, parameters such as fluid velocity, viscosity, magnetic field properties and more are considered in the model. The model includes the distribution of MO‘s concentration, magnetization or, conversely, the hydrodynamic radius as parameters for predicting the dynamic behavior of MOs. The model takes into account the interaction of magnetic fields, laminar fluid flow and convection-diffusion transfer. The simulation was carried out using finite element analysis (FEM) methods. The magnetic particle deviates from the straight flow and drifts due to the influence of two forces: the magnetic attraction force and the drag force of the fluid (Fig. 1a.). They counterbalance each other out for a given drift velocity, and if we integrate this over the elapsed time, we obtain the total displacement. The initial experimental data confirmed the applicability of the model, and the experimental parameters evaluated with the FEM (software "FreeFem") were predicted with a reasonable degree of reliability (Fig. 2). For example, an average displacement of 1.5 mm is predicted for the given experimental conditions using a neodymium magnet of 1T for particles with a diameter of 200 nm (Fig. 1c–d), which is in preliminary agreement with the experimental observations. To validate this model, we used the microfluidic chip (Fig. 1e) to assess the deviation of MO‘s flux under the influence of an external MF. While we disregard the impact of diffusion effects on this process as they are negligible, we consider the kinetic parameters of particle motion within the microfluidic device, together with the physical parameters characterizing the magnetic permeability and the distribution of the magnetic field gradient.

For this experiment, we utilized the standard microfluidic chamber ("Microfluidic ChipShop"). In this setup, we observed the displacement of the flux of MOs deviating from the central line and forming a region with a specific distribution of the density of MOs under the influence of an external magnetic field. We tested different MF configurations using experimentally adapted interposition of permanent magnets with a magnetization of 1T. This opens the possibility to define the most suitable conditions to achieve visually distinguishable (under standard microscopy) changes in laminar MO flow in the microfluidic chamber. In particular, it opens a way to customize the experimental conditions. Firstly, the application of multiple magnets to the flow can be explored. Secondly, reducing the flow velocity allows a longer duration of magnetic attraction. Thirdly, by adjusting the geometry by placing two magnets closer to the trajectory of the nanoparticles, the magnetic field gradient can be increased. Another variant is to significantly increase the concentration of MOs in the flow in order to avoid (or possibly support) the behavior typical of magnetic fluid.

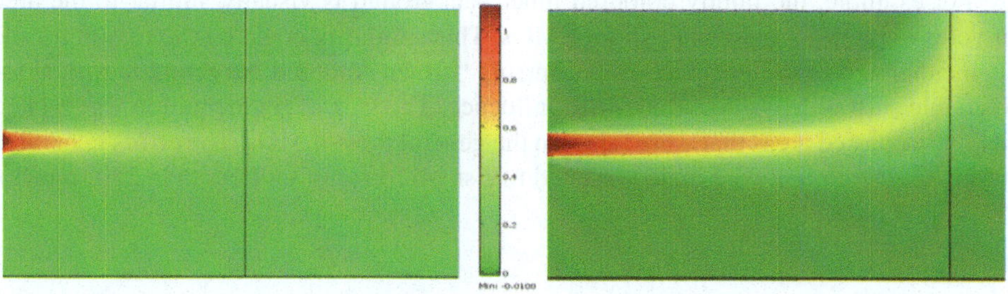

Fig. 2. Example of MOs kinetic visual simulation using FEM (left – MNPs 100 nm, right – MNPs – 500 nm).

Using FEA, we performed a series of simulations for various sizes of MOs. We used conventional equations to characterize "magneto-dynamic" and hydrodynamic parameters of MO's flow in a laminar stream [3, 5]. The relationship between the particle displacement in the defined region of the chamber and the size of the particles was derived from these simulations to create a calibration curve. It is therefore not surprising that the results for MOs larger than 100 nm were quite satisfactory, considering that the sizes of MOs are typically distributed over a range and the simulations assume conditions for largely uniform particles [6]. In the case of evaluation of cell magnetization, the standard conditions for cell loading were used [7].

The following step includes ML/AI processing to assess the specific behavior of magnetized nano-objects in a microfluidic chip to assess the hydrodynamic radius (or relative magnetization (RM) for magnetically loaded cells). The method helps to characterize the patterns (in defined regions of magnetically induced displacement) occupied in the microfluidic chip by MOs with relatively uniform magnetization and size under the influence of MF.

To test the effectiveness of the improved network in this work, various modules from [8, 9] such as ResAt-UNet(2SEAE), MLP, SVM, FCN8, Bilateral Segmentation Network (BisNet), Dual Attention Network (DANet), SegNet, PSPNet-101, DRNet, and Deep Feature Aggregation Network (DFA-Net). Spatial residual inception convolutional neural network UNet and ResAt-UNet modules were selected for building the image segmentation.

In particular, the videos from the experiments were utilized to generate the necessary time dynamics video-frame dataset for subsequent post-processing with machine learning. The video images captured with the microscope were used to input data into the neural network model. In a special experiment, we performed semantic segmentation of the analysed images using neural network architectures [8]. For training and validation, we used video-frame datasets extracted from the video using the Python OpenCV framework and reduced to 50 FPS and 500 × 500 pixels. The subsequent processing of the data largely followed the methods presented in [2]. After loading from LabelMe, the annotated JSON files were converted to the standard COCO data format, resulting in files annotated with the ranges. Semantic segmentation was chosen to analyse the frames representing the dynamics of the formation of structures that visually characterise the displacement kinetics of MOs in the given experiment.

For example, the faintly coloured inner background is visually similar to the area of yellow colour extrapolation shift of MOs. This similarity could be the reason for the segmentation error that occurs. The occurring "brown coloured" structure is typical for a reliable shift of MOs flow under MF influence. The lack of segmentation was related to faintly colored areas that were far from the general flow of MOs. The errors generated during segmentation were mainly related to a small learning set (less than 250 frames).

3 Results

Finally, 3890 images in 15 consecutive sets were extracted from the videos, manually labeled, and used as training sets. After a thorough analysis, it was found that the standard "UNet" network achieved the best validation metrics with the best mean intersection of unit and pixel accuracy among other associated networks [9]. To characterize the MOs accumulating into the framed visualized area of the microfluidic chamber, the trained U-net model was employed to analyse 6350 images (in 9 experimental sets).

Using the microfluidic chip, we estimated the drift of the magnetized MOs under the influence of the external magnetic field based on the corresponding evolving patterns. As we expected, the diffusion effects on this process were negligible, but we took into account the kinetic parameters of the particle motion in the microfluidic device, as well as the physical parameters characterizing the magnetic permeability and the distribution of the magnetic field gradient.

The first experimental data confirmed the applicability of the model and the experimental values/parameters evaluated with the FEM were mostly correctly predicted for magnetic nanoparticles from 150 to 500 nm, while the diameters of MOs with dimensions of 100–150 nm were estimated with significant measurement errors (Tab. 1). At the same time, the evaluation of cell magnetization using the AI application provides values that are more precise than in the case of FEA simulation for average MOs values exceeded 10^5 MOs/cell (Tab. 2).

In addition, the application of the UNet neural network and the corresponding AI analysis offers the possibility to improve the evaluation of the hydrodynamic radius for nanoparticles larger than 500 nm and for cells magnetized with MOs in the range of 10–100 μg/ml culture medium. We still have a problem with the evaluation of the diameter of MOs up to 100 nm and with the comparative analysis of slowly magnetized cells. The latter is possibly due to the heterogeneous loading of magnetic nanoparticles under conventional methodology.

Table 1. The results of HR evaluation using described approaches.

Hydrodynamic radius (HR),nm	Evaluated HR using FEM, nm	Approximation error (%)	Evaluated HR using AI, nm	Approximation error (%)
750	829	36,9	732	8,1
500	538	19,4	517	7,5
200	211	7,0	188	6,4
150	145	3,3	148	1,8
100	144	26,8	77	31,6
75	Non	applicable applicable	Non	applicable
50	Non		Non	applicable

Table 2. The results of relative magnetization evaluation using described approaches. * - average amount of MOs in single cell; ** - values were rounded to millions.

RM (MOs/ cell)*	Evaluated RM using FEM**	Evaluated RM using AI**
$5X10^6$	$3,8X10^6$	4,6X106
10^6	$1,7X10^6$	$0,9X10^6$
$5X10^5$	1,2X106	6X105
10^5	$7X10^5$	$2X10^5$
10^4	10^5	-
10^3	Non	applicable
10^2	Non	applicable

4 Conclusion

The proposed experimental methodology, backed by mathematical approaches, enables the evaluation of various parameters of magnetized objects in a wide range of values typical for use in biomedical research. This study can be conducted in conventional biological and medical laboratories using routine microscopy and simple microfluidic devices. The optimization and advancement of microfluidic devices for the production and characterization of magnetic micro- and nano-objects are crucial for the development of applications that rely on the precise control, movement, and manipulation of MOs in diverse fluid environments.

Acknowledgments. The authors acknowledge the support of the EU project BOW (ID:952183).

Author Contributions. Conceptualization, physical and biological experiments - V.G.; software and simulation, A.M.; validation, V.G., A.M.; formal analysis, management V.G.; writing—review and editing, V.G.

References

1. Kosuda, K., Bingham, J., Wustholz, K., Van Duyne, R.: Nanostructures and surfaceenhanced Raman spectroscopy. In: Handbook of Nanoscale Optics and Electronics, p. 309 (2010)
2. Applied Artificial Intelligence 2: Medicine, Biology, Chemistry, Financial, Games, Engineering; The Second Serbian International Conference on Applied Artificial Intelligence (SICAAI). Conference Proceedings, June 2024. https://link.springer.com/book/9783031608391
3. Hinderliter, P.M., Minard, K.R., Orr, G., et al.: A computational model of particle sedimentation, diffusion and target cell dosimetry for in vitro toxicity studies. Part. Fibre Toxicol. **7**, 36 (2010). https://doi.org/10.1186/1743-8977-7-36
4. Cao, Q., Han, X., Li, L.: Numerical analysis of magnetic nanoparticle transport in microfluidic systems under the influence of permanent magnets. J. Phys. D Appl. Phys. **45**(46), 465001 (2012)
5. Nimisha, W.: A systematic correlation of nanoparticle size with diffusivity through biological fluids. Thesis, Imperial College London (2010)
6. Van De Loosdrecht, M.M., et al.: A novel characterization technique for superparamagnetic iron oxide nanoparticles: the superparamagnetic quantifier, compared with magnetic particle spectroscopy. Rev. Sci. Instrum. **90**(2) (2019)
7. Goranov, V., Shelyakova, T., De Santis, R., et al.: 3D patterning of cells in magnetic scaffolds for tissue engineering. Sci. Rep. **10**, 2289 (2020). https://doi.org/10.1038/s41598-020-58738-5
8. Paszke, A., Chaurasia, A., Kim, S., Culurciello, E.: ENet: a deep neural net-work architecture for real-time semantic segmentation. arXiv preprint arXiv:1606.02147 (2016)
9. Fan, Z., Liu, Y., Xia, M., Hou, J., Yan, F., Zang, Q.: ResAt-UNet: a U-shaped network using ResNet and attention module for image segmentation of urban buildings. IEEE J. Sel. Top. Appl. Earth Obs. Rem. Sens. **16**, 2094–2111 (2023)

A Brief Survey of AI-Based Methods in Astrodynamics Problems with Disturbances, Noises and Uncertainties

Natalia Amelina[1], Oleg Granichin[1], Yury Ivanskiy[1(✉)], Kseniia Kopylova[1], Margarita Maksimenko[1,2], Vadim Sakharov[1], and Aleksei Tikhonov[1]

[1] Saint Petersburg State University, Universitetskaya Embankment, 7–9, 199034 Saint Petersburg, Russia
natalia_amelina@mail.ru, {o.granichin,y.ivansky,k.kopylova, v.sakharov,a.tikhonov}@spbu.ru, mig_sab@rambler.ru

[2] Empress Catherine II Saint Petersburg Mining University, 21st Line, 2, 199106 Saint Petersburg, Russia

Abstract. This survey is devoted to the analysis of the superposition of gravitational and magneto-electric effects in the problems of dynamics and control of the orbital and angular motion of spacecraft (SC) and SC constellations under the conditions of disturbances, noises and uncertainties. We consider spacecraft and constellations of spacecraft interacting with the geomagnetic field through Lorentz forces and moments, as well as moments caused by magnetic interaction. We briefly analyze a wide range of papers which describe the dynamics of the controlled motion of spacecraft and constellations of spacecraft moving in the gravitational and magnetic fields of the Earth with disturbances, noises and uncertainties and reduce their negative impact by applying such methods of artificial intelligence (AI) as randomized algorithms of optimization and estimation, genetics algorithm, fuzzy logic satellite attitude control, deep reinforcement learning applied to spacecraft attitude control and neural network-based method.

Keywords: Artificial Intelligence · Spacecraft · Satellite · Attitude Control · Attitude Stabilization · Disturbances · Uncertainties · Asymptotic Stability · Geomagnetic Field

1 Introduction

Spacecraft and spacecraft constellations are used in solving many problems in various areas nowadays. Modern geophysical, geodetic and meteorology problems involve the usage of artificial Earth satellites. Satellites are a necessary tool for navigation, astronomy and communication. During the flight, satellite positioning and navigation systems are influenced by disturbances of different nature: the drag affecting SC during the flight in the atmosphere results not only in force, but also in torque; the torque may be caused by near-Earth plasma influence on the reaction wheel; the actuators and sensors onboard the SC introduce a certain error into the estimates of motion and rotation vectors; the

© The Author(s), under exclusive license to Springer Nature Switzerland AG 2025
N. Filipović (Ed.): AAI 2024, LNNS 1446, pp. 419–427, 2025.
https://doi.org/10.1007/978-3-031-99201-8_39

inability to calculate the inertial matrix characterizing the SC in real time causes problems in control of the SC. Although the considered problems are usually solved in a restricted formulation, in general, the relationship between the translational and attitude motions of the SC should be taken into account. In this work, the main focus is on the attitude motions of the SC.

Various approaches based on artificial intelligence (AI) are used to solve complex problems in stochastic environment and under uncertainty conditions in order to deal with the effect of various disturbances or to estimate the system operating parameters. This paper provides a brief overview of AI methods used to calculate the spacecraft control strategy under the influence of uncertainty. The use of randomized algorithms of optimization and estimation, evolutionary algorithms and genetics algorithms, fuzzy logic satellite attitude control and neural network-based methods are considered in the paper.

The consideration of noise and model uncertainty, which included environmental torques, the gravity gradient and aerodynamic torques, representing the geomagnetic and magnetic hysteresis torques and the reaction wheel control torque, during the simultaneous model learning and optimization was made in [4]. In [5], a review of methods that use linearized models to solve the problem was provided and the algorithm which was proposed in [5] deals with the fully nonlinear system describing the rotational dynamics of a spacecraft as a solution to the problem. A summarized comparison between the different active and passive attitude control techniques was presented in [6]. Numerous control design methods, which have been adapted and applied for attitude control of flexible spacecrafts, were considered in [7]. In the presence of the inertia uncertainties and environmental disturbances, a decentralized sliding mode control law was developed to ensure that the rigid body attitude error dynamics can converge to a neighborhood of the sliding-mode surface in a finite time [8]. The hybrid controller's response is proportional-derivative (PD) based, and a body attitude and rate feedback were used in [9] to provide stability and robustness. Also, it was shown that the PD tracking controller is capable of stabilizing the system if it assumed to be rigid. However, it is possible for the system to be destabilized should the flexible modes of the satellite be excited sufficiently. Therefore, to avoid the excitation of these destabilizing flexible modes, a low jerk polynomial spherical-linear-interpolation profile was used and the PD control gains were tuned to ensure that the closed-loop response does not overly excite the flexible modes. The model predictive control problem based on satellite attitude control was considered in [12]. The original optimization problem is transformed into a multi-parametric quadratic programming problem with a bounded error.

2 Orbital and Attitude Control Model

The spacecraft rotational motion with respect to its center of mass is usually studied in one or another coordinate system, taken as the base one. The most widely used base coordinate systems are the Koenig coordinate system (sometimes called inertial coordinate system), and the orbital one, which is convenient for solving many problems related to observing the Earth's surface; in particular, a gravitational stabilization system based on the stabilizing effect of the gravity gradient torque $\vec{M}_G$. In some cases, when it is fundamentally important to take into account the interaction of an artificial satellite with the geomagnetic field through the moment of magnetic interaction $\vec{M}_m$ and the moment $\vec{M}_L$ of Lorentz forces, a coordinate system associated with the magnetic induction vector $\vec{B}$ of the geomagnetic field may be more convenient. One of the variants of such a coordinate system was proposed in [1, 10], where the torques $\vec{M}_m$ and $\vec{M}_L$ are used as control torques simultaneously in a unified control system called an electrodynamic attitude control system.

2.1 Model Without Taking into Account the Uncertainties, Disturbances and Noise

In some cases, the best model for research is the one that takes into account the maximum number of effects influencing the object. The satellite is assumed to have a controlled intrinsic magnetic moment $\vec{I}$ and a controlled electric charge Q. The dynamic equation of SC orbital motion is expressed as:

$$m\frac{d^2\vec{\mathrm{R}}}{dt^2} = -\frac{\mu m\vec{\mathrm{R}}}{\mathrm{R}^3} + \vec{\mathrm{F}}_L + \vec{\mathrm{F}}_C, \tag{1}$$

where m is the satellite's mass, $\vec{\mathrm{R}}$ is a radius vector of the satellite's center of mass, μ stands for universal gravitational constant, $\vec{\mathrm{F}}_L$ is the Lorentz force and $\vec{\mathrm{F}}_C$ is the control force.

A special case of simultaneous consideration of gravitational, magnetic and Lorentz torques was described in the model [1–3, 10]:

$$\frac{d}{dt}(J\vec{\omega}) + \vec{\omega} \times (J\vec{\omega}) = \vec{M}_m + \vec{M}_L + \vec{M}_G, \tag{2}$$

$$\frac{d\vec{\xi}_0}{dt} = \vec{\xi}_0 \times \vec{\omega} - \vec{\omega}_0\xi_0, \quad \frac{d\vec{\eta}_0}{dt} = \vec{\eta}_0 \times \vec{\omega}, \quad \frac{d\vec{\xi}_0}{dt} = \vec{\xi}_0 \times \vec{\omega} + \vec{\omega}_0\xi_0,$$

where $\vec{\omega} = \vec{\omega}' + \vec{\omega}_0$ is the absolute angular velocity of the satellite, $\vec{\omega}_0$ is the orbital angular velocity and $\vec{\omega}'$ is the angular velocity of the satellite relative to the basic coordinate system.

For any real spacecraft and especially for craft operating in modes close to the oriented motion, Lorentz torque $\vec{M}_L$ can be approximated by the expression [1]: $\vec{M}_L = \vec{P} \times A^T(\vec{v}_C \times \vec{B})$. Here, $\vec{P} = Q\vec{\rho}_0$, $\vec{\rho}_0$ is the radius-vector of the center of charge Q of the spacecraft with respect to its center of mass; the satellite orientation in the orbital coordinate system is defined by matrix A of direction cosines; $\vec{v}_C$ is the velocity of the satellite's center of mass with respect to the geomagnetic field; vectors $\vec{B}$ and $\vec{v}_C$ are considered given in the orbital coordinate system, $\vec{M}_m = \vec{I} \times \vec{B}$ is the magnetic interaction torque, $\vec{M}_G$ is the gravitational torque. The basis of the electrodynamic control system is the ability to control on the board of SC the magnitudes and directions of vectors $\vec{I}$ and $\vec{B}$, thereby creating the necessary control torques $\vec{M}_m$ and $\vec{M}_L$.

2.2 Disturbances, Noise and Uncertainties

High-precision inertial properties of the spacecraft are necessary for precise attitude control. Sudden and unpredictable changes during a mission cause unpredictable changes in the inertia tensor. To ensure accurate control, one needs to be able to detect these changes. Also, external disturbances with strong nonlinear dynamics affect the spacecraft or satellites. Detecting and compensating for these noises is important for nonlinear orientation tracking [23].

The dynamic equation of SC orbital motion taking into account disturbances, noise and uncertainties in external forces is expressed as:

$$m\frac{d^2\vec{\mathrm{R}}}{dt^2} = -\frac{\mu m\vec{\mathrm{R}}}{\mathrm{R}^3} + \vec{\mathrm{F}}_L + \vec{\mathrm{F}}_C + \vec{\mathrm{F}}_{\mathrm{d}}, \tag{3}$$

where $\vec{\mathrm{F}}_{\mathrm{d}}$ is the disturbance force. The SC attitude dynamics is governed by the following equation to account for disturbances, noise and uncertainties:

$$\frac{d}{dt}(J\vec{\omega}) + \vec{\omega} \times (J\vec{\omega}) = \vec{M}_m + \vec{M}_L + \vec{M}_G + d_e + d_i, \tag{4}$$

where d_e – external disturbances, d_i – internal disturbances.

External Disturbances. A lot of environmental forces and torques can influence significantly the orbital and attitude dynamics of SC. So, it may be important to take them into account alongside with the above-mentioned gravity gradient torque, magnetic interaction torque and Lorentz torque. As known, the aerodynamic drag becomes a nonnegligible factor, because of exponentially increasing atmospheric density with decreasing altitude [4]. Also, atmospheric drag caused by the aerodynamic forces arises in a low Earth orbit (the aerodynamic torques due to drag and lift). Moreover, atmospheric density and drag coefficients are known to be uncertain [24]. Solar radiation pressure is important for SC in high altitude orbits.

Internal Disturbances. The charge Q can change due to the interaction of the spacecraft with near-Earth plasma, causing inaccuracy in determining its value [26]. Reaction wheel is frequently used actuator in aerospace engineering, which is also used to provide attitude stabilization control torques [22]. Reaction wheels are vulnerable to the following four types of failures: F1, decreasing reaction torque; F2, continuous generation of reaction torque; F3, increasing bias torque; and F4, failure to react to the system control signals. To determine error dynamics, it is necessary to express errors based on quaternion algebra [23].

Noise. Vectors $\vec{P}$ and $\vec{I}$ may be noisy [28]. Uncontrolled interference in actuators and noise in measuring instruments usually occur in the practical implementation of vectors $\vec{P}$ and $\vec{I}$. Therefore, it is more correct to consider models of orbital motion and attitude motion in which such noise is included in advance. It is also important to investigate the stability of the implementation of the corresponding control algorithms relative to the possible noise level (investigate the permissible noise level).

Uncertainties. The inertia matrix cannot be accurately computed in real time [8, 24]. Let $J = J_N + J_R$, where J_N denotes the satellite nominal inertia matrix and J_R denotes the parametric perturbation [25].

2.3 On Methods for Constructing Controls, Observational Disturbances, Noise and Uncertainties

In the presence of inertia uncertainties and environmental disturbances, a class of decentralized adaptive sliding mode control laws was proposed in [8]. Article [24] proposes the method of the adaptive controller construction that exploits environmental torques for spacecraft attitude maneuvers in the presence of uncertainties in the time-varying CoM location, atmospheric density, drag and lift coefficients with guaranteed bounded state tracking through a Lyapunov-based stability analysis. Therefore, there are different techniques from classical control theory [29] to techniques, including Wave-Based-Control [30], LQR (Linear-quadratic regulator) [31], or algorithms which deal with the fully nonlinear system instead of relying on a linearization (with the base in variational techniques) [5], or AI. We made an effort to consider the latest techniques of AI in attitude control theory with disturbances, uncertainties and noise below.

3 Methods of AI

A significant level of disturbances, uncertainties and noise, described above, creates a problem of using traditional Eqs. (1) due to their sensitivity. Recently, it has become possible to reduce the negative impact of disturbances, noises and uncertainties by applying the AI methods.

3.1 Randomized Algorithms of Optimization and Estimation

Randomized algorithms of optimization and estimation allow to reduce the negative impact of disturbances, noises and uncertainties based on computationally simple and effective schemes and under weaker assumptions about the nature of disturbances, noises and uncertainties compared with many others (see, e.g., [11, 27]). An optimization-based fault detection filter is considered in [41] as a residual generator, which is designed to be robust against disturbance and sensitive to a fault. Also, a random forest (RF) algorithm is designed to obtain fault isolation with system input–output and residual signals.

3.2 Evolutionary Algorithms and Genetics Algorithms

A PD and a fuzzy controller for attitude control of a satellite model, with parameters obtained through the use of a simple genetic algorithm, were compared in [19] in terms of the sum of the errors yielded by each controller during the simulation of a single orbit. In [20] it was shown that the FOPID controller's performance with ITAE criteria in the genetic algorithm (GA) method for control of roll, pitch and yaw attitudes on nanosatellites has better performance in the form of a few of isolation waves. It is faster in reaching steady-state values than the classical PID method. The foundations of the most common evolutionary algorithms aimed at solving control theory problems for aviation and astronautics applications are presented in [18]. A Safety-Informed Evolutionary Reinforcement Learning algorithm (SERL), which combines Deep Reinforcement Learning (DRL) and neuroevolution to optimize a population of nonlinear control policies is presented in [42]. A nanosatellite optimal reorientation maneuver trajectory, taking into account external torques under control torque constrains, is explored in [43]. A linearized model is considered in [44] for adjusting the PID control parameters applies two different methods: "Model-Based Root Locus Tuning" and "Genetic Algorithm Based Tuning".

3.3 Fuzzy Logic Satellite Attitude Control

Attitude control models were linearized in order to apply a linear controller to the system in [13, 14]. The actuators for the maneuver are the reaction wheels and gravity gradient torque and a fuzzy logic controller (FLC) is used in [13] to control the attitude. The control law for magnetic wheel unloading is given in [14] with a description of a fuzzy gain-scheduler. The paper [15] describes the restrictions of using only "Electromagnetic" actuation for 3D stabilization and how to make the magnetorquer based attitude control feasible using Fuzzy Logic Control. In order to attain high-precision orbit, attitude and vibration stabilization a fuzzy-logic-based integrated orbit-attitude-vibration prescribed-time controller for large-scale flexible spacecraft is studied in [38]. A linear model-based controller was created to regulate CEACS attitude in [39]. An interval type-II fuzzy logic nonsingular terminal sliding mode and nonlinear extended state observer were used to create a finite-time chattering free attitude controller for rigid spacecraft without angular velocity measurement in [40].

3.4 Deep Reinforcement Learning Applied to Spacecraft Attitude Control Disturbances

A deep reinforcement learning algorithm successfully controlled the attitude of a spacecraft to reach the desired orientation in [16]. In [17] the performance of the attitude control system was assessed in terms of the pointing error between the satellite body-fixed and inertial coordinate frames. The response of a classical PD control law based on quaternion feedback is compared with the performance of the neural network controller. A critical scenario of actuator failure is also evaluated. A novel reinforcement learning-based attitude tracking control strategy for combined spacecraft takeover maneuvers with completely unknown dynamics is developed in [32]. An autonomous attitude control mechanism based on deep reinforcement learning is proposed in [37]. A near-optimal control strategy using reinforcement learning, which is applied to the six-degree-of-freedom (6-DoF) attitude control of dual-control aircraft, was proposed in [45].

3.5 Neural Network-Based Method

The intelligent sliding mode control for satellite dynamics based on neural learning and disturbance observer considering the dynamic uncertainty due to uncertain inertia matrix and the disturbance torque caused by orbit transfer engine was studied in [21]. A neural network is used to approximate the unknown dynamics, while an adaptive observer is utilized to handle the unknown disturbance. The prediction error is constructed by introducing the estimation model to understand the uncertainties better. Based on the finite-time design, an efficient controller is designed in [21] to achieve the attitude tracking.

The paper [33] proposes a robust generalized dynamic inversion (GDI) control system design with adaptive neural network (NN) estimation for spacecraft attitude tracking in absence of knowledge of the spacecraft inertia parameters. The robust GDI control system works to enforce attitude tracking, and the adaptive NN augmentation compensates for the lack of knowledge of the spacecraft inertia parameters. Neural network-based active control is applied to a model with external disturbance with bounded input in [34]. The article [35] presents a finite-time prescribed performance (FTPP) control approach based on a learning Chebyshev neural network (LCNN) for spacecraft attitude tracking with modeling uncertainties, actuator faults, and external disturbances. Spacecrafts can be controlled and measured with a novel inertial device called a VSMSCSG. An effort to improve attitude measurement and control accuracy of the spacecraft is made in [36]. A spacecraft attitude measurement and control method based on variable speed magnetically suspended control sensitive gyroscopes (VSMSCSGs) and the fractional-order zeroing neural network (FO-ZNN) steering law is proposed.

4 Conclusions

The article provides an attitude control model and reviews methods, which are used to eliminate the effects on the SC, including disturbances, noises and uncertainties. We tried to overview different approaches in recent papers that consider uncertainties in the model of SC orbital motion with the goal of subsequently creating a control algorithm that takes into account the maximum number of inaccuracies. It is important for achieving accurate orientation control. AI methods applied to the SC orbital motion control were briefly reviewed. There are almost no methods in the literature for using randomized algorithms to reduce errors associated with unknown-but-bounded disturbances, noise and uncertainties. In the future, we plan to demonstrate AI methods associated with randomized algorithms that can help to significantly improve the quality of satellite attitude stabilization and compare them with modern AI ones.

Acknowledgment. The research was carried out at Saint Petersburg University and supported by RSF (project No. 24-41-02031, https://rscf.ru/project/24-41-02031/).

References

1. Aleksandrov, A.Y., Tikhonov, A.A.: Natural magneto-velocity coordinate system for satellite attitude stabilization: dynamics and stability analysis. J. Appl. Comput. Mech. **9**(2), 513–520 (2023)
2. Tikhonov, A.A.: Natural magneto-velocity coordinate system for satellite attitude stabilization: the concept and kinematic analysis. J. Appl. Comput. Mech. **7**(4), 2113–2119 (2021)
3. Aleksandrov, A.Y., Tikhonov, A.A.: Application of a PID-like control to the problem of triaxial electrodynamic attitude stabilization of a satellite in the orbital frame. Aerosp. Sci. Technol. **127**, 107720 (2022)
4. Sabug, L., Jr., et al.: Simultaneous design of passive and active spacecraft attitude control using black-box optimization. Control. Eng. Pract. **135**, 105516 (2023)

5. Pedregal, P., Úbeda, V.: Optimal attitude control of satellites. SeMA (2022)
6. Bouras, M., Berbia, H.: Review of attitude control approaches for ADCS optimization and faults tolerance. In: Proceedings of the 8th International Conference on Modeling Simulation and Applied Optimization (ICMSAO), pp. 1–4. IEEE (2019)
7. Tkachev, S., et al.: Attitude stabilization of a satellite with large flexible elements using on-board actuators only. Mathematics **11**(24), 4928 (2023)
8. Xia, Y., et al.: Attitude control of multiple rigid bodies with uncertainties and disturbances, finite time and cooperative control of flight vehicles. Adv. Ind. Control **8**(10), 195–213 (2019)
9. Gordon, R., Ceriotti, M., Worrall, K.: Effects of model fidelity and uncertainty on a model-based attitude controller for satellites with flexible appendages. Acta Astronaut. **214**, 30–45 (2024)
10. Giri, D.K., Sinha, M.: Application of dynamic neural network model for magneto-coulombic attitude control of earth-pointing satellite. J. Aerosp. Sci. Technol., 118–135 (2015)
11. Granichin, O., Volkovich, Z. (V.), Toledano-Kitai, D.: Randomized Algorithms in Automatic Control and Data Mining. Springer, Heidelberg (2015)
12. Xu, Z., et al.: Lattice piecewise affine approximation of explicit model predictive control with application to satellite attitude control (2024)
13. Adunola, F., Sani, S., Nyitamen, D.: Comparative study of FLC and PID for satellite attitude control. 5 (2022)
14. Zapf, J.: Robust attitude control with fuzzy momentum unloading for satellites using reaction wheels (2006)
15. Asadabadi, A., Anvar, A.: Small satellite modelling and attitude control using fuzzy logic. Int. J. Aerosp. Mech. Eng. **6**(12), 2598–2601 (2012)
16. Enders, N.A.: Deep reinforcement learning applied to spacecraft attitude control and moment of inertia estimation via recurrent neural networks (2021)
17. Chagas, R.A.J.: Intelligent attitude control of satellites via deep reinforcement learning. CEP 12: 010 (2021)
18. Sands, T.: Advances in Spacecraft Attitude Control. Intech Open (2020)
19. Correa, C., Sandri, S.A., de Souza, L.C.G.: The use of genetic algorithms on a fuzzy controller for a satellite attitude control during the pointing phase. Revista Brasileira de Ciências Mecânicas **21**(Special Issue), 615–624 (1999)
20. Negara, A.T., Mustiko, S., Firmansyah, L.: Genetic algorithm based fopid controller for nano-satellite attitude control. Jurnal Informatika Polinema **9**(1), 59–66 (2022)
21. Wang, X., Xu, B., Pan, Y.: Neural network-based sliding mode control for satellite attitude tracking. Adv. Space Res. **71**(9), 3565–3573 (2023)
22. Li, B., et al.: Finite-time disturbance observer based integral sliding mode control for attitude stabilisation under actuator failure. IET Control Theor. Appl. **13**(1), 50–58 (2019)
23. Alipour, M., Fani Saberi, F., Kabganian, M.: Inertia-free nonlinear attitude tracking with disturbance compensation using adaptive-sliding control based on quaternion algebra. Simulation **96**(1), 43–54 (2020)
24. Riano-Rios, C., et al.: Aerodynamic and gravity gradient based attitude control for CubeSats in the presence of environmental and spacecraft uncertainties. Acta Astronautica **180**, 439–450 (2021)
25. Liming, F.A.N., Huang, H., Kaixing, Z.: Robust fault-tolerant attitude control for satellite with multiple uncertainties and actuator faults. Chin. J. Aeronaut. **33**(12), 3380–3394 (2020)
26. Tikhonov, A.A., Yakovlev, A.B.: On dependence of equilibrium characteristics of the space tethered system on environmental parameters. Int. J. Plasma Environ. Sci. Technol. **13**(1), 49–52 (2020)
27. Granichin, O., Ivansky, Y., Kopylova, K.: Polyak's Method based on the stochastic Lyapunov function for justifying the consistency of estimates produced by a stochastic approximation

search algorithm under an unknown but bounded noise. Comput. Math. Math. Phys. **64**(4) (2023)
28. Pope, C.: Calibration and uncertainty analysis of a spacecraft attitude determination test stand (2017)
29. Seyyed, A., Alireza, G.: Attitude control of a satellite using classical control methods. (2023)
30. Liu, F., Yue, B., Ma, B., Feng, C.: Wave-based attitude control of in-orbit spacecraft with large-amplitude slosh. J. Vib. Control **30**(1–2), 366–376 (2024). https://doi.org/10.1177/10775463221147339
31. Rouzegar, H., Ghanbarisabagh, M.: A novel on–off linear quadratic regulator control approach for satellite rendezvous. AS **6**, 613–620 (2023). https://doi.org/10.1007/s42401-023-00230-9
32. Liu, Y., Ma, G., Lyu, Y., Wang, P.: Neural network-based reinforcement learning control for combined spacecraft attitude tracking maneuvers. Neurocomputing **484**, 67–78 (2022)
33. Jafri, S.M.N., Aslam, M.I.: Spacecraft attitude control based on generalised dynamic inversion with adaptive neural network. Aeronaut. J. **128**(1321), 504–516 (2024)
34. Fu, Y.: Neural network–based active control of a rigid–flexible spacecraft with bounded input. J. Vib. Control **30**(5–6), 1117–1132 (2024)
35. Jia, Q., Li, G., Yu, D., Ahn, C.K., Zhang, C.: Learning Chebyshev neural network-based spacecraft attitude tracking control ensuring finite-time prescribed performance. Aerosp. Sci. Technol., 109085 (2024)
36. Li, L., Ren, Y., Wang, W., Pang, W.: Spacecraft attitude measurement and control using VSMSCSG and fractional-order zeroing neural network adaptive steering law. Sensors **24**(3), 766 (2024)
37. Sarmiento, J.A.R., Tan, V.H., Talampas, M.C.R., Naval, P.C., Jr.: Sample efficient deep reinforcement learning for Diwata microsatellite reaction wheel attitude control. Aerosp. Syst. **6**(1), 61–69 (2023)
38. Yue, X., Lyu, B., Liu, C.: Fuzzy-logic-based integrated orbit-attitude-vibration prescribed-time control for large-scale flexible spacecraft. In: Li, S. (eds.) Computational and Experimental Simulations in Engineering. ICCES 2023. Mechanisms and Machine Science, vol. 145, pp. 839–849. Springer, Cham (2024). https://doi.org/10.1007/978-3-031-42987-3_58
39. Aslam, S., Chak, Y.C., Jaffery, M.H., Varatharajoo, R., Ansari, E.: A, Model predictive control for Takagi-Sugeno fuzzy model-based Spacecraft combined energy and attitude control system. Adv. Space Res. **71**(10), 4155–4172 (2023)
40. Beyramzad, J., Daneshjou, K., Khanmirza, E.: Design a finite-time chattering free attitude controller for rigid spacecraft's without angular velocity measurement using interval type-II fuzzy logic nonsingular terminal sliding mode and nonlinear extended state observer. In: Proceedings of the 21st International Conference of Iranian Aerospace Society, pp. 1–8 (2023)
41. Chen, S., Yang, R., Zhong, M., Xi, X., Liu, C.: A random forest and model-based hybrid method of fault diagnosis for satellite attitude control systems. IEEE Trans. Instrum. Meas. **72**, 1–13 (2023)
42. Gavra, V., van Kampen, E.J.: Evolutionary reinforcement learning: hybrid approach for safety-informed fault-tolerant flight control. J. Guidance Control Dyn., 1–13 (2024)
43. Elisov, N.A., Kramlikh, A.V., Lomaka, I.A., Avariaskin, D.P.: An attitude control by the functional series in the problem of nanosatellite reorientation. Aerosp. Sci. Technol. **132**, 108038 (2023)
44. Sayin, E., Bitirgen, R., Bayezit, I.: Attitude control and parameter optimization: a study on hubble space telescope. Meas. Sci. Rev. **23**(4), 146–153 (2023)
45. Yuan, Y., Zhou, D.: Reinforcement learning for dual-control aircraft six-degree-of-freedom attitude control with system uncertainty. Aerospace **11**(4), 281 (2024)

Author Index

© The Editor(s) (if applicable) and The Author(s), under exclusive license to Springer Nature Switzerland AG 2025
N. Filipović (Ed.): AAI 2024, LNNS 1446, pp. 429–430, 2025.
https://doi.org/10.1007/978-3-031-99201-8

The manufacturer's authorised representative in the EU is Springer Nature Customer Service Centre GmbH, Europaplatz 3, 69115 Heidelberg, Germany. If you have any concerns regarding our products, please contact ProductSafety@springernature.com

Printed and bound by CPI Group (UK) Ltd, Croydon, CR0 4YY
30/09/2026
02225561-0013

Lecture Notes in Networks and Systems 1446

Series Editor

Janusz Kacprzyk, *Systems Research Institute, Polish Academy of Sciences, Warsaw, Poland*

Advisory Editors

Fernando Gomide, *Department of Computer Engineering and Automation—DCA, School of Electrical and Computer Engineering—FEEC, University of Campinas—UNICAMP, São Paulo, Brazil*
Okyay Kaynak, *Department of Electrical and Electronic Engineering, Bogazici University, Istanbul, Türkiye*
Derong Liu, *Department of Electrical and Computer Engineering, University of Illinois at Chicago, Chicago, USA*
Institute of Automation, Chinese Academy of Sciences, Beijing, China
Witold Pedrycz, *Department of Electrical and Computer Engineering, University of Alberta, Alberta, Canada*
Systems Research Institute, Polish Academy of Sciences, Warsaw, Poland
Marios M. Polycarpou, *Department of Electrical and Computer Engineering, KIOS Research Center for Intelligent Systems and Networks, University of Cyprus, Nicosia, Cyprus*
Imre J. Rudas, *Óbuda University, Budapest, Hungary*
Jun Wang, *Department of Computer Science, City University of Hong Kong, Kowloon, Hong Kong*

The series "Lecture Notes in Networks and Systems" publishes the latest developments in Networks and Systems—quickly, informally and with high quality. Original research reported in proceedings and post-proceedings represents the core of LNNS.

Volumes published in LNNS embrace all aspects and subfields of, as well as new challenges in, Networks and Systems.

The series contains proceedings and edited volumes in systems and networks, spanning the areas of Cyber-Physical Systems, Autonomous Systems, Sensor Networks, Control Systems, Energy Systems, Automotive Systems, Biological Systems, Vehicular Networking and Connected Vehicles, Aerospace Systems, Automation, Manufacturing, Smart Grids, Nonlinear Systems, Power Systems, Robotics, Social Systems, Economic Systems and other. Of particular value to both the contributors and the readership are the short publication timeframe and the worldwide distribution and exposure which enable both a wide and rapid dissemination of research output.

The series covers the theory, applications, and perspectives on the state of the art and future developments relevant to systems and networks, decision making, control, complex processes and related areas, as embedded in the fields of interdisciplinary and applied sciences, engineering, computer science, physics, economics, social, and life sciences, as well as the paradigms and methodologies behind them.

Indexed by SCOPUS, EI Compendex, INSPEC, WTI Frankfurt eG, zbMATH, SCImago.

All books published in the series are submitted for consideration in Web of Science.

For proposals from Asia please contact Aninda Bose (aninda.bose@springer.com).